Mine Investment Analysis

by

Dr. Donald W. Gentry
Dean of Undergraduate Studies and
Professor of Mining Engineering
Colorado School of Mines
Golden, Colorado

Dr. Thomas J. O'Neil
Manager, New Business Development
Amoco Metals Company
Englewood, Colorado

Published by

Society of Mining Engineers

of

American Institute of Mining, Metallurgical, and Petroleum Engineers, Inc.

New York, New York • 1984

Copyright © 1984 by the
American Institute of Mining, Metallurgical,
and Petroleum Engineers, Inc.

Printed in the United States of America
by BookCrafters, Inc., Chelsea, MI

Library of Congress Card Number 84-051346

ISBN 0-89520-429-0

Preface

Those familiar with the minerals industry are aware of the acute shortage of timely books and reference materials addressing some of the specialty areas in mining. Nowhere is this problem more pronounced than in the area of mineral engineering education where there is an obvious absence of appropriate texts available for instructional purposes in many segments of the undergraduate and graduate curricula within the minerals engineering disciplines.

One aspect of mineral engineering particularly in need of a modern textbook is the study of mine valuation. Significant changes in parameters such as state and federal taxation policies, project investment analysis, evaluation criteria, alternatives for project financing, spiraling inflation, and risk quantification have greatly impacted the mine valuation process within the last 10-15 years. As a result of the dynamic nature of these and other variables, many practitioners within the industry find themselves in need of more recent written material describing some of the more critical aspects pertaining to the procedures and eccentricities associated with analyzing mining-related investments. Thus, another reason for this book was perceived.

The authors recognized the need for a text which addressed the procedures and analyses associated with mineral investments a number of years ago and were encouraged to produce this book. It should be stressed that this is *not* another textbook in the field of engineering economy. Rather, this book addresses the concepts and principles associated with corporate investment decision-making and specifically the practice and procedures pertaining to mine property valuation. In general terms, this book relates more to finance (i.e., capital budgeting and the evaluation of investment opportunities) than to economics. As such, the business (as opposed to purely economic) aspects associated with the minerals industry are emphasized.

This book was written with the intent of satisfying three primary objectives. First, and perhaps foremost, was the desire to provide an up-to-date tool for teaching mine valuation concepts and procedures, as well as the principles of capital budgeting, to undergraduate minerals engineering students. Second, the authors attempted to present the material in a manner which would enable graduate students interested in specific aspects of mine valuation to appreciate the complexity, intricacies, and interrelationships of these topics. Hopefully this textbook will form the basis for graduate work in the area of mine valuation and stimulate graduate students to pursue work on some of the complex issues inherent in the process. The third and last primary objective was to provide a book which would serve as a timely reference document to the industrial practitioner working in the area of mine investment analysis. We trust this document accomplishes these objectives and represents a contribution to the minerals engineering profession.

Donald W. Gentry

Thomas J. O'Neil

May 1984

Acknowledgments

The writing of a textbook often begins with great energy and as a labor of love only to become an anchor of responsibility as time passes and professional obligations multiply. That we were able to see this project through to completion in spite of changes in residence and employment responsibilities is largely due to support and encouragement of Peggy O'Neil and Sheila Gentry. It seems like all authors thank their spouses, and now we know why. They understood better than we did how important this book was to us.

To Marianne Snedeker of the Society of Mining Engineers goes our sincerest gratitude for her patience and assistance. Her excellent performance on the production work for SME-AIME publications has become routine and has saved many poorly organized authors from paralyzing chaos.

Finally, we want to thank Bill Lampard, a good friend who reviewed the manuscript for administrative approval under severe time constraints. It goes without saying, of course, that he bears no responsibility for any errors that remain.

Contents

Preface • D.W. Gentry, T.J. O'Neil iii

Chapter

1   Introduction 1
2   Basic Valuation Concepts 11
3   The Time Value of Money 35
4   The Ore Reserve Problem 57
5   Estimating Revenues 81
6   Capital and Operating Cost Estimation 103
7   The Minerals Depletion Allowance 151
8   Mine Taxation 165
9   Project Evaluation Criteria 253
10   Inflation in the Mine Investment Decision 301
11   Selecting a Discount Rate 321
12   Special Evaluation Problems in Mining 347
13   Accounting for Risk in Mining Investment 375
14   Case Study 391
Appendix A: Interest Factors for Discrete Compounding . . . 435
Appendix B: Effective Interest Rates Corresponding to
    Nominal Rate 456
Appendix C: Interest Factors for Continuous
    Compounding Interest 457
Appendix D: Funds Flow Conversion Factors 472
Appendix E: Geometric Gradient Interest Factors 473
INDEX 489

1
Introduction

"A Western mine is a hole in the ground owned
by a liar..."
—Mark Twain

THE CAPITAL INVESTMENT DECISION

Decisions pertaining to a firm's proposed capital investments can have vital short- and long-term consequences on the organization's ability to compete, and even survive. Capital investment decisions, in general, center around two fundamental activities: (1) allocating capital funds to specific investment projects or assets, and (2) obtaining necessary financing in such proportion as to increase the overall value of the firm. In essence these activities or decisions describe the science of finance (as contrasted with economics).

Finance has been described in general terms as the study of how a present, known amount of cash is converted into a future, perhaps unknown, amount of cash (Archer, Choate, and Racette, 1979). Fundamental to understanding the science of finance and the capital investment decision process in general are the basic concepts relating to cash flow transactions, time, income generation (markets), expected returns, and risk. The functions of finance may be segregated into three fundamental decisions which a firm must address: (1) the dividend decision, (2) the financing decision, and (3) the investment decision. These decisions are interdependent and their joint impact on the objective of the firm must be considered. This book primarily addresses the concepts and components associated with the investment and financing decisions as they affect the practice of mine valuation.

Responsibilities of Management

A primary assumption incorporated in this book is that the fundamental objective of any firm is to maximize its value, or wealth of its owners (stockholders). Here, the term *wealth* refers to the total current market value of the firm's assets. In the case of corporations, the wealth or value is considered to be represented by the market price of the firm's common stock. The price of the firm's common stock obviously will be affected by its investment, financing, and dividend decisions. Consequently, an optimal combination of these three decisions should maximize the value of the firm to its stockholders.

It is important to point out that wealth maximization is a more appropriate and inclusive goal for the firm than profit maximization. Indeed, there is a difference between the two objectives in most situations. For instance, a firm can always raise profits by selling stock and investing the proceeds in treasury bills or certificates of deposit. This type of activity, however, would rarely cause net shareholder wealth to increase. Indications are that total profits are not as important to the investment community as other indicators such as earnings per share ratios. However, even the maximization of earnings per share is not an appropriate objective for the firm because it does not: (1) specify the timing of expected returns, (2) consider the risks associated with projected earnings, or (3) consider the dividend policy of the firm.

On the other hand, wealth maximization, as reflected by the price of the firm's common stock in the market place, does consider current and future prospective earnings, the timing of these earnings, and the impact of the dividend policy of the firm on its stockholders. As such, management may be evaluated indirectly by the stockholders because the market price of common stock serves as an index of the firm's progress as perceived by its owners. Over the long-run, the success or failure of a firm will be a function of how well the management of the firm handles investment and financing decisions, for the results of these decisions will directly impact the value of the firm in the marketplace.

Capital investment decisions ultimately relate to the allocation of scarce resources—land, labor, and capital. From the standpoint of mining-related investments, the resources of most concern are those associated with land (ore deposits) and capital. Fortunately, the principle of wealth maximization provides a very realistic format for the efficient allocation of resources. Consequently, if management adheres to the concept of wealth maximization, particular emphasis must be placed on the investment and financing decisions.

The investment decision is perhaps the most important decision facing a firm. It is a conscious decision to convert capital into some other asset with the anticipation of earning a positive return at some specified level. The investment decision, in effect, determines the composition of the asset portfolio held by the firm as well as the relative business risk of the firm. Obviously the synthesis of this information by the investment community will greatly impact the overall value (i.e., price of common stock) of the firm. A major component of the investment decision is the capital budgeting aspect. Capital budgeting is that element of capital investment dealing with the allocation of capital to projects in some optimal manner. The perceived benefits from these projects will be realized at some time in the future.

After the firm has reconciled the investment decision, the financing decision must be addressed. The financing decision concerns the selection of source and timing of new capital necessary to implement the investment decisions. Financing decisions must be concerned with formulating the best or optimal financing mix for the firm such that the value of the firm will be maximized. The topic of optimum capital structures is addressed in some detail in Chapter 11.

If a management hopes to efficiently allocate scarce capital resources and maximize the value of the firm to its stockholders—and thereby contribute to the long-run success of the firm—it must be capable of making consistent and *accurate* investment decisions. Management must, therefore, make every effort to develop criteria which lead to such consistency and accuracy. These criteria must be realistic, theoretically sound, and, above all, consistent with capital budgeting goals. Fortunately, financial decisions oriented toward the goal of wealth maximization lead to logical decision rules in virtually all situations. These rules, along with specific criteria applied to the capital budgeting process, should enable corporate management to accomplish its primary objective.

Capital Budgeting and Project Evaluation

The capital budgeting process refers to the sequence of decisions which ultimately lead to the firm's acceptance or rejection of investment proposals along with subsequent management of the proposals accepted. The entire capital budgeting process is normally considered to comprise the activities of planning, evaluation, selection, implementation, control, and, finally, continual reevaluation and auditing of results. As mentioned previously, capital budgeting refers to the current allocation of investment capital to investments whose perceived benefits will be realized at some time in the future.

In a general sense, the term *capital budgeting* involves the process of selecting among alternative investment projects. Capital budgeting addresses decisions pertaining to questions such as, "Is project A and/or B good enough to warrant investment recommendations?"; and "Is project A better than project B?" These are the kinds of decisions which must be predicated on specific criteria which yield consistent management decision-making.

The mine evaluation activity in capital budgeting is the main focus of this book. The term *evaluation* deals with assessing a single project. Estimates of costs, benefits, expected returns, and associated risks are made for each project or investment alternative available to the firm. All appropriate decision criteria are then calculated for each project. These individual projects, along with their respective evaluation criteria, are then incorporated into the capital budgeting process and ultimately some, or all, of the projects available to the firm may be selected for investment. This selection, of course, is based on the objective of maximizing the value of the firm to its stockholders.

In this book the term "investment analysis" refers mainly to the concept of project evaluation. Although emphasis is placed on evaluating mining-related investment proposals, some discussion of the overall capital budgeting process is also presented in Chapter 9.

The Engineer and Investment Analysis

Historically in the minerals industry there has been little interchange between individuals in the geological, mining, metallurgical, and financial disciplines during the evaluation stage of new properties. Typically each discipline has concentrated on its own unique set of problems and has ignored most, if not all, of the problems pertaining to the others. Unfortunately, this segregated approach has led to some poor investment decisions in mining.

There is no doubt that the evaluation of new mining projects in today's environment is much more complex than it was just a few years ago. There are typically myriads of variables which are directly or indirectly associated with the mine evaluation process. As such, mine valuation has become truly interdisciplinary in nature. An individual rarely is knowledgeable in all the areas involved in the evaluation process, particularly where major projects are being considered. Therefore, most organizations prefer to establish multidisciplinary groups which perform the evaluation function for new investment opportunities. These evaluation groups generally consist of individuals with expertise in each of the major aspects associated with the evaluation process (e.g., geology, mining, processing, economics, environment, regulations, etc.). This is the preferred approach to the problem, but many mining companies cannot afford the luxury of so many specialists and must rely on one or two individuals to perform all project evaluations.

The engineer's role in mining project investment analysis is to provide sound technical advice and information on parameters relating to design, extraction methods, production costs, recoveries, mining rates, and many other variables. In essence, the engineer must provide quantitative values for project variables based on sound technical analysis. Only when variables are quantified can project feasibility studies be brought to their conclusion and sound investment decisions made. The engineer's unique contribution to the investment analysis process comes in his ability to analyze information, which is inevitably incomplete, and generate a sound technical appraisal in such a form that lends itself to subsequent financial analysis.

One of the continuing controversies among engineering educators is the argument concerning whether a first-degree engineering graduate should receive a general engineering education with emphasis on the fundamentals, or whether he should be trained in one or two specific areas of specialization. Although this book does not seek to resolve this issue, the authors do believe that engineers choosing a career in the area of investment analysis should be capable of recognizing and ascertaining the importance of various key variables and associated trade-offs affecting investment decisions. Certainly engineers with expertise in specialty areas can and do contribute significantly in certain investment proposal analyses. However, someone who understands the interrelationships and trade-offs among key project variables must manage these experts on the evaluation team. This individual must have an appreciation for, and basic understanding of, the various disciplines involved and have the ability to view the project as a whole and not just the unique problems associated with one discipline. Because of the broad educational background which they receive, mining engineers are generally well suited for this role.

The process of evaluating mine investment opportunities is usually iterative in nature. The general process may be represented as follows:

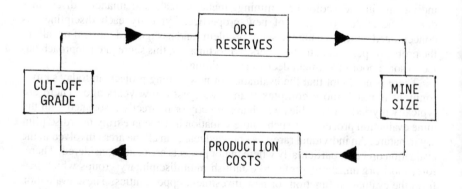

Each time a variable is changed, the analyst must assess the impact of this change on the other project variables and on the financial results. This iterative process must be repeated until the most economical design is achieved for the project being analyzed. This is indeed a time-consuming process, but represents the essence of engineering in the evaluation process.

MINING—A UNIQUE INVESTMENT ENVIRONMENT

Certainly the investment environment associated with the mining industry is unique when compared with the environment encountered by typical manufacturing industries. Some of the characteristics of mining which are often proclaimed as being unique are as follows:

1) **Capital Intensity:** Virtually every knowledgeable observer would agree that mining ventures are extremely capital intensive. Although the magnitude of capital required for a new mining venture will vary with the type of commodity, mining method, mine size, location, and other parameters, major new mines may require financial commitments ranging from $500 million to as much as $8-$10 billion. The infrastructure alone for mines in remote locations may cost several hundred millions of dollars. Even extremely small high-grade precious metal mines employing only a handful of miners can rarely be developed for operation for less than a million dollars.

2) **Long Preproduction Periods:** Once the occurrence of an ore deposit has been well established, it takes a number of years of intensive effort before the property is brought on-stream and ore is produced on a continuous basis. The amount of time required to develop a mining property for production can vary significantly. Until recently, a preproduction period ranging from 4-6 years was a good average estimate. However, more recently, environmental regulations and permitting requirements have caused some mining companies to experience time lags as great as 10-12 years from the time the investment decision was made until the mine was actually brought into production.

The significance of these long lead times is amplified when considered in conjunction with the capital intensity of the industry. Not only are companies committing extremely large capital resources to new mining ventures, but they are also financially exposed for a considerable period of time prior to project start-up. The longer the lead time, the higher is the probability of undesirable change in key engineering and economic parameters utilized in the initial investment decision.

3) **High Risk:** In addition to the obvious risks associated with capital intensity and long lead times, there are a number of other risks associated with mining ventures. Some of these risks may be under the control of the investor while others are clearly uncontrollable. In general, these risks may be placed under the general headings of geological risks, engineering risks, economic risks, and political risks.

Although numerous examples could be cited, one element of risk familiar to all mining people is that associated with mineral markets. Prices fluctuate significantly and precipitously for mineral commodities which are traded in international markets and when the economic distribution of a mineral commodity is rather limited on a worldwide basis. The Organization of Petroleum Exporting Countries (OPEC) amply demonstrated this feature for petroleum in 1973. The fact that mineral markets do vary significantly and are cyclic in nature can be easily checked by plotting company earnings for precious and base metal producers over the past 30 years.

Another example of risk in the mining sector which is often overlooked is political risk. Political risk is most dramatically illustrated by outright expropriation such as occurred in Chile in the early 1970s. However, there is an accelerating trend toward greater participation in mining projects by host governments throughout the world. This is occurring even in states, provinces, and nations that are otherwise committed to free enterprise. Therefore mining companies contemplating any new venture must assess these growing political risks to insure that the added financial exposure is warranted.

4) **Nonrenewable Resource:** Perhaps the most unique aspect of the minerals industry is the fact that it deals with the extraction of a nonrenewable resource. One result is that revenues from mining are derived from a piecemeal disposal of the project's major asset, the ore body. This gradual attrition of capital has been a justification for the minerals depletion allowance in the United States income tax code.

A second impact of the nonrenewable resource characteristic of mining is that all mines have finite lives determined by the size of the ore deposit and the mining rate. Investors must receive an adequate return by the time ore reserves are depleted, and new deposits must be continually discovered and developed.

Many people regard nonrenewable mineral deposits as assets created for the benefit of all mankind. This is particularly true where mineral deposits are contained on public lands. As such, these people believe that these mineral deposits should be mined only for the benefit of society as a whole. Under these circumstances they have little sympathy for profits required by private mining companies. This basic philosophy, often referred to as the "natural heritage theory," has led many states to enact tax policies which treat mining ventures differently—and typically more severely—than other industries. These tax policies are discussed in some detail in Chapter 8.

The mining industry occasionally creates tremendous growth in population in remote or rural areas with an attendant demand on public services. The development and operation of a mining operation require a large labor force as well as major support personnel and facilities. This sudden influx of people and support industry can have significant social and economic impacts on a small town or even an entire region within a state. Responsible mining companies plan to minimize such impacts, and the costs of infrastructure continue to rise as a percentage of overall project capital. This problem is perhaps best illustrated in some of the western states currently experiencing vast energy-related mining activity where entire planned communities are being constructed. Much of this mining development is located in fragile, arid environments which heal slowly and are difficult and extremely costly to reclaim.

The closing of a mine or groups of mines in an area can also have far-reaching impacts on local communities. One need only look at all the old abandoned mining camps, or ghost towns, scattered throughout the western US to realize that this is also a part of the mining business. Indeed, mine closures are inevitable due to the limited size of all mineral deposits. The closing of mining operations creates severe dislocations, affecting not only unemployment problems directly, but also often leading to the bankruptcy or closing of local businesses, significant decreases in local population as people seek jobs elsewhere, and plummeting real estate values. It is not surprising then that state and local planning commissions view mineral development in their regions with considerable mixed emotions. As a result many political leaders are insisting that mining companies themselves finance and assist in developing the infrastructure and social needs resulting from mineral development and be accountable for the depressed economics and unemployment in the future when the ore is gone. Again, these phenomena, although perhaps not unique, are very prominent in mining and create major investment challenges.

Mining is frequently cited for its historical limited environmental awareness, and there is ample evidence to support such criticism. However, many of these abuses can be traced to a time when society was far more concerned with economic growth than with the side effects of such growth. Although the attitude of the mining industry during the period did not differ markedly from that of other heavy industry, mining's environmental monuments from this era tend to be grander in scale and more visable to the public.

As a result of the national environmental movement, significant and restrictive environmental legislation was enacted in the late 1960s and 1970s which placed major operating constraints on the minerals industry. Although these regulations and constraints have been extremely costly, the mining industry has now established a good reputation for reclamation and environmental safeguards associated with mineral extraction in fragile environments.

Nonetheless, the very nature of mining itself requires that material be excavated and that the surface be altered in some manner, even if only temporarily. Consequently, mining will inevitably result in some environmental impact at the mine site, which usually occurs in rural areas and is aesthetic in nature. The environmental impact on the surface resulting from mining activities may take many forms. Open pits, highwalls, waste dumps, leach dumps, collection ponds, headframes, and the physical plant are examples of some of these features. Perhaps the features which are of concern to most people are those associated with waste products resulting from the mining and processing of mineral commodities. Often, these wastes contain toxic or potentially harmful chemicals which must be stored on the surface. In addition, the dust, noise, and other external factors typically associated with mining are environmental impacts which must be addressed.

Most rational analysts agree that trade-offs must occur between domestic minerals production and environmental regulations. It is well-documented that a country's per capita minerals consumption is directly correlated with a nation's standard of living. If the US populace chooses to maintain its standard of living, then raw mined materials must obviously be produced in increasing quantities. Clearly, there is a cost associated with this production, one portion being direct, while the other component is indirect and constitutes costs associated with conservation, environmental issues, and other nonproductive activities. The overall objective should be to produce the necessary mineral products while, at the same time, minimizing adverse environmental impacts from this production. The key to meeting this objective is to understand the relevant issues and then logically and realistically assess the trade-offs involved and the potential consequences of specific decisions.

Because of the unique features and characteristics of the mining industry already described, the financial analysis of mining projects has become a highly specialized field. Although the standard topics addressed in engineering economy texts are an important and essential component in the analysis of normal mine investment, these publications rarely treat the critical features of the analysis which result from the unique characteristics of the business. Furthermore, the standard economic approach to investment analysis typically does not focus on the financial aspects of investment decision-making nor on the unique business aspects of an industry. Therefore, this book seeks to bridge the gap between the well-developed fields of capital budgeting and finance and the peculiar and challenging investment characteristics of a single industry—mining.

PURPOSE OF VALUATION STUDIES

Valuation studies for mineral properties are conducted for many reasons.

Regardless of the specific purpose for the valuation determination, the ultimate objective of the study is to arrive at a monetary value or worth for the property in question. A specific value, or range in values, for a given property is often required for one or more of the following purposes.

1. **Acquisition:** The acquisition of mineral properties may transpire at any point in time between a raw prospect and an actual operating mine. Obviously the actual amount of data available on a property will depend upon where it lies within this spectrum. Depending upon the state of development of the property, the purchaser is acquiring assets with varying levels of risk. As such, the estimated value of the property must reflect not only the potential of the mineral deposit, but also the relative risks associated with these assets as well. Certainly the distribution of value estimates for an existing operating mine would be expected to have a rather low variance as contrasted to that for a raw prospect.

Regardless of the type of mineral property being considered for acquistion, it is essential that the seller and potential buyer have a property value in mind. Without such values, meaningful negotiations cannot occur. Alternative buyout provisions, royalty arrangements, bonus payments, production payments, advanced royalties, etc., must be assessed by both parties in terms of their perceived value of the property.

2) **Taxation:** Mineral properties must also be valued for taxation purposes. Perhaps the classic example here is with ad valorem property taxes levied by most state and local governments. The difficulty with value estimation of a mineral property for taxation purposes is that a *single* value is required for property worth.

Most states have enacted tax provisions which attempt to approximate the value of a mineral property through a formula or other mechanism which rarely serves as an adequate measure of property value for an actual sale. These mechanisms are rarely based on strong economic foundations and only serve as a convenient proxy for mineral property values. As a result, significant discrepancies can occur between the appraised value of a property for tax purposes and the value as perceived in the marketplace.The inability to resolve these differences in value estimates continues to result in considerable litigation.

3) **Financing:** The mode, mechanism, and magnitude of financing new mining properties or ventures are functions of the estimated property or project value. Certainly the risk of default must also be considered in mining and must be assessed in regard to the perceived intrinsic value of the property. This aspect is becoming increasingly important in view of the popularity of international joint ventures and project financing as a means of spreading project risks.

The fundamental concern of lending institutions is not whether a specific rate of return is achieved by the project owner, but rather that the project will generate adequate cash flow to service the debt. Thus, lenders approach mine valuation studies from a different perspective.

4) **Regulatory Requirements:** The federal government has also found it necessary to wrestle with the problems associated with estimating the value of federally-controlled mineral lands. For instance, according to the Federal Coal Leasing Amendments Act (FCLAA) of 1976, federal lands offered for coal leasing must be sold via competitive bidding and "no bid shall be accepted which is less than the fair market value, as determined by the Secretary of the Interior, of the in situ coal subject to the lease." Thus, the receipt of fair market value is a legal requirement in federal coal leasing. As a result, the federal government must determine a value for coal leases prior to competitive bidding in order to assure that

bonus bids and royalty provisions represent fair market value and are, therefore, acceptable. The federal government is faced with a similar valuation problem when determining or negotiating royalty provisions on other leased minerals.

TYPES OF VALUATION STUDIES

Valuation studies for mineral properties may be conducted for any number of reasons and may be as simple or sophisticated as the circumstances dictate. There are cases, for instance, where a highly successful exploration drilling program clearly suggests the presence of a major ore body. Under these conditions, rigorous economic analyses are not warranted to support the decision to continue the drilling project. Unfortunately, these types of projects are the exceptions rather than the rule in the mining industry. Most mining projects appear as economically marginal or borderline situations and no obvious decision is apparent. In these circumstances it is imperative that all variables affecting the project be assessed and incorporated in the decision-making process.

At this stage the normal procedure is to call for an initial feasibility study on the property. Feasibility studies are the heart of the valuation process. A feasibility study for a mining property is nothing more than an engineering/economic appraisal of the commercial viability of that property. The feasibility study is a relatively formal procedure for assessing the various relationships that exist between the factors which directly or indirectly affect the project in question. The objective of a feasibility study is to clarify the basic factors which govern the chances for project success. Once all the factors relative to the property have been defined and studied, an attempt is made to quantify as many of the variables as possible in order to arrive at a potential value or worth of the property.

As a mining property progresses from raw exploration through to the time when a management decision is made to develop and mine the property, a number of feasibility studies will be conducted on the property, each of which will be based on increasing amounts of data, require increasing amounts of time (and therefore expense) to prepare, and have increasing degrees of accuracy. Normally feasibility studies progress from order-of-magnitude studies to preliminary studies to definitive studies to detailed studies.

The effort expended on a feasibility study is dependent on the purpose of the evaluation and the required degree of accuracy. Obviously a complex, time-consuming feasibility study cannot be justified when only a crude indication of potential economic viability of the property is required by management. Also detailed feasibility studies are often self-defeating when dealing with exploration programs which have proven the existence of mineralization but have not delineated the entire mineralized zone or assessed the ore reserve potential. Care must be taken not to dignify a small amount of basic data with a detailed, sophisticated feasibility study. This is particularly true in the case of the economic analyses.

A more thorough discussion of the various types of valuation studies and the associated accuracy and cost trade-offs is presented in Chapter 6.

ADDITIONAL INVESTMENT GOALS

Throughout this book the focus is on calculating a monetary value for a mineral property based upon projection of future net cash benefits generated by the property. Thus, the subject matter deals strictly with mine valuation rather than the broader area of investment decision-making.

Capital budgeting theory implies that investment opportunities can be ranked numerically according to some investment criterion; an optimum mix of projects can be selected which will maximize shareholder wealth. In practice, of course, the process is far more complicated. Every investment opportunity is accompanied by a unique set of qualitative factors—sometimes referred to as irreducibles in engineering economy texts—and risks. In short, firms often do not select the project having the highest rate of return due to the level of risk involved and to other investment goals which management perceives to be in the long-range best interests of the firm. Some of these goals are:

1) **Degree of Necessity:** An expenditure may itself be unprofitable but may be necessary to permit a facility to continue operating to meet contractual obligations. Environmentally mandated expenditures occasionally fall in this category.

2) **Market Share:** In certain industries management perceives that maintaining a certain level of market penetration is essential and may weigh this aspect heavily in ranking investment opportunities.

3) **Customer Relations:** Some investments are undertaken to promote customer goodwill. In the mining equipment industry, for example, it is important to maintain regional parts depots and service staffs. It is difficult to assess the rate of return on such investments, but there is no question but that customer service facilities greatly enhance the marketing effort.

4) **Infrastructure:** Mines located in remote locations spend considerable sums to attract and maintain an experienced work force. Living accommodations, as well as shopping, recreational, medical, and other similar social expenditures are essential, though difficult to evaluate quantitatively.

5) **Strategic Investments:** Firms occasionally make major investments in new business areas, in high risk ventures, or in other activities from which the benefits are simply too speculative to accurately estimate. Good examples are research expenditures and a close analogy in the minerals industry, exploration. It is extremely difficult to attempt to calculate a rate of return on a basic research effort or a reconnaissance exploration program. Such expenditures are made in an attempt to assure the future well-being of the company, but they do not lend themselves easily to quantitative analysis.

Analytical financial models are powerful tools in the evaluation of prospective capital investments. However, it is naive to expect that the model can produce an investment decision. There are simply too many qualitative factors and unique risks associated with every mining investment opportunity to escape the need for sound managerial vision and judgment.

SELECTED BIBLIOGRAPHY

Archer, S.H., Choate, G.M., and Racette, G., 1979, *Financial Management*, John Wiley and Sons, New York.

Van Horne, J.C., 1974, *Financial Management and Policy*, 3rd ed., Prentice-Hall, Inc. Englewood Cliffs, NJ.

2
Basic Valuation Concepts

"There are two characters to the value of mining properties—one mine may have a value, owing to its real intrinsic worth; another (having no intrinsic value) may have a value by being so situated as to harass the working of the really valuable mine—in mining camps one is looked upon as much of ligitimate enterprise as the other."
— Victor Clement to Simeon Reed, Wardner, Jan. 11, 1890, Reed Mss.

APPROACHES TO VALUATION

Value and costs are two terms basic to financial analysis which cause, perhaps, the greatest amount of misunderstanding in mine valuation studies. When discussing the value of an asset or the costs of producing a product, it is absolutely essential that the speaker define the terms being used. There are many types of *value* and many categories of *costs*. Without specific accompanying definitions, value and cost data are not very meaningful.

Value refers to a measure of the desirability of ownership of property. Some of the types of value which might be encountered in a mine valuation study are:

1) Market value
2) Full cash value
3) Salvage value
4) Replacement value
5) Capitalized value
6) Book value
7) Assessed value
8) Insured value

Each of these has a specific meaning which can be applied to determine a monetary amount in a specific situation.

The major item of interest here is the broader question of "what is the value of the mine?" or "what is the mine worth?" In this case, we are speaking of *market value.* Market value is the value (price) established in a public market by exchanges between a willing buyer and a willing seller when neither is under duress to complete the transaction. Market value fluctuates with the degree of willingness of the buyer and seller and with the conditions of the sale. The use of the term *market* suggests the idea of barter. When numerous sales occur on a market, the result is to establish fairly definite market prices or values for specific assets.

The term *market value* is often used synonymously with the term *fair market value*. The courts have established the legal definition of fair market value as the amount in cash, or in terms reasonably equivalent to cash, for which in all probability the property would be sold by a knowledgeable owner willing but not obligated to sell to a knowledgeable purchaser who desired but is not obligated to buy. On numerous occasions the courts have stressed the importance of actual market values and the irrelevance of considerations and matters not directly affecting market values when determining *fair* market value.

Most property appraisers point out that the market value of a mineral deposit should be expressed as an estimated market value based on the time and conditions existing as of a specified date. Consequently, market value is as dynamic property which is constantly changing as market conditions and expectations change. At this point the important distinction between the *estimation* of market value as opposed to the *determination* of market value should be noted. Mineral economists, appraisers, and government tax officials, among others, are concerned with the estimation of market value for mineral properties. Ultimately the determination of market value for a specific mineral property can only be made by the market through an actual sales transaction. Note, however, that although a sale is a necessary condition for the determination of fair market value, it may not be a sufficient condition if either buyer or seller were acting under some actual or perceived duress.

To estimate the market value of any asset, most appraisers initially consider the three generally accepted approaches to value. These are: (1) the Cost Approach, (2) the Market or Comparable Sales Approach, and (3) the Income or Earnings Approach. All three approaches are based on the very important appraisal principle of substitution. A closer look at each of these approaches and their applicability to mining or mineral properties is provided in the following.

Cost Approach

With this approach one attempts to determine the depreciated replacement cost for the asset in question. That is, what would it cost to reproduce the asset with another of identical quality and state of repair? The fundamental concept with this approach is that a purchaser would not be justified in paying more for a property than it would cost him to acquire land and construct improvements having comparable utility, assuming no undue delay.

The cost approach is rarely applicable in mining because the correlation between construction costs and value of the property is very imperfect. If one were to build two 100 tpd mines—one on a very rich ore deposit and one on an economically marginal deposit—construction costs might be very similar, but fair market values of the two mines would clearly be substantially different. Similarly, the value of a concentrator when the ore deposit is depleted, or nearly depleted, is very limited regardless of the reproduction cost of the structure.

Another problem arises with this approach when applied to newly discovered mineral properties having no surface improvements or equipment of any kind. The very nature of mineral exploration and mining dictates that the discovery value of an ore deposit is generally greater than the cost incurred in making that discovery. If this were not true in the aggregate, money could not be justified for exploration. Furthermore, the notion of estimating the cost of acquiring a comparable asset (ore body) is not very useful. This cost could, for example, be infinite if nature failed to provide a duplicate for explorationists to find.

The cost approach is rarely applicable in valuing mining properties and is generally the least reliable method of valuation.

Market (Comparable Sales) Approach

The market approach is viewed by most appraisers and the courts as the best evidence of fair market value since it reflects the balance of supply and demand in the market place.

The market approach assumes a purchaser would not be justified in paying more for a property than it would cost him to acquire an equally desirable substitute property. In essence the analyst studies sales and purchases—the market—for similar assets in attempting to determine the market value for the item in question. The concept of market value also presumes conditions of an open market, exposure for a reasonable time, knowledgeable buyers and sellers, absence of pressure on either the seller to sell or the buyer to buy, and a sufficient number of transactions to create a stable market.

Although this method has been used extensively for estimating residential and agricultural property values, it encounters serious practical problems when applied to mining transactions. This is mainly due to two facts: (1) there are very few sales of mining properties and therefore few comparative data are available; and (2) since each mineral deposit is unique in quality, size, geographical location, degree of development, and many other parameters, any market data are of modest value at best. Even an actual previous sale of the property in question may not be a reliable indicator of its value if market conditions—commodity price expectations, production costs, ore reserve information—have changed. To be applicable the market data must not only be for similar assets, but also for a similar point in time.

In the absence of real or relevant data, the appraiser may estimate value using this approach by developing a model with supporting detailed calculations. Unfortunately, the appraiser's model, to a large extent, typically involves judgmental trade-offs rather than strict numerical relationships. As might be expected, many of the procedures incorporated throughout the model are judgmental in nature even though the procedures can be explicitly documented. As a result, significant variations in values may result from different appraisers following similar procedures.

Experience in the area of mineral property transactions suggests that arm's-length transactions and other assumptions previously mentioned in association with this approach are rarely present. In the rare instance when the criteria and assumptions previously delineated are met, it is often extremely difficult to ascertain the actual true value of the sale due to stipulations pertaining to production commitments, deferred payments, stock exchanges, production payments, and other subtle factors which can significantly affect value.

Income (Earnings) Approach

With the income approach, the value of an asset or investment-type property is estimated by calculating future annual net earnings generated from the producing property or asset and then discounting this earnings stream to the present time using an appropriate interest rate. Because of this procedure, many analysts refer to this approach as the capitalized income approach. The approach assumes that a purchaser would not be justified in paying more to acquire income producing property than the present value of the income stream to be derived from the property. Clearly, the presumption here is that the property is employed in business for the purpose of earning a profit.

In essence, the income approach is one step removed from the market (comparable sales) approach. If comparable sales data are unavailable or if one is estimating the value of a commodity in-situ, it is possible to arrive at a value estimate by combining the selling price of the commodity *produced* with the associated costs of producing the commodity from the property in question. By properly incorporating this data into a discounted cash flow analysis, it is possible to arrive at an estimate of property value even in the absence of actual production. The analyst must remember, however, that the value estimate thus obtained is not a direct estimate of the market value of a commodity in place, but rather it is an estimate of potential income generated from mining the commodity and selling the product.

Because mines have limited operating horizons and because there are well-established markets for mineral commodities, the income approach is used widely in valuing mineral properties. The approach is used commonly by the mining industry in assessing investment rates of return and determining appropriate purchase prices for mines or mineral prospects. Also, most taxing authorities employ some variation of the income approach when determining mine values for the purpose of ad valorem taxation.

Capitalized future income is a unit valuation method. That is, a single value is assigned to the ore deposit, surface and subsurface improvements, and all real and personal property used in the production process. Thus, the analyst does not partition the value among these assets. Because of their frequent remote location and specialized design, mining facilities have limited salvage values after the ore is exhausted. To a considerable degree, then, real property at mines has value only because of the presence of ore, and unit valuation is, therefore, appropriate.

This cursory discussion suggests that the estimating error could be large if either the market approach or the income approach were used to estimate the fair market value of mining properties. From a practical standpoint, the income approach has the capability of incorporating more obtainable, realistic data for analysis and, therefore, is the preferred approach. In addition, the income approach is consistent with the generally accepted definition for the value of a mineral property. That is, the value of a mining or mineral property, at a specific point in time, is simply the present value of all anticipated future net annual proceeds accruing from ownership. By the very nature of the income approach, one is necessarily projecting future costs and prices. The expectation of these various future events necessarily leads to variations in calculated *values* of the same mining property. In addition, the income approach is not without some other rather formidable problems, as will be discussed in the remainder of this book. In some valuation cases in the public sector, these difficulties become so troublesome that an alternative, less precise method is selected. Nonetheless, the preferred method for mining property valuation and the one unanimously used in commercial practice is the income approach. The basic element in the income approach to mine valuation is the pro forma income statement which is discussed in a subsequent section of this chapter.

Other Types of Value

As mentioned previously, there are a number of types of value which an analyst might encounter when performing a mine valuation study. Following are brief descriptions of some of them.

Salvage Value: Salvage is the net sum, over and above the cost of removal and sale, realized for a property or asset when it is retired from service. Salvage value and scrap value are synonymous when the property or asset retired from service is scrapped for the value of its materials.

Replacement Value: Replacement value refers to the existing value of a property or asset as determined on the basis of what it would cost to replace the property or its service with at least equally satisfactory and comparable property and service. The concept of replacement value is fundamental to the cost approach utilized by appraisers.

Book Value: Book value is the original investment in the property or asset as carried on the corporate books less any cumulative allowance for depreciation or amortization entered on the books.

Assessed Value: The assessed value of a property is the value entered on the official assessor's records as the value of the property applicable in determining the amount of ad valorem taxes to be paid by the property. Some of the various procedures and techniques used by state and local taxing authorities for determining assessed values are discussed in Chapter 8.

Insured Value: The insured value of a property refers to that value at which the property has been insured against loss or disaster. This value is generally associated with replacement value for tangible assets and earning capacity for property such as ore deposits.

Capitalized Value: The capitalized value of a property is the sum of discounted future annual net earnings generated by the property. The capitalized value concept is synonymous with the income approach to value estimation for properties.

BASIC COST TERMINOLOGY

Costs: Cost data are subject to even greater misunderstanding than are value data. The main reason for this is that, although the various categories of costs have precise meaning to the accountant, these categories often do not lend themselves to efficient cash-flow-based decision making. Furthermore, accounting interpretations can vary by a significant degree from company to company. As a consequence, when XYZ Copper Co. is reported to have production costs of, say, 75¢ per pound, very little useful information is communicated unless *production costs* is further defined.

An excellent classification of total costs of production is provided in Table 1, which is taken from Jelen (1970). The major breakdowns are:

I. Operating Costs
   A. Direct Costs
   B. Indirect costs
   C. Contingencies
   D. Distribution Costs
II. General Expense
   A. Marketing Expense
   B. Administrative Expense

With the possible exception of some industrial minerals, distribution costs would not be sufficiently large to jusify a separate heading and would be combined with the other operating cost categories.

Table 1. Production Cost Components*

I. Operating Cost or Manufacturing Cost
  A. Direct Production Costs
    1. Materials
      a. Raw materials
      b. Processing materials
      c. Byproduct and scrap credit
      d. Utilities
      e. Maintenance materials
      f. Operating supplies
      g. Royalties and rentals
    2. Labor
      a. Direct operating labor
      b. Operating supervision
      c. Direct maintenance labor
      d. Maintenance supervision
      e. Payroll burden on all labor charges
        (1) FICA tax
        (2) Workmen's compensation coverage
        (3) Contributions to pensions, life insurance, etc.
        (4) Vacations, holidays, sick leave, overtime premium
        (5) Company contribution of profit sharing
  B. Indirect Production Costs
    1. Plant overhead or burden
      a. Administration
      b. Indirect labor
        (1) Laboratory
        (2) Technical service and engineering
        (3) Shops and repair facilities
        (4) Shipping department
      c. Purchasing, receiving, and warehouse
      d. Personnel and industrial relations
      e. Inspection, safety, and fire protection
      f. Automotive and rail switching
      g. Accounting, clerical, and stenographic
      h. Plant custodial and plant protective
      i. Plant hospital and dispensary
      j. Cafeteria and clubrooms
      k. Recreational activities
      l. Local contributions and memberships
      m. Taxes on property and operating licenses
      n. Insurance—property, liability
      o. Nuisance elimination—waste disposal
    2. Depreciation
  C. Contingencies
  D. Distribution Costs
    1. Containers and packages
    2. Freight
    3. Operation of terminals and warehouses
      a. Wages and salaries plus payroll burden
      b. Operating materials and utilities
      c. Rental or depreciation

II. General Expenses
  A. Marketing or Sales Costs
    1. Direct
      a. Salesmen salaries and commissions
      b. Advertising and promotional literature
      c. Technical sales service
      d. Samples and displays
    2. Indirect
      a. Sales supervision
      b. Travel and entertainment
      c. Market research and sales analysis
      d. District office expenses
  B. Administrative Expenses
    1. Salaries and expenses of officers and staff
    2. General accounting, clerical, and auditing
    3. Central engineering and technical
    4. Legal and patent
      a. Within company
      b. Outside company
      c. Payment and collection of royalties
    5. Research and development
      a. Own operations
      b. Sponsored, consultant, and contract work
    6. Contributions and dues to associations
    7. Public relations
    8. Financial
      a. Debt management
      b. Maintenance of working capital
      c. Credit functions
    9. Communications and traffic management
    10. Central purchasing activities
    11. Taxes and insurance

---

*Jelen, 1970, p. 339.

Operating Costs are taken to be all expenses at the plant site, whereas *general expenses* are off-site management or corporate-level expenditures. General expenses may be directly related to plant-site activity or they may be indirect headquarters' items which are allocated across all production divisions.

Direct vs. Indirect Costs: Direct costs, or variable costs, are those items such as labor, materials, and supplies, which are consumed directly in the production process and which are used roughly in direct proportion to the level of production. On the other hand, indirect costs, or fixed costs, are expenditures which are independent of the level of production—at least over certain ranges.

In the limit, there are, obviously, few truly fixed costs. If the enterprise is liquidated, most *fixed* costs are eliminated; and in cases where production is severely curtailed or greatly expanded some indirect costs (e.g., insurance) would change. Nonetheless, the concept of fixed vs. variable costs is valid in a general sense and is useful in understanding some of the characteristics of a particular industry.

The mining industry, for example, is characterized by a high degree of capital intensity. In the category of assets per unit of sales, mining ranks near the top of all industrial sectors. In addition to depreciation, other items of indirect (fixed) costs, such as taxes, are also higher than average for mining. The result can be seen in the idealized graphs in Fig. 1. Here it can be seen that the relatively high level of fixed costs in mining usually means that the break-even production level for mining facilities is closer to capacity than for firms with lower fixed costs. This is a major contributing factor in why operators attempt to run mines at capacity, often employing three-shift, seven-day-per-week work schedules.

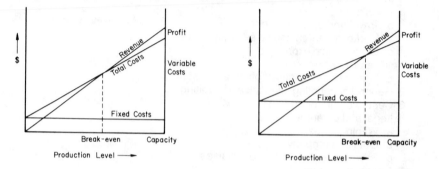

Firm with low fixed costs.   Firm with high fixed costs (e.g., mining).

Fig. 1. Idealized cost-revenue relationships for firms with relatively low and relatively high (e.g., mining) fixed costs of production.

A second characteristic of mining can also be observed in Fig. 1. The difference in slopes between the total cost and revenue curves is greater for mining than for less capital-intensive industries. That is, earnings in mining are highly leveraged by the level of production. A small increase or decrease in output yields a relatively large change in profit. This is simply another way of illustrating the well-known sensitivity of profits to metal prices in the metal mining business.

Capital Costs

In addition to *operating costs,* the mine investment decision clearly must also consider *capital costs.* Capital costs (or first cost, or capital investment) are those expenditures made to acquire or develop capital assets, the benefits from which will be derived over several years. The largest share of capital costs is incurred to initially get the project started, but some capital expenditures are made yearly throughout the life of the mine.

Capital costs fall into one of three classes, depending upon the treatment of the cost for income tax purposes.

1) **Depreciable Investment:** This is investment in a capital asset which is allocated over the useful life of the asset according to some formula acceptable to the tax authorities. All types of mining machinery and equipment fall in this category.

2) **Expensible or Amortizable Investment:** Expenditures in this class can, at the taxpayer's option, be either charged off against revenue immediately, or capitalized and amortized over some reasonable period. Mine development is a good example here where the amortization option can be exercised by charging off such development at the same rate as the ore which is thus exposed is mined.

3) **Nondeductible Investment:** Included here are capital expenditures which cannot be deducted for tax purposes. Examples are successful exploration and

property acquisition—which become the basis for the depletion allowance—and working capital, which is recovered at the end of the mine's life.

Obviously, the attractiveness of a mining investment is affected by the amount of capital investment involved. It may not be quite so obvious, however, that the *type* of capital expenditures involved can also be a very important factor in evaluating a prospective new project. This is primarily due to different tax treatments accorded different types of capital expenditures.

Other Cost Concepts

There are other cost concepts which arise frequently in investment analysis. Some of these are defined here for reference.

1) **Cash vs. Noncash Costs:** Cash costs are those which represent actual monetary outlays. Noncash costs do not directly represent such outlays but are permissible deductions from revenue, the sole impact of which is to reduce the income tax liability. Depreciation and depletion are two important examples of noncash costs.

2) **Sunk Cost:** A sunk cost is simply an expenditure that has already been made. Sunk costs are irrelevant to a capital investment decision which must weigh only future benefits against future costs. Although there may be a strong personal commitment to some previous capital investment, those funds have been irrevocably spent and that decision, therefore, should have a bearing in subsequent investment decisions only to the extent that some money can be recovered from these transactions in the future.

3) **Marginal Costs and Benefits:** Only those costs and benefits to be experienced by the firm due to the contemplated investment are relevant to an investment decision. These marginal cash flows do not include, for example, allocated corporate overhead which would be incurred regardless of whether or not the new project is accepted.

4) **Cost of Capital:** Capital costs, which were discussed previously, and the cost of capital are two entirely different concepts. The cost of capital is discussed in detail later in this book (Chapter 11). Basically, this term is used to refer to the cost of raising funds for capital investment. The cost is expressed as a percent and is usually determined by combining the costs of specific sources of capital (debt and equity) into a single value based upon the firm's relative use of the various sources.

5) **Opportunity Cost:** This cost refers to the yield or rate of return foregone on the most profitable investment opportunity rejected by a firm. These costs are generally experienced when capital rationing constraints are imposed in the capital budgeting process. When budget ceilings are imposed, projects which are otherwise profitable may be rejected. The resulting cost to the firm associated with rejecting these projects is the opportunity foregone on the most profitable investment alternative remaining unfunded.

DATA REQUIREMENTS

Nothing improves the output of an engineering/economic evaluation of a mining property more than good input data. Unfortunately, those preparing feasibility studies for mining properties or projects never have all the information they would like. In addition to inadequacy or unavailability of some needed data, care must be taken not to overlook any variable which may influence project viability. In this regard it is often helpful to compile a list of factors which should be considered when preparing feasibility studies on mining properties. Table 2 is an outline of some of the pertinent factors which must be considered, studied, and analyzed when evaluating mining properties. Obviously the significance of each factor will be a function of the specific property being investigated and the mineral commodity (metallic, nonmetallic, fuel) involved. For example, Table 3 illustrates the salient factors requiring consideration for feasibility studies in coal and clearly shows that the same variables are not of equal importance for all commodities. Nonetheless, all of these factors should be assessed to some degree during preparation of at least one of the feasibility studies conducted throughout the evaluation period.

Table 2. Salient Factors Requiring Consideration in a Feasibility Study*

I. Information on Deposit
  A. Geology
    1. Mineralization: type, grade, uniformity
    2. Geologic structure
    3. Rock types: physical properties
  B. Geometry
    1. Size, shape, and attitude
    2. Continuity
    3. Depth
  C. Geography
    1. Location: proximity to towns, supply depots
    2. Topography
    3. Climatic conditions
    4. Surface conditions: vegetation, stream diversion
    5. Political boundaries
  D. Exploration
    1. Historical: district, property
    2. Current Program
    3. Reserves: tonnage, distribution, classification
    4. Sampling: types, procedures
    5. Proposed program

II. Information on General Project Economics
  A. Markets
    1. Marketable form of product: concentrates, direct shipping ore, specifications
    2. Market location and alternatives
    3. Expected price levels and trends: supply-demand, competitive cost levels, new source of product substitutions, tariffs
  B. Transportation
    1. Property access
    2. Product transportation: methods, distance, costs
  C. Utilities
    1. Electric power: availability, location, ownership right of way, costs
    2. Natural gas: availability, location, costs
    3. Alternatives: on-site generation
  D. Land and Mineral Rights
    1. Ownership: surface, mineral, acquisition and/or option costs
    2. Acreage requirements: concentrator site, waste dump location, tailings pond location

E. Water
  1. Potable and process: sources, quantity, quality, availability, costs
  2. Mine water: quantity, quality, depth and source, drainage method, treatment
F. Labor
  1. Availability and type: skilled/unskilled in mining
  2. Rates and trends
  3. Degree of organization
  4. Local/district labor history
G. Governmental Considerations
  1. Taxation: federal, state, local
  2. Reclamation and operating requirements and trends
  3. Zoning
  4. Proposed and pending mining legislation

III. Mining Method Selection
  A. Physical Controls
    1. Strength: ore, waste, relative
    2. Uniformity: mineralization, blending requirements
    3. Continuity: mineralization
    4. Geology: structure
    5. Surface disturbance: subsidence
    6. Geometry
  B. Selectivity
  C. Production Requirements
    1. Relative production
    2. Development: methods, quantity, time requirements
    3. Capital requirements vs. availability

IV. Processing Methods
  A. Mineralogy
    1. Properties of ore: metallurgical, chemical, physical
    2. Ore hardness
  B. Alternative Processes
    1. Establish flowsheet, recovery grade
    2. Production schedule
  C. Production Quality vs. Specifications
  D. Recoveries

V. Capital and Operating Cost Estimates
  A. Capital Costs
    1. Exploration
    2. Mining
      a. Preproduction development (may also be considered operating costs)
      b. Site preparation
      c. Mine buildings
      d. Mine equipment
    3. Mill
      a. Site preparation
      b. Mill buildings
      c. Mill equipment
      d. Tailings pond
  B. Operating Costs
    1. Mining
      a. Labor

      b. Maintenance and supplies
      c. Development
   2. Milling
      a. Labor
      b. Maintenance and supplies
   3. Administrative and supervisory

*Gentry and Hrebar, 1978.

Table 3. Salient Factors Requiring Consideration in Coal Property Feasibility Studies*

I. Information on Deposit
  A. Geology
     Overburden
      stratigraphy
      geologic structure
      physical properties (highwall and spoil characteristics, degree of consolidation)
      thickness and variability
      overall depth
      topsoil parameters
     Coal
      quality (rank and analysis)
      thickness and variability
      variability of chemical characteristics
      structure (particularly at contacts)
      physical characteristics
  B. Hydrology (Overburden and Coal)
      permeability
      porosity
      transmissivity
      extent of aquifer(s)
  C. Geometry
     Coal
      size
      shape
      attitude
      continuity
  D. Geography
      location (proximity to distribution centers)
      topography
      altitude
      climate
      surface conditions (vegetation, stream diversion)
      drainage patterns
      political boundaries
  E. Exploration
      historical (area, property)
      current program
      sampling (types, procedures)

II. General Project Information
  A. Market
      customers

product specifications (tonnage, quality)
locations
contract agreements
spot sale considerations
preparation requirements

B. Transportation
property access
coal transportation (methods, distance, cost)

C. Utilities
availability
location
right-of-way
costs

D. Land and Mineral Rights
ownership (surface, mineral, acquisition)
average requirements (on and off-site)
location of oil and gas wells, cemeteries, etc.

E. Water
potable and preparation
sources
quantity
quality
costs

F. Labor
availability and type (skilled and unskilled)
rates and trends
degree of organization
labor history

G. Governmental Considerations
taxation (local, state, federal)
royalties
reclamation & operating requirements
zoning
proposed and pending mining legislation

III. Development and Extraction
A. Compilation of Geologic and Geographic Data
surface and coal contours
isopach development (thickness of coal and overburden, stripping ratio,
quality, and costs)

B. Mine Size Determination
market
optimum economic

C. Reserves
method(s) of determination
economic stripping ratio
mining and barrier losses
burned, oxidized areas

D. Mining Method Selection
topography
refer to previous geologic/geographic factors
production requirements
environmental considerations

E. Pit Layout
extent of available area

      pit dimensions and geometry
      pit orientation
      haulage, power, and drainage systems
  F. Equipment Selection
      sizing, production estimates
      capital and operating cost estimates
      repeated for each unit operation
  G. Project Cost Estimation (Capital and Operating)
      mine
      mine support equipment
      office, shop, and other facilities
      auxiliary facilities
      manpower requirements
  H. Development Schedule
      additional exploration
      engineering and feasibility study
      permitting
      environmental approval
      equipment purchase and delivery
      site preparation and construction
      start-up
      production

IV. Economic Analysis
      sections III and IV repeated for various alternatives

---

*Gentry and Hrebar, 1980.

    A quick review of these tables suggests that there are some fundamental areas of concern which are applicable to all mining property valuations. For instance, one of the first tasks associated with any mining property is estimating the magnitude and quality of the ore reserves. Ore is, of course, an economic term and is a function of commodity prices, production costs, mining method, recoveries, dilution, and a number of other variables. Because ore reserves are determined by ever-changing economic conditions, the exact amount of ore contained in a given deposit cannot be determined with certainty until the last truck-load or skip-load of ore has left the mine. Only then can one make an absolute statement regarding the ore reserves in a specific deposit. This uncertainty with respect to ore reserves has a significant impact on the valuation of mineral properties where long-term contracts at stipulated rates (for most or all of the reserve as determined from current conditions) are not available.

    Another area of fundamental interest in feasibility studies of mining properties which requires considerable data generation and analysis is projecting sales revenues. The timing and magnitude of mining revenues will depend upon factors such as ore reserves, production rates, commodity prices, markets, and metal-lurgical recoveries. These variables are often extremely difficult to estimate or predict—particularly for commodities traded in international markets. Even the prices of energy commodities have been historically difficult to predict in recent years. Note, for example, the wide variations between predicted and actual prices of coal and uranium from the mid-1970s through the early 1980s. For instance, the tremendous increase in uranium exploration, development, and production which occurred in the late 1970s was based on predictions of strong markets and high prices in the very near future. Although the price of yellow-cake did reach the

$41-$43-per-lb range in 1979, this was far less than the $60-$70-per-lb range anticipated by many companies. The irony is that the peak price of $43 per lb lasted only a few months before it began to slip downward. Following the serious accident at the Three Mile Island (PA) nuclear generating plant, the price of yellowcake dropped rapidly to the $20-$25-per-lb range. The net effect of these price, market, and subsequent revenue fluctuations was that many companies tabled existing projects, significantly reduced exploration activities, and closed down mines which had recently been brought into production.

Production technology is another key area of concern when performing mining feasibility studies. Technological advancements in equipment, mineral processing, and other areas can significantly impact projected operating and capital costs. A good example of the impact of technology changes on mining costs is the comparison between direct mining costs per ton of rock for underground and surface operations. While underground operating costs have been increasing at significant rates over the last ten years (mainly because of the lack of major technological advancements and the labor intensity which remains), unit operating costs at surface operations have changed far less. The technological advancements in mining equipment, yielding productivity increases, have helped keep direct operating costs in surface operations from escalating at the rates experienced by underground producers. The relative stability in surface operating costs has not been entirely free however. The technological advancements in mining equipment which contributed to this stability in operating costs carried with them significant increases in capital costs.

Operating and capital cost requirements must be determined separately when formulating the feasibility study. However, in the limiting case, it is the combination of the magnitude and timing of both of these costs which ultimately influence the analysis. Any changes in future production technology must be carefully analyzed and the impact assessed on overall operating and capital cost requirements.

The overall operating environment is another area of major concern. In recent years the operating environment of mining properties has been significantly impacted by environmental and other regulatory requirements. The net effect has been that various physical and legal constraints have been imposed upon mining operations. These constraints have invariably increased operating and capital cost requirements for the industry and have reduced or delayed the production of mineral commodities. The operating environment of mining operations is also affected by direct economic variables such as royalties and taxes mandated by federal, state, and local taxing authorities. All of these costs, whether direct or indirect, impact profit margins, ore reserves, mineral conservation, and ultimately project viability.

The basic problems and requirements set forth here regarding mining project valuation are addressed at some length in subsequent chapters of this book.

ACCOUNTING AND INVESTMENT ANALYSIS

A great deal of confusion has been created from time to time due to differences between accounting concepts and valuation concepts. To be sure, valuation, particularly of producing mines, relies heavily on the use of data collected for accounting purposes. However, the direct use of financial accounting statements in valuation calculations can lead to serious errors.

Accounting is a vital administrative function that serves many purposes in a commercial organization. The common summary financial statements prepared by the accounting department are:

1) **Income Statement (or Profit and Loss Statement):** This statement summarizes the financial *performance* of the firm. It matches the amounts received from selling the goods and other items of income against all the costs and outlays incurred in order to operate the company. Thus, it measures a flow—the total revenues, costs, and earnings over a given period of time—and shows how much the corporation made or lost during the time period.

2) **Balance Sheet:** This statement summarizes the financial *condition* of the firm. As such it measures a stock—what is owned by the firm (its assets), and what is owed by the firm (its liabilities) at a given point in time.

3) **Changes in Financial Position (or Sources and Uses of Funds):** This statement explains changes in the balance sheet from the end of the preceding period to the end of the current period. These changes result from the firm's financial performance in the intervening period (as measured by the income statement) as well as from capital transactions that occurred during that period.

Examples of these statements are shown in Table 4. These financial statements represent the summation of literally thousands of individual transactions completed during the period. Statements are prepared at least quarterly for all corporations in the United States.

In order to permit meaningful interpretation of accounting statements, a consistent classification and treatment of costs is obviously required. The Financial Accounting Standards Board (FASB) publishes a set of rules which public corporations observe in external reporting of the financial condition of the firm to their shareholders. Although these rules facilitate some comparisons between firms, considerable discretion still exists in their interpretation. Cost accounting, which is designed to assist internal management control, is less rigorously regulated and also results in frequent important inconsistencies among firms.

There are three areas where potentially major errors can occur in using financial accounting statements for evaulation purposes.

1) **Tax Accounting vs. Financial Accounting:** Every corporation keeps at least two sets of books—tax books, which attempt to conform to the ever-changing rules of the federal income tax code, and financial books, which comply with professional accounting practices as formally defined by the FASB. There are often major differences between the two procedures. For example, for tax purposes mining firms can benefit from accelerated depreciation and percentage depletion, whereas accepted financial accounting practice requires straight-line or unit-of-production procedures for capital write-offs.

Because of these tax benefits, a mining firm generally pays lower taxes and has a higher cash flow than is implied from the net earnings shown on the financial income statement. Although the differences between tax and financial accounting appear in the financial accounts as deferred taxes and deferred charges, the reconciliation is often difficult without considerable additional detail.

2) **Time Value of Money:** The proper method for evaluating a capital investment is to compare the present outlay with the anticipated positive net cash flows which will accrue from the project in the future. In making this comparison it is essential that the timing of the various cash flows be recognized by the use of an appropriate interest rate.

Table 4. Statement of Consolidated Income
(in Thousands of Dollars)

	1981	1980*	1979*
Revenues			
Sales and other operating revenues	$1,438,555	1,440,137	1,280,830
Equity earnings	33,781	23,280	16,311
Interest and miscellaneous income	46,530	39,239	8,341
	1,518,866	1,502,656	1,305,482
Costs and expenses			
Costs of products sold	1,230,930	1,191,922	987,997
Depreciation, depletion, and amortization	77,374	70,441	67,303
Selling and general administrative expenses	58,445	54,228	46,326
Interest expenses	41,921	43,747	40,117
Exploration and research expenses	31,075	22,994	20,768
	1,439,745	1,383,322	1,162,511
Income before taxes and extraordinary item	79,121	119,334	142,971
Provision for taxes on income	(20,629)	(28,020)	(32,200)
Income before extraordinary item	54,492	91,314	110,771
Extraordinary item.	10,829	—	—
Net income	$ 69,321	91,314	110,771
Per common share after preferred dividend requirement			
Income before extraordinary item . .	$ 2.61	4.20	5.06
Extraordinary item.	0.51	—	—
Net income	$ 3.12	4.20	5.06
Average number of shares outstanding	21,019,335	20,831,647	20,698,414

*Certain amounts have been reclassified for comparative purposes.

Statement of Consolidated Retained Earnings
(in Thousands of Dollars)

	1981	1980	1979
Retained earnings at begining of year	$ 876,083	821,112	741,248
Net income	69,321	91,314	110,771
Dividends declared and paid:			
Preferred shares	(3,775)	(4,038)	(6,062)
Common shares	(33,797)	(32,305)	(24,845)
Retained earnings at end of year.	$ 907,832	876,083	821,112

Consolidated Balance Sheet (in Thousands of Dollars)

	December 31*		
	1981	**1980**	**1979**
Assets			
Current assets:			
Cash and short-term investments, at cost	$ 15,428	19,576	12,034
Receivables, less allowance for doubtful accounts (1981, $7,559; 1980, $5,674; 1979, $3,874)	167,527	202,728	166,855
Inventories	177,039	176,257	152,046
Supplies, at cost or less	101,692	116,598	120,141
Prepaid expenses	6,782	4,682	6,774
Deferred income tax charges	4,961	5,274	4,330
Current assets	473,429	525,115	462,180
Investments and long-term receivables	190,881	171,904	176,635
Property, plant and equipment	1,473,333	1,408,244	1,354,772
Deferred charges	6,751	6,841	6,631
	$2,144,394	2,112,104	2,000,218
Liabilities			
Current liabilities:			
Short-term borrowings	$ 42,003	40,263	—
Current portion of long-term debt	11,208	11,195	45,958
Accounts payable and accrued expenses	198,995	199,639	165,338
Income taxes	38,585	34,421	35,197
Current liabilities	290,791	285,518	246,493
Long-term debt	579,493	626,749	605,554
Deferred income taxes	125,366	112,553	95,296
Other income taxes	11,451	9,561	9,092
	1,007,101	1,034,381	956,435
Redeemable Preferred Shares Redemption value $100, par value $1; 6,000,000 shares authorized; 470,000 shares outstanding (1980, 470,000; 1979, 750,000)	48,290	48,290	74,590
Common Shareholders' Equity Common shares, par value $6.25; 30,000,000 shares authorized; 21,739,162 shares outstanding (1980; 20,880, 971; 1979; 20,717,838)	135,870	130,506	129,486
Capital in excess of par value	45,301	22,844	18,595
Retained earnings	907,832	876,083	821,112
	1,089,003	1,029,433	969,193
	$2,144,394	2,112,104	2,000,218

Statement of Changes in Consolidated Financial Position
(in Thousands of Dollars)

	1981	1980	1979
Funds provided by:			
Operations:			
Income before extraordinary item $	58,492	91,314	110,771
Adjustment for items not involving working capital:			
Depreciation, depletion, and amortization	77,374	70,441	67,303
Deferred income taxes	12,813	17,427	13,953
Undistributed earnings of equity investments	(29,521)	(18,895)	(10,026)
Dry hole expenditures charged to income	6,448	2,193	1,002
Sales of preoperating production	277	2,566	16,061
Total from operations.	125,883	165,046	199,064
Long-term debt	—	31,285	1,000
Common shares issued in exchange for debt	25,377	—	—
Other increases in common shares outstanding.	2,444	5,269	748
Long-term receivables repaid or classified as current.	10,694	3,120	4,674
Equity in consolidated Aluminum Corp. at date of disposition	—	84,294	—
Less long-term receivable acquired	—	(67,088)	—
Book value of other assets sold	872	3,943	2,454
Other items, net	11,575	7,947	8,656
Total funds provided	176,845	233,816	216,596
Funds used for:			
Dividends	37,572	36,343	30,907
Capital outlays	147,653	126,668	106,807
Capitalized interest	12,311	9,572	11,750
Principal amount of debt acquired through issuance of common shares	35,630	—	—
Less extraordinary gain on exchange	(10,829)	—	—
Long-term debt repaid or classified as current	11,467	11,023	60,159
Purchase of preferred shares	—	26,300	—
Total funds used.	233,804	209,906	209,623
Increase (decrease) in net current assets. . . . . $	(56,959)	23,910	6,973
Change in net current assets:			
Current assets:			
Cash and short-term investments	(4,149)	7,542	(10,724)

Receivables, net	(35,201)	35,873	26,959
Inventories and supplies........	(14,124)	20,668	37,113
Prepaid expenses	2,100	(2,092)	3,088
Deferred income tax charges....	(313)	944	(2,639)
Increase (decrease)	(51,686)	62,935	53,797
Current liabilities:			
Short-term borrowing	1,740	40,263	—
Current portion of long-term debt	13	(34,763)	33,901
Accounts payable and accrued			
expenses	(644)	34,301	30,281
Uranium delivery obligations	—	—	(29,547)
Income taxes	4,164	(776)	12,189
Increase	5,273	39,025	46,824
Increase (decrease) in net			
current assets............. $	(56,959)	23,910	6,973

In contrast, a firm's financial accounts simply summarize monetary transactions and are not designed to judge the success of some past investment. For example, a large profit in the current year does not mean that the owner is receiving a high return on his investment. In order to assess the investment return on any project, the time profile of its cash flows must also be examined. An accounting income statement simply covers a single period and is ignorant of any previous or anticipated future profits or losses.

3) **Allocation vs. Valuation:** By recording only monetary transactions, accounting data do not include any change in the market value of capital assets after acquisition. For example, ore deposits are carried at discovery cost, which often bears little relationship to their value. Furthermore, accounting procedures allocate the capital cost of depreciable assets to operations over the estimated useful life of those assets. This allocation is purely a mathematical exercise and ignores the actual market value of the asset which could be declining at a different rate, remaining unchanged, or actually rising.

There is an imperative need for sound production cost estimates in any mine valuation exercise. To the extent that the valuation analyst often relies on data collected for accounting purposes as the basis for these cost projections, there is a clear need to understand accounting statements. However, the guiding principles of investment analysis are cash flow and time value of money, neither of which are of any direct concern to the accountant.

SELECTING A DECISION VARIABLE

The importance of the pro forma income statement in estimating the value of a mine or mineral property was alluded to previously. Also, examples of income statements were illustrated in the previous section along with a discussion on the fundamental differences between tax and financial accounting practices. Inasmuch as there are generally major differences between accounting profits and actual net cash benefits, investors are increasingly using cash flow as the primary measure of benefits produced by a capital project. Cash flow analyses and accounting concepts depict investments differently. The main difference between these approaches is the timing of costs, as mentioned previously.

A cash flow analysis relates the expenditures associated with investments to the subsequent revenues or benefits generated from these investments. Cash flows are frequently calculated on an annual basis for evaluation purposes and are determined by subtracting annual outflows from annual inflows resulting from the investment. Therefore a cash flow analysis may be made for any investment which has income and expense associated with it.

Annual cash flows resulting from an investment may be either positive or negative. Net cash flows for a new mining property will be negative during the preproduction years due to large capital expenditures. After production commences, the cash flows will usually be positive and an inflow of cash results from the investment in the project.

Net cash flow is basically a combination of two components. It consists of: (1) the return *on* the investment, and (2) the recoupment *of* the investment. In the mineral industry net cash flow is defined as net income after taxes plus depreciation and depletion minus capital expenditures and working capital. The net income after taxes represents the return on the investment while depreciation and depletion represent the recoupment of the investment.

The fact that depreciation and depletion are added back in the cash flow calculation often causes confusion. In a cash flow analysis, each investment receives credit for income taxes saved. Since depreciation and depletion allowances reduce the amount of taxable income (and therefore reduce the amount of taxes paid) they have the effect of saving the organization money. Therefore they are a credit to the cash flow calculation and are added to net income after taxes. It is important to realize that depreciation and depletion are noncash items and do not actually *flow* anywhere.

Table 5 illustrates the components and basic calculation procedure for determining annual cash flows for a mining property. Table 6 lists some of the more important factors relating to preproduction, production, and postproduction mining activities which need to be considered when preparing cash flow analyses. The appropriate use and manipulation of these input variables is an extremely important

Table 5. Components of an Annual Cash Flow Calculation

Calculation	Component
	Revenue
Less	Royalties
Equal	Gross Income from Mining
Less	Operating Costs
Equal	Net Operating Income
Less	Depreciation and Amortization Allowance
Equal	Net Income After Depreciation and Amortization
Less	Depletion Allowance
Equal	Net Taxable Income
Less	State Income Tax
Equal	Net Federal Taxable Income
Less	Federal Income Tax
Equal	Net Profit After Taxes
Add	Depreciation and Amortization Allowances
Add	Depletion Allowance
Equal	Operating Cash Flow
Less	Capital Expenditures
Less	Working Capital
Equal	Net Annual Cash Flow

facet of the cash flow analysis. The concept of cash flow analysis is a particularly useful technique for the evaluation of mineral-related projects because of the important impact of the depletion allowance in the United States.

In spite of the foregoing, some students find confusing the idea that cash flow is more important than profit. It is important to remember that profit is an accounting concept, subject to an extensive set of fairly rigid rules established by the accounting profession. In the final analysis, however, an investor is simply concerned with how much *cash* surplus a project will generate in relation to how much cash outlay it required. Unlike the accountant, the investor is not particularly interested in the method for determining the level of net cash flow from a project. His major concern is to estimate whether or not the cash in will exceed the cash out by a sufficient amount.

This is not to say that profit is irrelevant. Profit is often the largest component of cash flow, but depreciation, amortization, and depletion also often account for a large share of a project's cash flow. As mentioned previously, these three items are noncash expenses since they do not represent cash transactions during the current tax year. The financial impact of these noncash expenses is simply to reduce the amount of income taxes that would otherwise be paid.

The income statement presented on page 31 is designed to promote rapid calculation of annual cash flows. This is illustrated in the following example:

	Item	Amount	Cash Items Only
	Gross Sales (NSR)	$100	$100
less:	Royalties	2	2
	Gross Income from Mining	98	98
less:	mining costs	24	24
	beneficiation costs	20	20
	general expense	16	16
	Net Operating Income	38	38
less:	fixed charges S & A	8	8
	depreciation, amortization	10	
	depletion	4	
	Pretax Net Income	16	
	Income tax @ 50%	8	8
	Net Profit	8	
Add Back:	depreciation & amortization	10	
	depletion	4	
	Net Operating Cash Flow	22 =	22

The income statement format is needed in order to calculate the income tax liability, which is often one of the largest expenses of the venture. In this regard it might be worthwhile to reiterate the fact that in a cash flow analysis each investment receives a credit for income taxes saved. Therefore, for profitable organizations it is advantageous to maximize pretax deductions and thereby reduce the amount of taxable income and, consequently, income taxes paid. In order to take advantage of these tax savings as soon as possible, the firm would opt to expense all possible expenditures in the year incurred as opposed to capitalizing them followed by subsequent write-offs over the amortization period. Although the total amount of the pretax deduction would be the same in either case, by

Table 6. Factors for Consideration in Cash Flow Analysis of a Mining Property*

Preproduction Period

Exploration expenses	Land and mineral rights
Water rights	Environmental costs
Mine and plant capital requirements	Development costs
Sunk costs	Financial structure
Working capital	Administration

Production Period

Price	Capital investment—replacement and expansions
Processing costs	Royalty
Recovery	Mining cost
Postconcentrate cost	Development cost
Reserves and percent removable	Exploration cost
Grade	General and administration
Investment tax credit	Insurance
State taxes	Production rate in tons per year
Federal taxes	Financial year production begins
Depletion rate	Percent production not sent to processing plant
Depreciation	Operating days per year

Postproduction

Salvage value	Contractual and reclamation expenditures

*Laing, 1977.

expensing as soon as possible the firm will realize an earlier return of the resulting tax savings. This early return of tax savings enables the firm to utilize these dollars sooner than would otherwise be possible.

Normal cash flow analyses may be utilized to evaluate projects on the basis of maximizing benefits or minimizing costs. Most mining projects are analyzed from the standpoint of maximizing benefits, whereas equipment replacement decisions or production system alternatives are often evaluated from the standpoint of minimizing costs. When possible, investment decisions should be evaluated from the standpoint of maximizing benefits. Although there are a number of reasons why analyses based on cost minimization can be misleading, perhaps the most obvious is that minimum cost in any facility is usually achieved by shutting down, which is often not the best course of action. Only when the revenues from each alternative are identical can minimum cost be safely used as a proxy decision variable. This means that the output of each alternative must be identical in quality and quantity.

There are situations in which the analyst must work in terms of cost minimization simply because the only meaningful data pertinent to the investment decision are cost data. The equipment replacement problem for production machines in a mining operation is a classic example. Although it is perfectly acceptable to evaluate machine replacement on the basis of relative cost, the implicit assumption of identical output must be satisfied. In general, projects should be evaluated on the basis of benefits maximization whenever possible.

SELECTED BIBLIOGRAPHY

Anon., 1973, "Uniform Appraisal Standards for Federal Land Acquisitions", Interagency Land Acquisitions Conference Committee.

Gentry, D.W., 1981, "Valuation of Mines and Mineral-Bearing Lands", *Valuation*, Vol. 27, No. 1, American Society of Appraisers, Nov., pp. 28-38.

Gentry, D.W., and Hrebar, M.J., 1980, "Planning and Economic Aspects—Surface Coal Mining", short course notes, Colorado School of Mines, Golden, August, 250 pp.

Gentry, D.W., and Hrebar, M.J., 1978, "Economic Principles for Property Valuation of Industrial Minerals," short course at SME/AIME Fall Meeting, Nassau, Bahamas, September, 153 pp.

Gentry, D.W., and Hrebar, M.J., 1976, "Procedure for Determining Economics of Small Underground Mines", *Colorado School of Mines Mineral Industries Bulletin*, Vol. 19, No. 1, 18 pp.

ICF Inc., 1979, "Observations on Fair Market Value for Federal Coal Leases", submitted to Dep. of the Interior, December, 16 pp. plus appendices.

Jelen, F.C., 1970, *Cost and Optimization Engineering*, McGraw-Hill Book Co., New York, 490 pp.

Laing, G.J., 1977, "Effects of State Taxation on the Mining Industry in the Rocky Mountain States", *Colorado School of Mines Quarterly*, Vol. 72, No. 1, 126 pp.

O'Neil, T.J., 1980, "Inflation and the Capital Investment Decision in Mining", short course notes, presented in Tucson, AZ, Dec., 172 pp.

3
The Time Value of Money

"Money is like an arm or leg—use it or lose it."

—Henry Ford

INTRODUCTION

If it were not for the existence of interest, the analysis of investment opportunities would be greatly simplified. In the absence of interest, investors would be indifferent as to when cash outlays are made or cash benefits are received. It would, in fact, be irrelevant whether the outlays preceded or followed inflows as long as both amounts are known with certainty.

Of course, it does make a considerable difference whether, for example, a firm receives $1 million now or five years from now. The reason is that money does, indeed, have a value which is a function of time. Interest is how this time value is measured.

Interest is generally defined as money paid for the use of borrowed money. In other words, interest is the rental charge for using an asset over some specific time period. The rate of interest is the ratio of the interest chargeable at the end of a specific period of time to the money owed, or borrowed, at the beginning of that period.

In order to understand why interest exists, it is necessary to take the lender's viewpoint. The lender can justify charging interest for several reasons:

1) **Risk:** The lender is faced with the possibility that the borrower will be unable to repay the loan.

2) **Inflation:** Money repaid in the future will be in units of lower value due to inflation.

3) **Transactions Cost:** There will be expense incurred in preparing the loan agreement, recording payments, collecting the loan, and other administrative tasks.

4) **Opportunity Cost:** By committing limited funds to one borrower, a lender is unable to take advantage of other available opportunities.

5) **Postponement of Pleasure:** By lending money, an individual (or organization) is postponing the pleasure which that money could purchase.

Each of these is a valid reason for a lender demanding interest from a borrower. The level of interest is, like the price of other assets, determined by supply and demand. Nonetheless, from time to time various analysts have taken a then current average interest rate and broken it down into the components described previously.

Nelson (1975), for example, estimated that the 12% rate for high-grade commercial paper which prevailed at that time could be broken down as follows:

Opportunity cost	2%
Risk	2%
Transaction cost	$\simeq 0\%$
Inflation	8%
Total	12%

Note that there is nothing fundamental to the *definition* of interest which says that payment of interest is a legal obligation of the borrower. It is important to recognize the *concept* of interest applies in all investment transactions, not just those where repayment is legally enforceable. This is particularly relevant in regard to corporate equity financing of capital projects. Managers who treat equity funds as "free," thereby not providing an adequate return on the investor's capital, are not fulfilling their obligations to the owners. The result to the firm would be a slow attrition of present capital and the inability to raise new funds.

The history, philosophy, and theoretical underpinnings of interest are covered exhaustively in a large number of other text and reference books. This material need not be repeated here. It is sufficient at this point to recognize simply that money has *earning power*. That is, the timing of when payments are made and earnings are received in a capital project is very important.

INTEREST TERMS

Simple interest, often referred to as the *nominal rate of interest,* is the annualized percentage of amount borrowed (the *principal*) which is paid for the use of the money for some period of time (Hubbard and Hawkins, 1969). If the interest is not paid at the end of the period in which it is due but is added to, and becomes part of, the principal upon which interest is calculated, the interest is *compounded*. Interest rates in this text will always be compound interest rates unless otherwise specified.

The difference between simple and compound interest rates can be best illustrated by an example. Suppose $1000 is invested for one year at a simple rate of interest of 6%. At the end of the first year the investment would yield:

$$\$1000 + \$1000 \ (0.06) = \$1060$$

If this investment existed for a period of two years at a simple interest rate of 6%, the final amount of the investment would be $1120.

Now assume the same $1000 investment is made for a period of two years at an interest rate of 6% compounded annually. At the end of the first year the investment would be worth:

$$\$1000 + \$1000 \ (0.06) = \$1000 \ (1 + 0.06)$$

However, since interest is compounded annually, at the end of the second year the investment would now be worth:

$$\underbrace{\$1000 \ (1 + 0.06)}_{\text{1st year}} + \underbrace{\$1000 \ (1 + 0.06) \ (0.06)}_{\text{2nd year}}$$

or

$$\$1000 \ (1 + 0.06) \ (1 + 0.06) = \$1000 \ (1 + 0.06)^2 = \$1123$$

This compound interest relationship may generally be expressed as:

$$F = P(1 + r)^n$$

where F is the future sum of money which will accrue from investing a present sum, P, at an annual rate of interest, r, for n years. When the compounding frequency is annual, there is no difference between the *nominal* rate of interest, r, and the *effective* rate of interest, i. When compounding is performed more than once per year, the effective or true annual rate always exceeds the nominal annual rate. This is explained and discussed in more detail in "Varying the Payment and Compounding Intervals" later in this chapter.

INTEREST FORMULAS

Cash flows at different points in time are related by a series of six (6) basic interest formulas. These formulas in turn are based on five variables as follows:

F = a future sum of money
P = a present sum of money
A = a payment in a series of n equal payments, made at the end of each period of interest
i = effective interest rate per period
n = number of interest periods

Every interest problem is composed of four (4) of these variables; three are given and the fourth must be determined. The standard notation for describing the particular type of problem involved lists all four variables of concern in the following manner.

For example: $(F/P, i, n)$, means, "Find F, given P, i, and n"
Similarly, $(A/F, i, n)$, means, "Find A, given F, i, and n"

The notation is often shortened to simply F/P, or A/F, etc., where it is understood that i and n are given. This notation is used throughout this book and is widely accepted in the field of engineering economy.

The six (6) basic interest equations are developed and described below.

1) Single payment compound amount, $(F/P, i, n)$. This was developed intuitively in "Interest Terms" previously.

$$F = P(1 + i)^n \qquad (1)$$

Example: Find the amount which will accrue at the end of 7 years if $1250 is invested now at 8%, compounded annually. Given: $P = \$1250$, $n = 7$ years, $i = 8\%$

$$F = P(1 + i)^n = \$1250(1 + 0.08)^7 = \$2142.28$$

The quantity $(1 + i)^n = (F/P, i, n)$ has, like other interest factors, been tabulated for various values of i and n. A set of these tables for discrete compounding is located in Appendix A. For the foregoing example the correct factor (rounded to three decimal places) from Table A-12 is 1.714. Thus,

$$F = P (1 + i)^n$$
$$F = P (F/P, i, n)$$
$$F = 1250 (F/P, 8\%, 7)$$
$$F = 1250 (1.714) = \$2142.50$$

Due to rounding, the answers are slightly different. Many hand-held calculators can now solve such problems directly so that interest tables are becoming less important in investment studies.

2) Single payment, present worth, $(P/F, i, n)$. This is Eq. 1 solved for P.

$$P = F/(1 + i)^n \tag{2}$$

Example: If \$6500 will be needed in 5 years, how much should be invested now at an interest rate of $7\frac{1}{2}\%$, compounded annually? Given: $F = \$6500$, $n = 5$ years, i $= 7.5\%$

$$P = F/(1 + i)^n$$
$$P = \$6500/(1 + 0.075)^5 = \$4527.63$$

Using interest tables to solve this problem becomes somewhat more difficult because one must interpolate between 7% and 8%. For most interest problems, a linear interpolation is satisfactory. Thus, using interest tables A-11 and A-12, the solution is determined as follows:

$$P = F/(1 + i)^n$$
$$P = F (P/F, i, n)$$
$$P = 6500 (P/F, 7.5\%, 5)$$

for i = 8%

$$P = 6,500 (0.6806) = 4423.79$$

for i = 7%

$$P = 6,500 (0.7130) = 4634.50$$

Therefore, for 7.5%, $P \cong \$4529.15$

The widespread use of programmed calculators and computers for solving interest problems facilitates exact solutions so that interpolation is often unnecessary. However, the intellectual problem of calculating exact solutions from highly inexact data is of major concern in all investment studies and will be discussed in a subsequent chapter.

3) Uniform series, compound amount $(F/A, i, n)$.

Here, the concern is to determine the terminal amount when equal annual payments are made to an interest-bearing account for a specified number of years. It is important at this point to recall that A is defined as occurring at the *end* of the interest period. Therefore,

for $n = 1, F = A$
$$n = 2, F = A (1 + i) + A$$
$$n = 3, F = A (1 + i)^2 + A (1 + i) + A$$

or, in general,

$$F = A(1 + i)^{n-1} + A(1 + i)^{n-2} + \ldots + A(1 + i) + A$$
$$= A[(1 + i)^{n-1} + (1 + i)^{n-2} + \ldots + (1 + i) + 1] \tag{3}$$

Multiplying Eq. 3 by $(1 + i)$ yields

$$F(1 + i) = A[(1 + i)^n + (1 + i)^{n-1} + \ldots + (1 + i)^2 + (1 + i)] \tag{4}$$

Now, subtracting Eq. 3 from 4,

$$iF = A[(1 + i)^n - 1]$$

or

$$F = \frac{A[(1 + i)^n - 1]}{i} \tag{5}$$

Example: If payments of $725 are made at the end of each year for 12 years to an account which pays interest at the rate of 9% per year, what will be the terminal amount?

$$F = \frac{A[(1 + i)^n - 1]}{i}$$

$$= \frac{725 [(1 + 0.09)^{12} - 1]}{0.09} = \$14,602.02$$

Using Table A-13, approximately the same solution is derived as follows:

$$F = A(F/A, i, n)$$
$$= 725 (F/A, 9\%, 12)$$
$$= 725 (20.141) = \$14,602.23$$

It obviously makes a considerable difference if the annual payments are made at the beginning of the year (an *annuity due*) rather than at the end of the year (an *immediate annuity*). However, the end-of-year convention is most common, and nearly all interest tables and computer programs are constructed on this basis. To easily verify that any particular table or computer program adheres to this convention, note that for $n = 1$, $F = A$, regardless of the interest rate.

4) Uniform series, sinking fund, $(A/F, i, n)$.

Solving Eq. 5 for A will enable the analyst to determine what annual payments must be made to accumulate a specified amount by some future date at an interest rate, i.

$$A = i F/[(1 + i)^n - 1] \tag{6}$$

Example: With interest at 6%, how much must be deposited at the end of each year to yield a final amount of $2,825 in 7 years?

$$A = i F/[(1 + i)^n - 1]$$
$$A = 0.06 (\$2825) / [(1 + 0.06)^7 - 1]$$
$$= \$336.58$$

The concept of a sinking fund is well known, so that the A/F interest factor is often called the sinking fund factor. Solving the foregoing problem using Table A-10, gives the following answer:

$$A = F\,(A/F, i, n)$$
$$A = \$2825\,(A/F, 6\%, 7)$$
$$= \$2825\,(0.1191) = \$336.56$$

5) Uniform series, present worth ($P/A, i, n$)

This type of problem arises when the current value of a future series of cash flows is desired. This is often the case with investments in securities where an expenditure now will provide equal interest or dividend payments for several future periods.

For Eq. 5,

$$F = \frac{A[(1 + i)^n - 1]}{i}$$

Substitute Eq. 1, $F = P(1 + i)^n$

$$P(1 + i)^n = \frac{A[(1 + i)^n - 1]}{i}$$

and

$$P = \frac{A[(1 + i)^n - 1]}{i(1 + i)^n} \tag{7}$$

Example: An investment will yield \$610 at the end of each year for 15 years. If interest is 10%, what is the maximum purchase price (i.e., present value) for this investment?

$$P = \frac{A[(1 + i)^n - 1]}{i(1 + i)^n}$$

$$= \frac{610\,[(1 + 0.1)^{15} - 1]}{0.1\,(1 + 0.1)^{15}}$$

$$= \$4639.71$$

Solving the same problem using Table A-14, gives the following answer:

$$P = A\,(P/A, i, n)$$
$$P = 610\,(P/A, 10\%, 15)$$
$$= 610\,(7.6061) = \$4639.71$$

6) Uniform series, capital recovery, ($A/P, i, n$)

This is the reverse of the previous problem, and the equation is derived simply by solving Eq. 7 for A.

$$A = \frac{i\,P(1 + i)^n}{[(1 + i)^n - 1]} \tag{8}$$

Example: If an investment opportunity is offered now for $3500, how much must it yield at the end of every year for 6 years to justify the investment if interest is 12%?

$$A = \frac{i P(1 + i)^n}{[(1 + i)^n - 1]}$$

$$= \frac{0.12 \, (\$3500) \, (1 + 0.12)^6}{[(1 + 0.12)^6 - 1]}$$

$$= \$851.67$$

Similarly, using Table A-15,

$$\begin{aligned}
A &= P \, (A/P, \, i, \, n) \\
&= \$3500 \, (A/P, \, 12\%, \, 6) \\
&= \$3500 \, (0.2432) = \$851.20
\end{aligned}$$

SOLVING INTEREST PROBLEMS

Interest problems can usually be simplified if the two steps to be described are followed in order.

1) **Abstracting the Problem:** Interest problems are based upon the five variables—P, F, A, i, and n. The first step in solving any interest problem is to determine which three variables are given, and what is the fourth variable to be determined. Occasionally, interest problems become complex with many cash flows of various magnitudes occurring at different points in time. It is important in these cases to abstract the problem into terms of the five basic variables before proceeding.

2) **Drawing the Cash Flow Diagram:** It is usually helpful in visualizing the problem to construct a cash flow diagram, as illustrated in Fig. 1. This diagram is a plot of cash flow vs. time, where receipts are plotted to scale vertically upward and disbursements are shown as downward-pointing arrows. Such diagrams are partic-

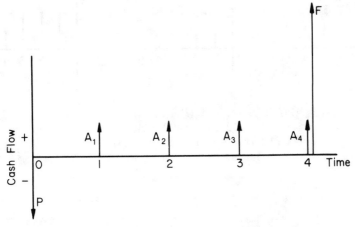

Fig. 1. Cash flow diagram with P, A, and F drawn to scale.

ularly helpful in problems involving A (equal payment series).

To illustrate this two-step approach to solving interest problems, the following example problem is presented.

Example: An investment opportunity is available which will yield $1000 per year for the next 3 years and $600 per year for the following 2 years. If interest is 12% and the investment has no terminal salvage value, what is the present value of the investment?

Step 1

The variables given in this problem are A, i, and n. The unknown to be determined is P. Actually, the problem must first be separated into subproblems because A is not constant over the life of the proposal. This is illustrated in the second step.

Step 2

The cash flow diagram, Fig. 2, shows the compound nature of the problem.

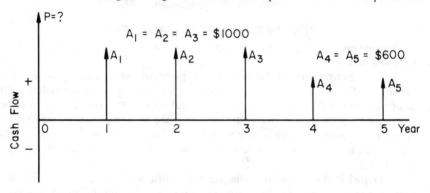

Fig. 2. Schematic of cash flow for example problem.

The problem can be solved rapidly in either of two ways:

$$P = A_1 (P/A_1, i, n_1) + A_2 (P/A_2, i, n_2)$$
$$= \$600 (P/A_1, 12, 5) + \$400 (P/A_2, 12, 3)$$
$$= 600 (3.6048) + 400 (2.4018) = \$3123.60$$

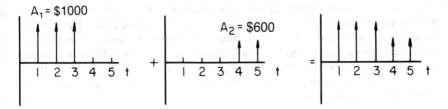

$$P = A_1 (P/A_1, i, n_1) + A_2 (P/A_2, i, n_2) (P/F, i, n_3)$$
$$= 1000 (P/A_1, 12, 3) + 600 (P/A_2, 12, 2) (P/F, 12, 3)$$
$$= 1000 (2.4018) + 600 (1.6901) (0.7118) = \$3123.61$$

The answers differ due only to rounding in the interest tables.

VARYING THE PAYMENT AND COMPOUNDING INTERVALS

All of the preceding discussion in this chapter relates to annual compounding and, where A enters the problem, to annual cash flows. In practice, these constraints are routinely violated as neither payments/receipts nor compounding need be on an annual basis. In fact, there is a continuous spectrum of possibilities with discrete annual compounding and discrete cash flows at one end of the continuum, and continuous compounding and continuous flow of funds at the other end.

Increased Compounding Frequency

Although annual interest compounding was used in the foregoing to introduce interest concepts, compounding can, of course, be performed at any other interval. It could be, for example, once every two years, twice a year (semi-annual), monthly, hourly, and so forth. When this occurs, there is an important difference between *nominal* and *effective* annual interest rates.

This difference can best be illustrated with an example. If one borrows $1000 from a finance company which charges interest at a compound rate of 2% per month, the *nominal* annual rate is 24%. However, it is clear that the true or *effective*, annual rate is greater than 24% due to the compounding effect that occurs every month. More precisely, if r = nominal annual interest rate and x = number of compounding periods per year the effective annual rate, i, is

$$i = (1 + r/x)^x - 1 \qquad (9)$$

or, in the foregoing example,

$$i = (1 + \frac{24}{12})^{12} - 1$$

$$= 1.268 - 1 = 26.8\%$$

The effective interest rate is the rate compounded once a year which is equivalent to the nominal interest rate compounded x times a year. Consequently, the effective interest rate is greater than the nominal rate. Also, as the frequency of compounding increases for any given nominal rate, the greater will be the difference between the effective and nominal interest rates. This is shown in Table 1. Appendix B provides the relationship between nominal and effective interest rates for various compounding frequencies.

Table 1. Nominal vs. Effective Annual Rates of Interest for Various Compounding Frequencies

Compounding Frequency	No. Periods Per Year	Nominal Annual Rate, %	Effective Annual Rate, %
Annual	1	12	12
Semiannual	2	12	12.36
Quarterly	4	12	12.55
Monthly	12	12	12.68
Weekly	52	12	12.73
Daily	365	12	12.75
Continuously	∞	12	12.75

The limiting case for increasingly frequent compounding is instantaneous or continuous, compounding. The effective annual interest rate when $x \to \infty$ is determined as follows:

$$i = \lim_{x \to \infty} (1 + \tfrac{r}{x})^x - 1$$

but since

$$(1 + \tfrac{r}{x})^x = [(1 + \tfrac{r}{x})^{x/r}]^r$$

and

$$\lim_{x \to \infty} (1 + \tfrac{r}{x})^{x/r} = e$$

therefore

$$i = \lim_{x \to \infty} [(1 + \tfrac{r}{x})^{x/r}]^r - 1 = e^r - 1 \tag{10}$$

Beginning with the relationship expressed in Eq. 10, a second series of six interest equations can be developed based upon continuous interest. For example for discrete compounding:

$$F = P (1 + i)^n$$

but now

$$i = e^r - 1$$

Thus, for continuous interest

$$F = P (1 + e^r - 1)^n$$
$$= P e^{rn} \qquad (11)$$

The term, e^{rn}, is the single payment, compound amount factor for continuous interest. It is designed as $(F/P, r, n)$ and is listed in the interest tables in Appendix C.

The six interest factors for continuous interest are listed for reference:

Continuous Interest Factors

Equation	Factor	Notation	
$F = P e^{rn}$	e^{rn}	$(F/P,r,n)$	(11)
$P = F/e^{rn}$	e^{-rn}	$(P/F,r,n)$	(12)
$P = A \left[\dfrac{1 - e^{-rn}}{e^r - 1} \right]$	$(i - e^{-rn}) / (e^r - 1)$	$(P/A,r,n)$	(13)
$A = P \left[\dfrac{e^r - 1}{1 - e^{-rn}} \right]$	$(e^r - 1) / (1 - e^{-rn})$	$(A/P,r,n)$	(14)
$F = A \left[\dfrac{e^{rn} - 1}{e^r - 1} \right]$	$(e^{rn} - 1) / (e^r - 1)$	$(F/A,r,n)$	(15)
$A = F \left[\dfrac{e^r - 1}{e^{rn} - 1} \right]$	$(e^r - 1) / (e^{rn} - 1)$	$(A/F,r,n)$	(16)

In recent years, nearly all financial institutions have abandoned annual compounding for most of their time deposits. Consumers have been deluged with advertising boasting *daily* or *instantaneous* interest paid by banks and savings and loans. As Table 1 shows, there can be a significant increase in the effective annual rate with more frequent compounding, but the increase becomes quite modest beyond monthly compounding.

Example: What is the worth at the present time of a promissory note which will yield $1200 in 5 years if interest in 10%, compounded quarterly?; continuously?

Quarterly Compounding

$$F = \$1200$$
$$r = 10\%$$
$$x = 4 \text{ compounding periods per year}$$
$$n = 5 \text{ years}$$
$$P = F/(1 + i)^n$$
$$= F/(1 + \frac{r}{x})^{nx}$$
$$= 1200 / (1 + 0.025)^{20} = \$732.33$$

Continuous Compounding

$$F = \$1200$$
$$r = 10\%$$
$$n = 5 \text{ years}$$
$$P = F/e^{rn}$$
$$= 1200/e^{0.5} = \$727.84$$

Thus, the difference between monthly and continuous compounding is usually not large.

Increased Cash Flow Frequency

In problems involving annuities, as long as the cash flows occur at the same frequency as compounding, the formulas developed in the preceding sections can be applied without difficulty. If, however, the cash flows occur more or less frequently than compounding, further examination of the problem is required. The two possible variations of this problem are considered in the following.

1) Cash flows less frequently than compounding. To handle this situation, the analyst must first convert cash flows and compounding to the same schedule. This can be done in two ways: (a) calculate an effective interest rate so that the compounding interval is increased to the cash flow interval; or (b) determine an equivalent intraperiod cash flow so that the cash flow interval is reduced to the compounding interval.

Example: If payments of \$250 are made every six months to a fund paying 8% per year compounded quarterly, how much will accumulate in 5 years?

(a) Change compounding interval.

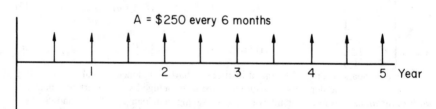

A = $250 every 6 months

Find the six-month effective rate.

$$i = (1 + \frac{r}{x})^x - 1 = (1 + \frac{4}{2})^2 - 1$$
$$= (1.02)^2 - 1$$
$$= 4.04\%$$
$$F = A\,(F/A, i, n)$$
$$= \$250\,(F/A, 4.04\%, 10) = \$3007.10$$

(b) Change payment interval

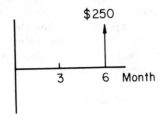

$$F = 250$$
$$i = 2\%$$
$$n = 2$$
$$A = ?$$

$$A = F\,(A/F, i, n)$$
$$= 250\,(A/F, 2, 2) = \$123.76$$

Now,

$$F = A\,(F/A, i, n)$$
$$= 123.76\,(F/A, 2, 20) = \$3007.10$$

2) Cash flows more frequent than compounding. The key to solving this type of problem is the intraperiod interest rate. That is, what interest rate—if any—applies to cash flows occurring between compounding points.

Obviously, there are many assumptions which could be made here. However, most financial institutions pay no interest on intraperiod deposits or withdrawals. Therefore, the most common procedure here is to assume that withdrawals were made at the *last* compounding period, and deposits are made at the *next* period.

Example: If deposits of $260 are made every month to an account which pays 12% per year, compounded semiannually, how much will accumulate by the end of three years?

$$i = 6\% \text{ per period}$$
$$n = 6 \text{ periods}$$

Assuming that no intraperiod interest is paid, deposits are assumed to be made at the next compounding period.

Thus,

$$A = 6 \times 260 = \$1560 \text{ every six months}$$
$$F = A\,(F/A, i, n)$$
$$= 1560\,(F/A, 6, 6) = 1560\,(6.975) = \$10,881.00$$

Continuous Flow of Funds

Analogous to continuous interest is the case where the flow of funds (receipts or disbursements) occur continuously. Thus, rather than monthly, weekly, daily, etc., cash flows, the same annual amount is received (or paid) but the funds flow on a continuous basis. The reader will note that, in view of the preceding discussion, this situation has no particular significance unless continuous interest also exists.

To solve funds flow problems, define $\bar{A}$ = the uniform flow rate of money per year.

Consider the future amount, ΔF which accrues from ΔP at a nominal annual interest rate, r, compounded continuously.

Since

$$F = Pe^{rn}$$
$$\Delta F = \Delta Pe^{rn}$$

But

$$\Delta P = \bar{A}\,\Delta t$$

So that

$$\Delta F = \bar{A}\,e^{rn}\,\Delta t$$

If

$$T \longrightarrow 0,\ dF = \bar{A}\,e^{rn}\,dt$$

or for the entire interval from 0 to n

$$F = \int_{O}^{n} \bar{A}\,e^{rt}dt$$

$$= \left[\frac{\bar{A}\,e^{rt}}{r}\right]_{0}^{n} \qquad = \bar{A}\left[\frac{e^{rn}}{r} - \frac{e^{o}}{r}\right]$$

or

$$F = \bar{A}\left[\frac{e^{rn} - 1}{r}\right] \qquad ,\ (F/\bar{A},\ r,\ n) \tag{17}$$

Similar equations may be derived from other situations as follows:

$$\bar{A} = F\left[\frac{r}{e^{rn} - 1}\right] ,\ (\bar{A}/F,\ r,\ n) \tag{18}$$

$$\bar{A} = P\left[\frac{re^{rn}}{e^{rn} - 1}\right] ,\quad (\bar{A}/P,\ r,\ n) \tag{19}$$

$$P = \bar{A} \left[\frac{e^{rn} - 1}{re^{rn}} \right] \quad , \quad (P/\bar{A}, r, n)$$

(20)

Funds flow problems are generally solved by converting $\bar{A}$ to an equivalent A and then solving as a continuous interest problem.
The necessary conversion factor can be determined by solving Eq. 17 when $n = 1$.
Thus

$$F = \bar{A} \left[\frac{e^{rn} - 1}{r} \right]$$

becomes

$$F = \bar{A} \left[\frac{e^r - 1}{r} \right]$$

when $n = 1$.

This simply gives the year-end equivalent for a single $\bar{A}$. Therefore, by multiplying $\bar{A}$ by the factor, $(e^r - 1)/r$, an equivalent A is calculated and the problem can be solved by the application of the previously developed continuous interest formulas.

Appendix D contains funds flow conversion factors for several values of r.

Finally, Table 2 summarizes the approaches to solving interest problems for all possible combinations of payment and compounding frequencies.

Table 2. Summary of Payment and Compounding Alternatives

Interest	Payments	Relative Frequencies	Solution
1. Discrete	Discrete	Same	General formulas, $i = r/x$ if not annual compounding
2. Discrete	Discrete	Compounding more frequent than payments	Determine either (1) inter-period i, or (2) equivalent A for compounding interval.
3. Discrete	Discrete	Payments more frequent than compounding	Usually no intraperiod interest applies. Therefore, deposits assumed at end of period; withdrawals at beginning.
4. Discrete	Continuous		Special case of 3.
5. Continuous	Discrete		$i = e^r - 1$ for annual payments; $i = \left[\frac{e^r - 1}{x} \right]$ for x payments per year.
6. Continuous	Continuous		$A = \bar{A} \left[\frac{e^r - 1}{r} \right]$, then solve as 5.

GRADIENT SERIES

A great many real problems in investment analysis involve unequal cash flow series and, therefore, cannot be solved with the annuity formulas developed in the foregoing. The general case of independent and variable cash flows can be analyzed only by repeated application of the single payment formulas. However, mathematical solutions have been developed for two special types of unequal cash flows—the uniform gradient series and the geometric gradient series.

A uniform gradient, G, exists when cash flows either increase or decrease by a fixed *amount* in successive periods. In this case, the annual cash flow consists of two components: (1) a constant amount, A_1, equal to the cash flow in the first period; and (2) a variable amount equal to $(n - 1)G$.

To solve such problems, a constant amount, A_2, is first calculated which is equivalent to the variable cash flow component. It can be shown through algebra (see Thuesen, et al., 1977) that

$$A_2 = G \left[\frac{1}{i} - \frac{n}{i}(A/F, i, n) \right]$$

The factor, $[\frac{1}{i} - \frac{n}{i}(A/F, i, n)]$, is called the uniform gradient factor; is designated $(A/G, i, n)$; and is tabulated for each value of i in tables in Appendices A and C.

The total A in a gradient problem then becomes:

$$A = A_1 + A_2$$
$$= A_1 + G(A/G, i, n)$$

and the problem can then be solved in a conventional manner.

Because receipts or expenditures in practice rarely increase or decrease every period by a fixed amount, gradient series problems have limited applicability and are covered in this book only briefly.

With a geometric gradient, the increase or decrease in cash flows between periods is not a constant *amount*, but a constant *percentage* of the cash flow in the preceding period. Like the uniform gradient, the geometric gradient also has limited applications, but may be useful in analyzing inflationary cost increases which may behave in a geometric fashion over a limited time horizon.

If j represents the percent change in cash flow between periods and A is the cash flow in the initial period, then the cash flow in any subsequent period, k, is:

$$A_k = A_1 (1 + j)^{k-1}$$

and the present value of the series is:

$$P = \sum_{k=1}^{n} A_1 (1 + j)^{k-1} (1 + i)^{-k}$$

or

$$P = A_1 (1 + j)^{-1} \sum_{k=1}^{n} \left(\frac{1+j}{1+i} \right)^{k}$$

This yields two solutions depending upon the relative values of i and j.

For $i = j$
$$P = \frac{n A_1}{(1 + i)}$$
(21)

for $i \neq j$
$$P = A_1 \left[\frac{1 - (1 + j)^n (1 + i)^{-n}}{(i - j)} \right]$$
(22)

In the nomenclature of engineering economy,

$$P = A_1 (P/A, i, j, n)$$

where $(P/A, i, j, n)$ is called the geometric series present value factor. Appendix E contains tables of both the present value and future value factors for geometric series for some representative values of i and j.

In the specific case where $i \neq j$ and $j \geq o$, Eq. 22 becomes

$$P = A_1 \left[\frac{1 - (F/P, j, n) (P/F, i, n)}{(i - j)} \right]$$

so that the geometric series problem can be solved using the interest tables in either Appendix A or C, whichever is applicable.

Example: Revenues from sulfuric acid sales from a copper smelter are presently $1,400,000 per year. For an annual interest rate of 8%, determine the present value of such revenue over the next ten years if: (a) sales rise by $150,000 per year; (b) sales rise by 10% per year.

(a) $i = 8\%$
$n = 10$ years
$G = \$150,000$
$A_1 = \$1,400,000$
$A_2 = G(A_2/G, i, n)$
$\quad = \$150,000 (A_2/G, 8\%, 10)$
$\quad = \$150,000 (3.8713) = \$580,695$
$A = A_1 + A_2 = \$1,400,000 + \$580,695 = \$1,980,695$
$P = A (P/A, i, n)$
$P = \$1,980,695 (P/A, 8\%, 10) = \$1,980,695 (6.7101)$
$\quad = \$13,290,662$ or roughly $13,300,000

(b) $i = 8\%$
$j = 10\%$
$n = 10$ years
$A_1 = \$1,400,000$
$P = A_1 (P/A, i, j, n) \qquad = \$1,400,000 (P/A, 8\%, 10\%, 10)$
$\qquad\qquad\qquad\qquad\qquad = \$1,400,000 (10.0702)$
$\qquad\qquad\qquad\qquad\qquad = \$14,098,280$
$\qquad$ or roughly $\qquad = \$14,100,000$

INTEREST FACTOR RELATIONSHIPS

When using interest tables it is often desirable to know the relationship between some of the interest factors. Following are some relationships based on given values of i and n.

1) The single payment compound amount factor $(F/P, i, n)$ and the single payment present value factor $(P/F, i, n)$ are reciprocals.

$$\frac{1}{(P/F, i, n)} = (F/P, i, n)$$

2) The sinking fund factor $(A/F, i, n)$ and the compound amount factor for an annuity $(F/A, i, n)$ are reciprocals.

$$\frac{1}{(A/F, i, n)} = (F/A, i, n)$$

3) The capital recovery factor $(A/P, i, n)$ and the present value factor for an annuity $(P/A, i, n)$ are reciprocals.

$$\frac{1}{(A/P, i, n)} = (P/A, i, n)$$

4) The capital recovery factor $(A/P, i, n)$ is equal to the sinking fund factor $(A/F, i, n)$ plus interest rate.

$$(A/P, i, n) = (A/F, i, n) + i$$

5) The present value factor for an annuity $(P/A, i, n)$ is equal to the sum of the first n terms of single payment present value factors $(P/F, i, n)$.

$$(P/A, i, n) = (P/F, i, 1) + (P/F, i, 2) + \ldots + (P/F, i, n)$$

6) The compound amount factor for an annuity $(F/A, i, n)$ is equal to 1.00 plus the sum of the first $(n - 1)$ terms of the single payment compound amount factor $(F/P, i, n)$.

$$(F/A, i, n) = 1.00 + (F/P, i, 1) + (F/P, i, 2) + \ldots + (F/P, i, n\text{-}1)$$

Also note that for large values of n which may not appear in interest tables.

$$(F/P, i, n) = (F/P, i, n_1)\,(F/P, i, n_2) \ldots (F/P, i, n_k)$$

where $n = n_1 + n_2 + \ldots + n_k$.

Thus, the compound amount of $100 invested today at 8% compounded annually for 125 years is,

$$P(F/P, 8, 125) = 100\,(F/P, 8, 100)\,(F/P, 8, 20)\,(F/P, 8, 5)$$
$$= 100\,(2199.761)\,(4.661)\,(1.469)$$
$$= \$1,506.178$$

EQUIVALENCE

The concept of equivalence is covered exhaustively in most textbooks on engineering economy, and, to be sure, it is of fundamental importance to invest-ment analysis. It is, however, not a difficult concept and can be grasped readily, given a clear definition and a few examples. That is the procedure which is followed subsequently.

Equivalence is nothing more than the fact that due to interest, two cash flow series, either receipts or payments, of entirely different magnitudes and timing may have the same present value. They are, therefore, equivalent, and a rational investor would be indifferent as to which series he or she would select.

The example in Table 3 was constructed to illustrate the notion of equivalence. In this example, an investor is presented with four riskless investment alternatives, each offering different total revenues. At a 10% rate of interest, however, all four projects have the same present value. They are, therefore, equivalent, and because they are worth the same, the investor would have no preference among the alternatives. Even though alternative D has the highest total cash flow, equivalence indicates that at 10% interest it is no more attractive than any of the other opportunities.

This situation could, of course, be easily changed by adopting another interest rate, by introducing risk into the analysis, or by altering the sequence or magni-tudes of the cash flows. Nonetheless, it is clear that interest can affect the selection among investment alternatives which have different magnitudes and timing of the cash flows.

Table 3. Equivalence of Four Investment Alternatives

Year n	Cash Benefits	(P/F,10,n)	Present Value	Year n	Cash Benefits	(P/F,10,n)	Present Value
Alternative A				**Alternative B**			
1	$ 1,000	0.9091	$ 909.10	1	$ 3,500	0.9091	$ 3,181.85
2	1,000	0.8265	826.50	2	3,250	0.8265	2,686.13
3	1,000	0.7513	751.50	3	3,000	0.7513	2,253.90
4	11,000	0.6830	7,513.00	4	2,750	0.6830	1,878.25
	$\Sigma = \$14,000$		PV~$10,000.00		$\Sigma = \$12,500$		PV~$10,000.00
Alternative C				**Alternative D**			
1	$ 3,154.71	0.9091	$ 2,867.95	1	0	0.9091	0
2	3,154.71	0.8265	2,607.37	2	0	0.8265	0
3	3,154.71	0.7513	2,370.13	3	0	0.7513	0
4	3,154.71	0.6830	2,154.67	4	14,641	0.6830	10,000
	$\Sigma = \$12,618.84$		PV~$10,000.00		$\Sigma = \$14,641$		PV~$10,000

i = 10%

THE BOND VALUATION PROBLEM

The bond valuation problem illustrates many of the concepts developed in this chapter. A bond is an instrument for long-term debt financing. It is a promissory note where the borrower agrees to repay the principal amount of the bond at a specified future date.

Bonds are characterized by a *face value* (usually $1000 for corporate bonds), a *coupon rate* (the interest rate applied to the face value which is paid to the lender), and a *maturity date* (when the principal amount will be repaid). Bonds, therefore, provide fixed income to lenders (bond purchasers), but as the market rate of interest falls or rises, this income becomes more or less attractive in relation to the face value. As a consequence, the price of a bond will change continuously with—and in the opposite direction to—the market rate of interest.

Suppose a 9% corporate bond, maturing on Jan. 1, 2008, is available for purchase on Jan. 1, 1983. Interest is payable semiannually. What is the present value of the bond if the market rate of interest is 12% (a) compounded semiannually, and (b) compounded continuously?

(a) Semiannual compounding

$$A = 1000 \times 0.045 = \$45$$
$$n = 50 \text{ periods}$$
$$i = 6\% \text{ per period}$$

Principal repayment = $1000 on 1/1/2008

$$P = A\,(P/A,\, i,\, n) + F(P/F,\, i,\, n)$$
$$= 45\,(P/A,\, 6\%,\, 50) + 1000\,(P/F,\, 6\%,\, 50)$$
$$= 45\,(15.762) + 1000\,(0.0543) = \$763.60$$

The bond just described is equivalent to (i.e., worth the same) as a present sum of $763.60, disregarding risk. Stated differently, an investor could afford to spend a maximum of $763.60 for this bond and still receive a nominal 12% return on his or her investment.

(g) Continuous compounding

$$A = \$45$$
$$n = 50 \text{ periods}$$
$$i = 6\% \text{ per period, continuous}$$
$$A = (P/A,\, r,\, n) + F(P/F,\, r,\, n)$$
$$= 45\,(P/A,\, 6\%,\, 50) + 1000\,(P/F,\, 6\%,\, 50)$$
$$= 45\,(15.367) + 1000\,(0.0498) = \$741.32$$

The higher effective interest rate per period, $e^{0.06} - 1 = 6.18\%$, as opposed to 6% in Part a, makes the fixed income of $45 every six months now look somewhat less attractive to the investor.

If the same bond is offered for sale for $850, what interest rate of return is available to the investor? Assume semiannual compounding.

$$A = \$45$$
$$n = 50 \text{ periods}$$
$$P = \$850$$

principal repayment = $1000 on 1/1/2008

$$i = ?$$

This problem really asks the question, "What interest rate will equate an $850 investment on 1/1/1983 with revenues of $45 every 6 months thereafter for 25 years, and $1000 on 1/1/2008?

$$P = A(P/A,\, i,\, n) + F(P/F,\, i,\, n)$$
$$850 = 45\,(P/A,\, i,\, 50) + 1000\,(P/F,\, i,\, 50)$$

Solving for i in problems where three or more unequal cash flows are involved is an iterative process.

Try 5%

$$850 \overset{?}{=} 45\,(18.256) + 1000\,(0.0872)$$
$$850 < \$908.72$$

Try 5.5%

$$850 \overset{?}{=} 45\,(16.932) + 1000\,(0.0688)$$
$$850 > \$830.74$$

By interpolation

$$
\begin{array}{ll}
\$\quad 908.72 & \$\quad 850.00 \\
-\ 830.74 & -\ 830.74 \\
\hline
\$\quad 77.98 & \$\quad 19.26
\end{array}
$$

$$i = 5\% + \left(\frac{19.26}{77.98}\right)0.5\% = 5.124\%$$

Thus, the *nominal* annual rate $= 2(5.124) = 10.25\%$, and the *effective* annual rate $= (1 + 0.05124)^2 - 1 = 10.51\%$.

REFERENCES

Hubbard, C.L., and Hawkins, C.A., 1969, *Theory of Valuation*, International Textbook Co., Scranton, PA, 247 pp.

Nelson, W.G., 1975, "Inflation—Hanging in There," *Financial Executive*, Vol. 48, No. 2, Feb., pp. 32, 37-41.

Thuesen, H.G., Fabrycky, W.J., and Thuesen, G.J., 1977, *Engineering Economy*, Prentice-Hall, Inc., Englewood Cliffs, NJ, 589 pp.

PROBLEMS

1. A savings and loan offers an annual interest rate of 10% compounded continuously for savings of $1000 or more deposited for at least 2 years. If $1642 is deposited on Jan. 1, 1984, how much would this amount to by July 1, 1992?

2. How much will a deposit of $575 now grow to in 15 years at an annual interest rate of 15%, compounded monthly?

3. A mining company holds an option to purchase some property in fee simple for $87,000 in 5 years. Recent exploration results have been favorable and the firm would like to purchase the property immediately. What would be a fair price if money is worth 12% compounded annually? (Assume that no other option or lease payments are involved).

4. If you participated in a Christmas Club which paid 10% per year compounded weekly, what must your weekly deposit be to accumulate $500 at the end of one year?

5. To accumulate the necessary $15,000 down payment in 6 years for a new home, a couple makes quarterly payments to an account which pays 9% compounded monthly. How much must these payments be?

6. A recent mining engineering graduate wants to purchase a new car in 5 years which presently costs $9500, but the price of the car is rising at an annual compound rate of 8%. If she makes semiannual deposits to an account which pays 10% compounded semiannually, how large must these payments be?

7. If income from the sales of an item occurs continuously at the rate of $17,000 per year and is deposited immediately in an account which yields 12% per year, compounded continuously, how much would accumulate at the end of 3 years?

8. Will annual payments of $4800 be enough to repay a $40,000 loan in 20 years of an

interest rate of (a) 10%, compounded annually?; (b) 10%, compounded continuously?; (c) 12%, compounded quarterly?

9. A crushed stone operation having a projected life of 12 years is offered for sale. If annual net cash flows are expected to be roughly $107,000, determine the appropriate purchase price for the property when interest is 9%: (a) compounded annually; (b) compounded continuously; (c) same as (b), but funds are also received continuously.

10. What is the value of a series of monthly royalty payments of $5,000 each for 8 years, if interest if 8% per year compounded annually?

11. To purchase a new home, a couple borrows $50,000 at 11% for 30 years. What will their monthly payments be if interest is compounded monthly? What is the principal remaining after 10 years (120) payments?

12. What must the annual earnings be from a new piece of equipment that costs $125,000 and lasts 15 years? Assume no salvage value, and interest is 14% per year.

13. Maintenance costs for a new piece of mining equipment are expected to be $20,000 in the first year, rising by $1000 per year every year thereafter. The machine has a life of 8 years and interest is 10% annually. To evaluate bids from outside firms for a maintenance contract, you need to know the present value of these costs. What is this value?

14. In the preceding problem, suppose the maintenance costs rise by 6% per year rather than by a fixed amount. Now find the present value of maintenance costs.

15. An investor brought $100 shares of a common stock 6 years ago for $24 per share. He just sold for $40 per share. What is the annual rate of return on his investment?

16. A $1000 corporate bond has a coupon rate of 8% per year, with interest payable semiannually. If the maturity date is in seven years, what is a fair price at the present time if interest is 12% compounded continuously?

17. Annual cash flow from a proposed small gold leaching operation is expected to be $360,000 per year for 8 years. Initial expenditure for plant and equipment is estimated to be $1.16 million. The salavage value of this investment is expected to be roughly $90,000. If the property is offered for sale for $450,000, calculate the approximate rate of return for this investment. Assume that the property payment would be made immediately but that the remaining capital investment would be divided equally over a two-year preproduction period.

18. How long will it take for an investment of $1200 to double in value if interest is 8%, compounded quarterly?

19. Find the net present value of the following series of cash flows: income of $6500 every six months for 10 years, income of 15,000 at the end of the 10th year, and expenditures of $5700 per year for 10 years. Interest is 12% compounded semiannually.

20. A certain coal stripping operation presently uses three dozers for reclamation work. To reduce costs, three alternatives are being considered for this job in the future: rebuild present equipment, purchase new dozers, and employ a contractor. Details for the alternatives are:

	Rebuild	**Purchase**	**Contract**
Number of units	3	2	N/A
Initial cost	$360,000	$920,000	0
Annual costs:			
Maintenance	140,000	85,000	
Labor	240,000	160,000	$525,000
Supplies	58,000	42,000	
Life	8 years	8 years	8 years
Salvage value	0	$120,000	0

If interest is 8% annually, which alternative should be selected?

4

The Ore Reserve Problem

"Pure gold was hidden in the quartz, they said, 'Twas proved by dreams and signs, and rods divining, By chemic tests, and spirits of the dead, In fact by everything—except by mining."

from R.E. White, "The Mining Town," *Alta California*, August 18,1873

Textbooks in mining engineering have traditionally covered ore reserve estimation and mine valuation in a single volume. Indeed, *Examination and Valuation of Mineral Property* by R. D. Parks (1957), having been the standard of the profession for over 20 years, is a classic example. However, the state of knowledge in these two fields has grown tremendously. Most mining engineering curricula now cover these topics in separate courses, and in practice, these tasks are usually performed by different individuals. Thus, this trend now suggests the need for separate, specialized publications rather than a single volume for both topics.

Because of the separate development of the science of ore reserve estimation, it is important to state at the beginning what this chapter does *not* cover. It does not address in detail the design of sampling programs, sampling methods, or the processing of samples or sample data compilation. There are a number of suitable references available on these topics, including Peters (1978), Gy (1979), as well as the previously cited book by Parks. Furthermore, this chapter does not cover in any substantial detail either the traditional methods for using sample data to calculate ore reserves or the more recently developed geostatistical methods. The traditional methods are thoroughly covered by Popoff (1966) and more concisely by Barnes (1980). Among the more useful references for statistical applications in ore reserve estimation are Barnes (1980), David (1977), and Clark (1979). A thoughtful review of the entire science and art of ore reserve estimation is provided by King, McMahon, and Bujtor (1982).

In spite of the fact that expansion of knowledge now calls for separate books on ore reserve estimation and financial evaluation, neither publication can ignore the important interrelationships between the two fields. In regard to this text, the unique and crucial role of ore reserves can not be omitted entirely; however, the field is now simply too extensive to treat here in any comprehensive manner.

The primary purpose of this chapter, then, is to define terms, cite pitfalls, and promote improved communication from the geologist acquiring sample information in the field, through the mining engineer designing production systems, to the financial analyst advising senior management in regard to the basic investment decision. Knowledge of the fundamental nature of mineral occurrence is vital to the analysis of mineral investments, but that knowledge must be based on a common set of definitions and assumptions. This chapter seeks to contribute to this goal.

NATURE OF THE PROBLEM

There is no topic in mining which has caused greater confusion or led to more misunderstanding than the term *ore reserves*. This is so even among experienced mining engineers and geologists. Communicating accurate and comprehensible information on ore reserves to nontechnical people requires unusual patience and skill.

Ore is defined as mineral that can be extracted from the earth at a profit. This economic constraint is the primary source of misunderstanding regarding ore reserves. Profit, in turn, is a function of production costs and sales prices, both of which change frequently—virtually continuously in some cases. As a consequence, ore reserves also change frequently, even though the physical endowment of the earth's crust remains constant.

The uncertainty created by the economic definition of ore is compounded further by the fact that prior to mining, the physical endowment is never known with certainty but is only estimated from sparse sampling data. Even after a major bulk sampling and testing program, at the very most only $\frac{1}{1000}$ of any mineral deposit is normally sampled prior to mining. The grade of the remaining 99.9% must be inferred from the known 0.1%. As more assay information becomes available and as the geologic interpretation of a mineral deposit changes, the ore reserves also change.

This fundamental dynamic nature of ore is one of the most challenging problems facing the mine investment analyst.

Impact on the Investment Decision

Ore, being an economic concept, is interrelated with other mine planning variables. In mine evaluations the analyst encounters the circular problem illustrated in Fig. 1. Here, the quantity of reserves is an important variable in determining the optimum mine size. Mine size, in turn, affects production costs, as economies of scale are enjoyed with larger production rates. Finally, the level of production costs determines what material can be mined at a profit and, therefore, determines the quantity of ore reserves.

The circular nature of this problem may or may not be significant in preliminary financial analyses, but the basic dynamic nature of ore reserves thoroughly pervades the mine investment decision. Mine management is continually analyzing new equipment and new methods hoping to either convert low-grade, submarginal mineral into ore or combat the seemingly inexorable rise in costs which reclassify low-grade ore into simply interesting mineralization. The ore-waste boundary is more sensitive to changes in economic conditions in some deposits than in others, but the principles described are applicable in all mineral evaluations, at least to some degree.

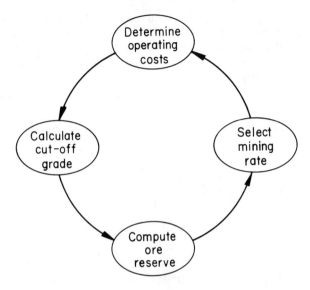

Fig. 1. The cyclic ore reserve problem.

BASIC CONCEPTS

As noted previously, ore is defined as mineral that can be extracted from the ground at a profit. The economic connotation is implicit in the word *ore*. Thus, whether or not a given volume of rock is ore depends upon the cost of extraction, the price of a particular mineral—both of which change frequently—and the amount of valuable mineral contained in the volume. Unlike the economic determinants of ore, deposit grade is fixed, but unfortunately it is unknown and must be estimated.

Mining engineers and geologists often conceptually depict ore bodies as shown in Fig. 2. These simple vertical sections transmit the notions of ore (valuable), waste (worthless), and the finite nature of the resource. To be sure, there are some stratabound deposits which can be reasonably represented in this manner. In most cases, however, such diagrams are vastly oversimplified and serve to reinforce the commonly held, but counterproductive *warehouse* view of ore reserves. That is, many people conceive of calculating ore reserves as analogous to taking inventory

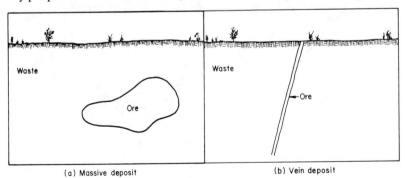

Fig. 2. Misleading conceptualizations of ore deposits.

in a warehouse. One simply counts the units on the shelves (or within the fixed ore-waste boundary) to find out how much of the commodity exists. These sketches are usually employed to illustrate mining concepts such as stripping ratio and various mining methods, but regretably they also communicate in a subliminal manner this misleading warehouse analogy of ore reserves.

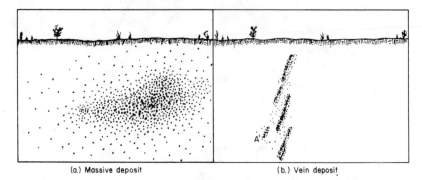

(a.) Massive deposit   (b.) Vein deposit

Fig. 3. Alternate representations of ore deposits.

, In reality, mineral deposits are more accurately represented by Fig. 3. Here, each dot represents desirable mineral, with the spatial density of dots being proportional to the concentration (grade) of the valuable mineral. With this representation one can gain far more insight into the nature of the ore reserve problem. It obviously is impossible to calculate the economic limits to mining and, therefore, the ore reserves in Fig. 3 without more information. Also, if one further acknowledges that the grade distribution in Fig. 3 is never known with certainty but is simply inferred from widely spaced samples, then the true dimensions of the ore reserve problem are clear.

The degree to which ore reserves vary with changes in production costs and mineral prices is decidedly site specific. In some deposits, there are distinct *structural limits* to the ore whereby one passes from ore to barren rock in a short distance over a relatively distinct boundary. Stratabound and vein-type deposits sometimes occur in this manner.

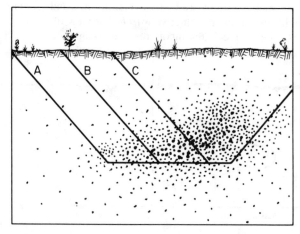

Fig. 4. Possible open pit limits in a massive ore deposit.

At the other extreme are massive low-grade deposits such as the copper porphyries where mineralization declines gradually from higher grade zones to barren rock over a considerable distance. In these instances ore is said to have *assay limits*, rather than structural limits, because mining ceases when the ore grade drops below some threshold value even though mineralization continues, albeit in nonprofitable concentrations.

Consider, for example, Fig. 4, which is intended to simulate a large, low-grade deposit such as a copper porphyry. Can mining proceed to limit *B* or limit *A*? Clearly this depends on the value of the rock excavated (price of the mineral, grade, and recovery of mineral during processing) as well as the costs of mining and processing. A change in any one of these variables can change the ore-waste boundary and, therefore, the ore reserves.

With stratabound or vein-type deposits, as depicted in Fig. 3b, there may be well-defined structural limits. Even here, however, the dynamic nature of ore reserves can be significant. For example, in Fig. 3b, the mineralized pod of material, *A*, has sharp boundaries, but whether or not it is ore is largely determined by whether the additional mine development necessary to gain access to the pod is economically justified.

In the final analysis, the mine planner must estimate whether a particular block of material in the earth's crust has a positive net value. There are many variables in this equation, each having a relatively high degree of uncertainty. Ore reserves contract or expand with each change in production costs, sales price, or new sampling data from the deposit.

Mineral Resource Classification

There are two principal sources of confusion regarding mineral resources. First, at any point in time quantitative estimates of mineral occurrence are accompanied by widely varying degrees of *geologic assurance*. Some deposits, for example, are well proven; others are highly conjectural. The second source of confusion is that mineral resources also exist with widely varying degrees of *economic feasibility*. A few deposits are virtual bonanzas; many others may only be attractive if and when mineral prices rise substantially.

An extremely useful diagram for classifying mineral resources was developed by McKelvey (1972). This graphical representation is now widely known as a McKelvey diagram, the original form of which is presented in Fig. 5. A number of modifications to the original model have been proposed, a popular one being that taken from USGS Circular 831 and reproduced in Fig. 6.

The original McKelvey diagram was viewed by most as applicable primarily to mineral resources in a macroscopic sense—e.g., to assist in preparing national and international mineral endowment estimates and mineral policies. The basic concept is, however, also applicable at the microlevel for discussing any particular mineral deposit. The modified version of the McKelvey diagram was developed with this extended application in mind.

With reference to the two axes of (1) degree of economic feasibility, and (2) degree of geologic assurance, the McKelvey diagram helps in defining several important terms.

Resources vs. Reserves: Resources is the more general term referring to the geologic endowment of minerals in the earth's crust in such concentration that commercial extraction is either presently or *potentially* feasible. *Reserves* is a subset of resources—that portion which can be extracted at a profit under the

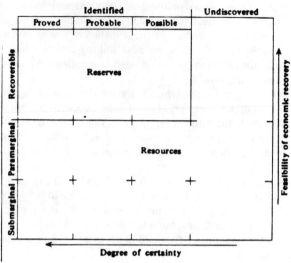

Fig. 5. Classification of mineral reserves and resources (from McKelvey, 1972).

present mineral price/production cost regime. In other words, the fundamental difference between resources and reserves is economics. Reserves can be presently extracted at a profit; resources need only offer the possibility of profit in the not-too-distant future. Fig. 6 illustrates the relationship between reserves and resources on a modified McKelvey diagram.

It is interesting to note that a United Nations special committee has recommended an international classification system for mineral resources intended to promote improved communication on this topic. The system (United Nations, 1979) recommends eliminating the term *reserves* and retaining only *resources* to reduce the confusion. However, *ore reserves* is so widely used in technical, financial, and legal circles that its demise in the United States is unlikely.

It is important to note that although resources are naturally occurring, reserves are generally created by human effort. That is, most reserves were known previously as resources until advances in technology, occasionally aided by market factors, permitted them to be reclassified as economically recoverable reserves. This fact runs counter to the view of economic rent in mineral resource development which is widely held in economic circles. In reality, the nature of income earned from the vast majority of mines differs little from that of most other economic activity. Only a small and ever-decreasing number of mines actually accrue windfall economic rent to any significant degree.

Identified Resources: Identified resources are specific mineral occurrences whose existence, location, and quality have been established. That portion of identified resources which is economically recoverable at the present time is specified as reserves.

Undiscovered Resources: Undiscovered resources encompass hypothetical resources and speculative resources. *Hypothetical resources* are those undiscovered deposits which can reasonably be expected to exist in known producing districts. *Speculative resources* imply a higher degree of risk and include conventional deposits in new districts as well as unconventional resources not previously recognized.

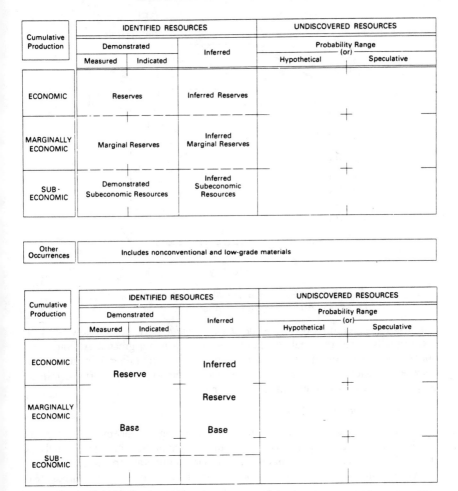

Fig. 6. Alternative versions of the mineral resource classification diagram (from USGS Circular 831, 1980).

The position of any particular mineral deposit is not static as depicted in Fig. 6. Exploration activity seeks to increase the degree of certainty of occurrence of the deposit, and advances in mining and mineral processing technology increase the feasibility of economic recovery. Thus, with continued technologic progress a mineral deposit tends to advance toward the upper-left corner of the diagram, although its rate of progress may be nil for long periods.

A third dimension to the McKelvey diagram has been proposed by Boyd (1979) to illustrate the substitutability of minerals. His point is that most minerals have little intrinsic worth and derive value only from desirable properties which they possess, properties which can usually be secured from other sources if the demand (i.e., price) is high enough. Thus, for example, whereas a two-dimensional McKelvey diagram for copper implies a limited geologic resource, a three-

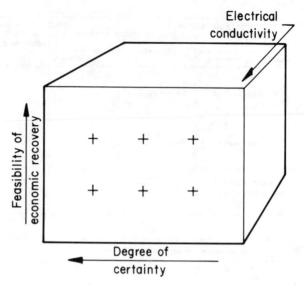

Fig. 7. Conceptual classification of mineral resources used for electrical conductivity.

dimensional diagram (Fig. 7) for *electrical conductivity* would show—at least conceptually—that the total functional resource is much greater. Although this concept is perhaps of limited interest to those performing mineral resource or ore reserve estimates, substitution is an important aspect to those performing mineral property valuations and investment decision making.

Ore Reserves for Mine Evaluation

The preceding discussion is useful in developing a conceptual view of the general ore reserve problem. However, in the context of mine investment analysis, the model now needs to be applied to an individual mineral deposit.

In evaluating a specific mineral deposit generally only identified resources are of concern, it being very difficult to place a value on undiscovered resources. The identified resources of a specific deposit can then be subdivided by (1) degree of geologic assurance, and (2) degree of economic feasibility (see Fig. 6).

Degree of Geologic Assurance: Two largely interchangeable classification systems have evolved over the years.

The *Proven, Probable,* and *Possible* ore reserve system is preferred by industry and is well entrenched in commercial and legal practice in the United States. Note that this system was also used by McKelvey in his original diagram.

The *Measured, Indicated,* and *Inferred* ore reserve classification is now being used widely in the geoscience/resource appraisal community. The US Geological Survey offers definitions of these terms as shown in Table 1, but notes that they are *loosely interchangeable* (USGS Circular 831) with proved, probable, and possible reserves.

In spite of the substantial effort that has been made to define these terms precisely, it is important to recognize that a considerable amount of discretion on the part of the analyst will probably always be required in the classification of ore reserves. For example, in structurally complex geologic environments, the amount of data necessary to classify some material as *proved* ore could be as much

Table 1. Reserve Definitions

1. **Measured (Proved) Reserves**. Quantity is compiled from dimensions revealed in outcrops, trenches, workings, or drill holes. Grade and (or) quality are computed from the results of detailed sampling. The sites for inspection, samples, and measurement are spaced so closely and the geologic character is so well defined that size, shape, depth, and mineral content of the resource are well established.
2. **Indicated (Probable) Reserves**. Quantity and grade and (or) quality are computed from information similar to that used for measured (proved) reserves, but the sites for inspection, sampling, and measurement are further apart or are otherwise less adequately spaced. The degree of assurance, although lower than that for measured reserves, is high enough to assume continuity between points of observation.
3. **Inferred (Possible) Reserves**. Estimates are based on an assumed continuity beyond measured and (or) indicated reserves, for which there is geologic evidence. Inferred reserves may or may not be supported by samples or measurements.
4. **Demonstrated Reserves** = measured reserves + indicated reserves.

Adapted from: US Geological Survey Circular 831, "Principles of a Resource Reserve Classification for Minerals."

as ten times greater than required for a similar sized block of material in a much simpler geologic setting. When evaluating an unfamiliar mine, a standard operating procedure should be to assume nothing in regard to ore reserve definitions. One should carefully examine the reserve calculation procedures to ensure that the results can be used with confidence.

Degree of Economic Feasibility: There are two calculations that are usually of interest in regard to the contents of a particular mineral deposit. First is the quantity and quality of material that is economically *recoverable* with present mineral prices and production costs. This is, of course, the ore reserve.

The second value is the total amount of ore-grade or potentially ore-grade material in the deposit. This is an in situ resource, a geologic endowment calculated at some cutoff grade. In the USGS system, the first quantity is called simply *reserves*, while the second equates roughly with *reserve base*.

These two resource calculations arise logically from the nature of the mining industry. Operating mines and mines in development have detailed mining plans and schedules, engineered systems and procedures, and projected costs. Under these conditions, it is perfectly logical to speak of ore reserves—economically recoverable mineral.

However, at any given time, most mining companies also control additional mineral resources for which ore reserve statements are not appropriate. This is because the resource is either not currently economic, or because it is still subject to such geologic uncertainty that it must be placed in the possible or inferred ore category. Material that is currently not economic would include (1) ore-grade material that is not recoverable in operating mines (e.g., isolated pods, pillars, material beyond the break-even stripping ratio, etc.); and (2) subore-grade material, either as entire deposits or as extensions in operating mines. It would be risky and perhaps misleading to publish *ore* reserves data for these resources, yet some quantitative measure of such assets must be determined by the firm for planning purposes. This is often accomplished by summing all material above a prescribed cutoff grade. The cutoff grade often used is the known cutoff grade for some operating mine in a similar geologic environment. Aside from this crude economic constraint, the resulting reserve does not assume any other costs or prices.

The terms *geologic reserve* and *mineral inventory* have sometimes been applied for such resource estimates. *Mineral inventory* has become popular with computer specialists who determine the inventory utilizing a block model of the deposit. A less ambiguous label is *in situ reserves*, which clearly implies that any constraints attributable to a mining system have not yet been imposed. Thus, *in situ reserves* and *reserve base* are roughly interchangeable.

Finally, it is interesting to note that one of the enduring sources of friction between geologists and mining engineers is the estimated loss in tonnage and grade when a deposit passes from the exploration department to the mining department, i.e., in-place tonnage and grade vis a vis mill heads. The difference between the two is due to percentage extraction—the portion of the deposit which is economically recoverable based upon the mining system selected—and dilution with waste rock as discussed subsequently. Such differences can be quite dramatic. For example, the unusual geometry of one large, deep copper deposit precluded the economic recovery of nearly two-thirds of the in-place reserves.

Dilution: One subtle, but extremely important, difference between in situ reserves and ore reserves is dilution. Dilution is the mixing of waste material with ore during the mining process. This may be either unintentional or a function of the mining system employed whereby minimum stoping dimensions require the mining of some waste. The result is that ore grade as mined, or mill-head grade, is lower, and the tonnage shipped is higher, than that calculated from mining plans.

Because the profitability of metal mines is extremely sensitive to changes in ore grade, dilution can be a critical variable in evaluating such mining ventures. An interesting example occurred with a block-caving copper mine wherein a thin high grade, but undulating ore zone with an in situ grade of 1.6% Cu became about 0.9% Cu at the concentrator, due primarily to dilution.

Like ore reserves, dilution is a term which has also been defined carelessly in mining. It is expressed as a percentage, although the figure has a slightly different meaning under different conditions. In stope mining, percent dilution usually means the amount of overbreak, or wall rock that is mined with the ore. If the waste material contains some mineral values, the mined ore grade is adjusted accordingly, and 100% extraction of the in situ reserve is assumed.

Example—Calculate the mill-head grade and tonnage for the following shrinkage stope:

In situ tons	:	9500
In situ grade	:	7.7 oz Ag per ton
Avg. grade of waste	:	2.9 oz Ag per ton
Dilution	:	20%
Total tons recovered	=	9500 (1.2) = 11,400
Mill-head grade	=	$\dfrac{9500\ (7.7)\ +\ 1900\ (2.9)}{11,400}$
	=	6.9 oz Ag per ton

The degree of dilution in stope mining is a function of many variables. It can be as high as 20% with incompetent wall rock in shrinkage stoping and as low as 5% with a highly selective mining system such as square set stoping. Generally, an allowance of at least 10% should be used.

In cave mining, waste dilution generally begins to affect ore grade when 60% to 70% of the ore is withdrawn. As the draw continues, dilution increases until the recovered grade drops to the cutoff grade, at which point mining is suspended. This may occur when more or less of the projected tonnage and contained valuable mineral is recovered. Recovery of the in situ resource is not assumed to be 100% and must, therefore, be stated.

Example—At a block-caving mine, determine the mined tonnage and grade for the following block:

Estimated in situ ore tons	:	1,300,000
Estimated in situ ore grade:	:	0.74% Cu
Ore recovery	:	104%
Dilution	:	15%
Copper content of block	=	1,300,000 × 0.0074
	=	9620 tons
Total metal recovery	=	9620 × 1.04 = 10,005 tons
Total ore mined	=	1,300,000 × 1.15 = 1,495,000
Grade of mined ore	=	10,005 ÷ 1,495,000 = 0.67% Cu.

The basic principles are the same, but only cave miners speak of recoveries in excess of 100%, which is sometimes confusing. Average values for block caving are 95–100% extraction and 15–20% dilution (Julin and Tobie, 1973). Thomas (1973) reports sublevel caving dilution as 20–30%, but these higher values are more attributable to ore deposit geometry than to the mining method.

Dilution is usually the reason cited for the difference in grade between the in situ geologic estimate and the actual delivered mill heads. However, some of this difference may be attributed to poor grade estimation rather than physical dilution with waste. Before costly operating procedures are adopted to combat dilution and, therefore, increase mill-head grade,* it is advisable to examine the ore grade estimation procedures. For porphyry coppers in particular a geostatistical ore reserve study might reveal that ore block grades have been overestimated and that physical dilution is less than indicated.

Reserve Reporting Practice

Most mining companies have more than one, and usually several, sets of ore reserve data. Particularly with firms that use computer-based reserve estimation systems, it is not difficult to generate a separate reserve estimate from each prescribed set of cost and price data. Most porphyry copper operations, for example, have detailed reserve estimates—and perhaps even mine plans—for cutoff grades of 0.2%, 0.3%, 0.4%, and 0.5% Cu.

With so many estimates issuing from the engineering office, it is easy for shareholders and other nontechnical observers to become confused. As mentioned previously, *ore reserves* implies material that is economically recoverable at the time of the estimate. This, in turn, implies that a production system with its attendant operational requirements and costs has been superimposed on the mineral deposit. Furthermore, the ore grade and tonnage reported should be fully adjusted for any dilution that is anticipated to occur during mining.

---

*Ingler (1975) shows that under certain conditions an increase in direct mining costs can be tolerated if a sufficient reduction in dilution is achieved.

Beginning in 1980 mining companies in the United States have been required by the Financial Accounting Standards Board (FASB) to report *mineral reserves* to their shareholders. Most companies have simply listed their ore reserves. Some (e.g., Kennecott and Asarco) used the term *mineral reserve* and at least one (Noranda) reported *mineral inventories*. Most firms have taken considerable care to provide as accurate information as possible. The excerpt in Fig. 8 from the 1979 Annual Report of Gulf Resources and Chemicals Corp. is a good example. For external reporting, it is always wise to provide adequate explanatory notes to minimize misinterpretations of the data.

Regardless of the name, however, the data reported appear to be the proven and probable ore reserves as defined previously. When mineral resource data from deposits of questionable profitability are reported, most firms take great care to indicate that *ore* is not involved. For example, in its 1980 Annual Report AMAX

Ore Reserve Summary.

All of Bunker Hill's ore reserves are based on estimates by Bunker Hill's geologists and mining engineers as of December 31, 1979, except that Star Unit Area ore reserve estimates were made by Hecla's geologists and mining engineers as of that date. The estimates given of current ore reserves relate only to that ore the existence of which is now reasonably established, but experience to date with the Bunker Hill, Pend Oreille and Star Unit mines has been that, as ore reserves have been mined, new ore reserves have been found to replace the mined ore and have been reflected in subsequent estimates of ore reserves. However, continued production beyond depletion of present reserves is dependent upon the discovery of additional reserves.

In the following table and elsewhere herein, ore reserves have been classified in accordance with the following definitions:

Proven Ore—The term "proven ore" means a body of ore so extensively sampled that the risk of failure in continuity of the ore in such body is reduced to a minimum.

Probable Ore—The term "probable ore" means ore as to which the risk of failure in continuity is greater than for proven ore, but as to which the assumption of ore continuity is deemed warranted.

Ore Reserve Summary December 31, 1979

Mine	Tons of Ore			Grade*		
	Proven	Probable	Total	% Lead	% Zinc	Oz. Silver Per Ton
Bunker Hill	964,033	1,113,771	2,077,804	2.5	3.6	1.6
Crescent	132,810	47,020	179,830	—	—	28.20
Star Unit Area (70% share)..	742,900	251,600	994,500	5.8	7.1	3.4
Pend Oreille—Yellowhead ..	72,460	37,570	110,030	0.7	6.0	—

*The percentages shown in this table may in some instances be higher than the percentages of metal per ton shown under "Mining" for the various mines. This is because the ore reserve tonnages and grades, although they do include allowance for normal mining dilution, do not include the extra tonnage of lower grade material that is produced by development work and by mining fringe (albeit economic) ores at stope limits. These extra tons are not part of the ore reserve, but once broken are economic to add to mill feed. The higher percentages which may be shown in this table do not indicate that the ore reserves are necessarily of a richer quality than the ores presently being mined. Bunker Hill production has been concentrated for several years in mine areas bearing greater proportions of lead-silver ores.

Fig. 8. External reporting of ore reserves by Gulf Resources and Chemicals Corp. in 1979 Annual Report.

reported over 900 million tons of *mineralized material* at its Mt. Tolman project. It is interesting to note that one recurring anomaly in reported reserve figures appears to be that *reserves* for flat-lying room-and-pillar/stope-and-pillar operations are not reduced by the material that remains tied up in pillars. This can be a sizable portion of the total reserves.

SAMPLING

Ore-body sampling is the crucial first step in ore reserve estimation. Sampling enables the engineer to estimate the quality and size of the ore body which are essential data in the financial analysis of any mining venture.

Estimating the tonnage and grade of an ore deposit is a classic example of applied statistics. The purpose of sampling is to determine, within some acceptable limit of error, certain properties of an entire population by measuring only a small portion of that population. In mine evaluation the property of the population (i.e., the ore body) which is usually of most interest is the ore grade. Therefore, a sampling program is designed to provide an estimate of ore grade which is reasonably close to the true grade of the deposit. The treatment of sampling data is briefly discussed in the section "Reserve Estimation," later in this chapter. This section focuses on the procedures involved in conducting a sampling program.

Sampling program design generally includes four major considerations:

1) *Sampling Methods*—Selection of the sampling unit (the type and size of the sample) and the actual mechanics of collecting the sample.

2) *Sampling Frequency*—Selection of the sampling density and pattern.

3) *Sample Processing*—Selection of procedures to accurately measure the mineral content of a sample.

4) *Program Management*—Developing procedures for continuously monitoring the preceding three steps as well as for gathering, marking, mapping, shipping, storing, and cross-checking samples in such a manner that errors are minimized.

As mentioned in the introduction, this chapter makes no attempt to treat sampling in any substantial detail. As a consequence, the foregoing four components of a sampling program are discussed in only a superficial manner. References are cited to direct the reader to a more complete treatment of the subject.

Sampling Methods

Sampling methods are either *direct*, where the sampler takes a sample directly from an accessible rock face or muck pile, or *indirect*, where the sample is obtained from a borehole.

Direct sampling methods include channel sampling, several varieties of chip sampling, and grab sampling. In channel sampling the dimensions of the sample volume are rigorously controlled and bias is thereby minimized. The method is, however, extremely laborious and costly. Chip sampling is faster and cheaper, but somewhat less accurate than channel sampling. Nonetheless, chip sampling with sufficient care can yield satisfactory results in most cases and is, therefore, now much more widely used than channel sampling. Both of these methods are described in more detail in Parks (1957), McKinstry (1948), and Peters (1978). An excellent description of sampling practice in South African gold mines is provided by Storrar (1977).

Grab samples of broken rock generally yield a relatively high error. However, satisfactory results have been achieved in using grab sampling for ore control in cave mining operations. Relating grab samples from a muck pile to in situ grade is not a recommended practice (Barnes, 1980).

Borehole sampling is more costly but often is the only practical way of sampling deposits at depth. A number of drilling methods are available for this purpose.

1) *Diamond Core Drilling*—The most commonly used method, diamond core drilling, provides the greatest amount of geologic information, is applicable in nearly all rock types, and has excellent depth capability. Furthermore, diamond drilling generally produces the most representative sample, particularly with the improved core recovery that has been achieved in recent years. The method is, however, costly and relatively slow.

2) *Rotary Drilling*—Although large oil-field equipment is more versatile, smaller rotary drill rigs used in mineral exploration are usually restricted to lower strength rocks and produce only finely ground cuttings for the sample. The method is, however, relatively fast and inexpensive and is often substituted for core drilling when only assay information is desired or when overlying barren rocks must be penetrated to reach the ore horizon.

3) *Percussion Drilling*—Drills which impart blow energy to rock through a long, sectional drill string have a serious depth limitation and produce only rock chips rather than core. However, down-the-hole hammers can drill straight and fast and can handle rocks of any hardness. However, they, too, become uneconomic at depth and in wet holes where it becomes costly to overcome back pressure on the drill exhaust.

4) *Churn Drills* are occasionally still used for sampling, particularly where labor costs are low or where extremely difficult ground is encountered at a relatively shallow depth such as with placer deposits (see Wells, 1973).

Rock drilling methods are explained in detail and their application in ore-body sampling is discussed by Payne (1973), and by Peters (1978).

Sampling Frequency

Ore-body sampling is an expensive, time-consuming process. Thus, in one sense every sampling program is an attempt to minimize total cost. This total cost, in turn, is the sum of the direct cost of conducting the sampling program plus the indirect cost of making an incorrect decision in regard to developing the deposit due to insufficient sample data. The cost of a wrong decision at this stage is, of course, not calculable except in a very crude manner, but most agree that this cost can be extremely high. Particularly in metal mining, a major contributing factor to nearly every troubled new mining project has been poor grade estimation or grade control. King et al. (1982) note that grade realization is almost always less than the expectation and suggest that to warrant development most deposits should have acceptable economics when the estimated grade is reduced by 10%.

Clearly, the more sample information available, the better will be the estimate of the true grade of the deposit, and the likelihood of a wrong decision will decline. However, the expected sampling error declines only in proportion to the square root of the number of samples. Therefore, the cost of achieving successive incremental improvements in grade estimation accuracy tends to increase very rapidly.

Because a sample value is used to estimate the grade of surrounding unsampled material, the appropriate sampling density is obviously a function of natural geologic variability. Complex geology may require dense sampling, whereas relatively uniform geologic conditions can tolerate much more widely spaced samples.

Prior to the widespread use of statistics in ore reserve estimation, sampling and reserve calculations were separate and discrete tasks. Geostatistics, however, permits a quantitative measure of geologic variability such that sampling frequency can be designed to achieve a certain level of precision in the subsequent reserve estimate. In other words, sampling and reserve estimation are now interrelated. This process is described in more detail by Barnes (1980, Chap. 9), and geostatistical methods are discussed briefly and referenced in the section on "Reserve Estimation" later in this chapter.

In regard to sampling patterns, Sinclair (1978) discusses the four general plans shown in Fig. 9. He shows that a regular sampling program generally minimizes the sampling error but acknowledges that the need for specific geologic information may require some deviation from this rule. A regular sampling pattern minimizes bias, and for the same reason the starting point for the pattern should be selected randomly. For example, if chip samples are to be taken along a drift, random selection of a starting point and the use of a fixed sampling interval tends to minimize the sampling error in most mineral deposits. Geologic anisotropies may prescribe different spacings in different directions, but spacings should remain fixed in any particular direction.

The appropriate sample spacing, or interval, is highly site-specific. Peters (1978) illustrates the wide range of sample spacings used in practice. Experience with similar deposits particularly within the same mining district may provide some guidance, but *rules of thumb* should be used with extreme caution. Barnes (1980) shows, for example, that a common drill-hole spacing for porphyry copper deposits (200 ft.) led to a serious overestimate of grade at a copper deposit in Arizona. Geostatistics can provide a good estimate of sample spacing required to

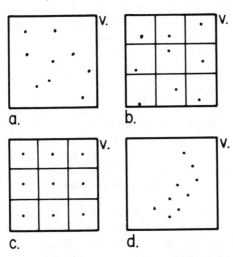

a. Random sample locations
b. Stratified random sampling
c. Regular sampling pattern
d. Biased sampling pattern

Fig. 9. Sampling patterns for locating nine drill holes in block V (from Sinclair, 1978).

yield a particular level of precision, but in practice the resulting high sampling cost often forces a compromise.

Sample Processing

Once a sample has been taken, the next task is to obtain an accurate measure of its contents. In the past this step has often received too little attention in sampling programs, but with recent large increases in precious metal prices, there is renewed interest in sample processing. Now, gold deposits having less than 0.05 oz per ton may be economic. This translates to an average of 0.00012 oz per 5 lb sample, or a gold concentration of less than two parts per million. With such a small amount of valuable mineral contained in a sample, proper processing becomes extremely important. Burn (1981) notes that sample processing is potentially the largest source of error in sampling.

The sample taken in the field usually must be modified in two ways before being assayed. First, the sample contains material of vastly different sizes, from micron-sized fines to large chunks. It is important to recognize that the concentration of the valuable mineral generally varies among the different particle sizes, often to a considerable degree. Consequently, some size reduction is often required to insure homogeneity in the subsample.

Second, the quantity of material contained in the original sample is usually too large to assay directly. The problem then becomes one of sampling the sample—extracting a smaller quantity of material which is representative of the original sample.

With the cutting of samples, the task is that of sampling particulate matter where particle size relative to sample size is of primary importance. The recognized authority on particulate sampling is Pierre Gy, and his recent publication, *Sampling of Particulate Materials* (1979), is a major effort to summarize the state of the art in this field. Practical problems in sample processing are also discussed by Lake and Perry (1973), Coleman (1978), and Barnes (1980). Also, the *Handbook of Mineral Dressing* by Taggert (1945) remains a good reference on this topic.

Program Management

Major sampling programs tend to generate a large quantity of data. Analytical testing often produces numerologic, petrographic, and geomechanics data as well as assay information. Also, frequent duplicate—even triplicate—assays are performed for quality control purposes, and the location of each sample must be accurately identified and accurately plotted on maps. Finally, sample storage is an important consideration so that check assays can be performed by potential joint venturers or other investors.

The magnitude of information generated in a well-organized and thorough sampling program for a medium-to-large mineral deposit virtually requires the use of a computerized data-based management system. In the absence of such a system, an enormous manpower effort is required to compile and utilize the expensive data which has been collected. Without a sound management plan and a firm set of operating procedures, significant errors are almost certain to be introduced. A consultant on computer-aided mine planning has remarked to the authors that with every major project on which he has worked, important errors in sample handling and data treatment have been discovered. In one case, a large copper deposit had been sampled with many thousands of feet of drilling. Some of the assay results were suspect, but checks were impossible as the limited drill core remaining was stored in a near-random fashion in an open shed in boxes that were badly water-damaged and torn.

The final investment decision is based heavily on data from a sampling program. With today's declining profit margins on bulk, low-grade ore deposits, errors in such data can have particularly disastrous impacts. As a consequence, sloppy handling of samples and sample data is less tolerable than ever. Barnes (1980) and Burn (1981) provide some good suggestions on this topic, and Peters (1978) is an excellent reference on exploration program management in general.

RESERVE ESTIMATION

The process of reserve estimation is one of extending sampling data to larger volumes of material and then usually converting the volumes to a weight measure. There are a variety of methods available to perform this task. The methods can be roughly separated into: (1) traditional methods and (2) the more recently developed statistical methods. These methods are described briefly in the following, but as with sampling, the purpose is simply to introduce important concepts and provide the interested reader with references for more thorough coverage.

Traditional Methods

The traditional methods of ore reserve estimation do not draw heavily upon the theory of mathematical statistics, but rather are based on the concept of the *area of influence*. That is, a sample value is extended to a surrounding area of influence in a manner based upon the judgement of the analyst. That area is then multiplied by some thickness, often arbitrarily established, to yield a volume of ore with an estimated grade. A summation of all such volumes produces an estimate of the in situ resource. If only those volumes which are economically recoverable are included, an ore reserve estimate results. A widely cited and comprehensive treatment of traditional ore reserve estimation methods is Popoff (1966).

In underground vein mines, ore is often blocked out by drifts on adjacent levels and raises between those levels. This gives rise to the classic *underground blocks* (influence-area-weighted assay) method described by Parks (1957), an example of which is shown in Fig. 10. Here the samples are assumed to be chip or channel samples, and each sample value is extended to an area of influence defined by the length of the sample and half the distance to adjacent samples.

Areas of influence may also be constructed around borehole samples using the well-known methods of *polygons* or *triangles*. In Fig. 11, the areas are polygons constructed by extending the perpendicular bisectors of lines connecting adjacent samples. Here the value of the sample is assumed to extend throughout its polygonal area of influence. This assumption is sometimes referred to as the rule of nearest points.

With the method of triangles (Fig. 12), the deposit is divided into a series of triangles with apexes at the sample boreholes. In this method, both the thickness and the grade of the ore are assumed to change linearly between boreholes so that the estimated tonnage and grade is calculated as indicated. This principle is sometimes referred to as the rule of gradual changes or the law of linear functions.

Fig. 10. Example of underground block method of ore reserve estimation.

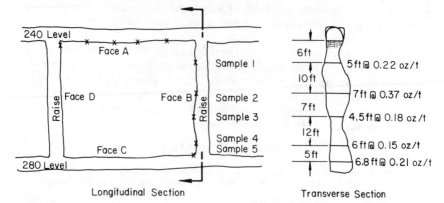

Longitudinal Section   Transverse Section

Assume minimum stoping width = 6 ft.

Average grade of face B, G_B $\dfrac{\sum\limits_{i=1}^{5} g_i a_i}{\sum\limits_{i=1}^{5} a_i}$

$$= \frac{(0.22)(5)(11) + (0.37)(7)(8.5) + (0.18)(4.5)(9.5) + (0.15)(6)(8.5) + (0.21)(2.5)(6.8)}{(6)(11) + 7(8.5) + 6(9.5) + 6(8.5) + 2.5(6.8)}$$

$$= \frac{12.10 + 22.02 + 7.70 + 7.65 + 3.57}{66.0 + 59.5 + 57.0 + 51.0 + 17.0}$$

$$= \frac{53.04}{250.50} = 0.21 \text{ oz per ton.}$$

Average grade of block $= \dfrac{\sum\limits_{i=A}^{D} G_i A_i}{\sum\limits_{i=A}^{D} A_i}$ Where g_i is grade of ith sample,
G_i is average grade of ith face,
a_i is area of influence of ith sample, and
A_i is area of ith face.

Fig. 11. Method of polygons for ore reserve estimation. Value of sample b assumed to extend throughout polygon B.

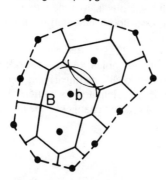

Note that the method of polygons yields a unique solution and is, therefore, more commonly used. There are many possible ways to connect a given set of drill holes to create a series of triangles. If this method is employed, it is generally recommended that the triangles be constructed as nearly equilateral as possible.

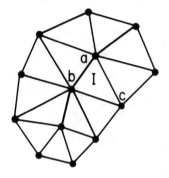

Fig. 12. Method of triangles for ore reserve estimation.

Thickness, t, of triangle I $= \dfrac{\sum\limits_{i=a}^{c} t_i}{3}$

grade, g, of triangle I $= \dfrac{\sum\limits_{i=a}^{c} g_i t_i}{\sum\limits_{i=a}^{c} t_i}$

Most ore reserve calculations in large mines are now done on a digital computer. With the introduction of the computer, however, modifications were required in the basic ore reserve calculation algorithms and refinements were introduced.

The basis of nearly all computer ore reserve systems is the *block model*, whereby the deposit is mathematically divided into a three-dimensional array of blocks, as illustrated in Fig. 13. To assign grades to individual blocks a number of methods have been used. These methods generally involve the block being assigned the grade of either (1) the nearest sample or (2) some weighted combination of several samples within a prescribed radius of influence around the block. For example, in Fig. 14, the ore block in question might be assigned the grade from the nearest sample (No. 4), which is really a variation of the method of polygons. Alternatively, the block could be assigned a grade by combining assays of several nearby samples in some manner. The following inverse distance weighting scheme has been used frequently:

where S_x is the assay of sample x; r_x is the distance of sample x from the center of the ore block; x is the weighting exponent, usually between 1.5 and 3, but selected by judgment and experience; n is the number of samples within the radius of influence, which is often based on experience; and G_a is the weighted average grade of block a.

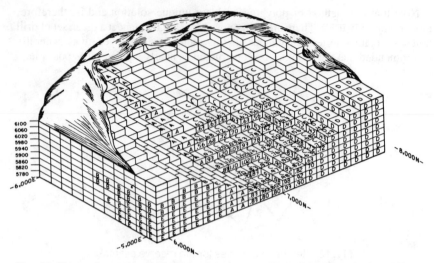

Fig. 13. Three-dimensional block model of an ore body (from Crawford and Davey, 1979).

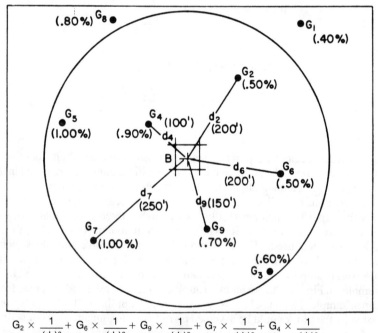

$$B = \frac{G_2 \times \dfrac{1}{(d_2)^2} + G_6 \times \dfrac{1}{(d_6)^2} + G_9 \times \dfrac{1}{(d_9)^2} + G_7 \times \dfrac{1}{(d_7)^2} + G_4 \times \dfrac{1}{(d_4)^2}}{\dfrac{1}{(d_2)^2} + \dfrac{1}{(d_6)^2} + \dfrac{1}{(d_9)^2} + \dfrac{1}{(d_7)^2} + \dfrac{1}{(d_4)^2}}$$

$$B = \frac{0.5 \times \dfrac{1}{(200)^2} + 0.5 \times \dfrac{1}{(200)^2} + 0.7 \times \dfrac{1}{(150)^2} + 1.0 \times \dfrac{1}{(250)^2} + 0.9 \times \dfrac{1}{(100)^2}}{\dfrac{1}{(200)^2} + \dfrac{1}{(200)^2} + \dfrac{1}{(150)^2} + \dfrac{1}{(250)^2} + \dfrac{1}{(100)^2}} = 0.77\%$$

Fig. 14. Assigning a grade to block B using the method of inverse distance squared (from Hughes and Davey, 1979).

$$G_a = \frac{\sum\limits_{i=1}^{n} \dfrac{S_i}{r_i}}{\sum\limits_{i=1}^{n} \dfrac{1}{r_i}}$$

Many refinements to the block model approach to ore reserve estimation have been made for specific applications. Variable block sizes and shapes are sometimes desirable for complex ores (Kim, 1979); elliptical areas of influence have been adopted in recognition of the existence of anisotropism (Barnes, 1980); and open pit limits generators—commonly based upon the floating cone concept (Lemieux, 1981)—are used to estimate true ore reserves by adding the economic dimension. The literature in this area is particularly rich, a good applied reference being Crawford and Hustrulid (1979).

Statistical Methods

The problem of estimating the ore grade of a deposit is, as mentioned earlier, a problem in applied statistics. The objective is to draw certain inferences about the population (the total number of samples which comprise the entire deposit) from a sampling distribution of the relatively small number of samples actually collected and assayed.

There is, of course, a large body of knowledge in classical mathematical statistics which deals with such topics. Much of the work in the application of this knowledge to ore reserve estimation was done by Hazen and is summarized in US Bureau of Mines Bulletin 621 (1967). Most models in classical statistics, however, require individual samples to be independent and the sampling distribution to be approximately normal so that the sample mean is an unbiased estimator of the population mean.

When the foregoing assumptions are satisfied, applications of classical statistics to ore reserve estimation are permissible. However, for many ore deposits, samples (1) are not independent but are geologically related and (2) produce a highly skewed (approximately lognormal) sampling distribution where the sample mean is a biased estimate of the population mean.

Credit for identifying and formalizing the problems associated with lognormal sampling distributions goes to H.S. Sichel and D.G. Krige for their work on South Africa gold ores. Their work is summarized well in Storrar (1977). Sichel (1966) developed tables to permit the calculation of both an unbiased estimate of the population mean and confidence intervals for the mean from a lognormal sampling distribution.

The theory of regionalized variables, or geostatistics as it is now commonly known, was formalized by G. Matheron who built upon Krige's work to include the spatial interrelationship of samples. Geostatistical ore reserve estimation was one of the most important technical developments in mining during the 1970s, although it had been used for specialized applications with certain minerals for many years prior to 1970.

The two basic steps in geostatistical ore reserve estimation are: (1) variogram development and (2) Kriging block and deposit grades.

The variogram is simply a graphical representation of the directional interdependencies of sample grades. Variogram development remains part art and part science, as the concept is fairly easy to describe and understand, but requires considerable judgment and experience in practice.

Kriging, named after D.G. Krige, is a mathematical procedure for assigning linear weights to samples such that the calculated estimation error for the block grade is minimized. This gives a mathematically justified distance-weighting scheme which compensates for intercorrelations among the samples.

The geostatistics literature is abundant but often mathematically intimidating to the practicing engineer. Among the more useful publications are David (1977), the most readable complete textbook on the subject; Barnes (1980), which emphasizes applications; and Clark (1979), which describes the basic procedures in clear language.

Geostatistics promises a *best* ore reserve estimate inasmuch as the grade estimation error is minimized. The question of whether and at what point the added analytical expense of geostatistics is justified is site-specific and always difficult to answer. However, an increasing number of firms are finding that the reduction of risk offered by geostatistics is well worth the extra expense.

Converting Volumes to Weights

Once grades have been estimated for individual blocks of material, the next step is usually to calculate the tonnage in each block. Given the density of the material in question, this is a simple mathematical exercise, but problems frequently arise in determining the correct density to use. Recently, for example, a southwestern uranium deposit in a relatively advanced state of evaluation suddenly became uneconomic when 25% of its reserves disappeared due to an error in estimating the density of the ore.

A separate density should be determined for every major rock type. In the exploration stage, a good estimate of density can be obtained from drill core averaged over many linear feet. On rare occasions, voids in the rock may distort the results, but generally discontinuities can be ignored. In an operating mine the best estimate of ore density is obtained by relating surveying results to actual tonnages milled over a period of time. In most mines this will differ from the value obtained by using truck counts or car counts reported by operators, but these latter figures are usually subject to larger errors.

Table 2. Calculating Tonnage Factor For a Massive Sulfide Ore

Constituent	(1) Weight %	(2) Specific gravity	(3) Pcf	(4) Lbs per ton ore	(5) (4 ÷ 3) cu ft per ton ore
Chalcopyrite	8	4.2	262	160	0.61
Sphalerite	10	4.0	250	200	0.80
Pyrite	18	5.0	312	360	1.15
Silica	48	2.7	168	960	5.71
Feldspar	16	2.6	162	320	1.98
Total					10.25 cu ft per ton
				or	195 pcf

Note *incorrect* result obtained by multiplying column 1 by column 3 and summing = 209 pcf.

In porphyry copper deposits and other mines where the ore grade is low, density is usually independent of ore grade. However, where ore grades are high and the ore minerals have densities that differ significantly from the gangue, there is an important correlation between density and grade. For example, a block of 15% lead ore will contain roughly 13% more tons than the same sized block of unmineralized waste rock. Note also that computing the weighted average density using weight percentages of the constituent minerals will give an erroneous result. The correct procedure is illustrated in Table 2.

SUMMARY

The value of any mine is ultimately derived from the quality and quantity of the geologic resources in the ground. Ore reserves give rise to mine plans and production schedules which, in turn, yield revenues and costs. An accurate appraisal of that reserve is, therefore, an essential element of any mine investment analysis.

With the continued development of statistical applications in ore reserve estimation, this topic can no longer be treated adequately in a book dedicated to mine investment analysis. Therefore, this chapter seeks only to outline the unique problems associated with ore reserve estimation and to cite appropriate references which deal with these issues in detail.

Perhaps the most important objective of this chapter is to instill in the reader the importance of precision when measuring and reporting resources and reserves. The entire valuation exercise is dependent upon the careful gathering, assaying, and recording of representative samples. No amount of sophisticated computer processing can compensate for fundamental errors in the data. The following three principal conclusions from the study by King et al. (1982) are generally consistent with these comments.

1) Ore reserve estimation is not a matter of more calculation. Calculations form only a part, and not necessarily the most important part, of the overall procedure.

2) The ore reserve statement should not merely be an estimate of what is in the ground, but should be a prediction of what will be fed to the mill.

3) Due to the limitations of sampling and the nature of ore, the accuracy of grade prediction rarely exceeds two significant figures.

Finally, the economic component of the definition of ore often creates important communication difficulties in mining. Few, if any, ore reserve statements should be accepted at face value. The analyst must further examine the procedures and definitions used in preparing the statement before the information can be used intelligently. Furthermore, a change in any one of several economic variables alters the amount and grade of ore, a relationship which is often important but frequently ignored in financial sensitivity studies. These unique characteristics of ore are largely responsible for the existence of mine investment analysis as a distinct field of study.

REFERENCES

Barnes, M.P., 1980, *Computer-Assisted Mineral Appraisal and Feasibility*, AIME, New York, 167 pp.

Boyd, J., 1979, Personal Communication.

Burn, R.G.,1981, "Data Reliability in Ore Reserve Assessments," *Mining Magazine*, October, pp. 289–299.

Coleman, R.L., 1978, "Metallurgical Testing Procedures," *Mineral Processing Plant Design*, Chap. 9, A.L. Mular and R.B. Bhappu, eds., AIME, New York, pp. 144–181.

Clark, I., 1979, *Practical Geostatistics*, Applied Sciences Press, London.

Crawford, J.T., and Davey, R.K., 1979, "Case Study in Open-Pit Limit Analysis," *Computer Methods for the 80's*, Chap. 3.3.3, A. Weiss, ed., AIME, New York, pp. 310–318.

Crawford, J.T., and Hustrulid, W.A., eds., 1979, *Open Pit Mine Planning and Design*, AIME, New York, 367 pp.

David, M., 1977, *Geostatistical Ore Reserve Estimation*, Elsevier Scientific Publishing Co., Amsterdam, Holland, 364 pp.

Gy, P.M., 1979, *Sampling of Particulate Materials: Theory and Practice*, Elsevier Scientific Publishing Co., Amsterdam, Holland, 423 pp.

Hazen, S.W., 1967, "Some Statistical Techniques for Analyzing Mine and Mineral Deposit Sample and Assay Data," Bulletin 621, US Bureau of Mines, Government Printing Office, Washington, D.C., 223 pp.

Hughes, W.E., and Davey, R.K., 1979, Drill Hole Interpolation: Mineralization Interpolation Techniques, *Open Pit Planning and Design*, Chap. 5, J.T. Crawford and W.A. Hustrulid, eds., AIME, New York, pp. 53–64.

Ingler, D., 1975, "Rock Dilution in Underground Slopes: Do you know How Much It Costs?," *World Mining*, August, pp. 54–55.

Julin, D.E., and Tobie, R.L., 1973, "Block Caving," *SME Mining Engineering Handbook*, Chap. 12.14, A.B. Cummins and I.A. Given, eds., AIME, New York, p. 12–166.

Kim, Y.C., 1979, "Open-Pit Limits Analysis—Technical Overview," *Computer Methods for the 80's*, Sec. 3.3.1, A. Weiss, ed., AIME, New York, pp. 297–303.

King, H.F., McMahon, D.W., and Bujtor, 1982, *A Guide to the Understanding of Ore Reserve Estimation*, The Australasian Institute of Mining and Metallurgy, Parkhill, Vic., Australia, 21 pp.

Lake, J.L., and Perry, J., 1973, "Sample Handling and Preparation," *SME Mining Engineering Handbook*, Chap. 5.4, A.B. Cummins and I.A. Given, eds., AIME, New York, pp. 5–70 to 5–74.

Lemieux, M., 1979, "Moving Cone Optimizing Algorithm," *Computer Methods for the 80's*, Sec. 3.3.5, A. Weiss, ed., AIME, New York, pp. 329–345.

McKelvey, V.E., 1972, "Mineral Resource Estimates and Public Policy," *American Scientist*, Vol. 60, No. 1, pp. 32–40.

McKinstry, H.E., 1948, *Mining Geology*, Prentice-Hall, Inc., Englewood Cliffs, NJ, 680 pp.

Mitke, C.A., 1930, *Mining Methods*, McGraw-Hill, New York, 195 pp.

Parks, R.D., 1957, *Examination and Valuation of Mineral Property*, 4th ed., Addison-Wesley Publishing Co., Reading, MA, 507 pp.

Payne, A., ed., 1973, Exploration for Mineral Deposits, *SME Mining Engineering Handbook*, Sec. 5, A.B. Cummins and I.A. Given, eds., AIME, New York, pp. 5–1 to 5–105.

Peters, W.C., 1978, *Exploration and Mining Geology*, John Wiley and Sons, New York, 696 pp.

Popoff, C.C., 1966, "Computing Reserves of Mineral Deposits: Principles and Conventional Methods," Information Circular 8283, US Bureau of Mines, Government Printing Office, Washington, D.C.

Sichel, A.S., 1966, "The Estimation of Means and Associated Confidence Limits for Small Samples from Log-Normal Distributions," *Mathematical Statistics and Computer Applications in Ore Valuation*, Johannesburg, March.

Sinclair, A.J., 1978, "Sampling a Mineral Deposit for Feasibility Studies and Metallurgical Testing," *Mineral Processing Plant Design*, Chap. 7, A.L. Mular and R.B. Bhappu, eds., AIME, New York, pp. 115–134.

Storrar, C.D., ed., 1977, *South African Mine Valuation*, Chamber of Mines of South Africa, Johannesburg, 472 pp.

Taggert, A.F., 1945, *Handbook of Mineral Dressing*, John Wiley and Sons, New York.

Thomas, L.J., 1973, *An Introduction to Mining*, Halsted Press, New York, 436 pp.

US Bureau of Mines and US Geological Survey, 1980, "Principles of a Resource/Reserve Classification for Minerals," Circular 831, US Geological Survey, Government Printing Office, Washington, D.C., 5 pp.

United Nations, 1979, "The International Classification of Mineral Resources," Economic Report No. 1, UN Centre for National Resources, Energy and Transport, New York, 8 pp.

Wells, J.H., 1973, "Placer Examination Principles and Practice," Technical Bulletin 4, Bureau of Land Management, Government Printing Office, Washington, D.C., 209 pp.

5

Estimating Revenues

Who steals our gold and silver, and copper,
zinc and lead?
Who takes the joy all out of life and strikes our
high hopes dead?
Who never wrote a schedule that to anyone
else was clear?
The sulphur-belching, miner-welching,
smelter engineer.
—Anonymous
From "The Engineer," *Engineering & Mining*
Journal, Apr. 13, 1918

INTRODUCTION

A striking characteristic of publications devoted to the principles of engineering economy and investment analysis is that the revenue side of the equation is generally either ignored entirely or addressed only in a superficial manner. This is understandable to a degree in the general case because methods of revenue estimation are highly dependent upon the specific type of project under consideration and the nature of the market served. An important contributing factor, however, is that developing estimates of future sales prices for most products is a risky activity which is less amenable to credible quantitative modeling than other topics in investment analysis.

The specific subset of capital intensive investments addressed in this book—mining projects—is particularly sensitive to projections of mineral prices, many of which are notoriously volatile. Furthermore, the unique nature of mineral markets, prices, and product specifications occupies a major role in mineral project evaluation. Even here, however, authors have tended to offer little assistance to the reader in estimating project revenues. For example, the most popular mining engineering textbook for four decades, *Elements of Mining* (Lewis and Clark, 1967), contains the following advice on estimating future mineral prices (pp. 359–360):

"Since the life of the mine will extend over a number of years the future price of mineral products must be estimated in order to calculate the probable profit to be realized. Current prices may be used, but it is preferable to take the average price for the past 25 to 35 years as it covers definite business cycles. Some engineers believe that a mine should make a small profit if its highest-grade ore is marketed at the minimum price during the past 35 years. Business experience and a sound understanding of economic conditions are valuable aids to the engineer's judgment in estimating future prices."

The preceding quotation is the extent of the authors' comments on revenue estimation in a 48-page chapter devoted to mine valuation. Most other contemporary publications contain little more on the topic.

Estimating mineral project revenue is, indeed, a difficult and risky activity. However, intractability is not a sufficient reason to ignore a problem of such critical importance to mine investment analysis. Therefore, this chapter was written to briefly describe some of the philosophies and methodologies used—past and present—to estimate mineral prices and revenues.

COMPONENTS OF REVENUE

Annual mine revenue is calculated by multiplying the number of units produced and sold during the year by the sales price per unit. In practice the number of units sold usually differs by a small amount from the number of units mined due to changes in inventory levels, but this difference is generally ignored in valuation studies. While the arithmetic involved in calculating annual mine revenue is trivial, determining the best value to use for each of these two critical variables is much more difficult.

Production

There are a number of important considerations in estimating the number of units produced and sold annually. One of the key variables associated with annual production is, of course, the **tonnage** of ore produced. Annual ore tonnage is derived from the mining schedule which, in turn, is a function of deposit characteristics, mining method, and many other factors. For preliminary studies mine schedules may only be prepared for multi-year increments. However, for major investment decisions mine schedules should be refined to an annual basis, or preferably, to a quarterly or semi-annual basis for at least the first three to five years of operation.

The second key variable associated with determining the annual production of salable units is the **grade** of the ore mined. The estimate of ore grade is obtained by employing one of the processes described in Chapter 4. Estimates of in-situ ore grade must first be adjusted for mining extraction and dilution before computing the annual number of salable units produced.

After arriving at appropriate estimates for annual tonnage and grade of ore produced, the analyst must then calculate the number of **payable** units produced annually. That is, the valuable mineral usually must be recovered from the gangue by some beneficiation process which invariably results in some loss of product in the tailings. Even under the unusual conditions of direct-shipping ore, smelter losses will occur which have the result of reducing revenues due to valuable mineral lost in slag.

Most ores require beneficiation before a salable product can be produced. The resulting milling losses must be estimated, and appropriate recovery percentages established. These recoveries are commonly estimated from a metallurgical testing program which is conducted at some point during the feasibility study. **Percentage recovery**, then, is the third basic variable which must be estimated to arrive at a final estimate of the annual production of salable units extracted from the mine.

In summary when estimating the production component of the revenue equation, the analyst must answer the following three fundamental questions for each

year in the analysis:
- what is the tonnage to be mined?
- what is the grade of the ore?
- what percentage of the valuable mineral in the ore will be recovered and sold?

Unit Price

The second major component of the mine revenue calculation is unit sales price. Estimating future mineral prices—particularly prices far enough into the future to be of use in mine investment analysis—is an exercise for which a high error of estimation invariably exists. The characteristically long preproduction lead times of mining projects mean that the success of these capital-intensive ventures will be determined by mineral prices five to ten years in the future. One need only reflect on the economic turbulence of the past ten years to appreciate the enormous uncertainty involved in these decisions.

Mineral prices are ultimately determined by supply and demand like any other product. However, there are major complications on both sides of the equation which seriously impair the value of quantitative econometric modeling for estimating mineral prices. On the demand side, one encounters the fundamental uncertainty surrounding the general level of economic activity that will exist in some future period. With the exception of fuels, consumption of most minerals is highly cyclical, being strongly related to the capital goods and construction industries. Furthermore, the demand for minerals lags most other economic activity, as consumers either accumulate or work off inventories before concluding that changes in business conditions will be sustained long enough to warrant changes in their raw materials (i.e., minerals) orders. Finally, demand for some minerals is based on speculation rather than intrinsic utility, and speculative demand is very difficult to anticipate.

On the supply side, mineral production curtailments or expansions are often not felt for several months in the marketplace due to the large amount of product in the "pipeline" en route to market. Also, mines have a high level of fixed costs and often operate in remote locations, both of which increase the resistance to shutdowns and start-ups when economic conditions change. Supply is also affected by new discoveries, new technology, and recycling. Many minerals are traded on world markets; therefore, in weighing supply and demand pressures the analyst must generally consider production and consumption in the entire free world along with any trade restrictions that might exist. Finally, net trade with Communist block nations must also be considered. This has been, for example, a significant factor in the nickel market in the early 1980s.

Clearly, the analysis of supply and demand of most minerals is a complicated matter. A further discussion of using supply and demand relationships in mineral price estimation is contained in the section of this chapter on "Projecting Mineral Prices."

MINERAL MARKETS

The topic of mineral markets and pricing is a complex one that is only briefly described in this section. A good reference for a more thorough exposition on mineral marketing can be found in Vogely (1976).

With respect to marketing and pricing, minerals can be discussed in two categories. The first category includes fungible commodities where there is little or no difference in product quality among producers. These minerals include most metals that are priced based upon negotiated settlements between buyers and sellers on one of the commodity exchanges. The most important exchanges are the London Metal Exchange (LME) and the New York Commodity Exchange (Comex).

The second category includes all other minerals. Here the production from every mine has a unique analysis which may significantly affect the price it will bring in the marketplace. Most industrial minerals fall in this category as do most coals, the prices of which are greatly affected by sulfur content and heating value. Sales prices of minerals in this category are generally determined through individual negotiations between buyer and seller, although there may be some published information regarding recent sales of similar material which can be used as a general guideline.

Commodity Exchanges

Commodity exchanges, as exemplified by the LME and the Comex, are formal free market auctions where buyers and sellers negotiate mutually acceptable sales prices. To facilitate such transactions the commodity traded must comply with a standard specification for quality, weight, and form (shape). In setting useful, market-clearing prices exchanges can only operate effectively if there are a large number of buyers, a large number of sellers, and the material traded complies with some widely accepted standard. For example, copper is traded today mostly in cathode form, and to be eligible for trading on the LME or the Comex, cathodes from any particular tank house must first undergo extensive quality control testing and be registered with the exchange.

Although the volume of minerals actually traded on the commodity exchanges is relatively small, the published transactions' prices form the basis for pricing most similar material throughout the free world. A copper producer may, for example, price at some premium to or discount from the Comex, depending upon the quality and shape of the copper to be marketed. Consumers will generally pay a slight premium over the Comex quotation to maintain good working relationships with, and secure good service from, a supplier. In the copper business, a further premium over the cathode price would be placed on continuous cast copper rod, the most common wire plant stock today.

Except for the modest price adjustments for varying quality and shape described previously, producers of commodities traded on exchanges have little flexibility in pricing. Because these minerals are fungible, a buyer will usually be unwilling to pay any more for Arizona copper than for, say, Zambian copper. The minerals traded on commodity exchanges are largely high unit value metals for which transportation costs do not comprise a significant share of the product's value. Thus, except where tariffs or import quotas distort the system, metals are truly world commodities, and significant premiums over commodity exchange prices are rarely attainable.

A final comment should be included on the importance of futures trading on the exchanges. Contracts are also traded on both the LME and the Comex that permit the holder to buy or sell fixed quantities of mineral at a specified price at a particular time in the future. Although speculators are active in the futures market with the hope of making highly leveraged profits, forward transactions also permit mineral producers and consumers to minimize their price risks through hedging. Any time, for example, that the forward price of a commodity rises to a level that will yield an acceptable return, a producer can purchase a forward sales contract to lock in this level of profit. For a consumer, acquiring a low-priced futures purchase contract will accomplish the same objective. Hedging transactions are described more fully in Labys and Granger (1973).

Purchase Contracts

Most mineral commodities are sold at prices determined in individual contracts negotiated between buyers and sellers. Most industrial minerals and coal are priced in this manner. There are no "going market prices" for these minerals as there are for base and precious metals. Most analysts can come to fairly close agreement as to what "the" price of, say, zinc is at any point in time. However, it is not possible to do the same for coal. Certain benchmark quotations for a wide variety of commodities are published in some trade publications such as *Engineering & Mining Journal* and *Coal Week*. These are, however, only rough guides, and actual sales prices can vary widely.

Some mineral commodities are traded primarily through long-term contracts, even though a smaller portion is sold in current markets at spot prices. Two classic examples of such commodities are coal and uranium.

During the mid-to-late 1970s when coal prices were escalating rapidly, utilities desperately sought coal on a long-term basis to meet current energy shortages as well as anticipated future growth requirements. The strategy of the utilities was simply to secure needed quantities of coal guaranteed to meet quality standards over reasonable time periods at prices below those anticipated on the spot market. The coal industry, not having experienced a dramatic increase in coal prices for a considerable period, saw the long-term contract as a means of procuring the financing needed to develop coal properties and of guaranteeing a given level of sales at a stipulated price, thereby substantially protecting its financial exposure and greatly reducing one of the primary risks associated with the mining industry.

Because of the mutual advantages cited, many long-term coal contracts were negotiated (5-, 10-, and 15-year contracts were typical). Most of these contracts contained provisions to protect the seller from inflation as well as from escalating property taxes, royalties, and severance taxes. For example, most contracts protected the seller's profit by indexing the selling price to the Consumer Price Index (CPI) or the Producer Price Index (PPI). Many contracts also contained provisions for passing royalty increases, severance tax increases, and additional bonus costs through to the buyer. Still other producers negotiated long-term contracts on a cost-plus basis. A typical arrangement was for the utility to pay the seller on the basis of 110% of actual production costs. Obviously, a contract of this type provided the producer with little incentive to minimize costs.

A number of coal producers chose not to participate in long-term contracts. Instead, they felt it was to their advantage to sell on the spot market and maximize returns for as long as the market would allow. The relative abundance of readily mined coal, the relative ease in bringing coal properties into production, the relatively low capital requirements for new coal properties, and the easing of the energy crisis in the late 1970s all contributed to the inevitable downward adjustment in coal prices. It is interesting to note that most of the coal producers still operating in 1983 were those fulfilling long-term contracts.

In the recent past uranium sold primarily via long-term contracts. In fact, just prior to the energy crisis in the mid-1970s uranium was being sold on long-term contracts for $20-$25 per pound. During the energy shortages of the mid-to-late 1970s, however, the price rose rapidly to a high of nearly $50 per pound. Because most traditional uranium producers had committed their production to long-term contracts, the high spot prices proved to be very tempting, and mining companies rushed to find and develop new uranium production to take advantage of these prices. At the time, this strategy seemed secure in view of the high projected growth of energy demand, uncertainties of energy supply, and the fact that existing capacity was largely committed to long-term contracts well into the 1980s. The result, of course, was that a significant amount of new production was brought on-stream just as the price of uranium plummeted to the $20 per pound range as a result of weakening demand for energy in general and nuclear energy in particular. Once again, most of the uranium producers who continued to operate were those supplying long-term contracts.

Smelter Contracts

Mining firms not having captive processing facilities may be able to market their products directly to custom mills or smelters. Custom smelting is common in the base metals industry where there is substantial worldwide trade in copper, lead, and zinc concentrates. Custom milling, however, is no longer a significant activity in the United States. The economic service radius around such a facility is very limited due to the high unit transportation cost of ores.

Custom smelting continues to be an important marketing mechanism in the metals business. Producers of concentrates often have the options of selling their concentrates to a broker or directly to a custom smelter, or tolling the concentrates. With tolling, the smelter—and in most cases the refinery—simply charges for its services and returns the refined metal derived from the miner's concentrates to the miner for marketing.

The largest volume of concentrate trade is in the copper industry, where a very large custom smelting industry was constructed in Japan in the past two decades. Although some smelters are dedicated exclusively to custom smelting, even most captive smelters accept some custom concentrates if capacity is available.

Transactions between a concentrate producer and a custom smelter are governed by the smelter contract. A hypothetical example is shown in Table 1, and an example is provided to illustrate the calculation of net smelter return. Whereas long-term, 10- to 20-year contracts were common during the 1960s, contracts today have much shorter durations, generally only 2 to 3 years. Economic uncertainty is the principal reason that contract lengths have been reduced. A common procedure employed for, say, a three-year contract is to annually renegotiate the terms for one-third of the tonnage involved.

Actual smelter contracts are much more lengthy and detailed than the simplified example shown in Table 1. Other important clauses which are generally included in a smelter contract are:

1) Weighing, sampling, and moisture determination. Describes procedures to be used, including resolution of disputes.

2) Assaying. Describes procedures to be used. Specifies the *splitting limits*, or maximum permissible difference between buyer's and seller's assays before requiring an umpire assayer to help resolve the difference.

3) Loading and unloading of concentrates. Specifies which party pays for the many costs arising from concentrate shipment. Penalties for shipping in nonstandard vessels or cars can be substantial.

4) Title and risk of damage or loss. Specifies responsibilities of the parties and procedures to be followed if cargo is lost or damaged in transit.

Table 1. Copper Concentrates—Japanese Smelter Contract

Quality:	Copper concentrates reasonably free of deleterious impurities and having the following approximate analysis: Cu 26% Ag 2.4 oz/dmt Fe 30% Au 0.15 oz/dmt S 33%
Quantity:	Approximately 23,000 dmt per quarter, plus or minus 10% at seller's option
Delivery:	F.o.b. vessel stowed and trimmed, Long Beach, CA
Price:	**Copper:** If final assay is less than 26% Cu, deduct 1.0 units; if greater than 26% Cu, deduct 1.1 units; and pay for 98% of the remainder at the LME Wirebar Settlement for the quotational period **Silver:** Deduct 0.8 oz, pay for 95% of remainder at the London Spot/US Equivalent for the quotational period, less 25¢ per oz **Gold:** Deduct 0.03 oz, pay for 95% of remainder at London Initial and Final gold quotations averaged for the quotational period, less $5.00 per oz
Treatment charge:	$85 per dmt, subject to cumulative annual escalation of 7% per year
Refining charge:	8¢ per lb of payable copper, subject to cumulative annual escalation of 7% per year
Price participation:	For every 1¢ per lb that settlement price for copper exceeds $1.00 per lb, refining charge increases by 0.1¢ per lb. This increase in refining charge shall be limited to a total of 4.0¢ per lb
Quotational period:	Either month of arrival or month following month of arrival at port of final destination, at buyer's option
Payment:	90% provisional payment 30 days after arrival. Balance upon receipt of final weights, assays, and quotations
Penalties:	Arsenic: $1.00/dmt for every 0.1% over 0.5% Bismuth: $1.50/dmt for every 0.1% over 0.1% Antimony: $1.00/dmt for every 0.1% over 0.5% Moisture: $0.50/dmt for every 1.0% over 8.0% Fractions in proportion
Length:	Two years beginning July 1, 1983

5) Force majeure. Specifies events beyond the control of either the seller or the buyer for which the failure of either party to meet the provisions of the contract is excusable.

6) Settlement of disputes. Describes procedures by which the parties agree to resolve any controversy or claim. Arbitration is often specified.

7) Environmental matters. Seller may be required to assume responsibility for disposal of sulfuric acid derived from his concentrates and may bear some of the risk for environmentally mandated expenditures in the future.

Example No. 1: Desperation Mines sells its copper concentrates to Honorable Smelting Co. in Japan according to the smelting schedule shown in Table 1. Determine the net smelter return per metric ton of concentrate shipped and per pound of copper.

Note: The term "net smelter return (NSR)," is the base against which royalties are commonly levied in mineral leases and is generally determined at the mill site (i.e., net of transportation costs to the smelter).

Given: Concentrate grade: 27.8% Cu   0.35% As
 3.1 oz Ag/t   0.20% Bi
 0.08 oz Au/t 0.15% Sb

Concentrate moisture: 9.2% shipped
 5.0% received

Transportation costs:
 Rail: mill to Long Beach $37.20/t
 Loading and stowing   15.10/t
 Ocean freight   35.40/t
 Transit losses   1.0%

Prices for quotational period: copper: $0.815/lb
 silver: $10.30/oz
 gold: $442.50/oz

Solution

A) Determine relevant weights.
 1) Shipped—9.2% moisture
 $2205 \times 0.092 = 203$ lb water/t
 $2205 \times 0.908 = 2002$ lb solids/t
 2) Average transit moisture
 $(9.2 + 5.0)/2 = 7.1\%$
 3) Received—5.0% moisture, 1% transit loss
 $2002 \times 0.99 = 1982$ lb solids
 $(1982/0.95) - 1982 = 104$ lb water

B) Payments (per wmt concentrates shipped).
 Copper: $1982(0.278-0.011) (0.98) (\$0.815)$ = $422.67
 Silver: $(1982/2205) (3.1-0.8) (0.95) (\$10.30-0.25)$ = 19.74
 Gold: $(1982/2205) (0.08-0.03) (0.95) (\$442.50-5.00)$ = 18.68
 Total payments $461.09

 Deductions
 Treatment charge: $(1982/2205) (\$85)$ = 76.40
 Refining charge: $1982(0.278-0.011) (0.98) (\$0.08)$ = 41.49

Penalties
Bismuth: (1982/2205) ($1.50) (0.2-0.1)/0.1 = <u>1.35</u>
Total deductions 119.24

C) Transportation Costs
Total cost/wmt: $37.20 + 15.10 + 35.40 = $87.70/wmt = 87.70
Net smelter return per wmt of concentrates shipped = $254.15

Net smelter return per dmt of concentrates shipped = $279.92
(2205/2002) $254.15

D) Summary

	Cost and Value ($ per lb Cu)	
	Payable Copper	Copper Shipped
Gross smelter return	0.585	0.545
less: transportation	0.169	0.158
Net smelter return	0.416	0.387
add: Au & Ag credits	0.074	0.069
NSR (incl. byproducts)	0.490	0.456
Copper price	0.815	0.815
Effective smelt./ref. cost	0.325	0.359

Administered Prices

With a limited number of minerals oligopolies exist where a few producers are responsible for most of the production in the Western world. From time to time, these producers have attempted to be price setters rather than price takers by marketing their production under a Producers Price. This price level was generally intended to promote orderly development of the industry—a price high enough to provide modest profits to the miner, but not so high as to cause substitution of other materials by the user.

This situation was best known in the copper industry in the 1960s and 1970s when a two-tiered pricing system existed. During this period the Producers Price was more stable and generally lower than the LME price for copper. With the overall depressed state of the copper industry since the mid-1970s and the loss by the United States of its dominant position among copper-producing nations, virtually all copper is now priced on the basis of LME or Comex prices.

Administered prices exist to varying degrees in other mineral commodities, aluminum and tin being two examples. Tin prices are affected by the open market operations of the International Tin Council, an organization established by major tin producers and consumers (excluding the United States) to stabilize prices.

Administered prices, then, cover a small group of mineral commodities where the producers' judgment of the appropriate price is substituted for a price derived in the marketplace, either collectively in a commodity exchange or singularly through a negotiation between a buyer and a seller. Over the long run producers are seldom able to correctly anticipate consumer behavior, so that administered prices must ultimately yield to market pressures. For this reason the Tin Council must periodically redefine its buying and selling price levels, and the US aluminum producers routinely sell substantial amounts at individually negotiated prices.

Sources of Information

Current and historical price information for mineral commodities is published in a number of periodicals and government documents. The most authoritative source of price information in the metals industry is *Metals Week*, which tabulates daily price movements for the principal metals. Monthly summaries of some of the data series included in *Metals Week* are found in *Engineering & Mining Journal*, which also publishes representative prices for a wide variety of industrial and fertilizer minerals. Prices for the nonmetallics can also be found in *Industrial Minerals*. Two weekly publications, that list prices for recent coal transactions are *Coal Week* and *Coal Outlook*.

Users of any published price information should take care to note the quality, quantity, and form of the mineral for which the price is quoted. It is also important to ascertain whether the price includes delivery costs to the consumer.

PROJECTING MINERAL PRICES

Mineral industry analysts have become increasingly equivocating on the question of mineral price forecasts. In fact, few analysts are willing to suggest that reliable forecasts of prices useful in mine investment analysis are possible. The currently popular approach is to occupy safer ground and issue price *projections*— or likely prices if certain assumed events actually occur.

Such was not always the case. The relative economic serenity and orderly development of the minerals industry in the 1960's encouraged the development of quantitative modeling for estimating future prices of minerals. During this period government leaders spoke of "fine tuning" the economy, and econometricians began to believe that prices of certain minerals (e.g. copper) could be reliably forecast by the use of statistical models.

By the early 1980s it had become painfully clear that, although supply and demand trends are important and must be studied, the world economy could not yet be accurately represented by a series of mathematical equations. Stung by a series of unsuccessful projects, investors lost faith in price forecasting and began adopting much more conservative investment criteria.

The following, then, is a brief review of the evolution and status of mineral price projection methodology.

Naive Methods

The best single estimate of tomorrow's price for any mineral commodity is today's price. This is the *no change* model, a naive method of price forecasting whereby the spot price at any given time is assumed to be as good as any other single point estimate of future prices. Another naive method is the *same change* model, derived by using standard regression techniques to fit a linear trend line to historical price data and extrapolating the trend into the future. The implication is that in the first case today's spot price would be used to evaluate a proposed new mine; whereas in the second case the required price estimates would be taken from the extrapolated trend line. In both cases, however, the basic assumption is the same: historic prices alone determine the level of future prices.

The mathematics employed in naive models can become more sophisticated. Weighted, moving, or logarithmic averages can be used, and higher order regression equations including Fourier Series can be fit to the data. However, the fundamental underlying assumption remains—future prices are determined by past prices.

As a general rule, these naive models are unsatisfactory for most investment analyses in mining. In preliminary scoping studies, however, more rigorous analyses of future prices may not be warranted, and a simple projection may be used with caution. Rarely is the application of a more complex model advisable in such situations. The analyst should strongly resist the temptation to disguise a weak forecasting premise through the use of a mathematically sophisticated model.

Econometric Modeling

A number of quantitative models have been constructed for various minerals relating price, as the endogenous variable, to a variety of lagged and unlagged exogenous variables. The statistically significant explanatory variables generally bring few surprises and nearly always relate to some measure of economic activity in the primary consuming sectors or to economic activity in the capital goods business in general. The magnitude of the coefficients in the resulting equations provides interesting insight into the historic formation of prices for the mineral under study.

Most econometric models for minerals have been of limited value for the evaluation of investment decisions due to timing. Most models of this type require knowledge of the level of explanatory variables one or two periods (usually quarters or years) prior to the date for which the forecast is desired. Investment analysis in mining generally requires price estimates for over five years into the future where the values of the explanatory variables are also unknown. Therefore, as long-range forecasting tools for mineral prices econometric models suffer from the inherent limitation that if reasonably accurate forecasts are desired, future values of explanatory variables must be known.

Two noteworthy efforts at econometric modeling of the copper industry were performed by Charles River Associates (1970) and by Fisher, Cootner, and Baily (1972). Charles River produced a complex linear regression model using annual data that yielded the following equation for the price of refined copper.

$$RP\,(t) = 0.580\,RP(t-1) + 0.072\,RPAL(t-1) + 0.245 \times 15^3 Q(t-1)$$
$$- 0.120\,\Delta S\,(t-1) - 0.184\,IR \qquad (1)$$

$$R^2 = 0.992$$

where $RP(t)$ is *Engineering & Mining Journal* weighted average price of electrolytic copper wirebars, at time, t; $RP\,(t-1)$ is the foregoing price at time, $t-1$; $RPAL\,(t-1)$ is the German price for 99% virgin aluminum ingot, at time, $t-1$; $Q(t-1)$ is the total domestic consumption of copper at time, $t-1$; $\Delta\,S(t-1)$ is the change in book value of manufacturers' durables at end of period, at time, $t-1$; and $IR(t-1)$ is the domestic producers' inventories of refined copper divided by domestic refined copper capacity, at time, $t-1$.

The Charles River work provided useful insight into the formation of copper prices, including the impact of inventories and aluminum substitution. Nonetheless, the limitations of Eq. 1 for forecasting copper prices in a time frame of value in investment analysis is clearly evident.

To summarize this section, statistical modeling can provide important quantitative information about the historical relationships between price and other variables. Furthermore, the modeling process can aid in prioritizing exogeneous variables for further study by the analyst. However, econometric modeling does not by itself yield credible long-run mineral price forecasts. Although econometric models examine underlying phenomena more carefully, they, like the naive models discussed in the preceding section, use only historical data to forecast the future.

Rational Pricing

During the 1970s when metal prices behaved chaotically considerable effort was devoted to projecting long-term prices on a so-called rational basis. The rational pricing approach is based upon the assumptions that (1) new productive capacity will be needed on a continuing basis to offset rising demand and/or depletion of present mines, and (2) rational investors will not invest in new mines unless the investment promises to deliver some minimum acceptable return. It then follows that the future price of a commodity must be high enough to attract the capital necessary to build the required new capacity. Conceptually, the problem then becomes one of defining a development schedule for specific new mines.

The price determined in the foregoing manner is a long-run normalized price and will generally differ from spot prices which reflect shorter-term phenomena. The magnitude of the difference between the rational price and prevailing spot prices is considered to be a measure of the pressure on prices once these short-term effects subside.

The rational pricing model is based upon the self-regulating nature of a market economy. In theory, when prices are low, investment will decline and supply will stagnate or even decline due to depletion of some mineral deposits; and in the face of rising demand, prices will recover. The opposite occurs when prices are high, spurring additional investment. Thus, under perfect competition in a market-driven economy, the actual price of a commodity will fluctuate around some *normalized* value. The normalized price, in turn, is that price which will cover, in the long run, the total costs of production, including some minimum acceptable return on the capital invested in the project.

Based upon the foregoing principle, most mining companies attempt to estimate the rational, or normalized, price for most minerals of interest to them. This price might be considerably above or below current spot prices but is intended to be representative of longer range equilibrium of supply and demand. Therefore, the rational price computed in this manner should be useful in mine investment decisions.

The rational pricing model for estimating future prices for minerals became most prevalent in the copper industry in the 1970s. Copper seemed to offer great promise for applying this theory because copper is a widely produced, broadly traded commodity that seemed to satisfy many of the economist's requirements for perfect markets. Furthermore, during much of the 1970s copper suffered from prices considerably below historical trends. Most copper producers in this period repeated many times the exercise of calculating the price of copper required to justify investment in new greenfield capacity. The calculations were usually based on bringing on line a new, low-grade, open-pit porphyry operation in the United States. The answer in the late 1970s was generally that a price in excess of $1.20 per lb was required. This led to the conclusion that the prices of $0.70 to 0.90 per lb that prevailed at that time were excessively low, and higher prices should be used in evaluating new copper projects.

There are some obvious shortcomings in the rational pricing approach to estimating future mineral prices due to the many imperfections in mineral commodity markets. Recent history in the copper industry has exposed many of these flaws as noted in the following.

1) *Specifications for the Model Deposit(s)*. Determining the normalized price for a commodity requires the financial modeling of an actual or hypothetical deposit(s) to determine the price at which this deposit(s) would be developed. In essence, determining the *normal* price requires the modeling of the *normal* deposit.

In the copper industry the "typical" low-grade porphyry deposit in the United States has generally been used in the model. However, it became clear in the early 1980s that these operations had become relatively high-cost producers during the preceding decade for a variety of reasons. Thus, they no longer represented the marginal new production. New mines in other countries—notably Chile—and the large expansion potential of many existing mines offered lower cost production and, therefore, lower "required" prices. Thus, the calculated rational price is very sensitive to the deposit assumed in the modeling process.

2) *Response to Market Signals*. The rational pricing approach assumes that producers will make investment and operating decisions based upon anticipated profit. However, in practice most producers make such decisions based upon many considerations, of which profit is but one component. For one class of producers— government controlled—providing employment and procuring foreign exchange can be greater motivating forces than profit. Again, this is a particularly important factor in the copper industry where roughly 60% of the newly mined copper produced in market-economy countries comes from government-controlled operations. These producers have continued to operate and to invest in new productive capacity even when prices were too low to offer commercially acceptable rates of return. The result has been that copper supply continued to expand even when inadequate profits were generated by the copper industry as a whole. Thus, the actions of individual producers may be guided by factors other than profit which creates abnormal investment behavior, at least in the short run.

3) *Demand Shifts*. In a market economy prices of mineral commodities are determined in the same manner as the prices of most other goods and services— through the interaction of supply and demand. The rational pricing model described previously ignores the demand side of the equation, implicitly assuming that demand trends will continue roughly as they have in the past.

Demand is probably the most important variable in determining the rate at which an abnormal commodity price—either high or low—will return to the "normal" level. Therefore, demand often controls the timing of when the normal price will be restored. If demand falls below expected levels price recovery from a recession will be slow, and the "normal" price may prove to be too optimistic for use in investment evaluation. It is small comfort to an investor to know that the price will *ultimately* recover when his investment is awash in red ink.

Referring again to the copper industry, the prolonged slump in the 1970s and early 1980s was caused by the compounding effect of many factors. However, an important element was the departure of copper demand from historical patterns. Increased substitution of plastics and aluminum for copper, as well as a generally lower demand in the industrialized countries, considerably extended the trough in copper's business cycle.

4) *Fluctuating Exchange Rates*. A factor that made the price of copper and other metals particularly difficult to forecast in the early 1980s was the strength of the US dollar relative to most other currencies. Whereas copper prices showed

some improvement in many currencies, the relative strength of the dollar resulted in copper prices, as measured in US dollars, showing little recovery. The appreciation of the US dollar relative to most other currencies was a significant cause of US copper producers suffering a serious loss of competitiveness in world markets (Commodities Research Unit, 1983a).

The rational pricing approach to projecting future mineral prices is rooted in basic laws of economics which will prevail over the long run. As a consequence, the estimation of rational prices is an exercise which contributes to a greater understanding of future supply, a crucial variable in the outcome of future prices. Successfully applying the model requires extensive knowledge of the particular industry, its cost structure, and undeveloped ore deposits.

There are, however, so many uncertainties surrounding the future that any estimate so produced should be considered to possess a large error term. In the late 1970s when copper producers computed a rational price of about $1.20 per lb at a time when spot prices were in the $0.70 to 0.90 per lb range, no price forecasts of $1.20 per lb were immediately forthcoming. However, most industry experts felt that the difference between the two prices was so large that stable prices of at least $1.00 per lb would evolve within one to two years. That this estimate was dramatically wrong highlights one the shortcomings of any bullish price forecast—they tend to be self-defeating because consumers substitute and producers build new capacity in anticipation of the new, higher prices.

Supply and Demand Schedules

All of the models discussed previously suffer from a serious fundamental flaw—they fail to explicitly take demand into account in estimating future prices. In this section, then, we briefly discuss some attempts that have been made to develop both supply and demand schedules for various minerals. It is both convenient and instructive to separate this section into discussions on *production cost schedules, short-run supply curves, long-run supply curves,* and *demand schedules.*

Production Cost Schedules: Due to the weak prices for minerals in the early 1980s, mine operators became increasingly cost conscious, and this attitude also extended into investment analysis. Mineral price forecasts during this period proved to be so notoriously inaccurate that investors sought more conservative investment criteria, focusing increasingly on the competitive cost rankings of prospective mines. Some companies required that a mine rank in the lower half or lower quartile of all producers of the particular commodity before being considered as a serious investment candidate (O'Neil, 1982). Low-cost mines benefit, at least in theory, from a cushion of higher cost producers that would be forced to curtail production earlier in a declining market, thereby tending to bring supply and demand back into balance.

It has now become fairly standard practice in the metal mining sector to compile production cost data for competitors and to plot this information in a cost schedule similar to that shown in Fig. 1. This figure is a snapshot of production costs for the free-world copper industry at a particular point in time, 1982. The only costs included are cash costs of production, usually including general and administrative costs but excluding depreciation, depletion, and financing costs. The relative positions of specific mines on the cost curve can and do change frequently for a variety of reasons including changes in the prices for byproducts.

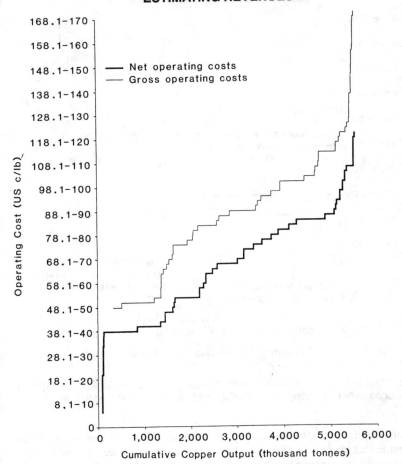

Fig. 1. Production cost schedule, total non-Socialist world. Cumulative output plotted against operating costs, 1982 (from Commodities Research Unit, 1983).

Two curves, gross operating costs and net operating costs, are plotted in Fig. 1, the difference between the two being byproduct credits which are relatively high for Canadian and African producers. An interesting variation to Fig. 1 is shown in Fig. 2. Here the data have been rearranged to more closely represent a supply schedule that reflects the unique conditions in the free-world copper industry. By-product copper has been entered at a $0.30 per lb cost, and mines that are unresponsive to market conditions (mainly government-controlled mines) are plotted at an arbitrary $0.50 per lb. Thus, during times of falling demand, the unpleasant task of reducing supply falls largely to the remaining 40% of the producers, many of whom are, therefore, more vulnerable than might be implied by the level of their production costs.

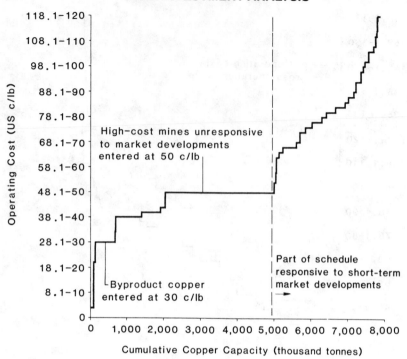

Fig. 2. "Real world" cost schedule, total non-Socialist world. Cumulative capacity plotted against net operating costs, 1982 (from Commodities Research Unit, 1983).

This production cost schedule represents short-run conditions and excludes new mines that may be developed in the near future. For this reason, and because *total* cash costs rather than variable costs of production are plotted, the production cost schedule has some important deficiencies in serving as a proxy for a supply schedule in estimating future price levels.

On the other hand, total cash costs of production for major mines can be estimated fairly accurately, so that a reliable data base can be developed. Also, in recessionary times management is keenly interested in defensive investment criteria such as the competitive cost ranking of a prospective new mine. A low-cost ranking, which carries little downside risk, can be of great comfort even if some profit on the upside must be sacrificed.

Short-Run Supply Schedules: A supply schedule indicates the amount of a commodity supplied to the market at various price levels. Supply schedules can be either short-run or long-run. Short-run schedules cover only currently producing and temporarily idle mines and exclude any prospective new mines.

A short-run supply schedule is the summation of all producers having marginal, or variable, costs of production below various price levels. The basic economic principle involved is that rational producers will continue to operate as long as revenues equal or exceed their variable costs of production. In theory, fixed costs are unavoidable and are, therefore, irrelevant in the production decision.

The short-run supply schedule covers only variable costs, whereas the production cost curve described previously includes *all* cash costs of production. Because mines will continue to produce even when losses are incurred—as long as their variable costs are covered—the amount supplied to the marketplace at a given price is generally greater than that suggested by a production cost schedule such as shown in Fig. 1.

An interesting attempt to develop a short-run supply schedule for copper is described by Foley (1979). She collected primary cost data for many US producers and developed linear regression equations to estimate costs for many other mines. The equations developed were as follows:

$$MNCOST = 36.19 - 23.27 X_1 + 2.11 X_2 - 0.21 X_3 \qquad (2)$$
$$R^2 = 0.975$$

where $MNCOST$ is the mining cost in 1972 dollars; X_1 is the ore grade mined, X_2 is the stripping ratio, and X_3 is the percent recovery of copper from ore.

$$MLCOST = 24.28 - 15.86 Y_1 - 5.20 Y_2 \qquad (3)$$
$$R^2 = 0.757$$

where $MLCOST$ is the milling cost in 1972 dollars, Y_1 is the ore grade milled, and Y_2 is the yearly milling capacity.

Foley's study covered only the US copper industry, and Eq. 2 applies only to open pit mining. Furthermore, important shifts in the costs of producing copper have occurred in recent years. Therefore, although the coefficients in Eqs. 2 and 3 may no longer be reliable, the study does illustrate that production costs in mining are amenable to systematic estimation.

In practice, estimating the fixed and variable costs of production for each producer in an entire industry is a staggering task. In the long run, all costs are variable, so some unambiguous set of definitions is needed to classify costs as fixed or variable. In fact, few operators have a clear idea even of their own fixed and variable costs, much less those of their competitors. As a consequence of the difficulty in estimating variable costs of production for each supplier, a production cost curve is often used as a proxy for a true supply curve in estimating price levels. It should be noted again, however, that this substitution tends to understate the supply that would actually be produced at any given price.

Long-Run Supply Curves: With short-run supply curves only presently operating mines are included. However, the competitiveness of an investment candidate must be measured not only against presently producing mines, but also against other mines that might be developed during the same period. The principal difference between the two sets of mines is that investment is a sunk cost for the first group, whereas investment costs are yet to be incurred for the second group. Clearly, a much higher price is required to call into production mines from the second group, but once in production these newer operations will be guided by the same shutdown and start-up considerations as the older mines. In theory, this point of operating indifference occurs when revenues equal variable costs of production. In practice, determining the shutdown/start-up point for a mine is much more complicated. In weak markets, mines exhibit a remarkable ability to control costs and minimize losses and, therefore, often continue to produce even when prices are exceptionally low.

As noted in the section on "Rational Pricing," revenues in the long run must cover the total costs of production, including a return satisfactory to attract the necessary investment capital. This is true not only for new mines but also for existing operations due to the continuing need for substantial sustaining capital throughout the life of a mine. Therefore, the long-run supply curve includes the total costs of production rather than just variable costs as in the case of the short-run supply curve, or cash costs as in the case of the production cost curve.

Through the Minerals Availability System (MAS), the US Bureau of Mines has developed the capability to produce minerals availability curves for a variety of minerals. These approximate long-run supply curves, and generally a rate of return on invested capital of 15% or 18% is included.

The Bureau has published many Information Circulars (IC) focused on long-run minerals availability schedules. A good example, again relating to the copper industry, is 1C 8809, "Copper Availability—Domestic" (Rosenkranz, et al., 1979) from which Fig. 3 is reprinted.

It is important to recognize that Fig. 3 simply represents the amount of domestic copper that could have been economically produced in 1978 at various price levels. It makes no statement about production costs for any property in subsequent years, does not consider the lengthy development period for those undeveloped deposits included in the reserve base, nor is there any assumption regarding demand. The publication does, however, include other curves showing the time profile of economic production at some copper prices.

Capital investment and recovery in mining is a long-term process, and investment analysis must consider the price trends affected by long-run supply. The problem, of course, is the high level of risk that accompanies any forecasts of the long run. The kinetics of the equation controlling the rate at which "normal" conditions are restored in the mining business are highly variable. Thus, although we may know that a certain price must ultimately prevail if mineral is to be produced, it is quite another matter to accurately predict the timing of this event. As a consequence, although long-run supply schedules should be examined in arriving at price estimates for the future, short-run supply schedules and production cost curves should also be studied carefully to ensure that the proposed investment can survive the near term.

Demand Schedules: Price is determined by the interaction of supply and demand, so price can be estimated by imposing a demand schedule on a supply schedule, for which the minerals availability curve in Fig. 3 serves as an acceptable proxy. However, accurately forecasting demand at a single price level is difficult enough without attempting to develop a demand schedule for all price levels. Nonetheless, a careful, detailed study of consumption trends in end-use markets can permit enlightened estimates to be made about likely levels of demand. In practice, it is rarely possible to quantify price elasticity of demand very accurately, although this is clearly an important factor with some minerals. There is little doubt, for example, that skyrocketing molybdenum prices in the late 1970s contributed to a dramatically lower molybdenum demand in the early 1980s. Unfortunately, the analyst must generally be satisfied with a demand forecast that is independent of price, particularly when the price estimate required for mine investment decisions is at least three years in the future.

The intersection of the demand estimate (or the range of estimates) with the minerals availability curve provides a price estimate (or estimates). For example, if average annual US copper consumption from domestic mines between 1969 and 1978 of 1.337 mt is plotted in Fig. 3, a price of $0.82 per lb (1978 dollars) would have been required for all producing mines to receive at least a 15% return on

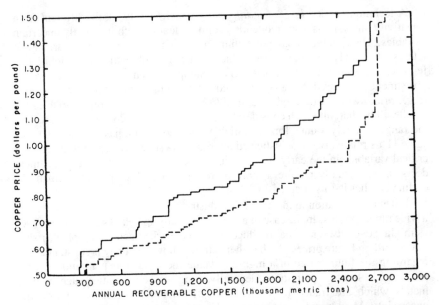

Fig. 3. Annual recoverable copper available from domestic deposits. Shows potential copper that could be produced at a specific price based on the 1978 copper reserve base and 1978 costs; duration of production not shown. Solid line, 15% rate of return; broken line, 0% rate of return (from Rosenkranz et al., 1979).

investment (Rosenkranz, et al., 1979). Based on a 0% return on investment (recover investment only with no profit), a price of $0.72 per lb would have been required.

A final note of caution should be repeated. As noted previously, one of the main difficulties in formulating demand projections is that, for practical reasons, demand is usually implicitly assumed to be independent of price. There is rarely sufficient empirical data to justify the quantification of any other assumption. In many markets, minerals comprise such a small share of the cost of the final product that the analyst can assume that, within reasonable bounds, demand is relatively price inelastic. In other markets there is strong competition (e.g., copper-aluminum, zinc-plastic) such that price changes may alter demand significantly. A mineral economics analysis of the industry often reveals these relationships in a qualitative manner and may permit the analyst to make some informed judgments about the price elasticity of demand for a particular mineral commodity.

ADDITIONAL COMMENTS

Probably the most common procedure employed for developing a higher level of comfort with respect to inherently uncertain mineral prices is sensitivity analysis. By the time funds are committed to a major new mining venture, a very large number of rate-of-return calculations have usually been performed to test the sensitivities of input variables. More often than not, price proves to be the most critical variable. Stochastic risk analysis is a powerful tool to use in this situation also, but this technique has met with only limited acceptance in industry. Both sensitivity analysis and risk analysis are described more fully in Chapter 13.

A technique called *level of indifference*, which is only subtly different from sensitivity analysis, is also frequently useful in dealing with inherently uncertain variables. Here, rather than prescribing a value for the particular variable of interest, the analyst computes the value for that variable which will yield indifference in the investment decision. For example, if an 18% rate of return on investment is required, the analyst would calculate the value of the input variable (e.g., price) which would provide an 18% rate of return. For some variables the result will be that, although considerable uncertainty exists concerning the future, the range of likely values for the variable does not encompass the indifference value. This relatively unsophisticated approach is especially useful in discerning critical variables in the early stages of investment analysis. The final investment decision will usually require a more thorough study of supply and demand as outlined earlier in this chapter.

Revenue estimation in the minerals industry is an activity which encompasses a wide range of risks. In the case of a long-term, fully specified sales contract such as might exist between a coal producer and an electrical generating plant, risk is low regarding future prices. At the other extreme, however, there is substantial risk regarding the future price of a high-unit-value metal such as platinum which is produced in limited quantities. Generally, fungible commodities such as most metals which are traded on exchanges suffer from the greatest future price uncertainty. Most metal markets are notoriously cyclical, and the amplitude and the period of the cycles defy accurate prediction.

In spite of the many problems cited in this chapter, estimates of revenue must be made. The recommended approach is not limited to the application of any one or two techniques, and is definitely not a mechanical process. Rather, it is a painstaking blend of economic theory, industry analysis, market analysis, and competitor analysis, combined with sound, experienced judgment. The situation is not fundamentally different from 1938 when Leith wrote (pp. 80-81):

"A forecast of future selling prices is perhaps the most doubtful and speculative of any of the elements of valuations. At best these can be dealt with only in very general terms of trend, and sometimes not even that. The first step obviously is to consider past trends and cycles, both for long and short periods, and to analyze as well as possible the various elements of which the past price trend has been the composite result. Some of these elements are the relative rates of growth of supply and demand, the effects of shortage and surplus of supply and plant capacity, the relations of mineral prices to commodity prices in general, the effects of public and private price-fixing efforts, the antitrust laws, fluctuations in currency and gold supplies, competition and substitution of other commodities, technologic trends, the effects of tariffs and many other restrictive measures affecting international competition.

With this historical background the appraiser then faces the almost impossible task of deciding which elements among the past trends are likely to dominate in the future. Human intelligence has not reached the stage of being able to integrate all the shifting variables to yield an accurate forecast. As difficult as the problem is, the appraiser has to attempt it."

REFERENCES

Charles River Associates, 1970, *Economic Analysis of the Copper Industry*, Report PB189927, National Technical Information Service, Springfield, VA, 336 pp.

Commodities Research Unit, Ltd., 1983, *Copper's Changing Cost Structure, 1980–1983*, Vol. 1, New York, 143 pp.

Commodities Research Unit, Ltd., 1983a, "Copper Production Costs and Currency Changes," *Copper Studies*, New York, Oct., pp. 1–4.

Fisher, F.M., Cootner, P.H., and Baily, M.N., 1972, "An Econometric Model of the World Copper Industry," *The Bell Journal of Economic and Management Science*, Vol. 3, No. 2, Autumn.

Foley, P.T., 1979, "A Supply Curve for the Domestic Copper Industry," M.S. Thesis in Materials Engineering, Massachusetts Institute of Technology, Cambridge, 115 pp.

Labys, W.C., and Granger, C.W.J., 1973, *Speculation, Hedging and Commodity Price Forecasts*, Heath Lexington Books, Lexington, MA, 320 pp.

Leith, C.K., 1938, *Mineral Valuations of the Future*, AIME, New York, 116 pp.

Lewis, R.S., and Clark, G.B., 1964, *Elements of Mining*, John Wiley and Sons, New York, 768 pp.

O'Neil, T.J., 1982, "Mine Evaluation in a Changing Investment Climate," *Mining Engineering*, Pts. 1, 2, Vol. 34, Nos. 11, 12, Nov., Dec.

Rosenkranz, R.S., Davidoff, R.L., and Lemons, J.F., Jr., 1979, "Copper Availability—Domestic," Information Circular 8809, US Bureau of Mines, GPO, Washington, DC, 31 pp.

Vogely, W.A., ed., 1976, *Economics of the Mineral Industries*, 3rd ed., AIME, New York, 863 pp.

6

Capital and Operating Cost Estimation

*The greatest of all gifts is the power to
estimate things at their true worth.*
LaRockefoucauld

INTRODUCTION

The primary reason for performing a feasibility study on a proposed mining venture is to investigate *all* phases of the proposal in as much detail as necessary to justify either dropping the project or continuing expenditures through the next stage. The economic evaluation component of the feasibility study, therefore, must ultimately be based on information which provides an answer to the question, "What is it going to cost?" Unfortunately engineers preparing feasibility studies never have all the engineering or economic information they would like or need. Consequently, the economic portion of the analysis can only be performed if estimates of the various costs associated with the project are made. Project cost estimates, in turn, require estimates of all the physical factors affecting cost components, including anticipated performance of plant and equipment as well as all pertinent geologic, mineralogic, and metallurgical variables.

Before useful cost estimating procedures can be applied considerable data must be collected, compiled, and organized. Preparatory work typically includes organization of cost data available for estimating purposes, development of techniques for updating historical costs, and establishing methods of forecasting prices and costs over the duration of the project.

Cost estimates for mining and mineral processing projects involve so many features and variables that it is absolutely essential to have some means by which the vast amount of detail can be organized. It is very important to identify, in as much detail as possible, all the cost data available (historical and developed) to the estimator. These costs, after being properly identified in detail, should be compiled, stored, updated regularly, and made readily available for use in developing new cost estimates.

Accessing files of current cost information requires a standardized numerical system of classifying cost data. When uniform definitions and procedures are used, every project record contributes to the common store of estimating data.

Such a standardized system is indispensable if accountants, estimators, designers, and managers are to have a common basis of understanding for allocation of project costs. In summary, good cost estimating is based on:
- Definition of cost elements so that cost records and estimates have a common basis.
- Collection of cost records based on the foregoing definitions.
- Classification and grouping of cost records.
- Analysis of the relationships among cost data.
- Utilization of cost estimating procedures based on these relationships.

If organized properly, cost estimating is one activity that gives, as well as receives, information in management information systems.

Sophisticated cost estimating procedures have been developed in the chemical engineering field. This is somewhat understandable because process plants treating homogeneous feedstocks generally consist of standard components. Detailed cost estimating techniques have been developed here based upon specific quantitative factors. Unfortunately mining projects mine and process unique, heterogeneous ores where there is rarely a useful body of historical data. As a result it is difficult to generalize with respect to cost estimating procedures, and the process remains more of an art than a science. In short, cost estimating in the mining portion of the minerals industry is a difficult and time-consuming process with few, if any, short cuts.

TYPES OF COST ESTIMATES

Many types and classifications of cost estimates have been proposed by various organizations and individuals over the years. The American Association of Cost Engineers (AACE) has adopted the following designations for cost estimates (Stackhouse, 1979):

1) Order of Magnitude: -30% to $+50\%$ accuracy range; use cost capacity curves and cost capacity ratios.

2) Budget: -15% to $+30\%$ accuracy range; use flowsheets, layouts, and equipment details; made for budgeting purposes.

3) Definitive: -5% to $+15\%$ accuracy range; use defined engineering data, specifications, basic drawings, and detailed sketches.

This classification replaced the earlier AACE classification of:

1) Order of magnitude (ratio estimate).
2) Study (factored estimate).
3) Preliminary (budget authorization estimate).
4) Definitive (project control estimate).
5) Detailed (firm estimate).

The names of the various types of cost estimates are not nearly as important as the identification of the estimate with the associated degree of accuracy of the estimate. There are four basic types of cost estimates used in evaluating new mining properties. They reflect various stages of progress in the project. The type and timing of these estimates can best be illustrated with an example.

A typical mining property moves to completion through a series of stages as follows (Gentry, 1979):

1) Discovery and indication of potential ore through exploration efforts.
2) ORDER OF MAGNITUDE cost estimate for preliminary feasibility.
3) Detailed exploration program including metallurgical testing of bulk samples and an indication of mineral processing requirements.

4) PRELIMINARY cost estimate for feasibility study.

5) Development of data necessary for engineering design of mine and mill including preliminary equipment selection.

6) DEFINITIVE cost estimate for feasibility study.

7) Detailed engineering design of mine and mill including site specifications, layout drawings, etc.

8) DETAILED cost estimate for feasibility study.

9) Plant construction and mine development.

10) Start-up operation.

11) Production.

This example illustrates that there are several key decision points during the life of a new mining property which require feasibility studies and associated cost analyses with varying degrees of accuracy. Table 1 illustrates the relationships among the four basic types of cost estimates and associated accuracies, time requirements, contingencies, percentages of preproduction engineering effort, and percentage of preproduction capital expenditures. The absolute values in the table will vary, of course, depending on the type and scope of the investment proposal being investigated. Nonetheless, the values do provide a relative comparison of various cost estimates and accompanying feasibility studies. It is interesting to note that a relatively small percentage of total capital expenditures occur through the definitive cost estimate stage of the project. At this point, if a decision is made to proceed with the project, serious preproduction activities begin to occur. Even so, most capital expenditures are typically made after the detailed cost estimate when equipment orders are confirmed and construction and development begin.

Before proceeding with a discussion of the various cost estimating techniques, it is worthwhile to examine in a qualitative sense some of the features and uses of these four basic types of estimates. The brief notes which follow are intended to place in perspective the characteristics of the four types of cost estimates proposed and their relationships to new mining property evaluation.

Order of Magnitude Estimates

These cost estimates are generally intended to assist management in making appropriate decisions regarding potential project feasibility and to justify a further expenditure of funds for the next stage of the project. Such estimates are sometimes suitable to reject a project, but are seldom adequate for positive project acceptance. The estimates are often based on known costs of similar projects and typically involve little or no design work for the mine and mineral processing facility in question. These estimates seldom become the basis for even conceptual design but may indicate the desirability of expanding work or efforts further.

Accuracy: The order of magnitude cost estimate provides a relatively low level of accuracy varying by as much as -30% to $+50\%$ and sometimes more when compared to true project cost. The sacrifice in precision, however, is often justified by the ability to screen a large number of investment proposals in a short time span.

Information Available for Estimate: For a new mining property information or *indications* will typically be available on project location, gross estimates of potential reserves and grade, probable mining method (surface or underground), probable processing method, possible production rates, and probable major mining and processing equipment required.

Table 1. Comparison of Various Cost Estimates with Pertinent Estimating Characteristics (after Gentry, 1979)

Type of cost estimate and associated stage of project development	Accuracy, %	Noncumulative time required for estimate	Contingency required, %	Percentage completion of preproduction effort, %	Percentage of preproduction capital expended at time of estimate, %
Order of magnitude	−30 to +50	1-7 days	20-30	5	<0.5
Preliminary	−15 to +30	1 week-2 months	15-25	15-25	2-5
Definitive	−5 to +15	3-12 months	5-15	50-60	10-20
Detailed	−2 to +10	2-9 months	4-8	90-100	50-60

Preliminary Estimates

The purpose of the preliminary cost estimate is to refine the order of magnitude estimate when additional data become available. These estimates are usually suitable to indicate or determine project feasibility and assist management in estimating a budget for the project. The estimate typically relates to a conceptual design of a mine or mineral processing facility.

Accuracy: The accuracy of a preliminary cost estimate is variable but will typically range between −15% and +30%, depending on the scope of the project.

Information Available for Estimate: At this stage better information is available from the exploration program relative to ore reserves and grade. More realistic estimates can be made about production rates, proposed mining method, and a proposed mine layout. Also, a proposed mineral processing method has typically been indicated along with preliminary flow charts, plant plans, recovery rates, and possibly plant layouts with varying degrees of information. Mine and mill equipment lists, indicating types and sizes, start to be developed. Development requirements can be estimated. The plant site location is specified, general site conditions are known, and building space requirements can be estimated.

Definitive Estimates

The purposes of definitive cost estimates are to: (1) provide for appropriation of funds or to establish a contract price, (2) provide the basis for project cost status reports, and/or (3) establish a format for final cost reports to aid accounting and provide feedback information on actual costs to use in future estimates and to improve existing estimating methods. An estimate of this type should enable management to authorize expenditures for completion of engineering specifications and drawings, design, and site surveys.

Accuracy: The level of accuracy associated with a thorough and well-planned definitive cost estimate should range between −5% and +15%.

Information Available for Estimate: By this time considerable data are available which describes the planned mine and mineral processing plant. Specifically, data of the following nature are available to the estimator: (1) definitive mine plans and layouts for a given mining method; (2) production capacities; (3) ore recovery, average grade, and dilution estimates; (4) definitive milling process with flow charts, plant plans, layouts, utilities, storage and handling requirements; (5) complete equipment lists with specifications for mine and mill; (6) nonequipment items are specified (access roads, paved roads, infrastructure, piping, etc.); (7) preliminary drawings for buildings and site development; and (8) the exact location of mine and plant site.

Detailed Estimates

The detailed cost estimate culminates the estimating procedure. It is based on complete engineering drawings, specifications, and site surveys. This type of estimate is normally suitable for accurate projections and funding for the project and provides a basis for authorization to proceed with construction and development. With particularly attractive projects, some construction may be authorized prior to completion of the detailed estimate, although this is not a recommended practice. A detailed cost estimate is seldom undertaken unless there is reasonable assurance as to project feasibility.

Accuracy: Detailed cost estimates should possess an accuracy between -2% and $+10\%$ of actual project costs.

Information Available for Estimate: A detailed cost estimate is predicated on detailed engineering drawings, a definite project schedule, and actual bids from manufacturers, contractors, and subcontractors.

The definition of a feasibility study implies that there are, or should be, varying degrees of quality and accuracy of the cost estimates that are prepared during project investigation. The preceding discussion on the four basic types of cost estimates addressed in this chapter has indicated that the "quality" of the cost estimate has to relate to the purpose of the study. In other words, what will be the end use of the study? Clearly, the effort expended on preparing the cost estimate, and the resulting quality of the estimate, must be consistent with the type of decision which is likely to result from the feasibility study being prepared.

Table 1 shows the very definite relationship which exists between effort expended and results achieved when preparing cost estimates. A relatively small amount of work in a short period of time results in an estimate of sufficient quality to give an indication of project viability. Further effort over a period of several months will rapidly improve the estimate quality. Beyond this point the law of diminishing returns is encountered with respect to estimate quality until final drawings and specifications are in hand for detailed costing and quotations. It is up to management to provide some guidance with respect to how much time and cost should be authorized for cost estimation and how accurate the results of the feasibility study must be at any given stage of project investigation.

In practice, an important estimating phenomenon usually appears: estimated project cost always increases with each successive improvement in the estimate. Rare is the project whose detailed capital estimate is less than it was at the preliminary or order of magnitude stages. Thus, some restraint is in order when a project looks extraordinarily profitable in the early stages.

The quality of a cost estimate is usually judged on the basis of the amount of contingency which is assigned to the total estimated cost. "Contingency," or "allowance for the unforeseen" in some engineering companies, is a misunderstood term. The American Association of Cost Engineers (AACE) defines contingency as a "Specific provision for unforeseeable elements of cost within the project scope; particularly where previous experience relating estimates and actual costs has shown that, statistically, unforeseeable events which will increase costs are likely to occur." Thus, contingency is money that will likely be spent, but can not be defined in any greater detail.

As a project progresses toward actual development, the quantity and quality of data becomes better and better. As a result contingency allowances become smaller and smaller as the cost estimates improve. Even though cost estimate accuracy and quality improve as the project moves closer and closer to completion, real world experience suggests that cost estimates will vary from actual project costs by some percentage. However, the actual values of variation about the mean are often unpredictable due to the multitude of project variables involved.

KINDS OF COST INFORMATION

Cost information may be classified into three primary categories: (1) historical costs, (2) measured costs, and (3) policy costs.

Historical costs are those collected from the literature, accounting records, governmental cost information sources, business reports, trade associations, technical publications, etc. These costs are often obtained internally within the organization and may represent data from projects previously funded by the firm.

Measured costs are defined as time-dollar relationships where direct observational processes and mathematical rules are followed. Measured costs are primarily limited to costs of work. Material quantities determined from drawings and specifications are also a kind of measured data. In general there are four methods normally employed for the determination of time, which are fundamental to determining measured costs.

1. Time Studies

Time requirements for unit operational components of the entire production cycle are observed, recorded, and analyzed with respect to $ per hour, $ per ton produced, etc. The following simple example illustrates the concept of measured costs derived from shift reports.

Example 1: A small fleet of 100-ton trucks was observed throughout the following cycle: load, flat haul of 3000 ft, spotting and dumping, return, waiting, spot and load. The average production for the truck fleet was 12,100 tons per shift.

Operating costs for the 100-ton truck fleet were as follows:

Operating labor	$ 800/shift
Fuel and oil	$1,000/shift
Repairs:	$ 600/shift
Tires:	$1,400/shift
Total	$3,800/shift

Therefore: the cost per ton of material moved is
$/ton = $3,800/12,100 tons = $0.314/ton.

2. Work Sampling

Observations are taken pertaining to specific activities of the person or machine at random intervals. Such a figure can be used in conjunction with other work samples to estimate the production capacity of machines in a given operation. The following example shows how work sampling can be used to determine production estimates for a given machine.

Example 2: A work sampling study of an electric shovel in an open pit mine was performed. Some 400 observations were made of the unit and the shovel was found to be in a waiting or nonproduction mode during 55 of these observations. How much of the total time is the shovel expected to be waiting for one reason or another?

Solution: Since work sampling is a statistical technique, the laws of probability must be followed. When n, the number of observations, is large the binomial distribution approaches the normal distribution. Therefore, the probability of event i occurring can be estimated from the number of observations (Ostwald, 1974d):

$$pi = \frac{n_i}{n}$$

where p is observed percentage of occurrence of an event i expressed as a decimal,

n_i is the number of observatins of event i, and n is the total number of observations. In this case: $pi = 55/400 = 0.14$ or 14%

Using the standard error of a sample percentage for a binomial distribution and the concept of confidence intervals, the following formula may be used:

$$I = 2 \left[\frac{pi\,(1-pi)}{n} \right]^{1/2}$$

where I is the confidence interval obtained from the study expressed as a decimal and α is the factor obtained from tables of probabilities for a normal distribution.

The problem can now be expressed as follows for a 90% confidence interval ($\alpha = 1.645$):

$$I = 2(1.645) \left[\frac{0.14\,(1-0.14)}{400} \right]^{1/2} = 0.057.$$

Therefore the electric shovel can be expected to be in a waiting mode 14% of the time $\pm$ 5.7%/2, or a range of 11.2% to 16.8%.

3. Man-Hour Reports

These reports include time card reporting of working hours by type of work, activities, etc. Cost distributions for labor can be obtained in this manner. It is important, however, that provisions be made in the recorded data for delays (controllable and uncontrollable), weather conditions, equipment malfunctions, work location, etc., if the information is to be of real value to the estimators.

4. Predetermined Motion-Time Data

This system stipulates times for various human motions, usually over very short intervals. Work performance (quantity) and subsequent cost distributions can theoretically be determined on this basis. These techniques are more applicable to assembly-line, mass-production situations and are rarely used in the mining sector.

Policy costs are unique in that they have the characteristic of being fixed for estimating purposes. Although the origins of these costs may be varied, they are accepted as factual and beyond control by the estimator. Examples of these types of costs are: (1) corporate specified administrative and overhead charges; (2) union-management wage agreements where predetermined policies dictate wages and wage escalator clauses; (3) social security tax required by the federal government; (4) mandated unemployment insurance; and (5) any other negotiated, contractual, or legislated costs. The interesting point about policy costs is that in most cases these costs can be used as absolute values for future cost estimates. This results, of course, from the fact that most of these costs stem from negotiated, contractual, or legislated actions covering some extended time period. Therefore, since policy costs are typically fixed for calculation purposes, they tend to eliminate, or at least reduce, one element of risk in the final cost estimate.

SOURCES OF INFORMATION

Cost estimates for use in feasibility studies may be based on information obtained from many sources. Often the information is obtained internally within the organization and represents data collected from technical publications and other literature sources, historical data on previous company projects, or accumulated data on similar projects funded by other companies. The major sources for this internal information are:

Accounting Department: General capital and operating costs on all phases of operations.

Personnel Department: Union contracts, wage rates, fringe benefits, etc.

Production Department: Direct costs per unit of production, maintenance, repair, supplies.

Purchasing Department: Current and past cost of materials and supplies, utilities, manufacturer quotations.

Marketing Department: Commodity prices, stability, supply-demand positions, sales contracts.

Legal Department: Interpretation of regulations with respect to operating constraints, requirements, etc.

Engineering and Geology Department: File data on in-house and other company properties relative to operations, costs, unique problems.

A great variety of basic economic data and trends can also be obtained from US government sources, perhaps the most important of which are:

Bureau of Labor Statistics: Provides elements of costs on the prices of materials and labor, wage surveys for particular industries, and overall movements of pay for various occupational groups. Compiles Consumer Price Index (CPI) and the Producer Price Index (PPI).

Department of Commerce: Domestic and international business activities, Gross Nation Product (GNP), etc.

Bureau of Mines: Studies on costs, commodity availability, resource and reserve assessments, etc.

Department of Energy (DOE): Costs associated with fuels, development scenarios, technology assessments.

Environmental Protection Agency (EPA): Monthly abstracts, data from studies performed on specific sites.

Various international agencies (e.g., the United Nations, World Bank), trade associations, and technical publications often provide valuable statistical information on measuring trade activities, flow of funds, international financing levels, etc., which may be critical in the final stages of project analyses.

Valuable information for cost estimates can also be obtained from sources such as (1) shareholders' reports, (2) manufacturing firms and agents, (3) consulting firm reports, (4) local banks, (5) Federal Reserve Bank, and (6) chambers of commerce. It should be stressed that most of the information necessary regarding operating aspects relating to equipment selection, performance, and costs is generally obtained from manufacturer and supplier representatives. Data from these organizations are essential for preparation of feasibility studies. This input greatly reduces the burden and time constraints often imposed on the evaluation groups.

There are many sources for deriving various capital and operating cost estimates for mining and beneficiation facilities. The key to the evaluation process is to determine where to obtain the best possible data in the shortest time possible, with the least expenditure of manpower, and at the same time, maintain a high degree of accuracy throughout the feasibility study.

COST ESTIMATING METHODS

There are almost a limitless number of techniques, procedures, and methods available for developing cost estimates. *The best cost estimating technique is the one that works.* It does not necessarily have to be sophisticated or broadly applicable to other situations, as long as it works for the project at hand. Some of the techniques are generally accepted methodologies while others are location specific or company specific.

Some of the major cost estimation methods commonly employed in association with the four basic types of cost estimates designated in this chapter are briefly discussed in this section. The level of sophistication and the type of base data required for each technique presented generally dictates the required level of estimation accuracy and the appropriate technique. In addition, every effort has been made to present the methods in order of accuracy (starting with order of magnitude and ending with detailed), although some overlap does exist.

At this point it is necessary to caution the reader that the numerical values presented or otherwise used in the following examples and throughout this section are intended for illustrative purposes only and *should not* be used as absolute values in calculating future costs. Every estimating technique is predicated on current and accurate base costs. Therefore, the estimator must continually adjust the various bases and factors utilized in the estimating procedure if quality estimates are to be achieved.

When developing cost estimates it is important to distinguish between the types of costs being estimated. The estimator is concerned with two primary types of costs for project estimation—capital costs and operating costs. The more general cost estimating methods are typically applied to capital cost estimates, because good operating cost estimates can be best developed after project specifics start to be defined.

Capital and operating cost estimating methods follow. Emphasis is placed on capital cost estimates for the following reasons:

1) Many of the estimating techniques are similar for capital and operating costs with the only difference being in the quality of the estimating basis.

2) Capital and equipment costs must be estimated prior to development of an appropriate operating cost estimate.

3) Fairly accurate operating cost estimates are typically derived from equipment lists, manning charts, production schedules, etc., which relate more to definitive and detailed cost estimates as opposed to the more general estimates.

Capital Cost Estimates

In the minerals industry, capital costs or capital investment generally means the amount of total capital dollars required to bring a mining property into production. A capital cost item to the Internal Revenue Service (IRS) is one which has its cost recovered over a period of time in the form of annual deductions from revenue before arriving at taxable income (see Chapter 8). Thus, it should be noted here that the capital associated with bringing a new mining property into production may not qualify as "capital" items for tax purposes.

Total capital investment consists of two primary components—a *fixed capital* portion and a *working capital* portion. The fixed capital costs refer to the total amount of money necessary to procure the site, purchase primary and ancillary equipment and facilities, and other expenses associated with project start-up. In the case of a new mining property, fixed costs are usually categorized into one of the following major headings:

1) Land Acquisition
2) Development
3) Preproduction Development (e.g., overburden removal)
4) Environmental Studies and Permitting
5) Mining Equipment (all), Buildings, Facilities
6) Milling Equipment (all), Buildings, Facilities
7) Supporting Facilities (roads, railroads, power lines, shops, etc.)
8) Design and Engineering Expense
9) Contingency

Working capital represents the amount of money beyond fixed capital needed to begin the operation and meet subsequent obligations during project start-up. The cost items typically associated with working capital are:

1) Inventories: (A) raw materials, (B) spare parts, (C) supplies, (D) materials-in-process, and (E) finished products
2) Accounts Receivable
3) Accounts Payable
4) Cash on hand (payroll, utilities, etc.)

Note that accounts receivable adds to the working capital required while accounts payable reduces the required working capital.

Working capital may be estimated by one of several techniques. Typically the estimate is based on 10–20% of the fixed capital investment or on a 1–3 month estimate of costs associated with the foregoing cost items. These specific types of estimates are perhaps more appropriate for process or milling facilites. Working capital for the mining operation itself is often determined as a portion of the annual operating cost:

$$\text{Working Capital} = \frac{\text{operating cost}}{\text{ton}} \times \frac{\text{tons mined}}{\text{year}} \times \frac{Y \text{ months.}}{12 \text{ months}}$$

The value of Y depends to a large extent on the length of the marketing "pipeline" for the facility (i.e., the length of time it takes product to reach its market and for payment to be received by the producer). Typically a period of three months is used in the calculation, although for remotely located mines, a higher value might be appropriate.

It should be noted that working capital is becoming an ever-increasing percentage of the total investment capital necessary to bring a new mining project into production. It is commonly assumed that the working capital account is established at the beginning of a new project, is "churned" throughout the project, and is finally recaptured at the end of the project. However, many organizations are now providing for estimates of working capital additions periodically throughout project life in order to adjust for inflationary effects on actual purchasing power.

The following major cost estimation techniques apply to fixed capital costs for mining and milling ventures.

Conference Method: This is a subjective estimation method providing a single value or estimate made through experience or direct comparison with a similar project. The procedure varies but generally involves representatives from various departments conferring on estimation in a round-table fashion and jointly estimating project cost as a lump sum. Major drawbacks are lack of analysis, lack of a trail of verifiable facts leading from the project to the estimate, and the assumption that the project in question is indeed sufficiently similar to some other project for which the costs are known.

Unit Cost Method: This method simply multiplies capacity by a unit cost which is typically expressed as installed capital cost per ton of annual production ($ per ton per year). Data for these estimates are collected from technical literature, private cost data, government sources, banks, or files on cost engineering. To illustrate the estimating procedure, assume capital costs are approximately $10,800 per annual ton of copper produced at typical open pit copper mines. If a new mine is anticipated to produce 90,000 tpy of copper, the capital cost of the project would be estimated at

$10,800 per annual ton of copper × 90,000 tpy = $972,000,000

Similarly, unit cost estimates for standard milling facilities are being estimated as follows:

Type of Mill (Standard)	Unit Cost (24-hr day)
Uranium	$10,000/tpd
Gold	$ 9,000/tpd
Flotation (base metal; 2 product)	$ 8,000/tpd

The strongest assumption necessary for the application of this method is that the mine or mineral processing plant to be estimated is similar to the mine, mineral processing plant, or design used to determine the cost relationship. Off-site or supporting facility costs may also be estimated by this method [i.e., buildings ($ per square foot), water supply systems ($ per gallon per minute), roads ($ per mile), etc.].

Perhaps the most common error committed with the unit cost method is assuming that a particular unit cost is essentially constant over a given range of capacities; this is simply not true. For instance, the unit costs for mill estimates given previously are for processing plants generally ranging from 250–10,000 tpd. Plant sizes above or below these values will have different values of unit cost. Because of this inadequacy in the estimating procedure, the so-called exponential adjustment for capacity method was developed.

Turnover Ratio Method: The turnover ratio method also utilizes historical data from similar plants or operations to calculate capital investment. The turnover ratio is equal to the product value per ton divided by project investment. For example, if the turnover ratio for a typical open pit porphyry copper deposit is estimated at between 0.30 and 0.35 per ton of capacity and if the product sells for $2000 per ton of copper metal, then the estimated investment in a mining complex anticipated to produce 100,000 tons of copper annually would be:

0.30 = $2000/Investment
Project Investment = $2000 ÷ 0.30 = $6667 per ton of annual capacity
$6667/ton of annual capacity × 100,000 tons annual capacity = $666,700,000.

Exponential Capacity-Adjustment (Scaling) Method: It is a well-known fact that the capital costs of a project can be expressed as

$$\text{Cost} = k \, (\text{Capacity})^x$$

where x is some exponential factor and k is a constant. Since capital costs do vary in some manner with capacity, it is important to incorporate this aspect into the estimating procedure. The exponential capacity-adjustment method enables the costs for a new project to be estimated from a known project cost and a ratio of their capacities. The exponent rule states that the ratio of costs is directly proportional to capacities raised to an exponential power. This relationship is expressed as follows:

$$\frac{(\text{Cost}) \, A}{(\text{Cost}) \, B} = \left(\frac{(\text{Capacity}) \, A}{(\text{Capacity}) \, B} \right)^x$$

where x is the exponential factor.

The most critical factor in this cost estimation calculation is the value of x, the exponential factor. Determination of this factor is based on the ever popular capacity-cost curves. The normal approach is to collect cost and capacity data for mining projects from technical literature, private cost data, government sources, banks, or files on cost estimating. The exponent is determined by graphically plotting capacity vs. cost from the data obtained. If these plots are made on log-log paper and a straight line is fit to the data, the slope of this line is the value for x, the exponential factor. Fig. 1 illustrates the concept.

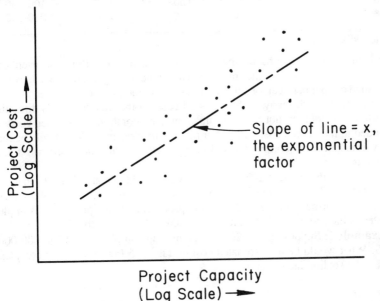

Fig. 1. Plot of capacity-cost relationships showing determination of the exponential factor.

Exponential scaling is employed extensively in the chemical processing industries. Table 2 from Jelen (1970) provides scaling factors for a number of common chemical plants.

Table 2. Scaling Factors for Common Chemical Plants*

Process	Capacity	Unit	Cost-capacity factor	Capacity range
Acetylene	10	Tpd	0.73	3.5-250
Aluminum (from alumina)	100 M	Metric Tpy	0.76	20 M-200 M
Ammonia (by steam-methane reforming)	100	Tpd	0.72	100-3 M
Butadiene	10 M	Tpy	0.65	5 M-300 M
Butyl alcohol	100 MM	Lb/yr	0.55	8.5 MM-700 MM
Carbon black	1	Tpd	0.53	1-150
Chlorine	100	Tpd	0.62	10-800
Ethanol, synthetic	10 MM	Gal/yr	0.60	3 MM-200 MM
Ethylene	100 M	Tpy	0.72	20 M-800 M
Hydrogen (from refinery gases)	10 MM	Cu ft/day	0.64	500 M-10 MM
Methanol	10 MM	Gal/yr	0.83	5 MM-100 MM
Nitric acid (50-60%)	100	Tpd	0.66	100-1 M
Oxygen	100	Tpd	0.72	1-1.5 M
Power plants, coal	100	Mw(elec)	0.88	100-1 M
Nuclear	100	Mw(elec)	0.68	100-4 M
Styrene	10 M	Tpy	0.68	4 M-200 M
Sulfuric acid (100%)	100	Tpd	0.67	100-1 M
Urea	250	Tpd	0.67	100-250
Urea	250	Tpd	0.20	250-500

* After Jelen, 1970.

It is important to emphasize that care must be taken in plotting consistent data. For instance, project capacities must be in the same units (e.g., tons per day or tons of refined copper per year, or some other unit, but not mixed), and the costs should be consistent. With many of the reported costs in the literature, it is difficult to ascertain whether or not milling, financing, exploration, and other costs are included in the quoted cost figure. For best results, separate plots should be made for various mineral commodities, mining methods, and domestic vs. foreign locations.

In general the exponential factor will vary from 0.1 to greater than 1.0. In mining projects typical values range between 0.5 and 0.9. Commonly used values are 0.6 to 0.7 for normal mining and mineral processing plant projects. Example 3 illustrates the exponential capacity-adjustment calculation procedure.

Example 3: Suppose a new 20,000-tpd mining complex costs $250,000,000 in 1975. What would be the estimated cost of a new 35,000-tpd mining complex in 1981 if the scaling factor is 0.7?

Solution:

$$\frac{(\text{Cost } A)}{(\text{Cost } B)} = \left(\frac{(\text{Capacity } A)}{(\text{Capacity } B)} \right)^{0.7}$$

$$(\text{Cost}) \, 35{,}000 = \$250{,}000{,}000 \left(\frac{35{,}000}{20{,}000} \right)^{0.7}$$

The estimated capital cost for the 35,000-tpd facility = \$370,000,000.

Note: This cost has not been adjusted for any inflationary effects from 1975 to 1981. This is addressed in the section on "Cost Indicies."

Although the assumption of a straight-line fit to the capacity-cost data is often made, a careful look at these data often indicates that real-world data are not adequately represented by a straight line. A curve turning slightly upward with increased capacity is more accurate. Fig. 2 illustrates the discrepancy which can occur between a conventional straight-line fit of the data and a curvilinear relationship. Fig. 3 shows a piecewise linear fit to the curve illustrated in Fig. 2. From these figures it is easy to see that when the approximate exponential factor was assumed to be 0.7, in reality it could have varied from 0.4 to 1.1. A straight application of the 0.7 value for the exponent would result in either overestimation or underestimation of costs. Obviously care must be taken to accurately determine the exponential factor at various capacities.

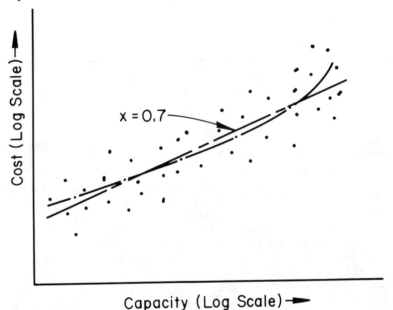

Fig. 2. Cost-capacity data represented by a straight line vs. some curve.

Figs. 2 and 3 illustrate another point which should be noted. As the capacity of the mining complex becomes progressively larger, the slope of the cost curve keeps increasing until at some point it reaches 45°. Once this point is reached, the exponential factor, x, becomes 1.0. From this point on there is no further reduction in the unit cost of the mine and plant with increasing capacity. This is known as the economic size of the plant, meaning that there generally is no economic advantage in terms of capital cost by increasing capacity. In other words, if the value of x increases with capacity, then the economy of scale decreases as one builds larger and larger projects.

There are, in fact, conditions under which x exceeds 1.0. One example is when the project reaches such a size that it exceeds the capacity of the region's infrastructure, thereby requiring added investment in such items as townsite and transportation facilities.

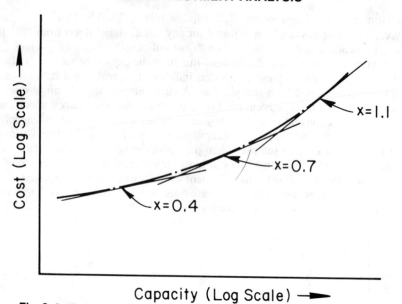

Fig. 3. Straight-line segments of curve in Fig. 2 showing hypothetical values of the exponential factor.

In practice these capacity-cost curves are usually not smooth. They often have discontinuities at various capacity levels or at the point of economic size. For example, in a mining operation the point of economic size may be reached at the capacity limit of the major equipment involved (e.g., a dragline in overburden removal at a coal property). A further increase in capacity means at least a major investment in equipment (perhaps another dragline or an auxiliary overburden removal system). This immediately creates an upward step in the curve and may place the curve just beyond the point of economic size.

Even assuming the mine and plant are at the economic size, they can still be influenced by technological and inflationary changes. Indeed, factors such as location, type of site, local economic conditions, labor rates, and regional productivity levels can cause substantial variations in the values of x. For example, consider the overall effect of labor escalation on capacity-cost relationships. Fig. 4 illustrates the cost pattern of a hypothetical underground mine (curve A) in 1965 when the exponential factor reached a value of 1.0 (45°) at 1000 tpd. Between 1965 and the present, labor rates have escalated greatly (i.e., the labor portion of mine costs at all capacities has greatly increased). However the labor content of the small mine will be a greater percentage of the total cost than for the large mine. It is therefore reasoned that a new cost curve B will develop for which the 1000-tpd mine no longer has an exponential factor of 1.0, but something less (say 0.84).

Although the preceding example is somewhat exaggerated, it does illustrate that rapidly increasing labor rates tend to flatten cost curves with time, forcing an economic-sized mine into larger and larger capacities. In view of the significant escalation in major equipment costs over the past five years, a reverse argument might be made for highly mechanized, equipment-intensive operations. The point is that the relative escalation in key cost components may have a significant effect on the overall capacity-cost curves for various types of mining ventures. Because

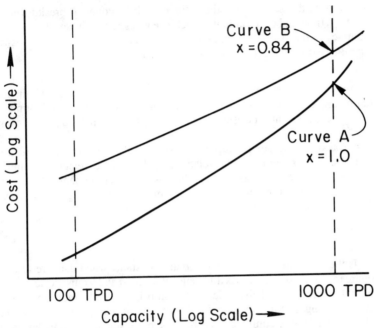

Fig. 4. Generalized effect of labor escalation on cost-capacity curve.

of this complicating feature, cost estimates are enhanced if the major cost areas of a mining project can be delineated and treated separately (i.e., mine, mill, storage and handling, utilities and services). A unique cost-capacity curve should be developed for each of these areas. Specific knowledge of these cost patterns would enable one to synthesize the total cost of a complete mining project at any given capacity.

The exponential capacity-adjustment method can, in theory, be modified to incorporate treatment of factors such as inflation, site location, and technological trends. The following illustrates how the basic formula has been modified to account for inflationary trends and project locations.

$$\frac{(Cost)\ A}{(Cost)\ B} = \left(\frac{(Capacity\ A)}{(Capacity\ B)}\right) \times \left(\frac{IA}{IB}\right) La$$

where IA is the cost index at the present time and location of the proposed project, IB is the cost index at the time of construction of the existing project, and La is the location factor for the proposed project.

There are some limitations associated with capacity-cost curves which should be remembered by the estimator. These may be briefly summarized as follows:

1) The estimate is only meaningful when comparable data are used for estimation. When constructing capacity-cost curves particular attention must be given to the mining and mineral processing method, the degree to which infrastructure is included, the commodity involved, and the similarity of cost data utilized.

2) A breakdown of the mining and mineral processing project into specific cost

patterns improves the accuracy of the estimate. The amount of breakdown is a function of the time and cost available to achieve a desired accuracy.

3) Accuracy of the estimate is more reliable when the capacity change is small. Scale-up ratios exceeding 3:1 are not recommended.

Equipment: Some capital cost estimating techniques for projects are built upon an initial estimate of equipment costs. Consequently some means of estimating major equipment costs must be established prior to developing total project capital costs.

One approach to estimating equipment costs is to develop *equipment variable curves*. These curves represent the relationship between a selected equipment variable (i.e., flywheel horsepower) and the cost of the equipment. Many of the major equipment items used in the mining and mineral processing industries have been shown to follow the form:

$$\text{cost} = a\,(X)^b$$

where X is the equipment variable and a,b are constants when data are fitted by least square methods. In most cases different values for a and b must be determined for different ranges in X. This basic relationship is generally more reliable for mineral processing equipment than for major mining equipment, the primary reason being that milling equipment has become more standardized in design and componentry. Table 3 shows data provided by Mular (1978a) for some basic mineral processing and mining equipment with appropriate values of X, a, and b.

To illustrate the procedure, suppose the cost of a 100-tph jig is desired. The equation developed from the data in Table 3 would be as follows:

$$\text{cost} = 1729\,(100)^{0.247} = \$5393.$$

Another method of estimating major equipment costs is to develop *unit cost* relationships. This information can be generated by dividing equipment cost by some suitable equipment parameter (e.g., $ per pound, $ per horsepower, etc.). Many prices of major mining equipment lend themselves to this form of estimating. For example the cost of walking draglines in 1981 could be estimated at between $2.40 and $2.70 per lb.

Obviously either of these approaches require continual updating and checking to ensure continued validity of the relationship. Technological developments or changes in machine design can significantly affect the estimating relationship.

Cost Ratio Method: This method can either relate to the project in total or to specific categories of major equipment. When estimating total project capital costs, the assumption is made that the total capital cost is proportional to the first or initial capital cost of the major equipment needed for the operation. The total project cost ratio method may be expressed as:

Total Project Cost = a factor (K) × total equipment cost.

The factor, K, is referred to as the Lang factor. These factors have been widely developed for chemical process plants, but are less common for mining projects. However, values of between 3.0 and 6.0 seem to apply for some mineral processing plants, depending on the mineralization involved. Obviously the Lang factor (K)

Table 3. Cost Estimating Parameters for Select Processing Equipment*

Equipment	Equipment variable, X	Range or capacity	a	b
Jaw crushers	area of feed opening, sq in.	15x24 to 32x40	227	0.807
		32x40 to 60x80	0.667	1.63
Gyratory crusher	receiver opening x mantle diameter, in.			
Primary		30x55 to 60x90	2.18	1.54
Secondary		16x50 to 20x70	83.5	1.10
Autogenous mills	horsepower	3,000 to 10,000	2,724	0.757
Ball mills (with liners and rubber backing)	horsepower	40 to 1,750	4,956	0.585
Rod mills (with liners and rubber backing)	horsepower	50 to 4,000	4,629	0.607
Wet cyclones	diameter, in.	1 to 15	227	0.75
Rake classifiers	length of tank, ft	20-36 (tank width = 4 ft)	11,750	0.27
		22-36 (tank width = 5 ft)	7,950	0.45
Spiral classifiers	spiral diameter, in.	24 to 28	69.8	1.69
Flotation cells	capacity in cu ft	10 to 600	472	0.52
Heavy media separator	capacity in dry tons/hr	15 to 200	92,898	0.23
Mineral jigs	max. capacity in dry tons/hr	35 to 200	1,729	0.247
Stainless steel autoclaves	capacity in US gallons	200 to 1,000	1,116	0.308
		100 to 1,000	2,734	0.612
Mixing tanks with agitator mechanism	capacity in gallons	168 to 1,540 (304 stainless)	27.5	0.602
		168 to 1,540 (316 stainless)	27.5	0.629
Rotary dryer (flue gas, direct (carbon steel))	dryer volume, cu ft	1,000 to 9,000	1,271	0.68
Rotary drum filter	filter area, sq ft	19 to 75	8,236	0.292
		75 to 113	2,662	0.649

Table 3. Continued

Thickeners	tank diameter, ft	10 to 200	1,110	0.965
Conditioners (excludes tank)	tank volume, cu ft	18,322 to 154,000	51.8	0.577
Vibrating grizzly feeder	width x length, sq ft	42 to 132	1,526	0.584
Cyclone dust collector	capacity in cfm	250 to 3,350	58.8	0.343
Power shovels	bucket size, cu yd	3 to 10	0.823	0.870
Off-highway trucks (diesel)	capacity, tons	22 to 170	5,292	0.932
Front-end wheel loaders	operating weight, lb	14,200 to 381,000	28.6	0.861
Rotary drills (truck mounted)	bit pulldown force, lb	30,000 to 120,000	5,292	0.932
Stationary compressors, 100 psi	horsepower	30 to 500	440.4	0.804
Scooptrams	tramming capacity, lb	5,000 to 39,000 (hard rock)	30.0	0.841
		5,000 to 15,000 (coal)	31.1	0.861

* Each cost corresponds to a Marshall and Swift cost index for mining and milling of 500 (base year 1926 = 100; costs in 1977 Canadian dollars). Source: Mular, 1978a.

will be greater than unity for mining situations because it embodies capital investment in the mine as well as off-site facilities.

The foregoing approach can be improved upon in terms of accuracy if the total equipment is subdivided into similar categories and a specific factor is applied to each group. The total project cost would then equal the sum of these categories each multiplied by an appropriate factor. This method is referred to as the equipment cost ratio method and may be represented as follows (Mular, 1978b):

$$\text{Project cost} = \sum_{i=1}^{n} F_i X_i$$

where X_i is the capital cost of the ith equipment category and F_i is the appropriate factor. Table 4 lists some proposed values for the factor F for several categories of mineral processing equipment.

Table 4. Equipment Cost Ratios for Select Mineral Processing Equipment*

Equipment category	Factor, F_i
Bucket elevator	2.0
Mixer	2.0
Furnace	2.1
Drum-type dryer	2.2
Kiln	2.2
Conveyor	2.3
Electrostatic precipitator	2.5
Blower, fan	2.5
Refrigeration unit	2.5
Boiler	2.8
Mills	3.0
Vacuum rotary dryer	3.2
Dry dust collector	3.5
Storage tank	3.5
Crusher	3.5
Process tank	4.1
Instrumentation	4.1
Heat exchanger	4.8
Wet dust collector	6.0
Pump	6.8
Electric motor	8.5

* Source: Mular, 1978b. Factors apply to costs in 1977 Canadian dollars.

To illustrate the general concept of the cost ratio method, the following example is offered (after Jarpa, 1977). Suppose a truck-shovel coal operation is being considered. The major cost centers would probably be categorized as overburden removal, drilling, blasting, loading, and hauling. Assuming there are no major pieces of equipment other than drills involved in the blasting unit operation, the principle units of equipment for the project will be scrapers, drills, shovels, and trucks.

If IC is Investment (Capital) Cost, then total project cost can be represented as:
Total Equipment IC = Scrapers IC + Drills IC + Shovels IC + Trucks IC.
Using the expression for the cost ratio method

$$\text{Total Project Cost} = K \text{ (Total Equipment } IC).$$

It is possible to calculate the investment cost for each classification of equipment by the previously discussed approaches. Assume the equipment cost is determined by the following equation:

$$\text{Cost} = a \, (X)^b.$$

It is therefore possible to calculate the total project capital investment for a project having a certain production capacity (A tpy) as illustrated in Table 5. With these parameters the total equipment capital investment is given as:

$$C = (Csc + Cd + Cs + Ct).$$

Table 5. Parameters in Cost Ratio Relationship

Equipment class	IC	% of total IC	Scaling component "b"
Scrapers	Csc	SC	sc
Drills	Cd	D	d
Shovels	Cs	S	s
Trucks	Ct	T	t
Total project	C	100	Y (unknown)

The total equipment capital investment required for the same or similar operation of different production capacity (B tpy) can be determined if it is assumed that the greater production will be achieved only by increasing the size or capacity of the equipment spread. If we let the capacity or size ratio be $Q = B/A$, the resulting expression may be written:

$$Q^y = Q^{sc}SC + Q^dD + Q^sS + Q^tT.$$

Solving the equation gives the value of the total project scaling exponent, y. Then the total project capital investment cost for the mining operation having production capacity B tpy is:

$$\text{Total project cost}_B = K \times Q^y \times C.$$

The following example (after Jarpa, 1977) illustrates the methods for an open pit copper property in Chile. The data used are representative of the early to mid-1970s. Note that the method does not, at this point, incorporate an adjustment for inflation. This is discussed in the section on "Cost Indices."

Example 4: From published data it appears that an appropriate value for K ranges between 4.0 and 6.0 for many medium- to large-sized open pit copper mines. Now suppose that the typical categories of equipment have been compiled and capital costs estimated on the basis of:

$$\text{Cost} = a \, (X)^b$$

where X is the capacity of the equipment.

Further suppose that the cost and technical input regarding the types and sizes of equipment can be summarized as follows:

Parameters of Cost-Capacity Relationships

Equipment class	Range of capacity	Constant "a"	Scaling exponent "b"
Trucks	50-200 tons	2,950	0.95
Shovels	6-25 cu yd	69,750	0.91
Drills	120-450,000 lb	6	0.90
Scrapers	9½-17¼ cu yd	2,340	1.27
Wheel loaders	6-24 cu yd	14,320	1.02

A compilation of some open pit copper mines from around the world indicates the following breakdown of percentages of equipment in various categories and the equipment capital investment for each mine.

Percentage of Investment in Mining Equipment by Category

Mining operation	Drills, D, %	Shovels, S, %	Trucks, T, %	Wheel loaders, W, %	Total equipment investment, $1,000
A	4.7	18.8	72.9	3.6	51,715
B	7.1	32.0	48.9	12.0	31,650
C	8.6	23.7	56.7	11.0	28,164
D	8.6	24.4	60.0	7.0	26,206
E	6.5	26.5	54.1	12.9	23,022
F	7.2	39.7	44.0	9.1	12,817
G	15.7	29.5	43.7	11.1	12,689
H	12.5	30.8	41.1	15.6	10,370
Average	8.8	28.2	52.7	10.3	

Therefore, for a given capacity of mine (say A tpd), the total equipment cost is calculated as:

Total Equiment IC = Trucks IC + Shovels IC + Drills IC + Wheel loaders IC
$$= 2950 (A)^{0.95} + 69{,}750 (A)^{0.91} + 6 (A)^{0.90} + 14{,}320 (A)^{1.02}.$$

Total Project Cost for the mine having a capacity of A tpd can then be calculated as:

Total Project Investment $= 5.60 [2950 (A)^{0.95} + 69{,}750 (A)^{0.91} + 6 (A)^{0.90}$
$$+ 14{,}320 (A)^{1.02}]$$

assuming $K = 5.60$.

In order to calculate project cost for a different mine having B tpd capacity, the scaling component, y, for the mining operation can be determined by substituting for the parameters in the following equation:
$$Q^d D + Q^s S + Q^t T + Q^w W = Q^y$$
where $Q = B/A$ is the capacity ratio.

From the table showing the average equipment spread for the mines given, the equation becomes
$$0.088 Q^{0.99} + 0.282 Q^{0.91} + 0.527 Q^{0.95} + 0.103 Q^{1.02} = Q^y$$
if the capacity increase of the mines in question went from 50,000 tpd (A) to 75,000 tpd (B), then $Q = 1.5$. Solving the equation for the total equipment cost yields $y = 0.94$. This implies that the total equipment investment cost for the new, larger mine (B) will be
$$(1.5)^{0.94} = 1.464 \text{ times the investment for the existing mine } (A).$$
Therefore, the total project capital investment cost for the mine with a capacity of

B tpd can be expressed as

$$\text{Total Project Cost} = K \times Q^y \times C$$
$$= 5.60 \, (B/A)^{0.94} \times C$$

where C is the total equipment investment cost for mine A. If the total equipment investment cost for mine A was \$90,000,000, the estimated total project cost for mine B would be:

$$5.60 \times (1.5)^{0.94} \times \$90,000,000 = \$738,000,000.$$

Component Cost Ratio Method: An extension of the previously discussed cost ratio method is the component cost ratio, sometimes referred to as the factored estimate based on installed equipment costs. This method involves a breakdown of fixed capital costs into components whose costs are a ratio of major equipment costs. After the cost of the basic equipment items has been determined, the next step is to find the cost relationship of the other components as a percentage, a ratio, or a factor of the basic equipment items. In a generalized sense the estimating process can be expressed as:

$$\text{Total Project Capital Cost} = [Ce + \Sigma \, (fi \, Ce)] \, (g + 1)$$

where Ce is the cost of selected major equipment items; fi is the factor for estimating buildings, installation, etc.; and g is the factor for measuring indirect costs such as engineering and contingency.

This method achieves improved accuracy by virtue of adopting separate factors for the various cost items. Also, equipment component adjustments can be made at this stage as well as conversion factors for location and time. Factored estimates should be based on proposed layouts, flowsheets, and equipment specification sheets for improved accuracy. The basic items provided in the layouts, flow charts, etc., should be major items such as mining equipment, milling equipment, structural shells for buildings, tons of concrete for foundations, roads, etc.

Jarpa (1977) discusses the application of the techniques previously discussed for open pit copper mining and has developed, on the basis of equipment cost-capacity data, relationships between production capacity and capital and operating costs. In general, developing good factored estimates in mining applications is quite difficult because of the extreme variations in development requirements, equipment spreads, and extraction techniques among properties—particularly underground properties. However, fairly accurate factored or component estimating relationships have been developed for mineral processing plants. Table 6 provides such an estimating approach.

Module Method: The module method is typically based on specific equipment lists, building sizes, flowsheets, etc., with accompanying list prices. This method is an extension of the factored (component) cost estimate and is much more precise and accurate, but requires more time and data. A specific piece of equipment is assessed and the cost of each auxiliary item attributable to that piece of equipment is obtained through the use of factors to get the final overall cost. The following example illustrates the concept.

Table 6. Mineral Processing Plant Component Cost Ratios

---

1. Purchased equipment costs from references
   and on current index basis $000,000
2. Equipment installation (0.17 to 0.25 times item 1) $000,000
3. Piping, material and labor, excluding service piping
   (0.07 to 0.25 times item 1) $000,000
4. Electrical, material, and labor, excluding building lighting
   (0.13 to 0.25 times item 1) $000,000
5. Instrumentation (0.03 to 0.12 times item 1) $000,000
6. Process buildings, including mechanical services and lighting
   (0.33 to 0.50 times item 1) $000,000
7. Auxiliary buildings, including mechanical services and lighting
   (0.07 to 0.15 times item 1) $000,000
8. Plant services, such as freshwater systems, sewers, compressed
   air, etc., (0.07 to 0.15 times item 1) $000,000
9. Site improvements, such as fences, roads, railroads, etc.
   (0.03 to 0.18 times item 1) $000,000
10. Field expenses related to construction management
    (0.10 to 0.12 times item 1) $000,000
11. Project management including engineering and construction
    (0.30 to 0.33 times item 1) $000,000
12. Fixed capital cost (sum of 1 + 2 + 3 + 4 + 5 + 6 + 7 + 8 + 9 + 10 + 11) . . . $000,000

Source: Mular, 1978 b.

A. Base machine price (fob manufacturing plant):	$18,000,000
B. Freight (1.5% of A):	270,000
C. Auxiliary extras not included in base price (1% of A):	200,000
D. Field erection (12% of A):	2,160,000
Total machine cost	$20,630,000

As an estimating system becomes more precise, however, its application becomes more specific, and thus a substantial amount of data is required to make it applicable to a wide range of different situations.

Detailed Estimates: Detailed cost estimates are the final and most accurate type of cost estimate. These estimates are based on detailed engineering drawings, layouts, flowsheets, and equipment lists showing model numbers, specifications, and so on. Typically at this stage actual quotations have been submitted by contractors, suppliers, and manufacturers.

These detailed estimates serve as a buying guide for purchasing and as a control and reference position during project development and construction. Detailed cost codes are itemized into the proper code (i.e., materials, labor, equipment, subcontracted job, etc.). Development of detailed cost codes also helps in finding and defining any cost items which may have been previously overlooked.

Contingency and Engineering Fee: As mentioned previously, the AACE definition of contingency is "specific provisions for unforeseeable elements of cost within the defined project scope; particularly important where previous experience relating estimates and actual costs has shown that unforeseeable events which will increase costs are likely to occur." Unforeseeable cost elements typically occur as a result of the absence of total quantity definition when estimating the project. Also contingency allowances must take into account the errors associated with the estimating procedures themselves as well as other errors which generally occur in the preparation of an estimate (e.g., errors in judgement, productivity rates, oversight of key elements, etc.). Obviously the amount of contingency assigned to the estimate will be a function of the type and accuracy of the estimate being prepared.

In mining projects contingency estimates are usually given more attention when definitive cost estimates are being prepared. Under these conditions the contingency allowance is understood to represent factors such as equipment price changes, omissions in minor pieces of equipment or attachments and extras, omission of support and/or ancillary equipment, and so on. A value commonly used is 15% of the total equipment capital cost. By obtaining firm quotations from suppliers, one can usually reduce the contingency on major equipment to ±5%.

Engineering expense should also be assigned to the capital cost estimate in order to account for those individuals working on equipment selection, mine layout, design, project development, contractor monitoring, etc. An average engineering cost estimate for mining situations is approximately 10% of the total equipment capital cost. However, some judgment must be used when determining this estimate. For instance, if the total capital costs for a given mining operation are primarily associated with one or two machines (i.e., a dragline coal operation), then it is probably unrealistic to assign a 10% engineering fee to the estimate when a value of 1–2% is more representative.

Engineering and contingency allowances are added to the equipment capital cost estimate in order to provide the total capital cost estimate for the project. The procedure is as follows:

	Investment, $1000	Total project capital cost, $1000
A. Draglines	55,000	
—accessories	1,500	
Subtotal	56,500	
Engineering @ 1%	565	
Contingency @ 15%	8,475	
Total		65,540
B. Scrapers	2,500	
Trucks	3,600	
Dozers	1,100	
FEL's	2,000	
Motor graders	800	
Misc. support equipment	6,200	
Subtotal	16,200	
Engineering @ 10%	1,620	
Contingency @ 15%	2,430	
Total		20,250
Total project		$85,790

Operating Cost Estimates

Preliminary operating costs are generally more difficult to estimate than capital costs for most mining ventures. The relative uniqueness of each mining operation with respect to equipment spreads, labor intensity, mining methods, location, maintenance philosophies, etc., makes the estimation of operating costs most difficult because of the general inability to develop consistent cost estimating relationships and techniques. For these same reasons, most estimators find it more difficult to develop cost estimates for underground mining operations than for open pit operations.

Operating costs in the US mining industry are typically expressed as "$ per ton." The common question that arises is, "What does the $ per ton number represent?" Does it include direct costs, indirect costs, taxes, depreciation, corporate overhead, and so on? Such a number really communicates very little useful information unless this cost is further defined and subdivided into components which can be updated and adjusted for future estimates.

Operating costs are defined as those ongoing, recurring costs incurred during normal functioning of the project. In general total project operating costs can be divided into three primary classifications—direct, indirect, and general (overhead).

Direct, or variable, costs are considered prime costs and can be traced directly to the product being produced. Direct costs primarily consist of labor and material charges. Although there are variations from company to company, the general cost headings typically associated with direct costs are as follows:

Direct Operating Costs:

1) Labor: (a) direct operating, (b) operating supervision, (c) direct maintenance, (d) maintenance supervision, and (e) payroll burden on the foregoing labor.

2) Materials: (a) maintenance, repair materials; (b) processing materials; (c) raw materials; and (d) consumables (fuel, oil, water, power, etc.).

3) Royalties.

4) Development (production area).

Indirect, or fixed, costs are expenditures which are independent of throughput. Although these costs often vary with the level of production, they do not vary directly with product throughput. In other words, they can not be traced to a unit of output. Some of the major cost components associated with indirect costs are:

Indirect Operating Costs:

1) Labor: (a) administrative, (b) safety, (c) technical, (d) service (clerical, accounting, general office), (e) shop and repair facilities, and (f) payroll burden on the foregoing.

2) Insurance (property, liability).

3) Depreciation.

4) Interest.

5) Taxes.

6) Reclamation.

7) Travel, meetings, donations.

8) Office supplies, upkeep, utilities.

9) Public relations.

10) Development (general mine).

General, or overhead, expenses may or may not be considered part of operating costs. Certainly they contribute, and are assignable, to total product cost even though they typically represent off-site charges. Some of these expenses could be incurred at a specific plant or mining unit, although in general they represent corporate level expenditures which are allocated to all operating units. General expenses typically include the following categories:

General Expenses:

1) Marketing/sales: (a) salespersons, (b) marketing analysis team, (c) supervision, (d) travel/entertainment, and (e) payroll burden on labor.

2) Administrative: (a) corporate officers and staff, (b) general accounting and auditing, (c) central engineering/geology, (d) legal staff, (e) research and development, (f) public relations, (g) financial staff, (h) contributions, and (i) expense accounts.

Because of the large number of cost components (many of which are very significant) which constitute overall operating costs and the relative uniqueness of mining operations and extraction techniques in general, the cost estimator finds there is very little middle ground in developing operating cost estimates. In the course of performing most feasibility studies, one is either working with rather crude estimates of operating costs or reasonably detailed estimates based on manning charts, equipment schedules, and so on.

Although there are fewer actual estimating techniques utilized for developing operating costs than for capital costs, those which are used are essentially the same as those applied to capital cost estimates. Because these methods were discussed in some detail in the preceding section, they are mentioned here only briefly.

Similar Project Method: One approach to estimating operating costs is to assume the project or process in question is similar to another project for which the costs are known. The similar or existing project may be one which is currently being operated by the firm, one which is operated by a competitor, one previously costed-out in detail by the firm, or simply one which is familiar to the estimator from previous experience.

This method inherits all the inherent problems associated with the use of similar projects. At best these are detailed estimates, but the geologic setting, equipment, staffing philosophy, and other variables often differ considerably from the project in question. Typically, rules-of-thumb start to appear in conjunction with these estimates. By estimating a small amount of information (say labor charges) from similar projects, the entire operating cost estimate may be developed. For example, some of the relationships frequently used for crudely estimating operating costs for underground mining operations are:

1) Labor = 50 to 55% of total operating cost.

2) Repair, maintenance, and supplies = 30 to 40% of total operating cost.

3) Miscellaneous = 5 to 20% of total operating cost.

Table 7 suggests the relationship between total production costs and the percentage of labor for select underground mining methods (Gentry, 1976). Obviously such relationships are predicated on some assumptions regarding degree of mechanization, labor intensity, development requirements, and so on. At this level of estimating the breakdown of direct and indirect costs is largely irrelevant.

Cost-Capacity Relationship Method: This method of estimating operating costs is essentially the same as that described for capital cost estimation. Operating costs are estimated by constructing cost-capacity graphs based on data compiled

Table 7. Labor as a Percentage of Total Production Costs for Selected Underground Mining Methods

Mining method	Labor as a percentage of total production costs
Room and pillar stoping	44
Cut and fill stoping	57
Sublevel stoping	60
Sublevel caving	64
Block caving	55

Source: Gentry, 1976.

from the literature or company records.

This approach is frequently infiltrated by errors. Use of data from the literature often results in significant errors because the estimator does not understand the basis for these data. For example, an article may describe a mining cost as follows: "Production cost is $12.50/ton." The estimator is uncertain whether or not this number includes both direct and indirect costs. Even if it is clear that both direct and indirect costs are included, the cost value does not provide any information on the cutoff points for these costs. Company data itself also is often unsuitable for direct use in the estimation procedure simply because of the manner in which the accounting system compiles and reports the cost components.

When plotting operating cost-capacity graphs one can expect considerable scatter of the data which tends to undermine confidence in the estimate. This scatter is not surprising when one considers the vast differences which exist in labor rates, utilities, development expenditures, inflation factors, etc., for mining operations around the world. For best results, cost-capacity plots should be developed for specific types of mining methods, preferably in similar geologic and geographic environments.

Component Cost Method: When the project has progressed to the point where crude estimates can be made for manpower requirements, required development footages, powder factors, approximate equipment lists, and so forth, then component cost estimates can be developed. These estimates are developed on the basis of *unit costs* and *factored costs*. For instance, unit costs such as the following are utilized:

$ per foot of drift development
$ per foot of shaft development
tons per man-shift
lbs explosive per ton broken
$ per ton overburden removal
$ per ton drilled
$ per operating hour for various equipment categories.

Factored costs are also utilized where appropriate. Usually these costs are expressed in terms of a percentage of some other major cost item. Some examples of factored estimates for operating costs follow:

Repair and maintenance   2-5% of equipment capital cost
General and administrative   2% of sales
Insurance   2-3% of capital equipment investment
Taxes (property)   2-3% of capital equipment costs
Indirect   10-30% direct labor + materials costs
Payroll burden   40% direct labor

From the various unit and factored cost relationships a component operating cost can be developed. It must be noted, however, that some preliminary design, equipment selection, and manpower estimates are necessary before component cost estimates are very meaningful.

Detailed Cost Breakdown Method: Ultimately, operating costs must be developed from a detailed breakdown of major cost items. Estimates on powder factors; footage drilled per bit; gallons of fuel consumed per operating hour per machine; quantity of detonation cord, primers, and delays per blast; and numerous other variables must be made in order to develop a detailed operating cost estimate. It is imperative that detailed manning tables be developed as well as specific equipment lists. A checklist should be developed for each job function in the operation along with estimates of labor, materials, supplies, services, etc., associated with each. Estimates of production rates, productivities, and advance rates must be made for each job classification where appropriate. These elemental costs are then combined into costs for various unit operations and subsequently summed into the operating cost for the facility.

This is indeed a time-consuming and laborious process. Yet, it is the only way to develop a reliable operating cost estimate for a mining project. There are few shortcuts! The estimate must be based on a final mine design and layout, drawings, estimates of operating characteristics, and actual calculated estimates of specific equipment operating costs. These equipment costs are often analyzed through computer simulation programs using haul-road profiles and equipment performance specifications based upon the mine design in question.

Table 8 illustrates the calculation procedure often utilized to estimate the hourly owning and operating cost for a front-end loader for a definitive detailed cost estimate. Similar approaches may be used for other pieces of equipment. Ownership costs relate to capital costs and these are treated through the depreciation deduction discussed in Chapter 8. It should be noted that the estimator must be careful not to double count for maintenance labor when performing these estimates. If labor is included in the maintenance and repair cost estimates calculated for a suite of machines, then it should not be tallied in the detailed manning table. It should be accounted for either in total labor or in the machine operating cost estimates but not both.

Table 8. Hourly Owning and Operating Cost Worksheet, Front-End Loader

Date:	4-15-81
Prepared by:	John Smith
No:	example

Location: _____

Application: _____

Machine: <u>12 cu yd</u>

Model: _____

OWNERSHIP COSTS

Amount in
$/oper.hr

A. Depreciation:

1. Purchase price (include attachments, extra, taxes)..... <u>$530,000</u>
2. Freight: <u>200,000</u> lb
   @ <u>$5.10/CWT</u> ....(+) <u>10,200</u>
3. Delivered price:.................................. <u>540,200</u>
4. Tire replacement costs:
   Front <u>$7,800 (2)</u>
   Rear <u>7,800 (2)</u> ....(−) <u>31,200</u>
5. Resale or trade-in value:
   (optional)..................................(−) <u>63,600</u>
6. Net depreciable value:........................... <u>445,400</u>
7. Depreciation period:
   a) service life (hours)   <u>11,000</u>   hr
   b) operating hours/year   <u>2,200</u>   hr
   c) years for write-off   <u>5.0</u>   yr
8. Hourly depreciation costs:
   $$\frac{\text{net depreciable value (line 6)}}{\text{depreciation period in hours (line 7a)}}$$ $ <u>40.49</u>

B. Interest, Insurance, Taxes:

1. Annual rates:
   a) interest   <u>14%</u>
   b) insurance   <u>2%</u>
   c) taxes   <u>2%</u>
   d) Total .... <u>18%</u>

2. Calculation procedure:

   a) $$\frac{\text{(line B1d)} \times \frac{N+1}{2N} \times \text{(line A3)}}{\text{line A7b}}$$

   $$\frac{\% \times \frac{yr}{yr} \times \$}{hr} \quad .... \$\_\_\_\_$$

   where N = line A7c
   or, if salvage is considered

b) (line B1d) x $\left(\dfrac{\text{line A3 (N+1) + line A5 (N-1)}}{\dfrac{2N}{\text{line A7b}}} \right)$

$18\% \times \left(\dfrac{(\$540{,}200 \times 6\ yr) + (\$63{,}600 \times 4\ yr)}{\dfrac{10\ yr}{2{,}200\ hr}} \right) \dots \dots \dots \underline{\$28.60}$

3. Hourly IIT costs (select either line 2a or line 2b above) . . . . $\underline{\$\ 28.60}$

C. Total Hourly Ownership Costs
   (line A8 + line B3). . . . . $\underline{\$\ 69.09}$

OPERATING COSTS

	Amount in $ Oper. Hr.

D. Hourly Tire Cost:
   1. Replacement cost:
      $\dfrac{\text{tire replacement cost}}{\text{estimated life (hrs)}}$
      $\dfrac{\$31{,}200}{3{,}200\ hrs}$ . . . . $\underline{\$\ 9.75}$
   2. Repair cost:
      tire repair factor (%) x hourly tire replacement cost (line D1)
      $\underline{15}\ \%/100 \times \$9.75$ . . . . $\underline{\$\ \ 1.46}$
   3. Hourly tire cost:
      (line D1 + line D2) . . . . $\underline{\$\ 11.21}$

E. Hourly Fuel Cost:
   est. consumption $\underline{20.0}$ gal per hr
   x unit price $\underline{\$1.04}$ per gallon. . . . . $\underline{\$\ 20.80}$

F. Service Costs:
   factor ratio x hourly fuel cost
   ratio $\underline{1/3}$ x line E $\underline{20.80}$ . . . . $\underline{\$\ \ 6.93}$

G. General Repair:
   repair factor x hourly depreciation cost
   factor $\underline{60}$ % x line A8 $\underline{40.49}$ . . . . $\underline{\$\ 24.29}$

H. Hourly Special Items: (cutting edges, bucket teeth, etc.)
   $\dfrac{\text{initial cost (\$)}}{\text{estimated life (hours)}}$
   $\dfrac{\$}{\text{hours}}$ . . . . $\underline{\$\ \ 1.25}$

I. Total Hourly Operating Cost:
   (exclusive of operating labor)
   (add lines D through H). . . . . $\underline{\$\ 64.48}$

J. Hourly Operator Cost:
   (including fringes, etc.) . . . . $\underline{\$\ 20.05}$

K. Total Hourly Ownership & Operating Cost:
(add lines C, I, and J) $153.62

Source: Martin, et al., 1981.

Table 9 shows a breakdown of reporting and costing detail which might be associated with an underground cut-and-fill mining operation. It illustrates the detail in which the cost estimator must be prepared to work in the final stages.

Table 9. Typical Cost Breakdown for a Cut-and-Fill Mining Operation

A. Labor
  Daily
    miners
    helpers
    service
    maintenance
    hoist man
    skip tender
    electricians
    billed in daily

  Salary
    superintendent
    general foreman
    foreman
    engineers
    geologists
    technicians
    office/clerical
    samplers

  Fringes
    holiday pay
    vacation allowance
    supplementary unemployment
    retirement/death/disability
    federal insurance contributions
    unemployment taxes
    workmen's compensation
    industrial accidents
    group life insurance
    group hospital insurance
    pensions wages
    pensions salary

B. Material/supplies
  Cable
  Electrical
  Explosives
  Hose, bits, drills
  Timber, lumber
  Machinery parts
  Pipe and fittings
  Rock bolts
  Stope fill (sand, gravel, cement)
  Sundry supplies
  Tools and utensils
  Track supplies
  Vent tubing
  Water
  Drills and parts

C. Other Indirect
  Power
  Draining
  Ventilation
  Development
  Tramming
  Hoisting and lowering
  Maintenance
    shaft, stations, level
    concrete equipment
  Auxiliary
    shops and services
    warehouse expense
    assay office
    pumping plant
    central sampling
    central heating
    compressor plant
    slime filling plant
    central pumping

D. Transportation/distribution

E. General expense

F. Total product cost

Contingency: A contingency factor is assigned to the estimated operating cost (direct + indirect + general) and is normally assumed to represent unexpected problems associated with operating conditions. These problems might include such items as extreme weather conditions, ground control problems, and excess dewatering. Contingency factor values ranging from 10–25% are typically assigned to operating costs, depending on the level of cost estimate generated.

Cost Indices

The cost estimating methods previously discussed for developing capital and operating costs were all based either on current unit costs or on comparisons with previous costs obtained elsewhere. None of these methods directly considered any variations or adjustments for cost values over time. Yet everyone recognizes that costs change continuously because of three primary variables:
• Technology.
• Availability of labor and materials.
• Changing value of the monetary unit (inflation).
Various cost indices are available to help the estimator keep up with the changes in cost and cost components over time.

A cost index provides a means of comparing cost or price changes from year to year for a fixed quantity of goods. In other words, the cost index is merely a dimensionless number for a given year showing the cost at that time relative to that base year. Therefore, if the cost of an item at some time in the past is known, the current cost may be estimated as follows:

$$(\text{cost now}) = (\text{cost then}) \times \frac{(\text{cost index now})}{(\text{cost index then})}$$

This method provides the cost estimator with a quick means of projecting the cost of a similar design or project from past to present. This is important because it provides a means of using actual historical cost values as a basis for determining current cost values.

There are many different types of cost indices published which cover practically every area of interest to an estimator. For instance, there are cost indices for building construction; types of plants or industries; wage rates for various industries; as well as indices for various types of equipment, materials, and commodities. Because there are so many different kinds of indices, it is most important that the estimator be aware of the more important ones, the various components of each index and their relationship to the project being estimated.

In general cost indices may be placed into two categories: (1) factor cost indices and (2) project cost indices.

Factor cost indices measure the cost trends for a specific type of product (e.g., fuel, bits and steel, mine and plant labor, etc.). Project cost indices provide the overall relative cost for an entire project which typically involves several individual factor inputs.

Perhaps the best known factor cost indices are the Producer Price Indices (PPI) which apply to the output of specific industries. The PPI (formerly the Wholesale Price Index) is designed to measure producer expenditure patterns. As such, PPI data are categorized on (1) an industry/commodity basis and (2) a stage of production basis. Thus the PPI catalogues prices for crude, intermediate, and semi-finished as well as finished products.

The PPI is compiled by the US Department of Labor, Bureau of Labor Statistics, which collects some 10,000 monthly quotations, primarily by questionnaire. As noted previously, these are aggregated by commodity and by stage of processing and finally into composite series on a monthly basis. The PPI commodity groups for annual reporting include farm products, processed foods, textiles, fuels, power, chemicals, rubber, plastics, paper, lumber, metals, machinery, household durables, nonmetallics, transportation equipment, and miscellaneous items. PPI data can be obtained from the monthly publication, *Producer Prices and Price Indices*, available from the US Government Printing Office.

PPI data are very useful in tracking the changing cost of factor inputs to various industries. Analyses of these data are particularly helpful in studying historical trends and projecting future levels of operating costs. A number of these indices relevant to mining are listed in Table 10.

Project cost indices are most applicable to capital cost analyses and projections. Perhaps those most useful in the mining area are as follows:

1) Marshall and Swift (formerly Marshall and Stevens) Cost Indices (M&S).
2) *Engineering News-Record* (ENR) Building and Construction Cost Indices.
3) *Chemical Engineering* (CE) Plant Construction Cost Index.
4) Nelson Refinery Construction Cost Index (NR).

The ENR and Nelson indices, both having large labor cost components, generally move in parallel, but they have increased at a more rapid rate than the M&S and CE indices. These later two indices also track closely but are influenced less by labor. Each of these indices are described in detail in the following and tabulated in Table 11.

Engineering News-Record Cost Indices (ENR): The ENR indices are compiled weekly for 20 cities and represent the costs of labor and building materials in the following proportions: 25 cwt of structural steel, 6 bbl of Portland cement, 1088 board ft of 2 x 4 lumber, and 200 hr of labor. (*Common* labor is used for the Construction Cost Index, *skilled* labor is used for the Building Cost Index). The ENR indices are published weekly in *Engineering News-Record* magazine.

Marshall and Swift Cost Index (M&S): The M&S series includes several indices. The index most often used and referred to is the all-industry equipment index. This index is the average of the indices calculated for 47 industries. Specific indices are calculated for eight process industries (cement, chemical, clay products, glass, paint, paper, petroleum products, rubber) and four related industries (electrical, power equipment, mining and milling, refrigerating and steam power). These indices are based on detailed equipment appraisals, installation labor, modifying factors, and judgments concerning current economic conditions. The M&S index reflects changes in installed equipment costs and therefore is an excellent index for monitoring cost trends for *equipment*. The M&S index for *mining and milling* is important to studies of mining costs. The M&S index is published in *Chemical Engineering* magazine.

Chemical Engineering Plant Construction Cost Index (CE): The CE Plant Cost Index was designed to reflect *plant* capital cost trends and is basically a materials-equipment-labor index. It is made up of 67 Bureau of Labor Statistics (BLS) price indices covering 155 individual commodities, three BLS indices covering 450 additional items, and BLS hourly earnings data for five groups of employees. The index also includes salary trends for engineers, draftsmen, and administrative workers. It is composed of four major components:

Table 10. Producer Price Indices Related to Mining

Year	Electric power	Refined petroleum products	Industrial chemicals	Explosives	Rubber & plastic products	Tires (tractor)	Wood products	Iron & steel	Construction machinery & equipment	Mining machinery & equipment
1967	100.0	100.0	100.0	100.0	100.0	100.0	100.0	100.0	100.0	100.0
1968	100.9	981.	101.0	102.2	103.4	102.7	113.3	101.9	105.7	102.9
1969	101.8	99.6	100.3	104.3	105.3	98.3	125.3	107.0	110.0	106.0
1970	104.8	101.1	100.9	106.3	108.6	105.4	113.7	115.1	115.9	109.9
1971	112.8	104.5	104.7	113.1	112.7	110.3	120.5	117.6	98.5	94.1
1972	120.7	106.6	103.9	115.2	112.8	111.6	136.9	123.9	102.0	96.9
1973	128.4	125.9	106.2	120.1	116.0	115.0	168.1	131.5	106.1	100.2
1974	162.0	218.6	155.7	150.0	140.6	143.8	174.2	172.4	123.6	118.8
1975	192.1	252.0	212.4	178.0	155.0	162.8	167.8	193.9	150.3	152.4
1976	206.1	270.6	225.2	187.1	164.3	178.3	195.1	208.4	161.4	175.4
1977	231.3	301.6	229.9	194.0	173.0	189.8	224.2	222.4	173.3	189.1
1978	248.9	314.1	231.6	208.6	180.4	196.3	261.9	244.8	188.8	206.5
1979	285.2	542.9	299.4	235.3	212.3	227.3	275.0	282.5	217.7	231.9
1980	337.9	716.3	334.6	262.3	223.5	262.6	299.4	316.0	301.9	320.4
1981	383.8	798.3	364.6	295.4	239.0	274.6	285.2	339.7	332.0	346.7
1982*	405.2	759.9	345.8	299.6	242.6	271.0	279.9	336.3	347.8	365.1
Base year	1967	1967	1967	1967	1967	1967	1967	1967	1967	1967
Index value	100.0	100.0	100.0	100.0	100.0	100.0	100.0	100.0	100.0	100.0
Average annual increase: 1975: 1982	11.3%	19.2%	12.5%	7.8%	6.7%	7.7%	7.9%	8.3%	13.2%	13.7%

Year	Bits & related steel	Mine repair parts	Transportation equipment	Average hourly wages, copper ore production workers	Mine & plant labor	PII all commodities
1967	100.0			100.0	100.0	100.0
1968	101.9			105.5	104.7	102.5
1969	107.0		100.0	112.0	112.8	106.5
1970	115.1		103.7	120.6	120.0	110.4
1971	121.8	100.0	109.4	127.6	126.9	107.4
1972	128.4	105.1	113.5	142.3	137.8	112.3
1973	131.6	107.7	114.2	150.3	147.5	127.0
1974	159.3	130.4	124.4	169.9	163.1	150.9
1975	147.8	121.0	140.4	195.1	184.1	164.8
1976	149.4	122.3	149.9	215.6	201.3	172.5
1977	162.0	132.6	160.0	229.8	216.9	183.0
1978	186.2	152.4	172.1	259.5	239.7	197.3
1979	188.5	154.3	193.6	292.3	265.6	235.1
1980	208.5	170.7	224.1	342.0	297.2	254.7
1981	187.6	NA	246.7	400.2	330.3	275.3
1982*	187.6	187.6	256.1	392.2	321.7	284.9
Base year	1967	1972	1968	1967	1967	1967
Index value	100.0	100.0	100.0	100.0	100.0	100.0
Average annual increase: 1975:1982	3.7%	6.6%	9.0%	10.7%	8.6%	8.2%

*November 1982.

Sources: US Department of Labor, 1978, *Handbook of Labor Statistics*.
US Department of Labor, *Monthly Labor Review*, various issues.
US Department of Labor, *Producer Prices and Price Indices*, various issues.

Table 11. Tabulation of Major Project Cost Indices (1967 = 100)

Year	Marshall & Swift mining & milling equipment cost	Engineering News Record		Chemical Engineering Plant	Nelson Refinery Index
		Construction cost	Building cost		
1967	100.0	100.0	100.0	100.0	100.0
1968	103.7	107.6	106.7	104.3	106.1
1969	108.0	118.3	117.0	108.1	114.8
1970	114.8	128.4	123.6	115.4	127.3
1971	121.9	146.2	139.7	120.9	141.6
1972	125.9	161.2	154.6	125.5	152.9
1973	130.1	176.6	168.6	131.8	163.2
1974	149.6	188.0	178.3	151.3	182.3
1975	171.3	205.7	193.3	166.9	200.7
1976	183.3	223.4	210.9	175.7	214.8
1977	194.1	240.0	228.6	186.7	227.8
1978	214.7	258.4	247.7	200.2	244.5
1979	235.3	279.5	269.3	218.4	263.9
1980	289.4	301.4	287.7	238.7	287.0
1981	349.6	332.4	312.7	261.2	315.3
1982	411.3	353.4	329.3	297.0	340.9
Base year	1967	1967	1967	1967	1967
Index value	100.0	100.0	100.0	100.0	100.0
Average annual increase: 1975:1982	13.5%	8.0%	7.9%	8.6%	7.9%

Sources: US Department of Labor, 1978, *Handbook of Labor Statistics.*
US Department of Labor, *Monthly Labor Review,* various issues.
US Department of Labor, *Producer Prices and Price Indices,* various issues.

Item	Weight, %
Equipment, machinery and supports	61
Construction labor	22
Buildings, materials and labor	7
Engineering supervision and manpower	10

The first component "equipment, machinery and supports," is made up of the following subcomponents:

Item	Weight, %
Fabricated equipment	37
Process machinery	14
Pipe, valves, and fittings	20
Process instruments and controls	7
Pumps and compressors	7
Electrical equipment and materials	5
Structural supports, insulation, paint	10

The CE Plant Cost Index is published in *Chemical Engineering* magazine.

Nelson Refinery Construction Cost Index (NR): The NR index is used primarily for cost estimates in the petroleum industry. The total index is based on material and labor in the following percentages: 20% iron and steel, 8% building material, 12% miscellaneous equipment, 60% labor (65% of which is skilled and 35% common). The NR index is reported quarterly in *Oil and Gas Journal*, which is published weekly.

Another index which often comes into play when negotiating sales or contract agreements for mining projects is the Consumer Price Index (CPI). The CPI measures the average change in prices over time of a fixed basket of goods and services. As such, the index is associated with consumer expenditure patterns as opposed to producer patterns discussed previously. Typically the CPI is used to adjust for inflation in labor rates, sales contracts, and other expenditures influenced by retail prices. It is discussed in this capacity in more detail in Chapter 10.

With the variety of indices available the estimator has a wealth of information to update a cost estimate from some previous period. The following example illustrates the use of cost indices.

Example 5: Suppose a 100-tph crusher was purchased in 1974 for $50,000. What would the expected price be in 1982 for the same basic crusher using the M&S mining and milling equipment cost index?

Solution:
Using the cost indices from Table 11, the following is obtained.

$$\text{cost (now)} = \text{(cost then)} \times \frac{\text{(cost index now)}}{\text{(cost index then)}}$$

$$= \$50,000 \left(\frac{411.3}{149.6} \right) = \$137,500$$

It is important to remember that most cost indices represent a composite of different factors having different relative weighting. The estimator must be aware of the components in each index and how the index is calculated. Only then can the proper index be selected for the estimate in question.

A few words of caution are necessary regarding the application of cost indices. Before selecting an index, it is important to ask the following questions:

1) What materials or items are included? How do these correspond to the cost data that will be updated?

2) What weighting method was used to construct the index or is the index an arithmetic mean?

3) What is the index intended to represent?

4) Do the prices used in the index reflect transaction or list price?

5) Are changes in the "mix" of commodities reflected in the index?

6) What changes are needed to account for changes in productivity of the labor and engineering components?

7) Are varying base periods used for different indexes?

8) Does the index reflect geographical areas?

It is also important to remember that cost indices measure historical cost trends. As such they may or may not be suitable for forecasting future costs. Projections of future cost index values may be made using any number of mathematical techniques. However, the calculation of future costs for investment projects based on projections of cost indices is a risky business.

Cost indices are very general and should be used with care. Many estimators choose to adjust the indices by adding a contingency factor of 5–25% for local variations in labor supply, transport facilities, materials availability, secondary equipment markets, competition among contractors, etc. The accuracy associated with updating costs via cost indices may be approximately ± 10–20% for periods not exceeding five years. The accuracy diminishes rapidly after five years and after ten years the use of cost indices should probably be abandoned in favor of the regeneration of an actual rigorous cost estimate. The accuracy of cost estimates is also improved if cost indices are applied to specific project cost components as compared with just one overall project cost. As in all cost-estimating techniques, the accuracy of the estimate is directly related to the amount of detail which can be provided about the project.

SYSTEMS APPROACH

Most organizations have developed formats, flow charts, or other formal procedures for determining capital and operating cost estimates for new projects. Unfortunately, most of these approaches to the problem are rather fragmented in that they only address major cost components associated with the project, and then in differing degrees of detail. Few of these estimating procedures address the cost estimating problem on a total systems basis. Indeed such a coordinated systems approach to the problem is difficult and requires considerable effort to develop.

MAS Approach

One such system approach is that used by *US Bureau of Mines Minerals Availability System (MAS)*. MAS is an activity of the USBM minerals intelligence division designed to conduct and maintain an inventory of minerals important to the nation. USBM contracted for the development of a method for preparation of order-of-magnitude-type estimates for capital and operating costs of mining and primary beneficiation of various types of mineral occurrences (except fossil fuels) in the US and Canada.

The resulting cost estimating handbook was developed for a user with knowledge and experience in both mining and estimating procedures. The curves, tables, and factors reflected in the estimating procedure represent average values; however, the estimator does have the capability to adjust the estimates by various factors if he or she is able to discern any differences between the estimating assumptions and actual deposit or cost characteristics. Various cost items can also be adjusted from base values used in establishing the estimating curves to current values by the use of cost indices.

Before an estimate can be obtained covering any phase of mining or beneficiation, the estimator should compile certain basic data in the following five areas of interest:

1) *General*
   Location
   Topography
   Climate
   Access to facilities and labor market
   Prevailing labor costs (including payroll burden)
   Daily or annual tonnage
   Applicable cost indices
   Electric power costs
   Transportation availability
2) *Surface and Underground Mines*
   Rock type and hardness
   Support or ground conditions
   Overburden (surface mines)
   Extraction and/or mining method
3) *Beneficiation*
   Crushability and/or grindability
   Extraction method (flow chart is useful)
4) *Exploration*
   Methods to be used
   Area (coverage)
5) *Capital Cost*
   Extent and method of preproduction development
   Freight rates
   Taxes

The actual estimating procedure is based on curves and equations depicting specific relationships. The estimator can work from either the curves or the supplied equations. Fig. 5 and accompanying descriptive material illustrate the cost estimating procedure for drilling and blasting overburden material. In the equation shown in the figure the value of x is for the horizontal scale and the value of Y is in dollars. The Y subscripts L, S, and E indicate labor, supplies, and equipment operation, respectively, for the cost component.

Perhaps the most significant advantage of this cost estimating manual or handbook is the fact that it treats the cost estimation problem on a total systems basis. It provides a consistent, straight-forward approach to estimating costs for the following general categories:

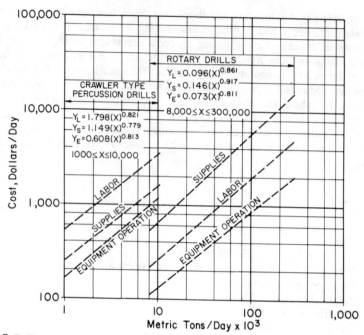

Fig. 5. Drill and blast, overburden and waste (STRAAM Engineers Inc., 1979). Scale: Logarithmic. The curves have been developed in two parts. The following distribution applies to equipment in this section.

Description	General Equipment Operating Cost Component Distribution		
	Repair parts	Fuel & lube	Tires
Drill equipment	50%	50%	
Trucks	32%	62%	6%

For mines excavating from 1000 to 10 000 t of overburden and waste per day, the curves reflect costs for drilling 6-m high benches with crawler-type percussion drills. Spacing of 2.5-in. holes is on a pattern of 1.5 x 2 m to a depth of 7 m. The powder factor is 0.30 kg/t.

For mines excavating from 8000 to 300 000 t of overburden and waste per day, drilling is performed with rotary drills having a down pressure of from 13,600 to 40,900 kg. The powder factor varies from 0.11 to 0.20 kg with an average of 0.14 kg/t of waste. Holes drilled average 12-¼ in. diam from a range of 6 to 13-¾ in. diam. Bench heights are 12 to 18 m, averaging 15 m. Drilling patterns and over-drilling varies with a range of 100 to 300 t of blasted material per linear meter of drill hole. Secondary drilling and blasting varies from 0 to 10% of blasted material.

The curves indicate average costs for a wide range of materials as can be noted by drill sizes given, bit sizes, powder factors, and drill pattern. To determine drilling and blasting costs consideration must be given to material hardness, abrasiveness, natural fractures and jointing, and maximum size fragments that can be loaded, hauled, and processed. Where the foregoing conditions are unfavorable, the costs shown on the curves can be increased up to 100%. For favorable conditions, the costs can be reduced up to 40%.

The labor cost for a typical rotary drill operation is based on an average labor rate for drill and blast crew of $8.00 per man-hour including rotary drill operators at $8.34. The labor cost for a typical percussion drill operation is based on an average labor rate for drill and blast crew of $7.91 per man-hour including percusssion drill operators at $8.17.

The supply costs for both curves include drill bits and steel related items at 24% of the total with the remaining 76% for blasting supplies.

Equipment operating costs include drills 75% and supporting equipment 25%.

I) Capital Costs
- A) Exploration and access roads
- B) Surface and underground mining
  - (1) Preproduction development
  - (2) Mine plant and buildings
  - (3) Townsite
  - (4) Restoration during construction
  - (5) Mine equipment
  - (6) Engineering and construction management fees
  - (7) Working capital
- C) Beneficiation
  - (1) Crushing
  - (2) Grinding
  - (3) Concentrating
  - (4) Waste and tailings disposal
  - (5) Site preparation
  - (6) Utilities and facilities
  - (7) Townsite
  - (8) Restoration during construction
  - (9) Engineering and construction management fee
  - (10) Working capital

II) Operating Costs
- A. Surface mining
  - (1) Production development
  - (2) Mining of ore
  - (3) Restoration during production
  - (4) General operations
- B. Underground mining
  - (1) Production development
  - (2) Mining of ore (by methods)
  - (3) Haulage of ore
  - (4) General operations
- C. Beneficiation
  - (1) Crushing
  - (2) Grinding
  - (3) Concentrating (by methods)
  - (4) Waste and tailings disposal
  - (5) Restoration during production
  - (6) General operations

III) Administration Costs
- A) Surface mining
  - (1) General expense
- B) Underground mining
  - (1) General expense
- C) Beneficiation
  - (1) General expense

Rather detailed cost breakdowns are utilized in this handbook and various factors can be adjusted to represent more closely the property being considered. Because the assumptions and cost component factors used in the development of each cost curve (equation) are provided, the estimation procedure can be adjusted for different physical parameters and inflationary trends rather easily. This is particularly important for any cost estimating method if any degree of longevity is desired.

It must be noted that this cost estimating approach *should not* be used to determine the cost of any single component of a mining or beneficiation system. This systems approach was not designed to determine component costs on a stand-alone basis; rather, it was designed to calculate these component costs such that they were properly incorporated into the total system cost calculation for a mining or beneficiation operation. Some users find this a limitation for day-to-day usage.

Some users also find it difficult to work with the cost components of labor, supplies, and equipment operation typically used in the handbook. In many cases these cost categories do not coincide with company cost accounting systems (for each cost category), nor do they coincide with the structure of many standard cost estimating formats typically used by estimators. Also, none of the curves or equations in this handbook has allowances for property and/or inventory taxes, general insurance or depreciation.

The manual is based on July 1975 cost values. Therefore, cost estimates must be adjusted for inflationary trends. This is accomplished by using cost indices for the various key cost items. As with any estimating technique, the curves and equations in this handbook require continued checking and updating to ensure that the basic cost relationships remain valid.

O'Hara Approach

One of the most complete factored-capacity cost estimating methods for mining and processing projects was compiled and published by O'Hara (1980). Capital and operating costs (including administrative and general services) for open pit mines, underground mines, and processing plants are estimated by developing relationships based on relatively few variables. The estimating technique is based on cost data from Canadian and foreign operations having similar or comparable unit costs. The cost relationships presented are in terms of 1978 Canadian dollars.

The relationships between mine project requirements and costs relative to plant capacity, operating costs of supplies and labor relative to daily tonnage mined or milled, and operating performance and mill recovery with respect to mill-head grade were determined by fitting data to an equation having the general form

$$Q = K T^x$$

where Q is quantities or costs and T is a variable causing changes in quantities or costs (tonnage rate, millhead grade, stope width, or some other physical factor).

The x values were determined to yield the lowest range of variation in K values across the widest range of T values for the reliable data available. Fig. 6 illustrates the approach for the cost of pit equipment for various mine sizes. The figure shows that the two major items in pit equipment cost are the fleet of shovels, each of which costs about \$230,000 $S^{0.73}$ (S is shovel size in cubic yards), and the fleet of trucks, each of which costs about \$9,000 $t^{0.85}$ (t is truck size in tons).

Capital cost estimates are related to sizing of mining equipment, mine development, plant-site topography, climate and accessibility, plant services, and personnel housing. Some of the cost estimating relationships developed for capital cost components are:

A. Open Pit Mines
   1. Preproduction stripping
      cost = $8,500 $To^{0.5}$ (rock)
where To is tons of overburden (millions).
   2. Pit equipment
      Total cost = $6,000 $Tp^{0.7}$ + $5,000 $Tp^{0.5}$ (truck/shovel operation)
where Tp is tons of ore and waste mined daily.
B. Underground mines
   1. Development
   cost = $\dfrac{\$40,000\,T}{W^{0.8}}$
where T is tons of ore mined per day and W is the average ore width.
   2. Equipment and installation cost = $30,000 $T^{0.6}$
where T is tons mined per day.
C. Process Plants
   1. Crushing plant, coarse ore storage and conveyors, cost = $45,000 $T^{0.5}$
where T is tons of ore milled daily.
   2. Concentrator building
      cost = $30,000 $T^{0.5}$ Fcl
where: Fcl = 1.0, mild climate; 1.8, cold climate; 2.5, severe climate.

Operating cost relationships at different daily tonnages were developed in a similar form and were found to be related to ore-body shape, mining method,

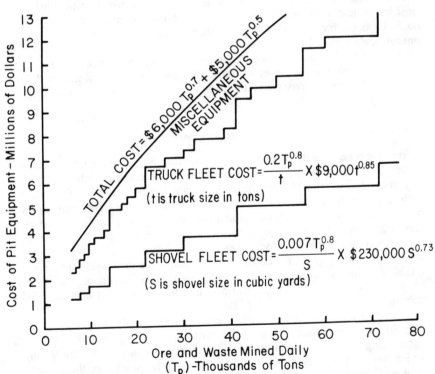

Fig. 6. Cost of pit equipment for various mine sizes (O'Hara, 1980).

milling process, and general plant services. The chief variable affecting all items of capital cost and operating cost was the daily tonnage of ore that is either mined or treated by the process plant. This greatly facilitates estimating procedures and reduces the size of the error while calculating the cost estimates.

One of the advantages of the relationships developed by O'Hara is that the overall project capital and operating costs can be made from a summation of cost items after judging the effect of specific local conditions on each item of capital and operating cost. These specific conditions or factors can be assessed from knowledge of the local topography, climate or accessibility, or from drilling results and tests on core samples, and incorporated into the cost estimate by virtue of the different relationships which have been developed. Also, operating costs consist of the standard subcomponents of labor, supplies, and administration and general services.

Like the USBM cost estimating handbook, the cost relationships developed by O'Hara offer a systems approach to estimating project costs, although certainly not in the same detail. The relationships offered by O'Hara are easier to use by the estimator than the intricacies of the USBM approach, but they also provide a rougher, or more general, cost estimate. These two techniques are really intended for two different levels of costs estimates. One major advantage of the O'Hara relationships, however, is that they can be used on a stand-alone basis for estimating certain cost components for projects in a general sense.

Obviously cost estimating relationships such as those developed by O'Hara should be used in conjunction with order-of-magnitude and/or early preliminary cost estimates. If a feasibility study using these types of cost estimating relationships indicates the project is viable, then a more detailed feasibility study should be performed utilizing more precise cost estimates before any financial commitments are made.

It is apparent that these types of relationships must be periodically checked and updated for accuracy. Nonetheless, these relationships represent the most complete factored-capacity cost estimating methods currently available for mining and processing projects.

SUMMARY

From the preceding discussions it is obvious that project cost estimates are only as accurate as the input data. As more definitive data become available on a project they should be incorporated into the cost estimate in order to enhance the accuracy.

Because feasibility studies are such a vital part of the investment decision process and because estimates of capital and operating costs are so crucial to the feasibility study, these estimates must be consistent with the purpose of the study. Certain recognized methods for cost estimating are accepted for specific types or levels of feasibility studies. The various cost estimating techniques should not be mixed in an analysis unless warranted by the data. The cost estimator must always keep in mind the reason for the estimate and the information it is intended to convey.

It is important to remember that cost estimates are simply that; they are simply estimates of what project capital and operating costs may be, based on numerous engineering and other variables. As such there is no one estimating technique which always gives the best answer, when best is defined as the percentage comparison between estimated and actual project costs. A good cost estimating method does not have to be a sophisticated one. A good technique is one that works within the required degree of accuracy.

Once the feasibility study reaches the definitive stage, the cost estimator must face the fact that there is no easy way out, and considerable time and effort must be devoted to preparing an appropriate cost estimate. These and future estimates will require manning tables, flow charts, design and layout drawings, and equipment lists including specifications. These cost estimates should incorporate data relative to unit operations, job functions, job requirements, timetables, etc. This task is perhaps even more difficult when estimating mining projects because of the uniqueness among ore deposits, mining options available, tonnage-grade relationships, production rates, and other operating parameters.

BIBLIOGRAPHY

American Association of Cost Engineers, 1980, *Transactions*, 24th Annual Meeting, Washington, D.C., AACE, Morgantown, WV.

Arnold, T.H., Jr., and Chilton, C.H., 1970, "New Index Shows Plant Cost Trends," *Modern Cost-Engineering Techniques*, edited by H. Popper, ed., McGraw Hill, New York, pp. 11–20.

Bettinger, D.S., 1978, *Selections From Transactions of the American Association of Cost Engineers*, Vol. 2, 1975–1978, AACE, Morgantown, WV.

Dubois, R., 1980, "A Study of Cost Indices," *Transactions*, 24th Annual Meeting, AACE, Morgantown, WV, pp. G–3, G–4.

Gallagher, J.T., 1970, "Rapid Estimation of Plant Costs," *Modern Cost-Engineering Techniques*, H. Popper, ed., McGraw Hill, New York, pp. 3–10.

Gentry, D.W., 1976, "Development of Deep Mining Techniques," *World Mineral Supplies—Assessment and Perspective*, Chap. 13, Elsevier Scientific Publishing Co., New York.

Gentry, D.W., 1979, "Mine Valuation: Technical Overview," *Computer Methods for the 80's In the Mineral Industry*, Alfred Weiss, ed., AIME, New York, pp. 520–535.

Gessel, R.C., *The Basic Principles of Estimating*, WABCO Construction and Mining Equipment Division, Peoria, IL, 23 pp.

Groggan, J., 1980, "Mass Property Valuation Using Cost Indexes," *Transactions*, 24th Annual Meeting, AACE, Morgantown, WV, pp. G–6, G–6.7.

Hackney, J.W., 1970a, "Capital Cost Estimates for Process Industries," *Modern Cost-Engineering Techniques*, H. Popper, ed., McGraw Hill, New York, pp. 43–58.

Hackney, J.W., 1970b, "Estimating Methods for Process Industry Capital Costs," *Modern Cost-Engineering Techniques*, H. Popper, ed., McGraw Hill, New York, pp. 43–58.

Haselbarth, J.E., 1970, "Updated Investment Costs for 60 Types of Chemical Plants," *Modern Cost-Engineering Techniques*, H. Popper, ed., McGraw Hill, New York, pp. 68–70.

Hoskins, J.R., and Green, W.R., eds., 1977, *Mineral Industry Costs*, Northwest Mining Association, Spokane, WA, 226 pp.

Jarpa, S.G., 1976, "Capital Investment and Operating Cost Estimation in Open Pit Mining," *14th APCOM Symposium*, R.V. Ramani, ed., AIME, New York, pp. 920–931.

Jelen, F.C., ed., 1970, *Cost and Optimization Engineering*, McGraw-Hill, New York, 490 pp.

Jelen, F.C., ed., 1979, *Project and Cost Engineers' Handbook*, American Association of Cost Engineers, Morgantown, WV, 260 pp.

Lawrence, B.W., 1977, "Underground Mine Development Cost Estimating," *Mineral Industry Costs*, J.R. Hoskins and W.R. Green, eds., Northwest Mining Association, Spokane, WA, pp. 119–146.

Martin Consultants, 1981, *Guidelines for Front-End Loaders Load-and-Carry Application*, Vol. 2, Sec. 6, prepared for US Bureau of Mines (Contract No. J0205045), May.

Miller, C.A., 1970, "New Cost Factors Give Quick, Accurate Estimate," *Modern Cost-Engineering Techniques*, H. Popper, ed., McGraw Hill, New York, pp. 59–67.

Mular, A.L., and Bhappu, R.B., eds., 1978, *Mineral Processing Plant Design*, AIME, New York, 883 pp.

Mular, A.L., 1978a, *Mineral Processing Equipment Costs and Preliminary Capital Cost Estimations*, Special Vol. 18, Canadian Institute of Mining and Metallurgy, Montreal, 166 pp.

Mular, A.L., 1978b, "The Estimation of Preliminary Capital Costs," *Mineral Processing Plant Design*, A.L. Mular and R.B. Bhappu, eds., AIME, New York, pp. 52–70.

Norden, R.B., 1970, "CE Cost Indexes: A Sharp Rise Since 1965," *Modern Cost-Engineering Techniques*, H. Popper, ed., McGraw Hill, New York, pp. 21–24.

O'Driscoll, J.J., 1979, "Cost Indices and Escalation," *Project and Cost Engineers' Handbook*, F.C. Jelen, ed., American Association of Cost Engineers, Morgantown, WV, pp. 7-1–7-17.

O'Hara, T.A., 1980, "Quick Guidelines to the Evaluation of Orebodies," *CIM Bulletin*, February, pp. 87–99.

O'Neil, T.J., 1979, "Procedures for the Preliminary Financial Evaluation of Metal Mining Ventures," *Computer Methods for the 80's in the Minerals Industry*, Alfred Weiss, ed., AIME, New York, pp. 556–580.

Ostwald, P.F., 1974a, *Cost Estimating for Engineering and Management*, W.J. Fabrycky and J.H. Mize, eds., Prentice-Hall, Inc., Englewood Cliffs, NJ, 493 pp.

Ostwald, P.F., 1974b, "Detailed Methods," *Cost Estimating for Engineering and Management*, Chap. 7, W.J. Fabrycky and J.H. Mize,eds., Prentice-Hall, Inc., Englewood Cliffs, NJ, pp. 195–196.

Ostwald, P.F., 1974c, "Forecasting," *Cost Estimating for Engineering and Management*, Chap. 5, W.J. Fabrycky and J.H. Mize, eds., Prentice-Hall, Inc., Englewood Cliffs, NJ, pp. 47–51.

Ostwald, P.F., 1974d, "Patterns of Cost Information," *Cost Estimating for Engineering and Management*, Chap. 3, W.J. Fabrycky and J.H. Mize, eds., Prentice-Hall, Inc., Englewood Cliffs, NJ, pp. 47–51.

Ostwald, P.F., 1974e, "Preliminary Methods," *Cost Estimating for Engineering and Management*, Chap. 6, W.J. Fabrycky and J.H. Mize, eds., Prentice-Hall, Inc., Englewood Cliffs, NJ, pp. 167–168.

Popper, H., ed., 1970, *Modern Cost-Engineering Techniques*, McGraw-Hill, New York, 538 pp.

Richardson, A.J., 1977, "Underground Stoping and Mining Costs," *Mineral Industry Costs*, J.R. Hoskins and W.R. Green, eds., Northwest Mining Association, Spokane, WA, pp. 147–162.

Soma, J.L., 1977, "Estimating the Cost of Development and Operating of Surface Mines," *Mineral Industry Costs*, J.R. Hoskins and W.R. Green, eds., Northwest Mining Association, Spokane, WA, pp. 67–90.

Stackhouse, E.E., 1979, "Cost Estimating," *Project and Cost Engineers' Handbook*, F.C. Jelen, ed., American Association of Cost Engineers, Morgantown, WV, pp. 3-1–3-19.

Stelly, W., and Wallace, A.J., 1979, "Cost Accounting," *Project and Cost Engineers' Handbook*, F.C. Jelen, ed., American Association of Cost Engineers, Morgantown, WV, 1979, pp. 2-1–2-19.

STRAAM Engineers, Inc., 1979, "Capital and Operating Cost Estimating System Handbook Mining and Beneficiation of Metallic and Nonmetallic Minerals Except Fossil Fuels in the United States and Canada," Contract No. J0255026, US Bureau of Mines, revised June, 384 pp.

Thuesen, H.G., Fabrycky, W.J., and Thuesen, G.J., 1977, *Engineering Economy*, 5th ed., Fabrycky, W.J. and Mize, J.H., eds., Prentice-Hall, Inc., Englewood Cliffs, NJ, 589 pp.

Weiss, A., ed., 1979, *Computer Methods for the 80's in the Mineral Industry*, AIME, New York, 975 pp.

Williamson, R.S., Browder, J.F., and Donnell, W.R., IV, 1980, "Estimate Contingency, Risk and Accuracy—What Do They Mean?," *Transactions*, 24th Annual Meeting, American Association of Cost Engineers, Morgantown, WV, pp. B–A.1–B–A.7.

7

The Minerals Depletion Allowance

It was a western truism that more money was made from selling mines than from buying them, just as it was accepted that many a good mine had been spoiled by working it.

from R. E. White, "The Mining Town,"
Alta California August 18, 1873

INTRODUCTION

One of the most important aspects of mineral enterprises of US corporations is the influence of the depletion allowance on cash flow. This allowance permits mineral producers to claim sizable federal income tax deductions not available to other industries, and these deductions significantly increase cash flows in mining.

In the past several years the minerals depletion allowance has come under increasingly heavy fire from many sectors as a tax loophole that deprives the federal government of large amounts of revenue. One estimate of the loss of tax dollars in 1968 was $2.25 billion due to depletion deductions and intangible expenses in the oil industry. Largely on the basis of such arguments, the Tax Reform Act of 1969 reduced percentage depletion rates for a number of minerals, and the Tax Reduction Act of 1975 totally eliminated statutory depletion for most oil and gas. In regard to percentage depletion for hard minerals, the federal government has estimated its resulting loss of revenue at $1.75 billion for 1980. The TEFRA legislation of 1982 further reduced the statutory depletion allowance available to iron ore, coal, and lignite beginning in 1984. Thus, it is clear that minerals engineers and managers should have a thorough knowledge of the origin, evolution, justification, and computational procedures relating to the depletion allowance.

ORIGIN AND EVOLUTION OF THE DEPLETION ALLOWANCE

While two temporary federal income tax acts were passed prior to 1900, the Corporate Franchise Act of 1909 marks the beginning of continuous corporate income taxation in this country. This act levied a corporation franchise tax of 1% on net income. Definitions of gross income, net income, allowable deductions and other critical items were not clear, however, which resulted in considerable confusion.

A group of metal miners were among the first to challenge the Act, contending that they were entitled, on constitutional grounds, to a tax deduction to recover their investments in mining properties (i.e., a depletion allowance). The Corporate Franchise Act of 1909 carried no such provision. The miners argued that if they were unable to recover their investments solely due to federal taxation, this would amount to unconstitutional confiscation. Although the Supreme Court ruled against the miners at that time, the first general Revenue Act of 1913, and each of the over 25 federal income tax acts since then, have contained provisions for depletion of wasting assets.

The Revenue Act of 1913 limited annual depletion deductions to 5% of the gross value of the minerals produced at the mine site. This was modified by the Revenue Act of 1916 to permit depletion charges up to the larger of (1) the capital invested in the property or (2) its fair market value on Mar. 1, 1913, the date of the first income tax act.

A major change was included in the Revenue Act of 1918 by permitting discovery *value*, rather than discovery *cost*, as the limit to depletion deductions. This act established the fair market price of new discoveries as the upper limit on total depletion deductions. While many arguments were offered in support of this change, the most defensible is that past discovery costs are frequently poor indicators of what future discovery costs will be when the original deposit is exhausted and a replacement is needed. This cost escalation is due to the disappearance of high grade and/or easily located deposits, and is entirely separate from the inflationary cost increases that prevent any industry from recovering replacement costs through depreciation. In this 1918 act, then, the right of (and justification for) permitting depletion recoveries in excess of capital costs first appears.

To eliminate alleged tax loopholes, the Revenue Act of 1921 limited discovery value depletion to the net income from the property, and a similar act in 1924 further restricted deductions to 50% of net income from the property.

As might be suspected, discovery value proved to be a very elusive quantity as anyone who works in the area of mineral property valuation can testify. Therefore, bearing the standard of computational simplicity, percentage depletion first made its appearance in the Revenue Act of 1926. At that time, the 27 1/2% depletion allowance was accorded to oil and gas producers, whereby annual depletion charges to oil and gas producers were limited to 27 1/2% of gross income or 50% of pretax net income calculated without depletion, whichever is smaller. Discovery value depletion remained applicable to mines until 1932.

In determining the percentage depletion rate for oil and gas, a comprehensive study was made of depletion charges granted to these producers in the years immediately preceding 1926. It was concluded that depletion deductions had been averaging 32% of gross income from a barrel of crude. With this guideline, the Senate recommended a 30% depletion rate, but subsequently compromised with the House, which had recommended a 25% rate.

Percentage depletion was first extended to mining in 1932 at the following rates:

Metal mines	15%
Sulfur	23%
Coal	5%

All deductions were (and still are) limited to 50% of net income before depletion. Note also that the number of mineral commodities named in the 1932 act was quite limited. Since that time, the list of minerals covered by depletion legislation has grown considerably, particularly during and shortly after World War II when the production incentive argument for national security was particularly strong. Table 1 lists percentage depletion rates for many common minerals for both domestic and foreign production. Since 1954, percentage depletion has been applicable to all minerals except those derived from inexhaustible sources such as air.

Table 1. Percentage Depletion Rates*

Mineral	Percentage rate	
	Domestic	Foreign
Antimony	22	14
Arsenic	14	14
Asbestos	22	10
Barite	14	14
Bauxite	22	14
Beryllium	22	14
Bismuth	22	14
Boron (borax)	14	14
Bromine (brine wells)	5	5
Cadmium	22	14
Cesium	14	14
Chromium	22	14
Clays:		
Kaolin, ball clay bentonite, fuller's earth, and fire clay	14	14
Clay and shale for sewer pipe or brick, and lightweight aggregates	7.5	7.5
Clay for alumina or aluminum compounds	22	14
Clay for drainage and roofing tile, etc.	5	5
Coal	10	10
Cobalt	22	14
Columbium	22	14
Copper	15	14
Corundum	22	14
Diamond (industrial)	14	14
Diatomite	14	14
Feldspar	14	14
Fluorspar	22	14
Garnet	14	14
Gemstones	14	14
Germanium	14	14
Gold	15	14
Graphite	22	14
Gypsum	14	14
Hafnium	22	14
Ilmenite	14	14
Indium	14	14
Iodine	14	14
Iron ore	15	14
Kyanite	22	14
Lead	22	14

Lignite	10	10
Lithium	22	14
Magnesium and magnesium compounds:		
Brucite	10	10
Dolomite and magnesium carbonate	14	14
Magnesium chloride	5	5
Olivine	22	14
Manganese	22	14
Mercury	22	14
Mica:		
Scrap and flake	22	14
Sheet	22	14
Molybdenum	22	14
Nickel	22	14
Oil Shale	15	15
Peat	5	5
Perlite	10	10
Phosphate rock	14	14
Platinum-group metals	22	14
Potash	14	14
Pumice	5	5
Quartz crystal, electronic grade	22	14
Rare-earth minerals and yttrium:		
Monazite	22	14
Other	14	14
Rhenium	14	14
Rubidium	14	14
Rutile	22	14
Salt (sodium chloride)	10	10
Sand and gravel:		
Common varieties	5	5
Quartz sand or pebbles	14	14
Selenium	14	14
Silicon:		
Quartzite	14	14
Gravel	5	5
Silver	15	14
Sodium carbonate	14	14
Sodium sulfate	14	14
Stone:		
For riprap, ballast, road material, etc.	5	5
Other	14	14
Strontium	22	14
Sulfur	22	22
Talc:		
Block steatite	22	14
Other	14	14
Tantalum	22	14
Tellurium	14	14
Thallium	14	14
Thorium	22	14
Tin	22	14

Tungsten	22	14
Uranium	22	22
Vanadium	22	14
Vermiculite	14	14
Zinc	22	14
Zirconium	22	14

*Adapted from Yasnowski and Graham, 1980.

Critics of the depletion allowance mounted a substantial assault with the Tax Reduction Act of 1969 and with the Tax Reduction Act of 1975. The most notable change was the reduction to 22% (in 1969) for oil and gas, and its elimination for major producers altogether in 1975—ironically, at the peak of the OPEC oil embargo. In 1969, sulfur, uranium, domestically produced molybdenum, lead, zinc, and a host of minor metals went from 23% to 22%. Copper, gold, silver, and iron remained at 15% except for foreign production, which dropped to 14%. Coal remained at 10% while many industrial minerals dropped from 15 to 7 1/2%. Among commodities of particular current interest are oil shale (15%), and certain geothermal steam production (see Breeding, Burke, and Burton, 1977), which is being gradually reduced from 22 to 15% by 1984. In 1982, TEFRA reduced percentage depletion allowances for coal, iron ore, and lignite by 15% of the excess of percentage depletion over the adjusted basis of the property. This change was to take effect in 1984. All depletion reductions are still limited to 50% of net income.

CALCULATING THE DEPLETION ALLOWANCE

At the present time the Internal Revenue Service (IRS) recognizes two types of depletion allowance calculations: (1) cost or unit depletion, and (2) statutory or percentage depletion. Each taxpayer must calculate both types and must use the *larger* of the two for tax purposes.

Cost or Unit Depletion

The unit or cost depletion method involves prorating the amount capitalized into the depletion account—the cost depletion basis—against the number of production units in the mineral deposit. The computations are analogous to unit-of-production depreciation calculations. The depletion basis is composed of acquisition costs and associated fees plus capitalized exploration expense.

To illustrate the cost depletion calculation consider a copper deposit estimated to contain 4.5 million tons of ore. Assume further that the purchase price for the property was $1,740,000, and that $600,000 in exploration costs was capitalized, for a total of $2,340,000 in the depletion account.

$$\text{Unit depletion rate} = \frac{\text{depletion basis}}{\text{reserve tonnage}}$$

$$= \frac{\$2,340,000}{4,500,000} = \$0.52 \text{ per ton.}$$

If 600,000 tons were mined in the year 1, allowable unit depletion would be:

$$600,000 \times \$0.52 = \$312,000$$

adjusted depletion basis, end of year 1 = 2,340,000 − $312,000 = $2,028,000

Now suppose market conditions improve causing a rise in the copper price such that ore reserves increase from 4,500,000 − 600,000 = 3,900,000 tons, to 4,300,000 tons. Now the unit depletion rate for the second year's operation is:

$$\frac{\$2,028,000}{4,300,000} = \$0.472 \text{ per ton.}$$

It is also possible that the depletion basis might increase if, for example, adjacent mineral-bearing lands are acquired and placed in production.

Unit depletion seldom results in a higher tax deduction than percentage depletion since the unit of depletion method limits total deductions to capitalized expenditures in the depletion account. The situation illustrated previously where ore reserves increase over time is a very common one. This is due to (1) increased prices for the mineral (2) and the completion of more detailed development drilling that enables the company to classify as ore material that was previously in the probable ore or possible ore categories. Consequently, as demonstrated, the unit depletion charge will decline whenever upward adjustments of ore reserves are made, thus making percentage depletion even more attractive. However, cost depletion, unlike percentage depletion, can be claimed whenever product is sold, regardless of whether or not any profit is earned.

Statutory or Percentage Depletion

As stated previously, the percentage depletion deduction is determined by selecting the smaller of (1) the product of the depletion rate for the mineral in question and gross income from mining, or (2) 50% of pretax net income calculated without a depletion deduction. The illustration presented in the following explains the percentage depletion calculation procedure in detail.

Consider the following simplified income statement for a mining operation:

	Gross revenue
less:	nonmining costs
	gross income from mining
less:	operating costs
	net operating income
less:	fixed charges
	net income before taxes & depletion
less:	depletion
	pretax net income
less:	federal income tax
	net profit

Using the preceding copper mining example assume the following revenues and costs for a domestic mining operation with no nonmining costs:

Gross income from mining	$10,800,000
Operating costs	6,600,000
	4,200,000
Fixed charges	1,900,000
Net income before tax and depletion	2,300,000

Percentage depletion is, then, the smaller of
a) Statutory percent of gross:
rate for domestic copper mining = 15%
$$0.15 \times 10,800,000 = \$1,620,000$$
b) 50% of predepletion net:
$$0.5 \times 2,300,000 = \underline{\$1,150,000}$$

The taxpayer then must use the larger of cost depletion and percentge depletion on the tax return. The entire calculation process is illustrated in a flowsheet in Fig. 1.

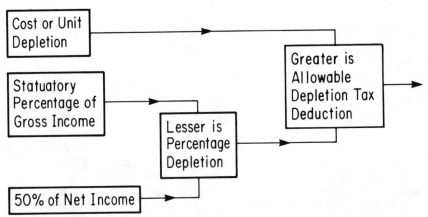

Fig. 1. Calculating the allowable depletion deduction for federal income tax purposes (adapted from Megill, 1971).

In those cases where one mine produces minerals having different depletion rates (e.g., copper and molybdenum), the gross income limitation is determined by applying the appropriate rates to the gross income from mining for each mineral. The sum thus obtained is compared to 50% of net determined on an aggregate basis to obtain the allowable percentage depletion.

The depletion basis is reduced annually by the amount of the depletion deduction claimed by the taxpayer. Whether that deduction results from the application of cost depletion or from percentage depletion is immaterial. As a consequence of this adjustment, the basis is eventually reduced to zero for most mines (the depletion basis cannot be negative), at which time no further unit depletion deductions can be claimed.

It is important to recognize, however, that the percentage depletion deduction is not constrained by the balance in the depletion account. This is an extension of the principle recognized in the Revenue Act of 1918 that discovery value, not discovery cost, is the appropriate limit to depletion deductions. Because the amount of the percentage depletion deduction is not affected by the basis, miners sometimes claim that property acquisition costs and successful exploration are not recoverable (i.e., the percentage depletion allowance is not affected by the level of these expenditures).

Finally, for financial accounting purposes in reports to shareholders, mining companies use unit depletion. This amount is generally different from unit depletion calculated for tax purposes (obviously both differ greatly from the amount of percentage depletion). This is because unit depletion for tax purposes is determined from the adjusted basis, and that basis may have been affected by *percentage* depletion deductions in previous years.

COMMON PROBLEMS

The minerals depletion allowance has, perhaps, resulted in more litigation than any other single item in mining tax law. It is impossible to cover this constantly changing topic in this chapter in any definitive sense, but some commonly occurring problems are discussed in a general manner. Solutions to specific problems obviously must be developed according to the specific circumstances surrounding each individual case.

Determining the Cutoff Point

In calculating percentage depletion, most problems arise in establishing the cutoff point for determining gross income from mining. Miners would like this item to be as high as possible since the percentage depletion rate is applied to this number. Further, many major metals producers in this country are fully integrated through to metal fabricating operations. Therefore, no income from external customers is generated until the ore is mined, crushed, concentrated, and smelted, and the resulting metal is refined and fabricated.

However, it was clearly not the intent of the law to permit depletion on manufacturing operations. Consequently, gross income from mining and the depletable costs are now defined by the courts to include costs and income derived solely from "mining" operations. Formerly, the cutoff point for depletion calculations was the "first marketable product," the form of which varied from mineral to mineral. This wording proved to be inadequate, and the IRS now lists specifically many processes which are deemed to be "mining," and others which are classified as "nonmining."

The allowable mining cost category includes: sorting, concentrating, sintering, loading, and substantially equivalent operations. However, for those minerals where there is substantial trade in crude ore, these costs must be deducted in determining gross income from mining. Processes which are designated as nonmining are electrolytic deposition, roasting, calcining, smelting, refining, and several other advanced processing or semi-fabrication steps (see Maxfield, 1975).

An additional complication arises if the ore must be transported farther than 50 miles to the plant where further "mining" processes are administered. If a distance less than 50 miles is involved, the IRS allows gross income from mining to include transportation costs. A special petition is required if this distance exceeds 50 miles.

Another potentially troublesome area concerns hydrometallurgical extraction processes where no marketable intermediate form is produced prior to the appearance of the refined metal. A good example of this is copper leach-electrowinning operations. Solvent extraction and electrolytic deposition are often considered not to be mining operations. However, solvent extraction is analogous to conventional concentration with sulfide copper ores, and the basic chemistry of electrowinning is the same as for precipitation, which is usually considered to be a mining process. This particular situation is still unresolved.

Allocation of Profit

Although the concentration phase marks the end of mining for depletion calculations in most metal mining operations, yet to be resolved is the problem of allocation of corporate profit. If the concentrates are sold to a custom smelter on an "arm's length" basis, no problems arise as the net smelter return (less royalty and rental payments, if any) is accepted as the gross income from mining. With vertically integrated firms, however, there are no conventional marketing transactions at the concentrate stage, so that another method must be used to partition total profit among the various production stages in order to derive gross income from mining.

The choice here has generally been between using a "representative field price" provided by the IRS or the "proportional profits" method described in the following.

The IRS has advocated allocating profit to production phases in proportion to their relative costs, but this "leads to the questionable conclusion that the profit from smelting concentrates from a low cost mine is greater than from smelting (identical) concentrates from a high cost mine" (Seery, 1971). Miners, who would like to assign as much profit as possible to mining to maximize depletion deductions, have advanced the rate of return concept where nonmining facilities would be allocated profit based on a fair rate of return on the capital invested in them. The IRS may accept methods other than "proportionate profits," but the burden of proof is on the taxpayer to show that the proposed method gives a better estimate of the gross income from mining.

Who is Entitled to Depletion?

Any taxpayer who has an *economic interest* in a property producing a mineral covered by depletion legislation may claim depletion deductions. Critics often cite this provision as one of the chief inequities of the mineral depletion allowance. The criticism focuses on the ability of property lessors to claim depletion on royalty income from mineral production. Frequently, the land owner paid little for the land and bears virtually none of the risk involved in bringing a new property into production, yet is entitled to claim depletion deductions to the fullest extent. Because lessors may claim depletion deductions on their royalty income, the operating lessee must deduct royalty payments from gross receipts when determining gross income from mining.

Example 1: Consider a copper mining company having a net smelter return of $51,530,000 per year. This would result in one of the percentage depletion limitations being:

$$\$51,530,000 \times 0.15 = \$7,729,500$$

However, if the mine were located on leased lands and were required to pay a 5% net smelter return royalty, the new gross income from mining would be:

$$\$51,530,000 - 0.05\,(51,530,000) = \$48,953,500$$

The new percentage depletion allowance would then be:

$$\$48,953,500 \times 0.15 = \$7,343,025$$

Therefore, at a 50% income tax rate, leasing costs the operator not only $2.58 million in royalty payments, but also possibly an additional ($7.73 million − 7.34) 0.5 = $193,000 in sacrificed tax benefits.

Aggregation of Properties

A provision of depletion legislation that permits mineral producers to maximize their depletion deduction is the aggregation of properties. Here, under certain conditions, a company may elect to combine two or more producing properties into one operating unit for the depletion calculation rather than determining individual deductions for each property. This may provide larger deductions since high cost mines that otherwise would have been restricted to 50% of net would then be able to borrow some unused depletion capacity from the company's more profitable operations.

Example 2: Aggregation of properties for depletion (from Seery, 1971):

	Deaggregated		Aggregated
	Mine A	Mine B	Mines A & B
Gross income	$100,000	$100,000	$200,000
Net income	40,000	10,000	50,000
Depletion			
15% of gross	15,000	15,000	30,000
50% of net	20,000	5,000	25,000
Depletion allowable	15,000	5,000	25,000
Taxable income	25,000	5,000	25,000
Tax @ 50%	12,500	2,500	12,500

It can be seen that it is never advantageous to deaggregate if all of the mines are operating at a profit.

To qualify for aggregation, the properties in question are subjected to several tests to verify that they are, in fact, operated as a single producing unit. Four criteria often used are:

1) Common field or operating personnel.
2) Common supply and maintenance facilities.
3) Common processing or treatment plants.
4) Common storage facilities.

Segments of the uranium industry in Colorado, Utah, and New Mexico have benefitted substantially from aggregation in the past.

Recapture of Exploration Deductions

Since 1969, the allowable depletion deduction in any tax year can be affected by the firm's prior treatment of exploration expenditures. Before 1969, mining firms could expense only a limited amount of exploration activity; the remainder was assigned to individual project accounts. If a project was abandoned the exploration costs in the account could be expensed, but exploration expenditures for successful projects were capitalized into the depletion basis.

Now, a mining company may elect to take all exploration expenditures as a current deduction, provided that if the project becomes a mine, the taxpayer must recapture these exploration deductions. The most common way to accomplish recapture is for the mine to forego any depletion allowance until the exploration expenditures are recaptured. This is consistent with the basic tenet of US tax law that successful exploration is not deductible. The details involved in computing recapture are covered in Chapter 8.

NON-US EXPERIENCE

For many years Canadian tax law contained many liberal provisions for the mining industry. One of the best known features of this law was the automatic 33 1/3% percentage depletion permitted to all mining firms. This percentage was applied to gross income with deductions permitted up to the full amount of the pretax, predepletion income. The depletion allowance and the three-year tax holiday were very helpful in building a strong stable mining industry in Canada and promoting the economic development of the vast northern regions of the country.

Recent changes in Canada's income tax philosophy, however, have greatly altered tax incentives for mining firms. The automatic 33 1/3% depletion rate has been discontinued in favor of *earned depletion*. In simplest terms, the concept of earned depletion permits depletion deductions only to the extent that a firm reinvests in exploration or productive capacity. Specifically, in Canada for each $3.00 of eligible expenditure, $1.00 of depletion will be permitted up to 25% of depletable income. Eligible expenses include a wide variety of items including capital investment, exploration and development expenditures, and the cost of acquiring mineral properties. See MacKenzie and Bilodeau (1979) for a more detailed description of Canadian mine taxation.

Although often named something other than minerals depletion allowance, the principle of exempting a portion of net mining income from taxation is recognized in many other countries. Legoux (1976) mentions such systems in Australia, South Africa, France, and the United Kingdom. Income tax regulations are constantly changing, but the basic idea embodied in the minerals depletion allowance has been adopted by many governments and is not unique to the United States.

IMPORTANCE OF THE DEPLETION ALLOWANCE

The minerals depletion allowance is extremely important to the US mining industry. During periods of normal profitability, percentage depletion often reduces the effective income tax rate for base metal, precious metal, and uranium mining to less than 30%. Table 2 shows the reduction in effective tax rates and corresponding tax savings for some major mining companies in 1978 and 1979. The early 1980's were generally years of poor earnings for US mining companies, so the effect of percentage depletion was much less significant.

Table 2. Impact of Percentage Depletion on Selected Mining Companies

Company	1978 Point reduction in tax rate	Tax savings, thousands	1979 Point reduction in tax rate	Tax savings, thousands
Hecla	14.0	$ 5,208	48.0	2,313
AMAX	21.3	92,630	30.6	62,050
Kennecott	14.9	24,778	0.0	0
Phelps Dodge	11.0	13,940	2.0	520
Homestake	16.6	14,255	18.9	8,008
ASARCO	5.9	18,690	10.5	7,538
Newmont	15.0	30,470	22.9	12,685
Pennzoil	10.3	37,559	8.1	15,683
		$237,350		$108,797

Because of these large tax savings it is not unusual for 20% of the cash flow from many mines to be directly attributable to the percentage depletion provision. However, because of the 50% of net limitation, percentage depletion affects operating mines differently than prospective mines.

Impact on Operating Mines

With percentage depletion restricted to 50% of predepletion net income, when there is not profit (i.e., no predepletion net income) there is no percentage depletion. In other words, the rescinding of percentage depletion would not cause any previously profitable mine to become unprofitable, nor would the granting of percentage depletion turn any existing mine that is losing money into a profitable venture.

One desirable effect that percentage depletion has with operating mines is to encourage more complete extraction of the resource. If additional investment is required to expand or to mine adjacent deposits, percentage depletion could provide the final increment of cash flow necessary to justify the investment. Generally speaking, however, although the loss of percentage depletion would make existing profitable mines less profitable, it would have little direct effect on mines already losing money.

The depletion allowance is often justified as an incentive needed to induce investment in the inherently risky mining business. However, it is clear that the greatest benefits from percentage depletion go to the most profitable mines, i.e., those that are least risky.

There are two responses relevant to the preceding observation. First, the important incentive effect of percentage depletion is not with existing mines but with prospective new mines. With new mines, percentage depletion can make a significant difference by (1) raising the rate of return on investment for a project from the marginal category into the acceptable range and (2) providing greater possible upside benefits so that investors will tolerate substantial downside risk and, therefore, proceed with the investment.

The second point to make here is that the incentive for risk argument is probably not the best justification for percentage depletion (see O'Neil, 1974). A more defensible position is that the minerals depletion allowance is analogous to the depreciation provision; that a mineral deposit is consumed in the production process in much the same manner as other capital assets. By recouping the value of the asset through noncash tax deductions the taxpayer can presumably afford a replacement when the original asset is used up.

However, whereas the investment cost and investment value are identical for most capital assets, they may be quite different for an ore deposit. Furthermore, it is the investment *value* which must be replaced when the deposit is depleted. Therefore, it can be argued that relating the depletion deduction to gross income from mining is a rational concept, even though the quantitative correlation is imperfect.

Impact on Prospective Mines

Although the existence of percentage depletion has little impact on the level of production in existing mines, it may have a profound effect on the mine investment decision with many prospective mines. At any given time a group of deposits exists for which projected cash flows are positive, but not high enought to warrant the associated investment risk. For many of these deposits, percentage depletion can provide the extra two- to four-point difference needed in the DCFROI (discounted cash flow-return on investment) to justify a decision to proceed.

A sudden discontinuation of percentage depletion would result in a major discontinuity in the mine investment cycle in the United States. A gradual phase-out of percentage depletion would make the impact less noticeable, but in either case investment in domestic new mine capacity would decline. Much of the reduced production would be exported to nations which could then offer substantial (although perhaps temporary) production cost advantages over the US.

SUMMARY

It might be inferred that the large loss of tax revenue to the federal government which has been ascribed to percentage depletion would have afforded high profitability in the mining industry. In fact, however, mining has had a mediocre earnings record for several years. This relationship, therefore, is not as clear as it might initially appear.

Percentage depletion raised profit rates at the time these deductions were initially conferred, which, in turn, attracted capital into minerals production to develop deposits that would otherwise have been submarginal. The net effect was modestly higher average returns for the industry although marginal rates of returns remained unchanged.

Subsequently, however, new mines have been developed whenever their projected net cash flows exceeded some minimum level. Whether or not that cash flow contained a depletion component was irrelevant to the investor. If the depletion allowance did not exist, investment in a given deposit might be delayed, but the deposit would still yield roughly the same rate of return upon development. Thus, once percentage depletion was implemented the major impact has been with the *timing* of mineral investments and not with the rate of return on such investments.

The elimination of percentage depletion would cause an important reduction in the level of domestic mineral investment for several years, requiring greater reliance on mineral imports. Increased federal tax revenues from existing mines would be at least partially offset by the loss of taxes from postponed new mines. It is very difficult to quantify this trade-off, but the annual cost to the public of maintaining a stronger minerals industry and more reliable sources of mineral raw materials appears to be far less than the $1.75 billion mentioned in the opening paragraph of this Chapter.

REFERENCES

Breeding, C.W., Burke, F.M., Jr., and Burton, A.G., 1977, *Income Taxation of Natural Resources*, Prentice-Hall, Inc., Englewood Cliffs, NJ.

Legoux, P. Ch. A., 1976, "Depletion Allowance," *Negotiation and Drafting of Mining Development Agreements*, Mining Journal Books, Ltd., London, pp. 34-42.

Mackenzie, B.W., and Bilodeau, M.L., 1979, "Effects of Taxation on Base Metal Mining in Canada," Centre for Resource Studies, Queen's University, Kingston, Ont., p. 190.

Maxfield, P. C., 1975, *The Income Taxation of Mining Operations*, Rocky Mountain Mineral Law Foundation, Boulder, CO, p. 365.

Megill, R.E., 1971, *An Introduction to Exploration Economics*, The Petroleum Publishing Co. Tulsa, OK, p. 159.

O'Neil, T.J., 1974, "The Minerals Depletion Allowance; Its Importance in Nonferrous Metal Mining," Parts I and II, *Mining Engineering*, Oct., Nov., 1974.

Seery, M.W., 1971, "Tax Planning for Exploration and Mining Projects," *Financial Analysis in the Mining Industry*, AIME, New York, NY, pp. H-1 to H-23.

Yasnowsky, N., and Graham, A.P., 1980, "Mineral Depletion Allowances and U.S. Import Dependence," Information Circular 8824, US Bureau of Mines, GPO, Washington, DC, p. 13.

8

Mine Taxation

"Who is the man who views the mines and promptly turns them down? Who is the one that thinks this is the short cut to renown? Who is it gives the bum advice to the innocent financier? The knowledge-feigning, theory-straining mining engineer."

—Anonymous

INTRODUCTION

Taxes levied against mining properties and operations are a critical cost in the economic evaluation of mining investments. Indeed, taxes represent a substantial cost of doing business in the minerals industry and often have a significant impact on corporate investment decisions. A good example was the postponement of mineral development in the state of Wisconsin, primarily because of what was perceived to be excessive taxes imposed by that state.

In many respects mining investments are no different from other industrial investments. Astute taxing authorities should recognize that geologic endowment is a necessary but clearly not a sufficient condition for mineral investments. The fact that a mineral deposit exists does not necessarily mean that it will ultimately be developed—a point that many taxing authorities fail to recognize. While it is true that ore deposits are not mobile in the sense that they cannot be physically moved to a district having more favorable taxes, corporate investment capital certainly is mobile and flows to ventures which maximize wealth to the firm. In short, higher taxes reduce project yields and tend to drive investment capital elsewhere.

The appropriate type and level of taxation imposed upon the mining industry continues to be a very controversial and emotion-charged topic. Mining activities have been taxed at various levels over the years due to widely differing taxation philosophies. Location has also influenced taxation policies, as evidenced by the diversity of tax laws and assessment procedures applied to mineral deposits by the various states. Indeed, mineral taxation varies from state to state, from county to county, and from one mineral commodity to another.

Whether or not a given tax is appropriate for a specific mineral deposit, or even an industry, is a difficult problem to assess. First, the type or kind of tax which should be imposed must be considered. Second, it is most important to define a fair and unambiguous base against which to levy the tax. Finally, one must consider the tax rate to be applied to this base. It is the combination of these two components which determines the level or magnitude of the resulting tax obligation. In addition the tax should be evaluated in the context of the entire tax program of which it is a part. Unless the entire tax package (federal, state, and local) is evaluated, it is difficult to assess the merits of a specific tax provision.

Mine taxation philosophies can often be traced to local attitudes toward mining activities during territorial times. These attitudes frequently led to mineral taxation policies which imposed excessive and unfair tax burdens on the minerals industry. The fundamental attitudes of proponents of heavy taxation on minerals can be summarized as follows:

1) The "natural heritage" theory is based on the concept that minerals are a free gift of nature for the benefit of all mankind, and therefore the benefits derived from resource extraction should be shared by all. Furthermore, since mineral deposits are exhaustible, society only has one opportunity to collect compensation for mining, and the mining company should pay handsomely for the depletion of the state's natural resources.

2) The "captive" theory assumes that since an ore body cannot be dismantled and moved to another location, any mine can be taxed with impunity. Further justification is offered for excessive taxation based on the fact that the corporate owner is typically an absentee owner that has little representation within the taxing jurisdiction. As pointed out earlier, this argument fails to recognize the mobility of investment capital.

Interest in mining taxation has increased recently as the result of a number of factors. Expanded environmental awareness has heightened criticism of some of the physical impacts from mining, and organized movements to oppose mineral development have strongly supported higher taxes to discourage such development. At the same time a shortage of some minerals and/or the significant increases in mineral prices (particularly fuels) have increased public awareness of the importance of minerals to our standard of living. As a result many individuals are having difficulty reconciling the need for expanded mineral production to maintain a given lifestyle and the goals and concomitant costs of preserving the environment. At the same time there is the ever-increasing need for revenue by taxing authorities to pay for growing social programs. As a result of these and other economic pressures there is a growing tendency to make mining operations prime candidates for increased taxation.

Certainly taxation policies, and particularly state taxation policies, can have a significant impact on mineral investment proposals as well as subsequent mine operating costs. Because taxes represent a substantial portion of total operating costs, the topic of mine taxation is discussed at some length in this chapter. However, the following discussion is in no way intended to represent a complete treatment of mineral-related taxation matters. Obviously such treatment is beyond the scope or intent of this book. Rather, the fundamental concepts, procedures, and effects of taxation policies are illustrated as they relate to mining and exploration activities. These are addressed from the standpoint of performing mine property valuations and do not describe all the peculiarities and unique procedures of actual tax accounting and corporate tax filings.

Those working in the area of mineral taxation and associated impacts on mine valuation appreciate the general level of uncertainty surrounding many of the taxation policies and statutes in existence. There are indeed many gray areas with respect to determining the "correct" procedure or method for handling a given situation in mineral tax matters. It is not possible in this chapter to anticipate and discuss all of the "what if" situations which may arise in performing mine valuations with multiple owners or carried interests, financial arrangements, income sharing, etc. When in doubt about handling a specific tax item in an economic evaluation, the analyst should consult the appropriate tax code or statute. If the "correct" procedure remains in doubt, the analyst should consult a tax lawyer either within the corporate legal staff or an individual specializing in corporate tax matters. Unfortunately, experience has shown that many of the really confusing issues cannot be easily resolved even by tax lawyers, because many of these issues have not been tested and resolved through litigation. When a decision is required, often the correct way to handle a given tax issue is to adopt a logical approach supported with substantial documentation which appears to be consistent with the intent of the tax statute. Even in situations where litigation has resulted in lower court rulings which are more favorable to taxpayers than the official position of the Internal Revenue Service, the Service refuses to officially recognize such practices until the interpretations are resolved by higher court decisions. Unfortunately those appeals can take years, whereas project evaluators must make decisions in a much more timely manner. As a result, most analysts tend to employ a conservative approach with respect to taxation matters when performing property valuations. This is the position advocated in this chapter.

Although federal tax statutes are relatively straightforward and rather well tested in most areas, the same cannot be said for state and local taxes. These taxes frequently are not based on sound theoretical or practical foundations and often exhibit a general lack of understanding of the fundamentals which apply to the minerals industry. These statutes are often drafted to address a specific situation—even a specific mine—without an understanding of the broader implications of the law.

To prepare a current list of state taxation procedures for a book of this type is not only difficult but also of limited value. Over the past five to ten years legislators at all levels have increasingly concerned themselves with tax "reform." As a result, state taxes applicable to the minerals industry may change in any given session of each state legislature. Because these taxes are so variable and often represent a significant uncontrollable cost to the mining operation, the project evaluator must make every effort to maintain current information on individual state tax statutes which affect mineral projects. This is not an easy task.

OBJECTIVES AND PRINCIPLES OF TAXATION

A tax may be defined as "any nonpenal yet compulsory transfer of resources from the private to the public sector levied on the basis of predetermined criteria and without reference to specific benefits received so as to accomplish some of the government's economic and social objectives" (Sommerfield, Anderson, and Brock, 1973). The key factors to remember with respect to taxes are the following:

1) They are an enforced (compulsory) contribution—usually monetary.
2) They are enacted by a power having taxing authority or power.
3) They are raised to provide revenue for public purposes.
4) They are a mechanism for diverting substantial wealth or income away from the control of the private sector to the control of the public sector.

Although taxes may be implemented for nonfiscal purposes (i.e., a mechanism of social control), tax programs are typically designed to generate revenue necessary to finance and support programs and services sponsored by the taxing authority. Presumably these programs and services are based on social, economic, and political objectives of the taxing authority involved (federal, state, or local government). The basic problem is the conversion of these various social, economic, and political goals into a meaningful tax policy which accomplishes the objectives in a reasonable and equitable manner.

Taxes are generally implemented to achieve one or more of the following objectives:

1) **Raising Revenue:** The purpose of most taxes is to provide revenue to governments without which they could not provide services such as public education, fire and police protection, public health and welfare, national defense, etc.

2) **Economic Development:** A major objective of taxation policy for most governments is to provide for economic development and full employment. The task of promoting economic growth can be divided into two basic components. First is the formation and attraction of capital. Certainly tax policies can effect the accumulation of capital in a given locality, effect overall savings or consumption of financial resources, and effect the attraction of capital to investment opportunities. The second component deals with the efficient use of all available resources (land, labor, and capital). Taxation policies can be designed to inhibit or promote the efficient use and development of natural resources, as well as influence efficient allocation of capital resources among competing uses.

Taxation, as a part of fiscal policy, is increasingly accepted as one of the more potent tools for achieving economic growth and full employment.

3) **Price Stability:** Price stability is another often-cited economic goal of most governments. If a government purchases considerable amounts of goods and services without taxing, total spending by the public and private sectors will soon generate considerable excess demand and, consequently, an inflationary bias in the economy. In order to maintain price stability, the government must either decrease its own or private spending. Although the government can employ monetary policy to help promote price stability, it can obviously also implement tax policies to decrease private spending. By controlling spending and taxing levels a government can either contribute to or inhibit inflation and effect price stability.

4) **Wealth Redistribution:** Taxation policies may be designed to effect a redistribution of wealth among individuals (private wealth) or from private to public control (presumably to achieve and maintain a desired level of economic health). The decision to redistribute wealth is generally predicated on a social value judgment or on the basis of an economic judgment which believes one pattern of wealth distribution is healthier than another. The latter is a public policy decision.

5) **Regulatory Medium:** Taxes may be imposed for nonfiscal purposes, such as to influence social behavior that is considered detrimental to society. For instance taxes may be placed on the purchase and consumption of specific goods, such as alcohol and tobacco. Alternatively, the objective may be to enforce or induce industry to perform in a certain manner or meet given objectives (e.g., minimize air and water pollution by placing taxes on waste emissions). Examples which relate specifically to the mining industry are the "reclamation fees" associated with the Surface Mining and Control Reclamation Act of 1977, and the Black Lung Revenue Act (March 1978). In fact some state governments have deliberately imposed extraordinary taxes—usually severance taxes—on minerals production to curtail mineral development in the state. These tax measures can be utilized to slow down development and provide the state an opportunity to consider the impacts and trade-offs associated with mining activities.

After a taxing authority has decided upon and prioritized its objectives and also decided to help accomplish these objectives through the imposition of a tax, it must then decide upon a tax base and a tax rate structure. The tax base is the foundation or basis upon which the tax is applied. Taxes are generally imposed on one of the following bases:

1) **Income:** Income implies the process of receiving wealth. Income taxes are typically applied to ordinary net income (revenues less costs of production) or to excess profits.

2) **Wealth:** This refers to actual ownership of wealth rather than a transaction in which wealth is exchanged. Property taxes on real and personal property are classic examples of taxes imposed on the wealth base.

3) **Expenditures:** A tax base may be associated with any transaction involving spending. Sales taxes are good examples of this type of tax base.

4) **Activity:** A tax base may be associated with a specific activity or profession. Such tax bases are often applied to the severance of natural resources and are sometimes referred to as a "privilege" tax. Typical taxes which fit in this category may take the form of transactions taxes, excise taxes, and franchise taxes.

SPECIAL FEATURES OF MINING

Every industry believes it has unique or distinguishing characteristics which make it different from other industries. In this regard mining is no different. Most will agree that responsible taxation of minerals must recognize the unique characteristics associated with the minerals industry, for these characteristics determine the economic impact of a specific tax policy on the operating and investment decision-making of individual firms. Some of the unique features of the mining industry as they relate to taxation policies were identified in Chapter 1 and are discussed briefly here.

Bonanza Image

The history of mining during the frontier era conjures up visions in the minds of legislators and their economic advisors of fabulously rich, high-grade ore bodies. As a result the natural inclination is that this windfall wealth and good fortune should be shared by the state through heavy taxation. Occasionally an ore discovery does bring back visions of the bonanzas of early Butte, Virginia City, and Tombstone. Unfortunately for everyone, these occurrences are indeed rare in today's mining industry.

Typically, today's ore bodies consist of low-grade mineralization, explored and delineated by high-cost drilling programs. Margins are smaller and satisfactory profits are generally only attainable by the investment of enormous sums of capital. Profit levels rarely achieve the lofty levels of the famous bonanzas of the past.

Even considering these facts, the old bonanza image of mining persists. The notion that a mine—particularly a metal mine—is a virtually bottomless source of tax dollars dies very hard indeed with the public.

High Risk

Mining ventures have long been recognized as inherently risky investments. This reputation is well deserved in view of the inherent geological and technical risks encountered in mines, as well as the normal business factors associated with all industries. There is ample evidence to suggest, however, that normal business and financial risks are more pronounced in mining ventures than in most other industries. The capital intensity and long lead times coupled with volatile markets are perhaps the most important parameters which affect the mining industry's risk profile.

Economic Rent

A currently popular term associated with mine tax policy is economic rent. Borrowed from agriculture, economic rent refers to the payments that a knowledgeable investor would be willing to make for the right to cultivate richer soil (or mine higher grade ore). The owner of such scarce resources can extract these rents due to the inherent higher productivity of such soil (ore) in comparison to that of the lower quality resources used by price-setting marginal producers.

Economic rent is a term derived from classical economic theory and may be defined as that portion of a rent to land, labor, or capital which could be expropriated *without* altering its mode, intensity, or location of employment. Thus, by definition, redistributing economic rents will not alter the allocation of resources. From a theoretical standpoint, economic rent represents excess producer rent, producer surplus, or "windfall" profits which are an unearned increment from a natural resource that can bear any level of taxation without discouraging use of that resource. In the mining industry economic rent refers to the present value difference between the revenue from the mined commodity and the total production costs, including the cost of initially attracting sufficient capital to the project. In short, this excess producer surplus is perceived to be unearned in the sense that it is in excess of the amount necessary to pay all factors of production.

Although the theory of economic rent is an interesting academic concept, its value in practice is quite limited due to problems of definition and measurement. From a practical standpoint, the concept of economic rent is extremely difficult to apply in the minerals industry because:

1) There is no general agreement that such rents do, in fact, exist in today's mineral industry.

2) There is no general agreement as to the level of economic rent in those cases where rent does apparently exist.

3) There is no general agreement as to how to divide such rents if indeed they do exist and if they can be measured.

To illustrate the foregoing difficulties with economic rents and minerals, consider the fact that a minimum required return on invested capital is clearly a cost of production, but how much is "minimum?" How does the mining firm get compensated for the risky nature of exploration which may have required millions of dollars before ore was finally discovered? If the price of a mineral commodity is subject to wide fluctuations, where does rent start and the effect of market aberrations stop? Certainly a high level of earnings may occur temporarily due to sharp cyclic swings in commodity prices, temporary increases in ore grade or other factors that do not promise excessive earnings over the long-run. Economic theory is of little assistance here, and a practical solution must be developed.

At this point the political process normally takes over with the result being a tax compromise which may or may not bear any resemblance to the original objective. It would appear that a tax which is imposed in the name of economic rent should be levied against that rent. The tax should really be applied to net proceeds, perhaps in the form of a graduated income tax, to promote rent-sharing between the producer and the taxing authority. In contrast, a tax on gross proceeds (i.e., a severance tax) would be an inappropriate mechanism to compensate for alleged excessive profits on the part of the miner.

In summary, pure economic rent is: (1) a fascinating topic for academic discussions among economists, (2) reasonably easy to define, (3) difficult to identify, (4) extremely difficult to quantify for any particular mine, and (5) virtually impossible to collect efficiently through taxation.

Exhaustibility

The fact that the principal asset of a mine—the ore deposit—is consumed in the course of production and ultimately exhausted has historically created difficult problems for taxing authorities. Once mined and processed the ore is gone and can never be renewed or replaced. As discussed in Chapter 7 this unique feature led to the development of the controversial depletion allowance in federal income tax law as early as 1913. At that time miners argued that a special tax deduction was warranted due to the *capital* nature of minerals. The logic offered was that the value of a capital asset consumed in the production process—be it natural or man-made—is a prepaid operating expense that is rightfully deductible from gross revenue. Congress ultimately accepted the concept of the depletion allowance in order to permit the miner a return for his eroding capital—the ore deposit.

It should be noted that an important consequence of depletion is that the best deposits tend to be discovered, developed, and exhausted first because mineral exploration tends to be a systematic process over the long term. As a result, generally lower quality deposits remain for the future which are more costly to find, and ultimately the cost of mineral supply rises with time. This is an important concept when considering the replacement cost of an ore deposit.

Ultimate exhaustibility of mineral deposits is of concern to government leaders who must deal with the social problems which may arise upon final closure of the mine. Mines are often located in remote locations where they are virtually the only sources of economic activity. Although responsible mining companies are eager to help mitigate economic and social dislocations created by mine closures, public officials occasionally look to taxation to bridge this difficult period. The closure of any industrial facility and the dissipation of the surrounding community is rarely joyful, but it sometimes is unavoidable in mining.

Capital Intensity, Long Lead Times

Mining is a very capital intensive industry. Tremendous sums are required for exploration programs, definitive drilling, sampling, testing, permitting, mine development, and capital equipment acquisitions long before any production takes place. In addition to the large capital requirements, these ventures are characterized by long lead times—8 to 12 years being common—before the mine is brought into production. Therefore, capital investment decisions on new mining ventures involve the commitment of significant sums of capital over long time periods. Only the electric utility industry faces problems of a similar magnitude. Unlike the utility industry, however, where revenues can be adjusted to offset risks during this period, mines have no guarantee to profitability. Any number of things can happen during the preproduction period (i.e., price fluctuations, operating cost escalations, increased tax burdens, etc.) which subsequently impact project success.

Uncertainty in Determination of Ore Deposit Value

The concept that the value of a mine is equal to the capitalized stream of earnings which the mine will yield in the future is almost universally accepted. However, the procedures for determining this value are the basis for almost universal disagreement. Estimation of future earnings is the key, but the level of such estimates unavoidably involves personal judgment.

Like most other firms, mining companies face marketing risks. However, unlike other firms mining companies are faced with considerable uncertainty in the measurement and valuation of any given ore deposit. Mining firms can not accurately determine, in advance and in detail, the quantity and quality of material contained within the ore deposit. This is particularly true for metallic deposits where structural geologic controls are closely associated with mineral concentrations. It should be noted, however, that the limits of an ore body are not only geological, but are also functions of economic factors such as market prices, operating costs, and taxes. In addition to these economic parameters, mining firms must also deal with other uncertainties associated with the ore deposit such as: (1) geologic features (e.g., ore discontinuities), (2) geomechanics features (e.g., rock strength, water inflow), (3) decline in metal content of ore.

The perceived level of these risks has a pronounced effect on the investment decision to develop and exploit any given mining property. These uncertainties also complicate the problems of appraising a mining property at the fair market value for tax purposes.

Indestructibility of Metals

While metals vary significantly in the extent to which they are indestructible, they all possess a permanency that is absent in the products of manufacturing and agriculture. This imperishable nature of metals results in a constantly increasing stock of secondary materials (scrap) which serves as a reservoir for industrial needs and places a limit on the price to which new metal can rise before coming into competition with recycled metal.

Recycled metal does have an impact on metal demand-supply relationships at the present time for commodities such as lead, copper, and aluminum. It is easy to envision a significant impact on primary metal markets if a massive recycling effort were economic. Such economics might be promoted through government tax subsidies.

Rental Payments

Rental payments associated with mineral lands or leases often encompass many objectives. Typically, however, rental payments are imposed on lessees for the right to explore or have access to potential mineral-bearing lands. These rental payments are intended to represent nominal, but adequate, compensation to the lessor for (1) environmental damage to land surfaces during exploration activities (drill roads, pads, etc.), (2) access to the lease area, and (3) sole exploration privileges for the lease area for some specified time interval.

In theory rentals are related to surface values only with royalties covering the extraction of underlying minerals. In practice this may or may not be the case. Normally, however, rental values are nominal and are levied on a per-acre basis. Rental rates typically escalate over time in order to promote or encourage diligence on the part of the lessee. This is not unreasonable in view of the fact that the leasing process itself requires the lessor to forego the next most valuable use for the leased land. Therefore, the rental payment is often considered to represent compensation for the opportunity foregone on the leased land.

Practical Limitations

The economic principles associated with the preceding special features of mining represent some difficult and complex issues. For instance the problems associated with determining appropriate economic rents, ore deposit value, and depletion deductions on mineral deposits are not easy to resolve. Although each of these special features has considerable theoretical economic merit, practical limitations often preclude the adoption of statutes which clearly address these issues. Most taxation policies are the result of compromise and practicality and often do not specifically reflect unique features of any individual industry. When unique features of a specific industry are represented in tax policy, it is usually to impose special taxes or rates on this industry as a result of its uniqueness.

Clearly legislators would have great difficulty in structuring tax policies which treated each industry within their state in accordance with what appears to be unique features of that industry. Even if this were possible, these individuals would be subjected to considerable criticism from constituents for inequity in taxation. However, many state legislators are becoming increasingly concerned with appropriate and fair taxation of specific industries within their states, as evidenced by the growth of state tax commissions and study committees.

In general states opt for tax policies which insure stability of income, generate sufficient revenue to run state government and programs, tend to maximize revenue or promote other objectives (i.e., environmental protection, etc.), and are reasonably easy to administer. These practical trade-offs do not always respond to the unique or special features associated with mining, resulting in some tax policies which are unintentionally wasteful of both the state's natural resources and finite capital resources of the private sector. At the federal level the primary recognition of mining's uniqueness—exhaustibility of the asset—is recognized through the minerals depletion allowance.

Impact on Mine Operations

Every tax is composed of a tax base and a tax rate which is applied to that base. Furthermore, the tax can be levied against wealth, against income, or simply triggered by some transaction. The magnitude and fundamental structure of the tax frequently affect operating decisions at a mine. These impacts are discussed in the following.

Types of Taxes

The minerals industry is subjected to the normal types of taxes imposed on most businesses and industries as well as some which are unique to the extraction of natural resources. Following is a brief discussion of the primary types of taxes which affect mining firms.

1) **Income Taxes:** Imposed on individuals and on corporations. Levied by the federal and most state governments, the tax is usually based on net income, or gross income less certain defined deductions. The tax rate is generally either fixed or progressive (i.e., higher levels of net income pay higher tax rates).

2) **Property Taxes:** Typically an ad valorem (according to value) tax based on appraised value of real and personal property. In the case of mining it is extremely difficult to accurately determine the fair market value of property—particularly if the ore deposit is taxed.

3) **Severance Taxes:** A tax unique to the extraction of natural resources (renewable as well as nonrenewable) and commonly considered to be an excise tax. It

is distinguishable by the fact that it is related in some manner to the actual removal of natural resources (fish, timber, minerals).

4) **Transaction Tax:** A tax imposed upon the consummation of a retail sale. A good example is the common retail sales tax which is levied against all retail sales of certain commodities.

5) **Excise Tax:** A tax imposed on the manufacture, sale, or consumption of specific, selected commodities and/or activities. They are not imposed on persons or property as such. All license taxes are considered to be excise taxes.

6) **Other Miscellaneous Taxes:** Other taxes commonly paid by mining companies are: (a) unemployment taxes, (b) import and export taxes, (c) franchise taxes, and (d) user taxes on various activities. The distinction between these different types of taxes is not always well defined.

Of the foregoing listed types of taxes, income, property, and severance taxes are probably the most significant to the mining industry. These taxes will be discussed in more detail in subsequent sections of this chapter.

Evaluating Taxes as a Cost of Operation

All taxes obviously contribute to the overall level of operating costs in a business. In mining these costs can be substantial indeed and may even exceed the costs of labor and material in some situations.

All taxes, however, do not affect operating costs in the same manner or at the same level. As a result the relative merits of a particular type of tax must be assessed carefully. As mentioned previously, it is more appropriate to judge the adequacy of an entire tax program and the role of a single tax within that program as opposed to trying to assess the merits of a single tax provision against some set of criteria. In fact it is difficult to establish a set of universally acceptable criteria against which to evaluate a particular tax. Adam Smith (1904) set forth four criteria which he referred to as "cannons of taxations" for a "good tax." He suggested that a good tax should be:

1) *Equitable*—referring to equal treatment of similarly situated taxpayers. He did not, however, discuss the tax base or basis to be used for determining under what conditions taxpayers are "similarly situated," nor did he distinguish between horizontal equity (all purchasers of the same commodity pay the same tax) and vertical equity (unequally situated taxpayers being taxed on their ability to pay as per progressive taxation philosophies). Equality in tax issues is, to a large extent, in the eyes of the beholder. The concept of tax equity is generally related to one of two major tax policy goals. In the case of the mining industry the goal is considered to be maximization of resource utilization, whereas the goal for the taxing authority is often maximization of social and economic benefits to the individual state or taxing power.

2) *Convenient*—referring to a tax that can be readily and easily assessed, collected, and administered.

3) *Certain*—referring to consistency and stability in the prediction of taxpayers' bills and the amount of revenue collected over time. This is particularly important to local and state governments who look solely to tax revenues as sources of income.

4) *Economical*—referring to the fact that compliance and administration of a tax should be minimal in terms of cost. One measure of efficiency is the difference between the amount of money collected by the tax and the cost of collecting the tax.

Three recent additions to Smith's four cannons which are receiving considerable support are:

5) *Adequacy*—referring to the fact that a tax should have the ability to produce a sufficient and desired amount of revenue to the taxing authority.

6) *Achievement of Social and Economic Effects*—referring to the use of taxes to reallocate resources in order to achieve various specific social and economic objectives.

7) *Neutrality*—recognizing that a tax should not encourage inefficient allocation of resources by being so extreme that taxpayers make counterproductive economic decisions,

From the viewpoint of the minerals industry, the key features of any tax code pertaining to mining investments should incorporate (1) uniformity, (2) stability, (3) simplicity, and (4) recognition of mining's special features (Byrne, 1968). Tax laws should not unfairly discriminate between taxpayers either through specific statutes or by administrative procedures. All taxpayers should be treated uniformly with respect to property valuations, tax rates, etc. In regard to tax stability, investors' fear of the unknown is especially strong. A high level of taxation will not necessarily preclude capital investment in mining. Significant instability in taxation will, however, be an even stronger investment disincentive. US federal corporate income tax is a good example of historic relative stability and the fact that it affects all taxpayers uniformly. The severance tax, however, is notoriously unstable and inherently nonuniform.

The need for simple, straightforward, and understandable tax codes and statutes is real, but perhaps unattainable in today's complex society. The major features of the statute should be unambiguous and tested in the courts. One reason that seemingly attractive tax laws enacted by some nations to attract foreign investment are so ineffective is that there is no comforting history of jurisprudence covering the interpretation of the law.

The call for tax laws which recognize the special features associated with mining may appear to threaten the requirements of stability and uniformity previously advocated. In general, this recognition of mining's special features is a plea for a tax policy which provides an incentive to the minerals industry for continuously channeling new investment capital into the exploration, development, and exploitation of ore deposits. The key factor here is the distinction between the need for an incentive as opposed to a request for a periodic subsidy. Properly applied, incentives are a much more stable form of motivation than the subsidy approach. Nonetheless, the call for special incentives in a specific industry rests on theoretically shaky ground in that many believe the free market mechanism (if allowed to function) should provide incentive enough for industrial investment.

In retrospect it is safe to say that tax policies have a direct influence upon the development of mineral resources—particularly at the state level. Taxes represent a cost of operation that, unlike most other costs, can not generally be controlled by improved engineering, technology, or management. Any increase in taxes adversely impacts future mineral development.

Effects on Mineral Conservation

The imposition of a tax that raises the cost of mining has the effect of raising the cutoff grade of ore that can be profitably mined. The net result is that lower grade material becomes unprofitable rock rather than valuable ore, and deposit reserves diminish—possibly to the point of having a nonviable project. Consequently, a tax can affect (1) initial investment decision on a property, (2) indicated rate of return, (3) level of mineral reserves, (4) resource conservation, (5) rate of production at a property, and (6) incentive to explore for minerals by mining companies.

Perhaps the best way to illustrate the impact of taxes which increase production costs, raise cutoff grades, and subsequently affect resource conservation is through the use of tonnage-value relationships for idealized ore deposits. For any given ore deposit, or mining district, there exists some tonnage-value curve. This relationship is typically represented by an exponential curve (Fig. 1) and in general indicates some limited tonnage of ore having high values and an increasing tonnage with much lower ore values. Lacy (1969) points out that value distribution within a mineralized body might vary, however, and might be (1) confined, (2) locally enriched, or (3) dispersed. Therefore, an idealized tonnage-value curve would not represent all deposits equally.

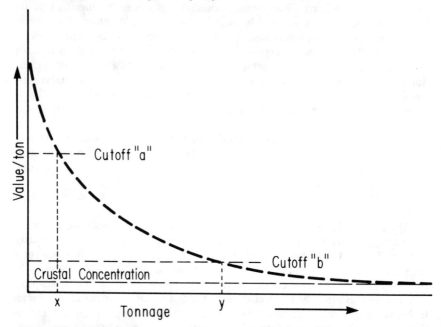

Fig. 1. Idealized relationship between reserve tonnage and metal value per ton in an ore body or mineralized district. A cutoff at "a" would give "x" tons of ore; dropping the cutoff to "b" would increase ore tonnage to "y" (Lacy, 1969).

From an operational viewpoint ore deposits may range from bonanza-type, high-grade, selectively mined ore bodies at one end of the spectrum to the more common disseminated, low-grade, bulk mining occurrences at the other end. There is an almost infinite variety of deposits occurring between these extremes with the moderate-grade, selectively mined deposits occupying a position somewhere in the middle. The high grade vein-type deposits are often characterized as being relatively small to moderate-sized with high unit value and offering a wide margin of operating profit. Initial capital investment as well as ore reserves are typically small to moderate. These deposits frequently have a sharp boundary between ore and waste (i.e., ore is high-grade and wallrock immediately adjacent to it is essentially barren). Fig. 2 illustrates a tonnage-value relationship for an idealized ore deposit of the bonanza type. Notice that raising the cutoff grade within normal limits does not appreciably affect deposit reserves. As a result these types of deposits can often survive levels of high taxation, mismanagement, and high overhead rates because of the wide profit margin. Unfortunately, such deposits have virtually disappeared from the mining scene and are of little consequence in today's total production picture.

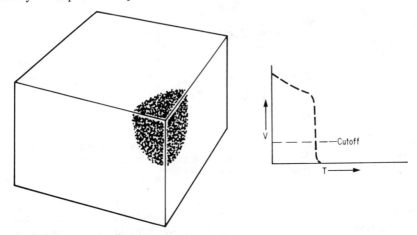

Fig. 2. An idealized deposit with confined values. Variations in cutoff, within limits, have little effect on reserves (Lacy, 1969).

Fig. 3 represents a situation illustrating a more irregular high-grade deposit. Because of the discontinuous and irregular nature of the ore body, mining costs are higher due to the selectivity required, necessitating a higher cutoff grade. Thus, any increase in production costs would create more serious consequences than in Fig. 2.

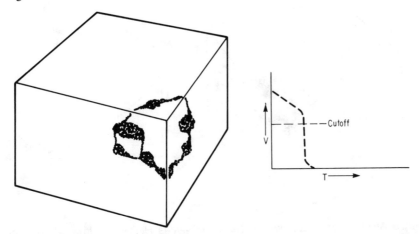

Fig. 3. Irregular, pockety deposit with confined values. Though the tonnage:value graph is the same as in Fig. 2, the discontinuous character of the ore body results in a higher mining cost and higher cutoff. Any additional costs would result in a marginal economic condition (Lacy, 1969).

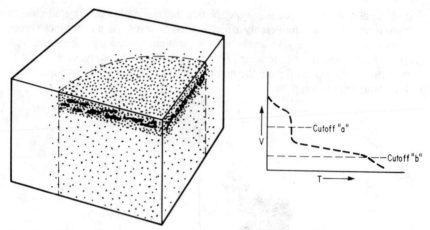

Fig. 4. A disseminated deposit with dispersed values. An enriched zone caps the deposit. As improved technology enables the cutoff to be lowered from "a" to "b," a great increase in tonnage is realized (Lacy, 1969).

Fig. 4 illustrates what may loosely be termed the typical porphyry copper model. Initially only the supergene enrichment zone or blanket was mined, but this material represented only a modest reserve. However, as the costs of surface mining declined as a result of technological developments, the cutoff grade was lowered and vast amounts of mineralized material entered the ore reserve category. This phenomenon results in the unusual shaped tonnage-value curve shown in Fig. 4. In these cases ore boundaries are not easily defined except through careful analysis of contained values and inherent total production costs. The shape and size of the ore body may be significantly influenced by minor fluctuations in the cutoff value. Fig. 5 illustrates an idealized disseminated deposit with lower-grade values

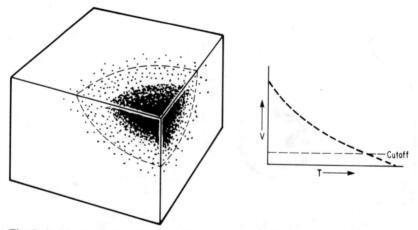

Fig. 5. An idealized deposit with disseminated values. High values toward the center grade uniformly outward into lower grade material. The deposit is exposed at the surface so that its position, shape, and continuity make it amenable to low-cost mining methods. The tonnage:value relationship is very sensitive to changes in cutoff value (Lacy, 1969).

radiating uniformly outward from the center, higher-grade values. Such a deposit is very sensitive to changes in cutoff value.

Fig. 6 represents a more common type of deposit which contains disseminated or dispersed mineralization with higher-grade material isolated in pockets, or perhaps in veins. In this situation a bulk, low-cost mining operation may be capable of extracting virtually the entire deposit while maintaining a satisfactory level of profit. However, if the operation is faced with rising production costs, or declining metal prices, the operation might have to revert to the extraction of only the higher-grade material and abandon the low-cost bulk method in preference to a higher-cost, selective mining system. These increases in the cutoff grade significantly shrink reserves to that quantity of material contained in the higher-grade pockets or veins.

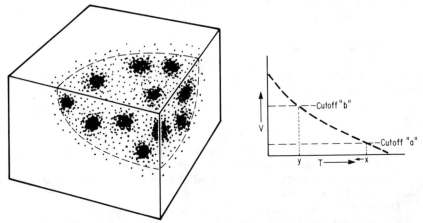

Fig. 6. A disseminated or dispersed type of deposit with higher values in isolated pockets. Though the tonnage:value graph is identical to that shown in Fig. 5, a minor adjustment in cutoff may have a marked effect upon both mining method utilized and mining costs, and would precipitate a great increase in cutoff ("a" to "b") and serious reduction in ore reserves ("x" to "y") (Lacy, 1969).

As Lacy (1969) points out, during the frontier era taxes were relatively low and deposits relatively rich. Minor shifts in cost resulting from taxes, labor, and mismanagement were often of minor consequence. Deposits fell in the steep portion of the tonnage-value curve, and thus minor cost fluctuations had little effect upon reserves. However, at the present time with low cutoff grades and a low margin of profit, a small rise in costs may result in a large reduction in reserves, thereby economically sterilizing a large portion of the in situ resource. Therefore, one goal of an ideal taxation policy should be that no ton of mineralized rock should remain unmined due solely to taxation. Taxes which raise cutoff grades are inherently wasteful of the nation's natural resources.

State Mine Taxation

All states impose taxes on mining operations within their borders. Some states impose no special taxes on mineral commodities, some impose special taxes only on specific minerals, and some impose special taxes on all minerals but perhaps in differing amounts. The primary types of special taxes imposed on mining operations are severance or production taxes and gross income or sales taxes. Severance taxes are distinctly unique to the extractive industries.

Not many years ago analysts performing mining property evaluations handled the costs associated with state and local taxes by simply adding 2-3% to the federal income tax rate and ignored specific costs resulting from individual state taxes. Recently, however, state and local taxing authorities have increased tax bases and rates to the point where these taxes now represent a significant cost of production. As such these tax obligations can no longer be adequately represented by assuming some effective tax rate for the property in general. The astute analyst will recognize that because of their complexity and magnitude, specific state tax provisions should be used in mining property valuations. Because of the volatile and ever-changing nature of these taxes, considerable time and effort are required to collect and interpret current tax statutes for all states in which a company may be performing business or exploration activities.

Basic Tax Definitions

Before proceeding with dicussions on specific tax statutes and policies on the state and federal level, consider some basic tax definitions which apply to material presented throughout the remainder of the chapter. For purposes of discussion in this chapter, the following definitions will apply (Gentry, Hrebar, and Martin, 1980):

Revenue: total income derived from product sales

Deduction: an amount deducted from revenue to arrive at taxable income

Taxable Income: the amount on which a tax is calculated

Depreciation: a deduction over a period of years for cost of plant and equipment (real and personal property)

Depletion: a deduction allowed as a mineral deposit is mined

Amortization: any deduction other than depreciation or depletion allowed over a period of years

Expense: an expenditure treated as a deduction in the year it is made

Defer: to save an expenditure for deduction in one or more later years

Capitalize: to establish an account for an expenditure which, depending on the type of expenditure, may or may not qualify as a deduction

Recapture: the act of foregoing a deduction until some amount of money is recouped. Recapture is generally associated with an amount of money which was previously expensed, but which should have been capitalized

Basis: the amount in a capitalized account at any point in time. The original basis is usually the cost of an asset

Expenditure: an outflow of money for the purchase of goods or services. The three types of expenditures for tax purposes are as follows:

(1) Expenditures for items which are consumed in one year and are non-recoverable. These expenditures can be expensed resulting in a deduction of the full amount in the year of the expenditure (e.g., operating labor, supplies, power, lubricants, etc.).

(2) Expenditures for assets which are consumed more gradually and are nonrecoverable. These expenditures can be capitalized with a fraction of the expenditure treated as a deduction annually over the life of the asset. Deductions are in the form of depreciation or amortization (e.g., mining equipment, buildings, etc.).

(3) Expenditures for assets which are not consumed and are recoverable

after the end of the project. These expenditures are considered recoverable and no deduction is allowed on them over the life of the property (e.g., land).

Types of State Taxes

Property, income, and severance taxes are generally considered to be the most important taxes imposed on mining companies by state taxing authorities. These three types of taxes are briefly discussed in the following sections.

Property Taxes: Property tax, which is nearly always an ad valorem tax, is one of the oldest taxes imposed on mineral property. Ad valorem property taxation is often the most important source of revenue to local and state taxing districts and is a highly developed art form in many states. Property tax is levied against the value of tangible property (real and personal) that is located within a given tax jurisdiction. As such it is intended to be a tax on wealth as opposed to income. Historically property taxation has been justified on the basis that property ownership inherently indicates ability to pay as well as requiring benefits and services provided by government. In the case of mining operations the amount of benefits and services received from local governments is usually minimal since the mine typically provides its own security, fire, water, sewage, and other necessary systems.

In theory, and according to most state constitutions, property taxes are based on equity and uniformity for each class of property. On this basis each taxpayer should contribute to government expenses in fair proportion to his wealth, and all property should be treated equally, regardless of type. It is sometimes difficult to recognize this principle in practice.

There are three basic steps involved in the determination of property taxes: (1) the property is valued at its fair market value (or full cash value which is generally considered to be synonymous); (2) the property is assessed at a percentage of this value; and (3) the property tax liability is calculated by applying a predetermined tax rate to the assessed value.

Market Value—Ad valorem property tax statutes normally require that a mine be taxed at its fair market value. The standard recognized test for fair market value is that price which a willing buyer would pay and what a willing seller would accept where both are well informed and under no compulsion (Brightwell, 1969). Appraisers generally employ three approaches for determining such values: (1) the market approach, (2) the cost approach, and (3) the capitalized earnings approach.

If there is an active market in the assets being considered (e.g., used cars), a relatively low risk value for the particular asset can be determined using the *market approach*. Unfortunately there is no organized market for mineral properties nor could one be devised that would provide comparative sales values upon which appraisals could be based (Starch, 1979). Even on the rare occasion when mineral properties do change hands, it is often extremely difficult to ascertain what the true value of the sale actually was due to common stipulations pertaining to production commitments, carried interests, deferred payments, stock options, and other complicating factors.

With the *cost approach* the value of an asset is presumed to be the cost of reproducing that asset with one of identical design and state of repair. This procedure is used frequently in valuing common commercial buildings but finds little applicability in valuing mining property. Employing the cost approach would yield the conclusion that two 100-tpd mining operations of identical design and age—one mining very rich ore; the other, very lean ore—have the same value. Clearly, highly variable ore deposit quality renders the cost approach to valuation of little value in most mining situations.

This leaves the *capitalized earnings approach* as the only generally applicable procedure for producing a reliable indicator of value for a mining property. Because the value of a mine is generally recognized to be equal to the capitalized stream of earnings which that mine will yield in the future, it is not surprising that experts will usually disagree on the value of a given mining property. Estimated mine value will be a function of anticipated ore reserves, future metal prices, production costs, investment capital, and a host of other variables. Because of the difficulty in arriving at an agreement on mine value between the taxing authority and the taxpayer, several simpler, but indirect, statutory methods have also evolved which attempt to produce a proxy valuation for property tax purposes. These methods retain the basic premise that the earnings approach is the appropriate appraisal technique, but make some simplifying assumptions to minimize disagreements with the taxpayer.

The three basic methods employed by states for determing mine property value are as follows:

CAPITALIZATION OF FUTURE NET PROCEEDS:

This method estimates the value of a mining property by calculating the present worth of future net proceeds of the mine during its life. The discounting procedure is specified by the state and usually employs a single discount rate, although some states continue to use the two-rate Hoskold premise. The application of present value formulas requires assumptions regarding mine life, discount rates, mineral prices, and many other variables. Obviously the prediction of future variables such as metal prices, inflation, and interest rates, production costs, ore reserves, foreign competition, technologic changes, and geologic changes will have an impact on value determination.

The capitalization method, as applied in Arizona is a classic example of this procedure. The Arizona system was described thoroughly by Kearns (1970) with recent changes noted by Petrick (1976). It is important to note that by estimating future production and earnings this method of valuation imposes a tax on the mineral reserves in place. This is a disincentive to exploration drilling—and, therefore, efficient mine planning—because additions to ore reserves increase the mine life, creating greater future earnings and a greater tax liability.

The relative advantages and disadvantages associated with this approach to determining property value are as follows:

Disadvantages

1) Ore reserves are taxed in the ground which discourages efficient exploration for, and development of, reserves far in advance of mining.

2) The method encourages increased production rates and reduced mine life.

3) Litigation is encouraged insofar as tax administrators and tax payers rarely agree on estimates of mine life, future costs, and prices which are necessary to the determination of full cash value.

4) The Hoskold valuation premise, which is still employed in some states, is not representative of modern business practice and can yield incorrect answers when applied to unequal annual earnings.

Advantages

1) The method is rational and follows the basic procedure employed in commercial practice for estimating the value of mining properties.

2) The method provides a *unit value*, a valuation that includes both the ore deposit and all surface values and improvements.

GROSS PROCEEDS:

Mine valuation by a gross proceeds method typically designates all or a fraction of the annual gross proceeds from a mining operation as the assessed value of the mine. Value may be determined from gross proceeds for the year in question or based on an average of several previous years.

States employing a gross proceeds approach to mine valuation, such as Wyoming and Colorado, normally base assessed value on the previous year's mineral production or ore sales. Real and personal property (machinery, equipment, etc.) in addition to improvements other than those from mining are valued separately for assessment at some percentage of actual value.

Gross proceeds valuation avoids the problem of discretionary selection of valuation variables inherent in the future income capitalization method. This is accomplished by establishing value by reference to historic rather than future earnings froms the property. Although this approach is fundamentally flawed, it does eliminate most of the disagreements concerning the calculation of full cash value.

Some of the more important characteristics of the gross proceeds method applied to mine value determination are as follows:

1) Only minerals which have been extracted are taxed; mining properties not in production must be assessed by some other means.

2) No distinction is made between profitable and unprofitable mines.

3) Provides no discouragement to explore since ore reserves in the ground are not taxed.

4) Administration of the tax is simple, inexpensive, and convenient.

NET PROCEEDS:

With the net proceeds method of property valuation, the value of a mine is determined on the basis of some formula which designates a percentage of the annual net proceeds of the mining operation. This value may be based on the previous year's net proceeds or an average of the past several years. For example, Utah averages net proceeds for the preceding three years. Total property value may or may not include a separate assessment for real and personal property associated with the mining operation.

Net proceeds are defined as gross proceeds resulting from sales during the preceding calendar year of all products from the mine, less certain statutory deductions which vary from state to state, but usually include mining, processing, and administrative costs. Obviously, this method requires that the taxing authority have greater access to company records than with gross proceeds valuation to ensure that deductible costs are legitimate and properly calculated.

Some characteristics of the net proceeds method to full cash value determinations are as follows:

1) The procedure does not tax ore is the ground and, therefore, requires no verification of ore reserves.

2) It recognizes the concept of ability to pay because it values property according to profitability.

3) It offers no incentive to accelerate production, thereby aiding resource conservation.

4) Revenue fluctuations to the taxing authority due to fluctuations in metal prices, shutdowns, or increasing production costs are greater than with the gross

proceeds valuation. This can be tempered by using an average of several years' proceeds which occasionally results in inequitable tax burdens on a year-to-year basis.

5) Both gross and net proceeds methods suffer from a basic flaw in logic: the value of a mine should be determined by anticipated *future* earnings and not by *historical* profits or sales.

6) Administrators have little discretion in market value calculations thereby minimizing disputes with the tax payer.

Assessment Rate—Assessment rates are highly variable among states, but one generalization should be noted. Property used in mining is rarely—if ever—assessed at a lower rate than other property. For example, the state of Arizona historically assessed mining property at 60% of full cash value while assessing general industrial property at 27% and residential property at 18% of full cash value. In 1980, however, the state began to gradually eliminate this differential such that assessment rates will be equalized by 1990 at 25%.

Tax Rate—Property tax rates are infinitely variable, being dependent upon the particular budgetary needs of the taxing authority. These rates are generally uniform among types of properties within a taxing district, with the combined rate being the sum of many state and local tax rates. The largest component is often the public school district tax, but it is not unusual to have a dozen or more separate taxes represented in the combined rate.

Income Taxes: Most state governments impose income taxes on corporations in addition to the federal income taxes levied on these firms. State income tax procedures are generally patterned closely after federal tax law and, in general, they affect mineral producers no differently than any other corporation. The range of rates among the states is large, but the effective rate is generally rather modest (5-10%) when differences regarding allowable deductions are taken into account. Either a flat rate or a progressive rate structure may be applied to the tax base for calculating the state income tax liability. In the past, states limited taxable income to income derived from business activities within that state only. However, an increasing number of states are enacting unitary taxation statutes whereby all earnings of a corporation are taxable by the state regardless of the geographic location of the source activity. The existence of such legislation has become an important consideration in the siting of new facilities.

States imposing an income tax usually list specific deductions which a firm may deduct from revenues to arrive at the tax base. These deductions normally include the costs of mining and processing, general and administrative costs, state and local taxes, insurance, and depreciation. Some states also allow federal income taxes and the depletion allowance as a deduction in determining the state income tax base. The case where the depletion allowance is allowed as a deduction for state income taxes presents an interesting situation because state income taxes are also allowable deductions for calculation of the depletion allowance when working with the 50% of income limitation. In this case, a procedure offered by Hrebar (1977) utilizing simultaneous equations should be employed. The procedure is illustrated:

Let: N = Net after Costs and Depreciation
         D = Depletion based on 50% Net Income
         ST = State Income Tax Based on $R\%$ x $(N - D)$
Therefore: $D = (N - ST) \times 0.50$
         $ST = (N - D) \times R$

Solving for ST: $D = (N - ST) \times 0.50 = 0.5N - 0.5ST$
$$ST = (N - (0.5N - 0.5ST)) \times R$$
$$ST = 0.5RN/(1 - 0.5R)$$

A similar situation arises when federal and state income taxes are each deductable in the calculation of the other.

When compared with other forms of taxation, income taxation encourages mineral conservation, exploration, and long-range planning since ore reserves in the ground are not taxed (Gentry and O'Neil, 1974). Income taxation encourages exploration and development of future reserves, as these costs can be deducted from annual gross income, thus reducing the tax base and taxes paid. Income taxes are generally less onerous to the taxpayer because they are proportional to profitability. Also income taxes do not adversely affect or penalize a mining operation during periods of unfavorable economic conditions due to depressed metal prices, strikes, or rapidly rising production costs.

Income taxes are relatively easy to assess and collect and are equitable, being based on the ability to pay. However, they do not provide very predictable tax revenues to the taxing authority.

Income taxes have been criticized on the basis that they promote operational inefficiencies due to the fact the higher a firm's costs of production, the lower its net income and consequently the lower the taxes paid. This would have a tendency to penalize the more efficient firms and reward or subsidize those less efficient. Such disincentives do, in fact, emerge at high rates of income taxation which is an important consideration in establishing a suitable tax policy.

Severance Taxes: Severance taxes represent the only tax which is uniquely applied to the extraction of natural resources, renewable (e.g., fish and timber) as well as nonrenewable (i.e., minerals). As such, the tax is clearly discriminatory in nature. A severance tax may have some other official name, but the distinguishing characteristic is that the tax liability is based directly on the number of units extracted. Other names for severance taxes in some states are excise, license, privilege, production, or occupation tax. Regardless of the name used, it is important to remember that the operational impact of the tax is the same.

Severance taxes are generally treated as excise taxes which are paid for the *privilege* of mining and are not considered to be property taxes. As such, severance taxes are not subject to state constitutional requirements of uniformity and equality as are property taxes. Also, when imposed in addition to property taxes, severance taxes do not constitute double taxation.

Severance taxes vary not only from state to state, but also vary from commodity to commodity, and, sometimes, by method of mineral extraction within a state. Severance taxes, like other taxes, are determined by establishing: (1) a specific tax base and (2) a tax rate applied to that base. There are generally two specific types of severance taxes—specific severance taxes and ad valorem severance taxes. These classifications refer to the base upon which the tax is calculated. Specific or "true" severance taxes are based on the physical or unit volume of production from the mine. Ad valorem type severance taxes are based on the value of production or product mined (either gross or net value). Severance taxes based on gross value tend to discriminate against less profitable operations since the tax does not vary in relation to costs. Marginal producers and newly developed mines also will operate at a relative disadvantage.

Severance tax rates applied to the tax base may be either a flat rate per unit of production or a percentage of the value of the resource produced. These tax rates vary considerably from state to state and commodity to commodity, and the resulting tax burden can be very misleading if not analyzed in the context of the entire tax system. Many states allow certain credits and exemptions which are applicable to severance taxes such as credits for some or all of the ad valorem taxes paid, exemptions for minimum levels of production during a specified time period, exclusion for "small" producers, etc.

Severance taxes are similar to royalties except that, in the former case, payment need not necessarily be made to the owner of the mineral rights. The concept of ownership, however, has often been used as a basic argument for severance taxes in that it serves to compensate citizens of the particular state for the exploitation and irretrievable loss of the natural resources. This "natural heritage" rationale persists today, although the apparent loss or cost associated with the extraction of an ore deposit to the citizenry is not quantifiable. In practice, it is much easier to measure benefits associated with the mining of a mineral deposit than it is to quantify the loss of future wealth.

Another common justification for severance taxes is to extract a greater share of the economic rents that are perceived to accompany some mining ventures. As discussed earlier, severance taxes are not well suited for this role because economic rents are found in net income, whereas severance taxes are typically levied against gross income.

In recent years one of the most quoted rationales (primarily by western governors) for increased severance taxes is that concerning the socioeconomic impact which mining has on new or expanding communities in proximity to mineral or energy development. The increasing responsibility of state and local governments to provide expanded programs, social goods, and services requires increasing state revenues. Requiring that developers pay a large share of infrastructure costs associated with their developments is a notion that is not restricted to mining. Rarely, however, are severance tax revenues earmarked primarily for this purpose. Regardless of the rationale offered in support of severance taxes, the underlying fact is that they are very adaptable and can usually be enacted to meet the need for more state revenue without incurring the wrath of the electorate. Starch (1979) provides an excellent discussion on the various arguments in support of, and the criticisms leveled against, severance taxes imposed on mineral production.

Some of the important features associated with severance taxes as they pertain to the mineral industry are as follows:

1) Severance taxes do not discourage development of ore reserves since these reserves are not taxed.

2) Severance taxes only apply if production takes place; therefore, the mine is not taxed during periods of strikes and shutdowns.

3) Severance taxes can be used by taxing authorities to accomplish specific objectives such as: (a) encourage or discourage mineral producers to process minerals within state boundaries, (b) discourage or delay undesirable mineral activities.

4) Severance taxes are relatively easy to administer.

5) Severance taxes are objectionable to the minerals industry because (a) they are limited to the extractive industries; (b) property plus severance taxes constitute a disproportionately large share of the state's revenue needs from the mining sector; and (c) they are extremely easy taxes to manipulate and/or change.

6) Severance taxes do not create a steady or predictable amount of revenue to the taxing authority.

7) Severance taxes have the effect of increasing mining costs, raising the cutoff grade, and subsequently reducing overall reserves which negatively impacts resource conservation.

Taxation and Mineral Conservation

A desirable feature of any form of mineral taxation is that it not encourage operating practices which are harmful to mineral conservation. Therefore, taxes which provide economic incentives to raise cutoff grades, minimize long-range planning and increase risk are generally counterproductive. Property, severance, and income taxes on mineral properties have different impacts on mineral conservation which are assessed in general terms in this section.

Property Taxes: Ad valorem property taxes represent a fixed cost to the operation, because they are levied whether or not the mine is operating. Fig. 7 illustrates the impact of ad valorem taxes on ore reserves for two types of deposits. When ad valorem taxes are levied on a marginal mining operation, the figure shows that these increases in fixed costs could result in no profit—even at full operating capacity. To again achieve the break-even point, the operator must mine higher unit value ore, and the higher operating leverage results in a higher critical operating rate (Gentry and O'Neil, 1974).

Under these conditions the mine operator is usually faced with two alternatives (1) close down the operation because it is unprofitable, or (2) raise the cutoff grade and resort to selectively mining only the higher-grade material. This latter alternative is illustrated in Fig. 7. As ore cutoff grade is raised, costs associated with increasingly selective mining methods also increase, which in turn tends to increase cutoff grade even further. This has the effect of reducing ore reserves— particularly in bulk, low-grade deposits. Note that as cutoff grade is raised from r to s to maintain the same total profit, tonnage of ore reserves on curve A of a disseminated type of deposit decreases substantially from x to y. However, in the case of the bonanza-type of deposit (curve B) the decrease in tonnage from p to q is much less (Lacy, 1969).

Unfortunately, if a mining operation must adopt a high-grading posture, it often becomes neither technically feasible nor economic to reenter the lower-grade areas of the deposit once they have been bypassed. Therefore from the standpoint of resource conservation, property taxes have the undesirable effect of reducing the amount of economically recoverable resource.

Severance Taxes: Severance taxes are incurred as long as there is production, even though the operation may not be profitable. Severance taxes represent a direct variable cost of production to the operator and, as such, have the effect of raising cutoff grades and decreasing ore reserves as discussed with property taxes.

Fig. 8 shows the impact of severance taxes on two distinct types of mineral deposits. It is ironic that severance taxes actually oppose one of the primary goals for which they are often imposed—mineral conservation. Note, however, that after the initial adjustment, the required critical mining rate remains unchanged when the operator restores the previous level of profitability.

There is no question but that a severance tax based on production output, gross value, or income represents an additional cost of production. As stated previously, an important effect of increasing costs is higher cutoff grades and lower resource extraction than would have been produced in the absence of the tax. Therefore,

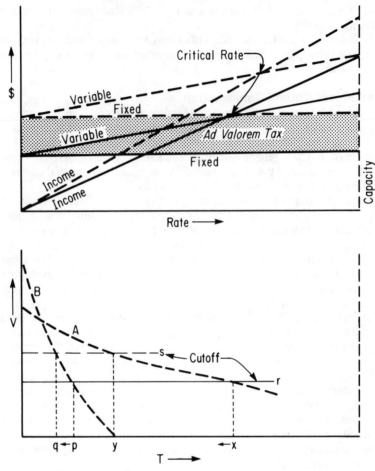

Fig. 7. Graphs illustrating the effect of ad valorem tax in increasing costs and necessitating an increase in cutoff from "r" to "s." Such a tax would have only minor effects upon an ore body illustrated by value:tonnage curve B, but would eliminate more than half of the reserves in ore body A (Lacy, 1969).

there is no doubt but that severance taxes encourage high grading of mineral deposits. However, some observers view this as a positive aspect of severance taxes, because the resulting increased production costs tend to reduce mineral production rates, retaining more mineral in the ground for future years. Unfortunately, the operational reality is that once bypassed, lower grade materials may never be mined. The long-run stability of mining is rarely, if ever, enhanced by the application of a severance tax.

Income Taxes: Income taxation has a lesser effect than property or severance taxes on the direct and indirect costs of a mining operation until the break-even point is reached. Fig. 9 shows income tax to be a variable production cost that is incurred only after a profitable operating position is achieved. As a result, the cutoff grade and the critical mining rate remain unchanged. Also, because operat-

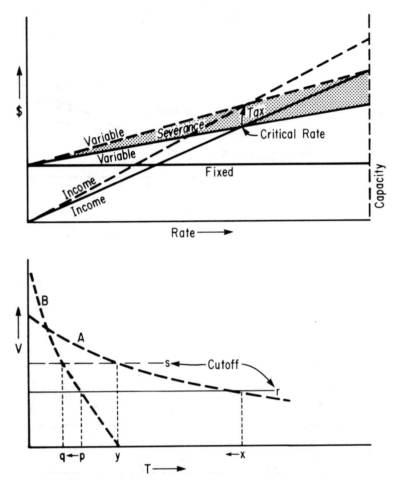

Fig. 8. Graphs illustrating the effect of severance tax in increasing costs and necessitating an increase in cutoff from "r" to "s." Its effects are similar to that of an ad valorem tax (Fig. 7) in causing a rise in cutoff and in the decreasing of ore reserves; however, it does not penalize a mine unable to operate at full capacity (Lacy, 1969).

ing costs below the break-even point are not affected, ore reserves remain unchanged. Since income taxes are based on profitability and do not affect ore reserves, they are perhaps the most rational means of taxing the minerals industry from a conservation standpoint.

Effects on Mine Valuation

Because of the significant impact of state taxes on minerals production, it is imperative that the evaluation of prospective new mining ventures incorporates current state tax statutes into the economic analysis of the project. The analyst should look for any subtle features in the statutes which may escalate total tax payments, as well as for mutual credits, interrelationships among the various taxes, deductions, and allowances. It is also important to become aware of

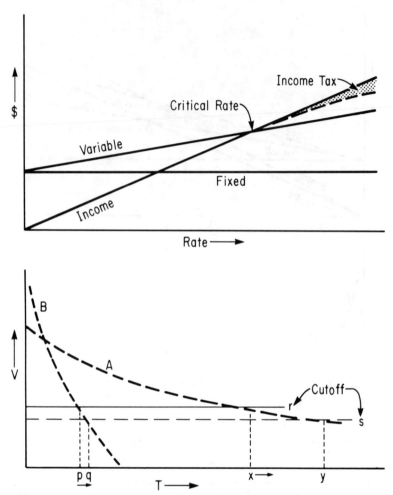

Fig. 9. Graphs illustrating the possible effect of income tax. Income tax does not alter ore reserves; on the other hand, income tax deduction incentives may encourage development of additional reserves that justify expanded operations, lowering costs and cutoff, and increasing ore reserves. Such encouragement increases tax yield (Lacy, 1969).

potential changes in tax statutes and the impact these may have on project viability.

Since many states have taxes which vary according to commodity, it is important to assess how these specific taxes will impact the market realizations for a particular product. For instance, many sales contracts pertaining to energy commodities contain provisions for the pass-through of any future escalations in severance taxes which may be imposed on the producing property by the state. Such contractual agreements recognize: (1) the general fickle behavior of state taxation policy—particularly as it relates to severance taxes, (2) the state's relative ease in increasing tax rates, and (3) the state's ability to export or shift the tax burden to consumers in other states. Although on the surface the ability to pass-through some state tax increases may seem beneficial, the project evaluator must consider the relative competitiveness of a project in a state having high taxes with one in a state having lower taxes when the time arises for contract renewal. One cannot lose sight of the fact that a buyer must still be found before tax increases can be passed on to the consumer.

In contrast to energy commodities, most hard-rock minerals are traded on international markets, with producers being price takers rather than price setters. Thus, tax increases cannot be readily passed on to the consumer. The net effect of increased state tax obligations on producers of these commodities is to reduce the mine's competitive position with respect to other producers. On a national level this can lead to increasing imports of raw materials from foreign sources.

A recent study performed by the US General Accounting Office [GAO (Comptroller General, 1981)] on the impact of federal and state taxes on domestic mineral profitability and production concluded that:

1) State taxes can have a significant effect on the profitability of existing mines, the development of known deposits, and the level of exploration activity within a state.

2) State taxes can have a significant effect on the level of domestic mineral production.

3) The state tax burden is a substantial portion (40%) of the total taxes paid by all mines in the study.

4) Many states, in order to obtain stable revenue streams or for other purposes, often enact taxes that discourage efficient mineral production.

5) Adequate analysis by state taxing authorities has not previously been undertaken to determine the effects that state taxes have on mineral development, production, and investment decisions.

6) Changes in bases, rates, and timing of state taxes can significantly affect the investment potential of nonproducing mineral deposits.

Because state taxes are allowable deductions for determining federal taxable income, the GAO noted that "changes in state taxes have a rippling effect that transcends state revenue streams and affects Federal tax receipts, allowances, and credits and have a major impact on the overall profitability of a mine." It was also concluded that many states were unaware of federal mineral policy objectives (Mining and Mineral Policy Act of 1970). In this legislation it was suggested that given the critical interaction of federal mineral policy, state tax policy, and the profitability of domestic mining, institutional means should be devised to try and better harmonize state tax policy with national mineral production objectives.

Perhaps one of the best indications of the impact and magnitude of state mineral taxation—particularly severance taxes as applied to energy minerals—was the recent introduction of federal legislation which would place limits on the level of severance taxes which could be imposed on coal sold to utilities. Although not enacted, it does indicate the federal government's awareness of state taxation policy impact on minerals and its willingness to delve into states' rights issues which have long been held sacred.

Summary of Tax Practices in Selected States

Because of the dynamic nature of state taxes, it is difficult to provide a complete, current, and definitive list of state property, severance, and income taxes which affect the minerals industry. Therefore, some examples of these state taxes are provided for illustrative and comparative purposes only.

Table 1. Property (Ad Valorem) Taxation for Selected States

State	Mineral	Valuation method		Basis	Assessment rate, %
		Mineral estate	Improvements		
Alabama				20% of fair market value	65
Arizona	All	Capitalized net future proceeds, unit valuation		Full cash	52—1980-82 44—1983-85
Colorado	Industrial minerals	Capitalized earnings, unit valuation		Market value	30
	Metal mines	Higher of 25% of gross proceeds for the preceding year or 100% of net proceeds for producing mines; 30% of actual value for nonproducing mines.		Previous year	100
		Also all metal mining properties are taxed an additional amount based on 0.1% of assessed value.	Appraisal	Book value	30
Missouri	Nonfuels		Appraisal	Market value (original cost + improvements)	33⅓

State	Category	Base	Valuation	Basis	%
Montana	All, except coal & metal	Net proceeds minus specified costs		Annual net proceeds	100
	Metal	Net proceeds		Annual gross proceeds	3
	Surface coal	Gross proceeds		Previous year	45
	Underground coal	Gross proceeds		Previous Year	33⅓
Nevada	All	Net Proceeds	Appraisal	Previous year	35 / 35
New Mexico	All, except potash	Annual net profits	Appraisal	Previous year	300 / 33⅓
	Potash	Gross proceeds	Appraisal	50% of market value of production year	33⅓ / 33⅓
Utah	Metal mines	Net proceeds	Appraisal	Previous 3-yrs. average × 2 + $5/acre	100 / 30
	All other	Net proceeds	Appraisal	Reasonable full cash value	30
Wyoming	All	Gross proceeds		Previous year	100

Sources: Laing, 1976; O'Neil, 1977; US General Accounting Office (Comptroller General, 1981); Commerce Clearing House, 1981b, *State Tax Guide*, 2nd ed., Vols. 1, 2.

Table 1 illustrates some of the approaches taken for property or ad valorem taxation in selected states.

Table 2 shows the rates, bases, and some of the allowable deductions associated with income taxes for selected minerals-producing states. In most cases the tax rate is based on taxable income as defined for federal income tax purposes. Although the rates are fairly uniform, the allowable deductions vary. In most states federal income taxes are not accepted as allowable deductions for state income taxes; however, the federal minerals depletion allowance often is an allowable deduction for state income tax bases.

Tables 3 to 5 list rates, bases, and special features associated with severance taxes on coal, metals, and nonmetals for select states. These tables illustrate the extreme variability in severance taxes with respect to rates, bases, commodities, special allowances, etc., from state to state. The tables include only taxes that are defined strictly as severance taxes. Other taxes which have the same impact as severance taxes, such as Arizona's Transaction Privilege Tax, are excluded.

In addition to property, severance, and income taxes, states may collect mineral royalties for mining activities conducted on state lands. Table 6 lists examples of mineral royalties for select state lands. It is difficult to generalize regarding the financial impact of these royalties because of the variation in rates and bases. Nonetheless, these royalties are typically considered to be less onerous than normal state taxes applied to the minerals industry.

Federal Income Taxation

This section describes, in general terms, some of the more important federal income tax rules which affect mining property valuations. Although not as volatile as state tax regulations, federal income tax codes undergo changes in definitions, allowable credits, deductions, tax rates, and procedures. As such, the specific information contained in this section is subject to change. However, the fundamental concepts associated with federal income tax law are well tested and do not change radically from year to year. It is the overall procedures associated with these concepts which are addressed in this section.

Again, the analyst must recognize that it is imperative that current information on tax changes be maintained and incorporated into project evaluations. The approach taken in this section is to expose the evaluator to the basic taxation regulations which are treated when performing mining property valuations. The evaluator should remember, however, that virtually every statute is fraught with vagaries and subtleties which often require the assistance of legal tax counsel for special or unique situations.

Special Features of Mining

Early in this chapter some of the unique characteristics of mining were briefly discussed as they related to tax policy matters. That discussion will not be expanded upon here; however, there are some rather special features of mining which specifically relate to federal income tax regulations. These features are presented in the following.

Depletion: The depletion allowance is perhaps the most significant and controversial feature of federal income tax law relating to the minerals industry. This unique aspect was discussed in detail in Chapter 7 and will not be repeated here. However, since other tax regulations do affect, or are interrelated with, the depletion allowance, this tax deduction will appear at various times throughout the subsequent discussions on federal income taxation as it relates to mining ventures.

Table 2. Comparison of State Income Taxes for Selected States

State	Rate and basis of tax	Allowable deduction of minerals depletion	Federal income tax deductible
Arizona	Maximum of 10.5% of net income above $6000 with lesser percentages on a sliding scale down to 2.5% on net income under $1000	No	Yes
Colorado	5% of taxable income	Yes	No
Idaho	6.5% of taxable income	Yes	No
Missouri	5% of taxable income	Yes	Yes
Montana	6.75% of net income	Yes	No
New Mexico	4% 1st $1,000,000 5% 2nd $1,000,000 6% over $2,000,000	Yes	No
Tennessee	6% of federal taxable income	No	No
Utah	4% of taxable income	No	No

Sources: O'Neil 1977; US General Accounting Office (Comptroller General, 1981); Commerce Clearing House, 1981b, *State Tax Guide*, 2nd ed., Vol. 1.

Table 3. Mineral Severance Taxes—Rates, Bases, Other Features—Coal

State	Tax	Base	Rate	Other features
Alabama	Coal and severance	Per ton	$0.135	In addition to Coal Severance Tax and $0.50 per ton Coal Severance Tax levied in De Kalb County.
	Coal and lignite severance	Per ton	$0.20	
Arkansas	Natural resources severance	Per ton	$0.02	Coal and lignite. In lieu of all other privilege or excise taxes.

Table 3 cont.

Colorado	Severance	Per ton (surface)	$0.60	Rate changes 1% for every 3-point change in the wholesale price index; 8000 tons per quarter exempted; 50% credit for lignite coal and coal produced underground.
		Per ton (underground)	$0.30	
	Mine license	Per mine		Minimum tax $0.50 per ton.
		500 tons or less	$10	
		500–1000 tons	$25	
		More than 1000 tons	$50	
Kentucky	Coal severance	Gross value	4.5%	
Louisiana	Natural resources severance	Per ton	$0.10	
Montana	Coal severance	Per ton or value (whichever is greater).		Applies to production over 20,000 tpy on the basis of Btu per lb heating quality of coal and whether mined by surface or underground techniques. Tax is on per ton or a value basis, whichever yields the greater amount of tax. Value means contract sales price.
		Btu per lb	Surface	
		Under 7000	$0.12 or 20%	
		7000–8000	$0.22 or 30%	
		8000–9000	$0.34 or 30%	
		Over 9000	$0.40 or 30%	
		Btu per lb	Underground	
		Under 7000	$0.05 or 3%	
		7000–8000	$0.08 or 4%	
		8000–9000	$0.10 or 4%	
		Over 9000	$0.12 or 4%	
	Mineral mining	Gross value	0.5%	Applies to gross value over $5,000, plus fee of $25.

State	Tax	Basis	Rate	Remarks
New Mexico	Resource excise	Gross value	0.75%	Imposed as a resources, processors, or service tax.
	Severance	Per ton	$0.38	Steam coal.
		Per ton	$0.18	Metallurgical coal. On July 1, 1978, and each succeeding July 1, the tax rate on coal will be increased by a surtax based on changes in the consumer price index.
	Oil and gas conservation	Value	0.19%	Royalty payments to the United States, state and Indian interests exempted. Rate may be raised or lowered depending on amount in conservation fund. Some expenses deductible.
North Dakota	Coal severance	Per ton	$0.65	Tax may be increased $0.01 per ton for every 1-point increase in the wholesale price index. Expires June 30, 1979.
Ohio	Resource severance	Per ton	$0.04	
	Coal excise	Per ton		Applies to any consumer using more than 200 tpy for generating steam or electric power. Tax based on sulfur content of coal delivered. Tax adjusted for processing to remove sulfur.
		Sulfur content		
		− 0.50%	$0.40	
		0.5-1.0%	$0.35	
		1.0-1.5%	$0.30	
		1.5% +	$0.15	
South Dakota	Oil and gas severance	Gross value	4.5%	In lieu of all other mineral taxes except sales, use, and property taxes.
Tennessee	Coal severance	Per ton	$0.20	
Wyoming	Mining excise and severance	Gross value	10.5%	In lieu of taxes on land.

Source: Starch, 1979.

Table 4. Mineral Severance Taxes—Rates, Bases, Other Features—Metals

State	Tax	Base	Rate	Other features
Alabama	Iron ore mining	Per ton	$0.03	Iron ore.
Arkansas	Natural resources severance	Per ton	$0.15	Barite, bauxite, titanium ore, manganese and manganese ores, zinc ore, cinnabar, and lead ore.
			$0.02	Iron ore.
Colorado	Severance	Gross income	2.25%	All metals except molybdenum. On gross income exceeding $11 million only. Credit allowed for ad valorem assessed, not to exceed 50% of Severance Tax.
Florida	Solid minerals	Per ton	$0.15	Molybdenum ore.
Idaho	Ore severance	Market value	5%	Includes rare earths.
Louisiana	Natural resources severance	Net value	2%	All ores.
Minnesota	Iron severance	Per ton	$0.10	All ores.
		Net value	15.5%	All ores except taconite, semitaconite, and iron sulfides.
		Net value	15%	Taconite, semitaconite, and iron sulfides. Credits for underground mining and beneficiation.
	Ore royalty			Applies same rates as Iron Severance Tax to all royalties.
	Taconite, iron sulfides, and agglomerates.	Per ton	$1.25	Taconite and iron sulfides. Beginning in 1978 will be indexed to steel production and not less than $1.25 per ton. Tax is in addition to Iron Severance Tax and Ore Royalty Tax, but in lieu of all other taxes; 1.6% of total tax per ton added for each 1% over 62% iron content.
			$0.05	Agglomerates.
		Per ton	$0.10	An additional tax imposed on tailings not disposed of in accordance with permits.

	Tax	Basis	Rate	Notes
	Semitaconite	Per ton	$0.10	A tax of up to $10 per acre is levied on certain reserve lands. Tax is $0.05 per ton if agglomerated or sintered in Minnesota. Add $0.001 per ton for each 1% of iron content above 55%. Tax is in addition to Iron Severance Tax and Ore Royalty Tax, but in lieu of all other taxes. A tax of $1 per acre is levied on certain reserve lands.
	Copper-nickel	Net value	1%	Occupational tax of 1% of value and an additional mining, quarrying, and production tax of $0.025 per gross ton. Additional tax for higher grade ores. Certain expenses deductible from value. Tax on royalties is 1% + an additional 1% of royalties paid on gold, silver, platinum, and other precious metals. In addition to other taxes, but in lieu of Iron Severance Tax. All metals and precious and semiprecious stones.
		Per gross ton	$0.025	
Montana	Metalliferous mines license	Gross Value		
		Up to $100,000	0.15%	
		$100,000-$250,000	0.575%	
		$250,000-$400,000	0.86%	
		$400,000-$500,000	1.15%	
		Over $500,000	1.438%	
Nevada	Mineral mining	Gross value	0.5%	$25 plus 0.5% on gross value over $5000. Applies local property tax rate to net proceeds of all mines in state.
	Net proceeds of mines	Net proceeds	Local property tax rate	
New Mexico	Resources excise	Gross value	0.75%	All metals except molybdenum. Imposed as a resources, processors, or service tax. Molybdenum.
			0.125%	

Table 4 cont.

	Severance	Value	0.5%	Copper. Some production expenses deductible.
		Per pound	$0.05-$3.24	Uranium. Rate varies with value of U_3O_8. For example, if value is $40 to $50 per lb, tax is $1.99 per lb plus 12.5% of excess over $40. Surtax on ore valued over $50 based on consumer price index.
			0.125%	Thorium, rare earth metals, gold, silver, lead, zinc, molybdenum, manganese, and all other metals. Gross value is defined for various commodities.
	Oil and gas conservation	Value	0.19%	Uranium. Value is 25% of resources excise value.
Oklahoma	Oil, gas, and mineral gross production	Gross value	0.75%	Lead, zinc, jack (zinc), gold, silver, and copper.
South Dakota	Mineral severance	Gross value Net profits	5% 4%	Uranium. All minerals including gold, silver, other precious metals, iron ore, and uranium; $100,000 market value minimum.
Utah	Mining occupation	Gross value	1%	Gold, silver, copper, lead, iron, zinc, tungsten, uranium, and other valuable metals; $50,000 exemption each mine.
Wisconsin	Metalliferous minerals occupation	Average net proceeds		Metalliferous minerals. Average net proceeds is based on preceding 3 years. Deductions allowable to reach net proceeds are specified.
		$100,000-$4 million $4 million-$10 million	6% 12%	

State	Tax	Base	Rate	Other features
Wyoming	Mining excise and severance	$10 million-$20 million	16%	
		$20 million-$30 million	18%	
		Over $30 million	20%	
		Gross value	2%	Any valuable deposit. In lieu of taxes on the land.
		Gross value	5.5%	Uranium.

Table 5. Mineral Severance Taxes—Rates, Bases, Other Features—Nonmetals

State	Tax	Base	Rate	Other features
Arkansas	Natural resources severance	Per ton	$0.01	Crushed stone, including chert, granite, slate, novaculite, limestone, construction sand, gravel, clay, chalk, shale, and marl.
		Per ton	$0.015	Gypsum (sold for out-of-state use) chemical grade, limestone, silica sand, and dimension stone.
		Per 1000 bbl	$2	Salt water used as raw material for bromine. Plus $25 per well Conservation Tax.
		Market value	5%	Diamonds, fuller's earth, ochre, natural asphalt, native sulfur, salt, pearls, other precious stones, whetstone, novaculite, and all other natural resources.
Colorado	Severance	Gross proceeds	4%	Oil shale. Graduated to 4% by fourth year after production reaches 50% of design capacity; first 15,000 tons or 10,000 bbl per day exempted; 25% credit for in situ methods.

Table 5 cont.

State	Tax	Basis	Rate	Description
Florida	Solid minerals	Market value	5%	Includes clay, gravel, lime, shells (excluding live shellfish), stone, sand, and rare earths. Limited credit allowed for ad valorem taxes paid on property. No tax on minerals extracted to improve site upon which Florida sales tax is paid or which is sold to government agencies in Florida. Refund allowable for reclamation program.
Idaho	Ore severance	Net value	10%	Phosphate rock.
			2%	All ores.
Louisiana	Natural resources severance	Per ton	$0.03	Sand and gravel and stone.
		Per ton	$0.06	Salt.
		Per ton	$0.005	Salt from brine used in manufacturing.
		Per ton	$0.04	Shells.
		Per ton	$0.10	Ores.
		Per ton	$0.20	Marble.
		Per long ton	$1.03	Sulfur.
Mississippi	Salt severance	Value	3%	Salt.
Montana	Micaceous minerals license	Per ton	$0.05	Vermiculite, perlite, kerrite, maconite, or any other micaceous minerals.
	Cement license	Per ton	$0.22	Cement.
		Per ton	$0.05	Cement, plaster, gypsum, and gypsum products.
	Mineral mining	Gross value	0.5%	All minerals; $25 plus 0.5% on gross value over $5000.
Nevada	Net proceeds of mines	Net proceeds	Local property tax rate	Applies local property tax to net proceeds of all mines in state.
New Mexico	Resources excise	Gross value	0.75%	All taxable nonmetals (except potash). Imposed as a resources, processors, or service tax.
		Gross value	0.5%	Potash (if taxed under Resources Tax; 0.125% if under Processors Tax).

State	Tax	Base	Rate	Commodity / Definition
	Severance	Gross value	2.5%	Potash. Gross value defines as 0.333% of proceeds less deduction for expenses.
		Gross value	0.125%	Clay, gravel, gypsum, sand, fluorspar, pumice, and all other nonmetals. Gross value defined for individual commodities.
Ohio	Resource severance	Per ton	$0.01	Limestone, dolomite, and sand and gravel.
		Per ton	$0.04	Salt.
Oklahoma	Oil, gas, and mineral gross production	Gross value	0.75%	Asphalt.
South Dakota	Mineral severance	Net profits	4%	All minerals including limestone, soda, saline, trona, bentonite; $100,000 market value minimum.
Texas	Cement distributors	Per 100 lb	$0.0275	
	Sulfur production	Per long ton	$1.03	
Wyoming	Mining excise and severance	Gross value	2%	Any valuable deposits (except trona and oil shale).
			5.5%	Trona.
			4%	Oil shale.

Source: Commerce Clearing House, Inc., 1967, *State Tax Guide*, all states, New York, Chicago, and Washington, (with updated supplements to March 1978).

Table 6. Mineral Royalty Systems for State Lands

State	Rate	Base	Definition of Base
Alaska	None	None	Mineral location system used.
Arizona	5%	Mine mouth value	"...net value of minerals produced from the claim. Net value equals gross value less cost of transportation from place of production to place of processing, less costs of processing and taxes levied and paid on the production thereof." —27-234 Arizona Revised Statutes

Table 6 cont.

State	Royalty	Value basis	Definition
California	Min. 10% or negotiate %	Mine mouth value Net profits	"...gross revenue less approved costs for transporting and processing such products."
Colorado*	Sliding scale 4%–10% (based on ore value)	Gross purchase price	"...gross purchase price received by lessee for ores delivered to buying station or mill."
Idaho*	Sliding scale 2¾%–10%	Ore value per ton	"...F.O.B. mine price received by lessee for ores delivered to a mill or buying station." For captive market: "...assay value of ore and applicable market price quotation less any transportation and processing costs after departure of the ore from the mine property." —State of Idaho Mineral Lease
Missouri	5%	Gross value	"...value at point of shipment of metalliferous concentrates." Excludes all transportation, smelting, refining, handling, and selling charges. —State of Missouri Mineral Lease
Montana	Min. 5%	Full market value	"...value of metalliferous minerals recovered by the lessee from the state lands." —77-3-116 Montana Revised Statutes
Oregon	5%	Gross purchase price	"...gross purchase price paid by buyer for value of minerals mined or extracted and sold from said Leased Premises." Captive Markets: "...amount being received from independent miners for ores of like character and quantity." —State of Oregon Mining Lease

State	Rate	Basis	Definition
Texas	Min. 6¼%	Production value	"...gross value at the mill or buying station less reasonable transportation from mine to mill and less reasonable processing costs if gross value of ore is enhanced." —*Leasing State-Owned Minerals*, B.J. Beard
Utah	Sliding scale 3%–12½%	Gross ore value	"...gross value of ores at receiving or processing mill or plant." —Utah State Lease for Metalliferous Minerals
Utah (revision of 1980)	8%, fissionable; 6%, precious metals; 4%, all others	Mine mouth value	"...proceeds from the sale of metalliferous minerals as crude ore F.O.B. at the mine site." —Staff recommendation, Utah Board of State Lands
Wyoming	Sliding scale 5%–25%	Ore value per ton	Not further defined as there is no mineral production on state lands in Wyoming under this provision. Other minerals (uranium, sulfur, trona, phosphate and potash) pay royalties of 5% of ore value.
New Mexico	Min. 2% (5% for certain minerals, primarily uranium)	Net smelter return	"...gross returns from smelter, mill, or other sale, less reasonable transportation, smelting, and reduction charges, if any." —New Mexico Statutes 19-8-22

*Does not apply to energy minerals.

A slight change in the depletion allowance for iron ore, coal, and lignite resulted from enactment of the Tax Equity and Fiscal Responsibility Act (TEFRA) of 1982. Starting in 1984, the statutory depletion allowance must be reduced by 15% of the excess of statutory depletion over the adjusted basis of the property. The adjusted basis is defined as the initial yearly value determined for the cost depletion account. The depletion rates applied to revenue after royalties remain at 15% for domestic iron ore and 10% for coal and lignite.

Exploration: One of the characteristic features of the minerals industry is the need to explore for new mineral deposits. Companies involved in mining generally spend considerable sums of money in order to find and secure raw material assets so that their business activities can be continued. These expenditures are necessary if a firm intends to replenish its reserves which are continually being depleted by mining, and they constitute very significant sums of money. The tax treatment of these expenditures can be very important to the financial health of mining firms.

For tax purposes exploration expenses are defined as those expenditures incurred in ascertaining the existence, location, extent or quality of any deposit of mineral before the beginning of the development stage of the deposit. In general, the development stage begins when deposits of ore are shown to exist in commercially marketable quantities. Thereafter all expenditures attributable to that specific deposit are either development or operating expenditures.

Exploration expenditures also include those expenses normally associated with prospecting such as preliminary geological surveys, obtaining the right of entry permission from landowners, and geochemical surveys. Exploration expenditures do not include the costs of acquiring or replacing depreciable fixed assets; however, the annual depreciation allowance on equipment used in exploration activities is considered an exploration expenditure to the extent that it is so used.

The costs associated with the *acquisition* of mineral lands (e.g., claim staking) is not considered an exploration expense. This aspect may cause confusion when considered in terms of a typical arrangement whereby a company acquires a percentage interest in a property for performing certain exploration activities. Under these conditions are the exploration expenditures truly exploration or expenditures for property acquisition?

Consider the situation where mineral property owner A conveys to B an undivided 2/5 operating interest in the mineral property on the condition that B incurs all exploration expenditures thereon. Under these conditions the taxpayer who pays or incurs exploration expenditures in connection with the acquisition of a fractional interest in the operation may elect to classify the expenditures to the extent of his fractional share as exploration costs. Therefore, in the example, B may elect to classify 2/5 of his exploration expenditures as exploration, but he must capitalize the other 3/5 of the expenditures into the depletable basis of his 2/5 operating interest. In essence, a taxpayer can elect to declare exploration expenditures only to the extent of the fractional share of the operating interest (so acquired) in a mining property.

The financial impact of the preceding situation can be quite considerable because the difference between the two classifications can be immediate tax savings (expensing) vs. no tax savings. Because percentage depletion deductions are not restricted to the depletion basis, additions to that basis may have no impact on the firm's income tax liability. Therefore, if percentage depletion prevails over cost depletion, additions to the depletion basis for aquisition costs are, in reality, never recovered.

Another situation which sometimes arises and becomes confusing from a tax standpoint is the case where small quantities of mineral may be recovered and sold during the exploration stage. Under these circumstances, exploration expenses for the year are considered to be the excess of exploration expenditures over the sale of mineral in that year.

The careful definition and classification of allowable exploration expenditures are important because of their tax impacts. Prior to 1951, mining exploration costs were considered to be capital expenditures recoverable solely through the depletion allowance rather than current expenses which are deductible in the year paid or incurred. Since 1951, there have been several changes in the tax treatment of mineral exploration expenses, the most significant being associated with the 1969 Tax Reform Act. Since 1969 a taxpayer may elect either to capitalize his domestic exploration expenditures or deduct them currently under Section 617 of the Code. If he chooses to deduct these expenditures, they are subject to later recapture if the mine reaches the production stage. The basic premise remains, however, that exploration expenditures which are associated with unsuccessful properties (i.e., those which never reach production) are a currently deductible expense, whereas exploration expenditures associated with successful properties (i.e., those which reach the production stage) are capitalized and recovered—in theory, at least—through the depletion allowance.

Because of this tax treatment, it is obviously important to accurately and properly allocate exploration expenditures to specific property units, and subsequently to specific mines or deposits. The exact method of allocating exploration costs to specific properties has been the source of considerable controversy. Some companies allocate exploration costs associated with exploration reconnaissance programs to "areas of interest." As these areas of interest, or acreages within large tracts, are abandoned or released, the associated exploration expenditures, which were deducted currently for income tax purposes under Section 165, are not subject to future recapture. However, the exploration expenditures associated with areas from which a mine is developed must be capitalized, so that to the extent these expenditures were taken as current deductions at an earlier date, they must be recaptured as defined below.

The Internal Revenue Service (IRS) has addressed the treatment of geological and geophysical exploration expenditures incurred for the purpose of obtaining data that will serve as a basis for the acquisition or retention of a mineral property. The IRS has taken the position with respect to such expenditures, that to the extent they are allocable to an indentifiable project area, they must be capitalized, subject to the election under Section 617(a), as part of the cost of acquiring any property within such project area. A project area is defined as a territory that the taxpayer determines can be explored in a single integrated operation. Since costs deducted under Section 617(a) of the Code are subject to recapture, the allocation of exploration costs can be significant, and it behooves the mining company to give this matter careful consideration since exploration costs deducted under Section 165 of the code are not subject to recapture.

EXPLORATION AS A CURRENT DEDUCTION:

The reason that this discussion on expensing, capitalizing, and recapture, is important is simply that appropriate tax treatment of exploration expenditures can significantly affect overall cash flow. Under Section 617 of the Code and prior to 1983, a taxpayer could elect to deduct currently all exploration expenditures paid or incurred prior to the development stage of the mine. The election does not apply to costs incurred in the exploration for oil or gas. If the taxpayer chose not to deduct exploration expenditures currently, he must capitalize exploration expenses and either write them off under Section 165 when the property is abandoned or include them in the basis of the cost depletion account if the property comes into production. If the taxpayer wished to deduct exploration expenditures currently under Section 617(a), he simply deducted such expenses on the return for the first tax year for which such treatment was desired. A Section 617 election to deduct currently was applicable to all exploration expenses paid or incurred in the year of election and for all subsequent years, unless it was timely revoked with permission from the Commissioner. The election may not be revoked without permission at any time after Sept. 30, 1972. The taxpayer electing to deduct exploration expenses under Section 617 had to clearly specify on all such tax returns the amount of the deduction which was attributable to each specific property or area of interest with sufficient identification of the latter. This identification is necessary and important for application of the recapture provisions if and when the property results in a producing mine.

The Tax Equity and Fiscal Responsibility Act (TEFRA) of 1982 affected the current deduction for exploration provision slightly. In the past all exploration expenditures could be fully expensed in the year incurred. Now, only 85% of such expenditure can be expensed; the remaining 15% must be written off over a five-year period. The five specified yearly depreciation percentages are 15, 22, 21, 21, and 21%, respectively. The first deduction (i.e., 15% of the 15% exploration expenditure) is declared in the year the cost is incurred, while the remaining deductions are declared in the following four years. It should be noted that the 15% of the exploration expenditure so capitalized is eligible for investment tax credit.

RECAPTURE:

If the taxpayer has elected to deduct exploration expenses currently under Section 617 and exploration and development efforts subsequently result in a mine's reaching the production stage, Section 617(b) provides for the "recapture" of exploration expenditures previously deducted under Section 617. This recapture must be performed for all previously expensed exploration expenditures.

At this point the Code provides the taxpayer two alternatives for recapture:

1) The taxpayer may elect to include in gross income for the year an amount equal to the "adjusted exploration expenditures" (defined subsequently) for all mines reaching the production stage that year. The production stage is generally considered to commence when the mine is primarily being operated for production rather than for development. The amount included in income is also added in the year of recapture to the depletable cost basis of the properties for the respective mines. However, the amount included in income (recaptured) is *not* added to gross income from the property for percentage depletion purposes, presumably because the original Section 617 deduction could only affect percentage depletion in the 50% of net limitation, and any reduction in the depletion deduction there is accounted for in the "adjusted exploration expenditure" requirement (Maxfield, 1975).

2) If the taxpayer does not elect to include the amount of adjusted exploration expenditures deducted under Section 617 in gross income for the year in which the mine reaches the production stage, he must forego taking any depletion deduction on the mine until the amount of the depletion foregone equals the amount of the adjusted exploration expenditures assigned to that mine. The depletable cost basis is not reduced by the amount of depletion that is foregone. This method of recap-

ture must be used if the taxpayer does not elect option 1. The net effect of recapture option 2 is often to spread the recapture amount over several years. In general a taxpayer with a large amount of gross income would opt for the second alternative whereas a taxpayer with large amounts of other deductions for the taxable year may profitably include the entire amount in gross income. Considering inflationary pressures and the time value of money, most organizations choose option 2 for required recapture in order to maximize the present value of annual cash flows. The same method of recapture must be used for all mines reaching the production stage in a given taxable year, however a taxpayer may choose a new election (option 1 or 2) each year.

ADJUSTED EXPLORATION EXPENDITURES:

The term *adjusted exploration expenditures* is defined in Section 617(f)(1) as the excess of (1) total exploration costs deducted under Section 617(a) in all taxable years which would have, if capitalized, been included in the depletable cost basis, over (2) the amount, if any, of reduction in the depletion allowance for the property or mine because such costs were expensed (under Section 617(a)) rather than capitalized. Consequently the taxpayer need recapture only to the extent that he derived tax benefit from the original deduction thereof. Thus, to the extent that the 50% of taxable income limitation resulted in a reduction of the depletion allowance because of the deduction of such exploration expenditures, there need be no recapture.

The House Ways and Means Committee in its technical exploration of the foregoing offered the following example of computation to illustrate the procedure involved (Burke and Bowhay, 1978).

Example 1:

Assume that A owns the working interest in a large tract of land located in the United States. A's interest in the entire tract of land constitutes one property for purposes of Section 614. In the northwest corner of this tract is the operating Mine X, producing an ore of beryllium, which is entitled to a percentage depletion rate of 22%. During 1971, A conducts an exploration program in the southeast corner of this same tract of land, and he incurs $400,000 of expenditures to which Section 617(a)(1) applies in connection with this exploration program. A elects to deduct this amount as expenses under Section 617(a). During 1971, A's gross income from the property was $1 million, with reference to the property encompassing Mine X and the area in which exploration was conducted. A's taxable income from the property, before adjustment to reflect the deductions taken with respect to the property during the year under Section 617, was $400,000. The cost depletion deduction allowable and deducted with respect to the property during 1971 was $50,000. The amount of adjusted exploration expenditures chargeable to the exploratory mine (hereinafter referred to as Mine Y) at the close of 1971 is $250,000, computed as follows:

Expenditures allowed as deductions under Section 617(a) $400,000
    Gross income from the property $1,000,000
    22% thereof 220,000
    Taxable income from the property, before adjustment to
      reflect deductions allowed under Section 617 during
      year 400,000
    50% thereof, tentative depletion deduction 200,000
    Taxable income from the property after adjustment to
      reflect deductions allowed under Section 617 during

year ($400,000 minus $400,000)...............   0
Cost depletion allowed for year   50,000
Amount by which allowance for depletion under Section 611
was reduced on account of deductions under Section 617
($200,000 minus $50,000)   150,000

Adjusted exploration expenditures at end of 1971 ....$250,000

In the course of performing mining property evaluations most analysts do not, as a practical matter, distinguish between straight exploration expenditures and "adjusted exploration expenditures" for recapture purposes. Most analyses are performed simply on the basis of recapturing the actual exploration expenditures incurred or associated with the mine under consideration. The net impact of this approach is to introduce some conservatism into the analysis which is not perceived to be a significant concern in view of the reliability of other income and cost estimates contained in the analysis.

The following example illustrates the two recapture provisions for exploration expenditures previously expended under Section 617 and does not include the "adjusted exploration expenditure" complication.

Example 2:

A feasibility study on the Big Rock copper mining property suggests the following cost and operating data:

Exploration expenditures previously expensed: $8.0 MM
Property acquisition costs (cost depletion basis): $3.0 MM
Annual production: 2,500,000 tpy
Annual gross sales: $6.5 MM
Reserves: 37,500,000 tons
Annual royalty: $0.5 MM
Percentage depletion rate: 15%
Estimated annual operating costs (inclusive of depreciation): $3.5 MM

Calculate the depletion allowance for Big Rock Mine for the first four years under both recapture provisions for previously expended exploration ($8.0 MM).

Option 1 (amounts in 1000's)

	Year 1	Year 2	Year 3	Year 4
Revenue: (6,500 + 8,000) =	$14,500	$ 6,500	$ 6,500	$ 6,500
Royalty:	500	500	500	500
Net after royalty:	14,000	6,000	6,000	6,000
Costs:	3,500	3,500	3,500	3,500
Net after costs:	10,500	2,500	2,500	2,500
Depletion claimed:	900	900	900	900
Taxable income:	9,600	1,600	1,600	1,600

	Year 1	Year 2	Year 3	Year 4
Depletion basis:	11,000*	10,100	9,200	8,300
Reserves:	37,500	35,000	32,500	30,000
Unit depletion:	0.293	0.289	0.283	0.277
Production:	2,500	2,500	2,500	2,500
Cost depletion:	733	721	708	692
50% net:	1,250†	1,250	1,250	1,250
15% net after royalty:	900‡	900	900	900

Depletion earned:	900	900	900	900
Depletion recapture:	8,000	—	—	—
Recapture balance:	0	—	—	—
Depletion claimed:	900	900	900	900

*(3000 + 8000) = 11,000
†(6500 − 500 − 3500) × 0.50 = 1,250
‡(6500 − 500) × 0.15 = 900

The reader will note that option 1 is not entirely accurate because the "adjusted exploration expenditures" were not recaptured (the entire $8.0 MM was recaptured). The recaptured amount was not added to gross income from property for percentage depletion purposes. However, the example does represent the concept of the acceptable procedure for recapture under this option.

Option 2 (amounts in 1000's)

	Year 1	Year 2	Year 3	Year 4
Revenue:	$ 6,500	$ 6,500	$ 6,500	$ 6,500
Royalty:	500	500	500	500
Net after royalty:	6,000	6,000	6,000	6,000
Costs:	3,500	3,500	3,500	3,500
Net after costs:	2,500	2,500	2,500	2,500
Depletion claimed:	0	0	0	0
Taxable income:	2,500	2,500	2,500	2,500

	Year 1	Year 2	Year 3	Year 4
Depletion basis:	3,000	3,000*	3,000	3,000
Reserves:	37,500	35,000	32,500	30,000
Unit depletion:	0.08	0.086	0.092	0.10
Production:	2,500	2,500	2,500	2,500
Cost depletion:	200	214	231	250
50% net:	1,250	1,250	1,250	1,250
15% net after royalty:	900	900	900	900
Depletion earned:	900	900	900	900
Depletion recapture:	8,000	7,100	6,200	5,300
Recapture balance:	7,100	6,200	5,300	4,400
Depletion claimed:	0	0	0	0

*Basis not reduced by amount of depletion foregone.

As can be seen in this example, option 1 results in substantially more taxable income (and therefore taxes paid) in year 1 and is therefore generally considered to be the less desirable option. However, generalizations must be regarded with some skepticism, and each case should be evaluated on its own merits in order to ensure that the corporation maximizes annual net cash flows on an after-tax basis.

Development: When a mineral property progresses from the exploration stage to the development stage, the taxpayer is allowed under Section 616 of the Code to currently deduct expenditures incurred in the development of the mine. As with exploration expenditures, TEFRA (1982) specifies that only 85% of the development expenditure can be expensed in the year incurred. The remaining 15% must be written off over a five-year period in accordance with the percentage rates specified previously for exploration expenditures. Similarly, the 15% capitalized is eligible for investment tax credit.

Development expenditures are all expenses paid for development of the mine after the existence of ores or minerals in commercially marketable quantities has been confirmed. The IRS considers development expenditures to include all costs, resulting "directly from the mining process" of making the mineral accessible by the driving of shafts, tunnels, and similar processes or activities. The courts have held that development expenditures are those which are incurred for the *specific purpose of developing the mine or deposit.*

The point in time when a mine leaves the exploration stage and enters the development stage is, of course, dependent upon the circumstances of each particular case. For example, core drilling would be treated as an exploration cost until it has been established that a deposit contains commercially marketable quantities. However, subsequent drilling related to delineation of the extent (perimeters) and continuity of that deposit generally would be treated as a development cost and expensed under Section 616 without recapture. Thus, although core drilling is sometimes casually called "exploration drilling," it often is more precisely "development drilling."

For tax reasons the transition from exploration to development is important. The corporation involved should make every effort to establish this transition point in terms of a factual determination as early as possible in the life of the project. This point is usually substantiated through a preliminary financial analysis to show "commercial quantities." Further documentation which is helpful in determining when development commences may include (1) affirmative action by the board of directors, (2) approval or acceptance of development status through normal established corporate procedures, and (3) appropriate public and agency notification.

Development expenditures may be incurred during the development stage of the mine or during the production stage. While the development stage is considered to be that period in which a deposit is prepared for extraction by providing access to the deposit, the mine is considered to be in a production stage when the principal activity of the mine becomes the production of developed ore rather than the development of additional ores for mining. In other words, the production stage is that period when the major portion of production is obtained from workings not in the development stage or when the principal activity is production rather than development. The specific point in time when a mine leaves the development stage and enters the production stage depends on the facts and circumstances of the case. From a tax standpoint it makes little difference because such expenses would be considered either development or operating, and both can be expensed currently.

Development expenses often occur concurrently with production (operating) costs, although the precise distinction is sometimes unclear. For instance, the IRS has ruled that preproduction stripping is indeed a development expenditure, but the periodic removal of overburden in a strip mining operation is a production activity and not developmental. Another complicating situation which often arises is the case where drilling is being conducted at a producing property in the hopes of finding additional reserves. Is the ore delineated by this activity the result of exploration or development efforts and expenditures? If these expenditures are considered to be exploration, they are subject to recapture; if development, they can be currently expensed. This is obviously a gray area, although the criterion seems to be deposit continuity. If the ore drilled out is a logical extension or continuation of the ore body being mined, the expenditures are considered to be developmental. On the other hand, if the ore drilled out suggests a new discovery which can not be shown to be part of the existing ore body being mined, the expenditures are considered to be exploration expenses and subject to recapture.

Not all expenditures incurred in the development stage constitute development expenditures. For example, expenditures for depreciable assets used in development activities are not considered development expenditures. However the annual depreciation allowance on such assets is considered a development expenditure to the extent the asset is used for development. Installation costs for such equipment are also considered as development expenditures as are day-to-day type repairs of depreciable development equipment. On the other hand, costs incurred for assets which have no independent physical life of their own apart from the mineral deposit (i.e., roads, slopes, shafts, etc.) are legitimate development expenditures. It makes no difference if the road or other assets will subsequently be used in the production stage for haulage or some other function.

If a taxpayer incurs development expenses in association with the acquisition of a fractional share of a working interest, then these development expenditures are deductible only to the extent of the fractional interest so acquired. For instance, a taxpayer who acquires a 40% interest in a gold mine by agreeing to pay for all development costs for the mine may deduct only 40% of the incurred development expenses. The remaining 60% must be capitalized as depletable cost of his mineral interest.

DEVELOPMENT DEDUCTIONS DEFERRED:

Rather than deduct development expenditures currently under Section 616(a), the taxpayer may elect to defer such deductions under Section 616(b) for each specific mine or deposit. These deferred expenditures are not capitalized as mineral property cost for depletion purposes, nor do they become part of the depletable basis. The amortization of these costs is allowed as an ordinary deduction in the years benefited in addition to the depletion deduction. The election to defer is applicable to development expenses incurred both in the development and production stages. When the mine is in the development stage, however, the election to defer applies only to the excess of development expenditures during a taxable year over net receipts received from the sale of mineral from the mine during the year. Such expenses not in excess must be currently deducted. The taxpayer is thus prevented from timing the deduction of his development expenditures to avoid the effect of the 50% limitation on the amount of the percentage depletion deduction.

When the election is made to defer depletable expenditures these expenses are, in effect, capitalized and subsequently deducted on a ratable basis as the units of mineral benefited by the development are sold. The amount of the deduction may be calculated using the following relationship (Maxfield, 1975):

$$\frac{A}{B} = \frac{C}{D}$$

where: A is amount of deduction, B is total development expense for the particular mine, C is the number of benefited units sold during taxable year (tons, lb, oz, etc.), and D is the number of benefited units remaining as of taxable year.

The number of units benefited by such expenditures is the number of units remaining to be recovered at the end of the taxable year plus the number of units sold during the taxable year.

The election to defer development expenditures is made on a mine-by-mine basis and applies only to development expenditures incurred in the year of election. Consequently, a taxpayer may elect to defer expenditures incurred in one year and currently deduct similar development expenditures in the same mine the following year. However, the taxpayer must treat all development expenditures consistently for each mine during a given taxable year (i.e., some expenditures may not be deducted and some deferred in the same taxable year at the same mine). Once made, the election to defer is binding with respect to the year made and cannot be revoked.

When development expenditures are deferred they are deducted on a ratable basis as described previously. Under these conditions the "amortized" development deduction should be treated similarly to the depreciation and depletion allowances for cash flow determinations. The cash flow calculation incorporating these deferred or amortized development deductions may be represented as follows:

$$
\begin{array}{l}
\text{Revenues} \\
- \text{ Operating costs} \\
- \text{ Depreciation} \\
- \text{ Deferred (amortized) development} \\
- \text{ Depletion} \\
\hline
\text{Taxable income} \\
- \text{ Federal income taxes} \\
\hline
\text{Net profit} \\
+ \text{ Depreciation} \\
+ \text{ Deferred (amortized) development} \\
+ \text{ Depletion} \\
\hline
\text{Operating cash flows} \\
- \text{ Capital expenditures} \\
\hline
\text{Net annual cash flows}
\end{array}
$$

Profitable firms usually prefer to expense development expenditures as opposed to deferring them because expensing currently reduces taxable income and provides an earlier return of cash to the firm. However, some taxpayers may not have that luxury if profits from other activities are not available from which to deduct development expenditures. The tax ramifications which should be considered when deciding whether to currently deduct or defer development expenditures include the following:

1) The impact of deducting such expenditures on the net income limitation for percentage depletion. This seems to be the most important consideration.

2) The minimum tax impact (discussed in a later section).

3) The utilization of net operating losses and investment tax credits (discussed in a later section).

Obviously the decision to deduct or defer development expenditures has to be based on long-term projections for each property and the resulting financial impact on the taxpayer.

Finally, the reader should note that a mining company makes an entirely separate election of whether to capitalize or expense development expenditures on the financial books. If different elections are made for the financial books and the tax books, the difference will be recorded as deferred charges in the financial accounts.

Production Payments: Burgeoning capital costs associated with financing new mining ventures have added to the financial complexity of the minerals industry. Few, if any, traditional mining companies can afford to finance new property development directly and must rely on other companies or financial institutions for the needed capital. These financing arrangements can be very complex and their detailed treatment is beyond the scope of this section.

One such means of financing, typically utilized by the banking community, is through so-called production payments. For federal tax purposes, a mineral production payment is a right to a specific share of production from a mineral property (or a sum of money in lieu of the production) when that production occurs. The payment is secured by an interest in the minerals. The right to the production is for a period of time *shorter* than the expected life of the property, and the production payment usually bears interest.

The tax treatment of production payments was changed considerably by the Tax Reform Act of 1969 and now depends on the type of production payment and the use to which the resulting money is put by the mineral developer. For instance prior to August 1969, all production payments were considered to be economic interests in the mineral in place. However under present law a production payment is an economic interest only in the following situations: (1) when the proceeds from the assigned production payment are pledged to the exploration and/or development of the mineral property burdened by the production payment; or (2) when the production payment is retained by a lessor in a leasing transaction.

Production payments are typically created in one of two ways. (1) If the owner of any interest in a mineral property assigns his interest and retains a production payment, payable out of future production from the property interest assigned, the production payment is said to be *retained*. (2) If the owner of any interest in a mineral property assigns a production payment to another person but retains his interest in the property from which the production payment is assigned, the production payment is said to be *carved out*. In mining situations the carved-out production payment is most common where an operating company carves out a production payment to a financial institution in exchange for capital.

Under current law a production payment carved out of a mineral property after Aug. 7, 1969, is to be treated as a mortgage loan on the property and does not qualify as an economic interest in the property. The Reform Act of 1969 provides that the assignor (mining company) of a carve-out production payment does not recognize income in the year of the assignment, even though the company then receives a consideration in cash or its equivalent. Rather, the company reports income as minerals are produced even though part of the proceeds from sale of production are applied to reduction of the carved-out production payment. For tax purposes the company accounting will be exactly the same as though it had made a loan secured by production. Thus, if company XYZ, the owner of the working interest in a mining property, assigns a production payment of $1,000,000 plus some amount for interest to a financial institution in exchange for $1,000,000 in cash, the company reports no income at the time of assignment. In future periods if sales of production from the mine are $30,000 annually of which $25,000 is applied on the production payment, company XYZ must report the entire $30,000 as income subject to depletion (Burke and Bowhay, 1978). During this time operating or production expenses are deducted in the year that gross income is generated, so that there is a matching of income and expenses. It is important, for tax purposes, to remember that in the preceding illustration the financial institution does not possess a depletable economic interest in the mining property but rather, in effect, holds a mortgage loan.

Although the foregoing example of a carve-out production payment is correct in the general case, some complications arise if the production payment is assigned for exploration or development of the mining property. Since this is the normal type of arrangement in the mining industry, some clarification is required. As pointed out previously in this section, when a carve-out production payment is assigned to the exploration or development of the property, then an economic interest is involved in the transaction. Simply stated, if the owner of a working interest assigns a carved-out production payment to another party who makes a contribution to the exploration for or development of a mineral property, or assigns a production payment for a cash consideration which is pledged to be used in the exploration for or development of a mineral property, he does not realize taxable income (Burke and Bowkay, 1978). If the assignee of the production payment (a financial institution) is responsible for the exploration and/or development activities or expenditures, then the assignor (the mining company) has not realized depletable gross income and is entitled to no deduction for expenditures incurred by the assignee in the development of the property. Under these conditions the subsequent production payments to the financial institution would constitute a depletable economic interest.

In the normal case where the mining company receives a cash contribution in exchange for a carved-out production payment and performs the required exploration and development activities itself, then the company is required to offset the cash received against appropriate exploration and development costs. Under these circumstances the carved-out production payment would constitute an economic interest in the property and the holder of the production payment would be taxable on the income and allowed depletion on such income.

Financing mining ventures through the use of production payments can be a very complicated matter. Considerations such as: who uses the money, how is it used, what kind of production payment is involved, the implications of the Tax Reform Act of 1969, and many others must be considered before the appropriate tax consequences for the assignor and the assignee can be determined. Although some of the issues and specifics remain to be resolved, it is clear that the Tax Reform Act of 1969 has substantially restricted the availability and utility of both kinds of production payments.

Historically, the biggest tax effect on the mineral industry with respect to production payments was the very reason given by the Senate Finance Committee for the changes incorporated in the 1969 Tax Reform Act. First, with carved-out production payments, taxpayers were able to advance the time at which income was reported, with the result that the 50% limitation on taxable income for depletion purposes was avoided. Prior to 1969, contributions received from the sale of a carved-out production payment were considered to be ordinary income to the assignor (mining company) and subject to depletion allowance. In essence, the assignor was held to have realized currently the value of future income. Production applied in satisfaction of the production payment was considered taxable to the donor at the time that it was paid by the donee. Second, by means of a retained production payment, taxpayers were able to finance and pay off the purchase of a mineral business with before-tax dollars. The committee report criticized the treatment of production payments prior to the Reform Act as creating a large disparity between the mineral industry and other industries in the tax treatment of their respective methods of financing (Maxfield, 1975).

Bonus Costs: A typical situation in the mineral industry is one in which the owner of a mineral property executes a lease with another party and retains a nonoperating interest (i.e. a royalty, net profits, or some other interest) in the property over the life of the lease. As a part of this transaction it is not uncommon for the lessee to pay the lessor-owner an initial payment for the execution of the lease. This payment is referred to as a *bonus* and is commonplace in the private sector as well as in the public sector (i.e., the leasing of federally owned mineral rights). Also, a cash payment made to a land owner for an option to lease mineral lands is regarded for tax purposes as a form of bonus.

PAYER'S TAX TREATEMENT:

The regulations provide that bonus payments must be capitalized and recovered through the depletion allowance. As such they are considered to be acquisition costs and go into the cost depletion account. In the normal case where statutory depletion exceeds cost depletion, the payer of the bonus receives no tax advantage from capitalizing these expenditures as long as the lease is maintained.

Bonus payments received by the payee, or property owner, are considered for tax purposes to represent advance royalties and are therefore depletable. Because the bonus payment is depletable for the payee, the payer is required to deduct a prorata portion of the bonus from gross income from property in each taxable year before computing percentage depletion. The magnitude of the deduction is determined by taking that percentage of the bonus that is the percentage of total estimated mineral reserves sold during the tax year. The following example illustrates the procedure.

Blister Mining Co. pays property owner A $500,000 for a lease on a particular mineral property. During the tax year, 5000 tons of ore are produced and sold out of an estimated 250,000 tons of recoverable reserves. The amount of bonus to be excluded by Blister in the current year from gross income from property for the purposes of computing percentage depletion is:

$$\frac{5,000 \text{ tons sold}}{250,000 \text{ tons recoverable}} \times \$500,000 \text{ bonus} = \$10,000$$

At the present time it appears as though the amount of bonus excluded from gross income from property may not be added back in determining taxable income from property for purposes of the 50% limitation on statutory depletion.

PAYEE'S TAX TREATMENT:

As mentioned, a bonus is considered to be in the nature of advance royalty to the payee for tax purposes. As such it is depletable even though there is no assurance of any future production from the lease. If the payee takes a depletion deduction on the bonus payment and the lease subsequently expires, terminates, or is abandoned in a later tax year and there has been no production under the lease, the payee must restore the depletion deduction previously taken to income in the year in which the lease was terminated or abandoned. Restoration of these depletion deduction monies may be avoided by making a complete disposition by sale or gift of the property (i.e., disposition of the retained mineral interest prior to termination of the lease). Also if some *marginal* production is obtained and sold, the regulation appears not to require restoration. What constitutes marginal production is unclear.

It should be pointed out that a taxpayer who has deducted exploration expenditures under Section 617 for a mining property, and who subsequently receives a bonus or royalty payment from the same property, must forego depletion on the bonus or royalty until the amount of depletion which otherwise would have been allowed equals the amount of adjusted exploration expenditures associated with the property.

The regulations cite an example of a taxpayer who, having elected to defer development expenditures, subsequently leases the property for a royalty consideration. Under these conditions, the deferred development expenditures should be allocated in relation to the expected production accruing to the retained royalty. Such deductions should be ratably applied to royalties as they are received. If the taxpayer had received a bonus or advance royalty payment, an appropriate part of the deferred expenses should be allocated in proportion to the amount of the payment and the amount of future expected royalties.

The payee may take either cost or percentage depletion (whichever is larger) on bonus income. Percentage depletion is determined by multiplying the amount of the bonus by the proper percentage depletion rate with the resulting amount not to exceed 50% of the taxable income from the property. Cost depletion is more difficult because the landowner often has little or no basis in his mineral ownership. The bonus payment is considered an advance royalty which represents a part of the payee's share of income from units to be produced and sold in the future. However, because these future units to be sold cannot be readily estimated, the IRS has allowed the use of dollar equivalents in computing the cost depletion on a lease bonus. The prescribed formula is suggested (Burke and Bowhay, 1978):

$$\text{Cost depletion} = B \left(\frac{A}{A + R} \right)$$

where B is adjusted basis of the depletable property just before the lease, A is advance royalty or bonus payment, and R is royalties expected to be received in the future.

The problem, of course, with this formula is that there is no cost depletion unless there is a depletable basis. The taxpayer has the burden of proving the appropriate basis allocable to the minerals. If the land was acquired primarily for its surface value, little or no cost can be allocated to any mineral content subsequently found. Similarly the burden of proof is on the taxpayer when estimating royalties to be received in the future. If the taxpayer fails to carry the burden of proof in these areas, he may be limited to percentage depletion.

The depletable basis in a property for cost depletion purposes is less difficult in the case of a sublessor. The following example illustrates the depletion determination on a lease bonus under this condition.

Example 4

Mr. Hustle obtained a lease on a mineral property for the amount of $50,000 cash. He subsequently negotiated a sublease agreement with Jippo Mining Co. for $100,000 plus an overriding royalty of $0.02 per ton.

The estimated recoverable reserves in the property are 3,000,000 tons. Therefore, Mr. Hustle's overriding royalty interest would be estimated at (3,000,000 tons × $0.20 per ton) = $60,000.

Mr. Hustle's cost depletion deduction on the bonus payment is calculated using the $50,000 basis as follows:

$$\$50,000 \left(\frac{\$100,000}{\$100,000 + 60,000} \right) = \$31,250 = \text{cost depletion}$$

Mr. Hustle's statutory depletion (assuming the mineral qualifies for the 15% rate) is calculated as follows:

$$\$100,000 \times 0.15 = \$15,000$$

As a result, Mr. Hustle would report the $100,000 bonus payment as income and take from this income the cost depletion deduction of $31,250 since it exceeds the statutory rate.

The following example is offered in an attempt to summarize some of the more important tax considerations associated with common mineral property agreements.

Example 5

Assume that A owns some land having mineral potential but which has never been leased, explored, developed, or otherwise exploited for minerals. A's land just happens to be a part of a larger area which Faultless Mining Co. wishes to explore for mineral occurrences. Faultless has engaged an exploration group to perform some preliminary geological and geochemical surveys in this area to the tune of $40,000. These surveys indicate that mineral potential exists on A's land and encourage Faultless to gain access to the land in order to perform some exploration drilling.

After negotiations, Faultless agrees to pay A $5000 in cash for damages to the surface and thereby obtains the right of access to A's land and performs core drilling. In a separate transaction Faultless agrees to pay A $4000 for an option to select for future lease any portion of A's land which may be of interest to Faultless, plus a 10% royalty to A. The option specifically states that Faultless will make a lease bonus payment of $20 per acre for any acreage selected from A's land.

As a result of core drilling, Faultless chooses to exercise its option on 500 acres of A's land even though the survey made by the exploration group covered a much larger area. Faultless therefore pays A $10,000 (500 acres × $20 per acre) for the mineral lease on the 500 acres. In addition Faultless owes the exploration group $60,000 for the core drilling that led to the selection of the 500 acres of A's land.

For tax purposes, the following is the relative status of each party in the activity described.

1) Faultless has incurred the following expenditures:

$60,000 for exploration core drilling.

$40,000 for the preliminary reconnaissance surveys.

$5000 for surface damages and access right.

$4000 for option to select portions of A's land.

$10,000 for mineral lease on 500 acres.

The first three items ($105,000) represent exploration costs and may be currently expensed. Should the property subsequently become a producing mine, these expenditures would be subject to recapture. The $14,000 paid for the mineral lease and the initial option are considered property acquisition costs and are capitalized into the cost depletion account.

2) The exploration group received $100,000 for its services and must treat this amount as ordinary income.

3) A has received $19,000 as a result of owning the property. Of this amount, $14,000 is treated as a lease bonus and is subject to depletion since it is considered an advance royalty for tax purposes. The $5000 received for surface damages and exploration access represents ordinary rental income.

The Depreciation Allowance

Depreciation is an allowable deduction when computing taxable income which represents the exhaustion, wear, and tear of property used in a trade or business, or of property held for the production of income. The purpose of the depreciation deduction is to provide a means by which a business or trade can recapture the capital needed to keep itself in business. Therefore depreciation allowances for capital assets are deducted from taxable income in an orderly manner such that the property owner has deducted the initial investment in the asset by the time it wears out or becomes exhausted. Having recaptured the initial asset cost from the annual tax deductions, the owner can, in theory, replace the worn-out piece of equipment with a new one and keep himself in business. At present, however, depreciation methods do not take into account the escalation in capital costs which has occurred since the last machine was purchased, so that the amount recovered is rarely sufficient to purchase a replacement. The concept of depreciation applied to real or personal property is similar to the concept of the depletion allowance on ore deposits, another type of asset.

In order for a taxpayer to be eligible for the depreciation deduction, he must have an investment in the depreciable property. A taxpayer who owns a royalty, production payment, or net profits-sharing interest in a mineral property cannot declare a depreciation deduction since he does not own an interest in the equipment on the property.

Depreciable property may be either *tangible* or *intangible* property that is allowed a depreciation deduction or can be amortized. Tangible property is property which is physically employed in the production process but does not include inventories, stock-in-trade, a depletable natural resource, or land apart from its improvements. Intangible property includes items such as patents, copyrights, franchises, licenses, contracts, or similar assets having a limited useful life. Goodwill is not, therefore, depreciable because its useful life cannot be determined. Because there is generally no physical deterioration or obsolescence in intangible assets, the more general term, amortization, is used rather than depreciation.

Depreciable property may also be classified as to real or personal property. Real property is considered to be land and generally anything that is erected on, or attached thereto. In the case of mining, real property is represented by concentrators, preparation plants, silos, offices, shops, warehouses, etc. Land itself, however, is never depreciable. Personal property is property other than real estate and includes machinery and equipment. Equipment representing personal property often is housed in a real property building (i.e., a classifier in a concentrator).

Property is considered to be depreciable if it meets the following requirements:

1) It must be used in business or held for the production of income.

2) It must have a useful life that can be determined, and its useful life must be longer than one year.

3) It must be something that wears out, gets used up, becomes obsolete, or loses value from natural causes.

Depreciation begins when the property is placed in service and stops when the asset is retired from service. Property is considered placed in service when it is first available or is in a state of readiness for service. Even if a machine is ready for service in January, but is not actually used until September, depreciation may start in January. Depreciation may also be claimed on an idle asset if it is usually used in business but is temporarily idle. If there is a drop in the price of a given commodity causing the temporary shutdown of a group of assets, the assets are still treated as being used in business or trade for Federal income tax purposes.

In order to calculate the annual depreciation deduction for an asset, it is necessary to determine: (1) the basis for depreciation, (2) the useful life of the asset, (3) the estimated salvage value at the end of the asset's life, and (4) the method for allocating the depreciable amount over the useful life of the asset.

BASIS:

The basis in the property to be depreciated is the amount from which depreciation will be deducted. The original basis is typically the purchase price of the asset, although in the case of major pieces of equipment the basis may also include freight and erection expenditures. The basis is generally considered to be the total cost associated with placing an asset into service.

The basis of a piece of property may change with time. For instance, normal repairs and replacements on an asset which do not increase the value of property, make it more useful, or lengthen its useful life may be expensed currently. However, if any repair or replacement increases the value of property, makes it more useful, or lengthens its life, these expenditures must be capitalized into the basis and recovered through annual depreciation deductions. The capitalized amounts of such repairs and overhauls include the expenditures for labor as well as parts.

USEFUL LIFE:

Depreciation on an asset is to be calculated over the estimated useful life of the asset to the taxpayer, and not over the longer period of the asset's physical life. The useful life of an asset must be an estimate by the taxpayer as to how long it will continue to be useful in trade or business. Factors affecting the estimate of useful life for an asset include (1) frequency of use, (2) maintenance schedules and philosophy, (3) asset age when acquired, and (4) the job conditions to which it is subjected. The useful life of an asset can also be affected by technological changes, economic changes, regulatory laws, and many other factors.

Whenever possible the taxpayer should estimate the useful life for an asset based on specific job conditions and previous personal experience with similar assets. When this is impossible the experience of the industry can be used until better property-specific data become available. In the case of mining, some of the guideline lives provided by the IRS are:

1) Equipment used in mining	10 years
2) Transportation equipment:	
a) ore trucks under 13,000 lb	4 years
b) ore trucks over 13,000 lb	6 years
c) railroad cars	15 years

Use of the guideline lives is optional and does not preclude the use of longer or shorter lives. However, shorter lives must typically be justified by the taxpayer's replacement policies. Taxpayers may employ one method for determining the useful life for one asset and another method for other assets.

It is important to note that some expenditures for mining equipment may be expensed currently even though the equipment may have a useful life in excess of one year. Expenditures for equipment which are required to maintain the normal output of production solely because of the recession of the working face of the mine may so qualify. This is the well-known *receding face theory*. A good example of such equipment might be the auxiliary ventilation fans used when headings or entries reach the point where additional air is required in excess of that produced by the primary ventilation system. To qualify for current deductions, this equipment must not increase the value of the mine nor decrease the cost of production per unit of minerals. According to Maxfield (1975), vertical as well as horizontal recessions qualify, i.e., the opening of a new or deeper working face in the same deposit constitutes a qualifying recession where the old face is almost exhausted. However, the reason for the expenditure must be the recession of the face, and not something else such as a manpower shortage, requirements of law, or safe mining practices.

As certain facts and parameters change, the taxpayer may adjust his estimate of the remaining useful life of an asset. The change may be made regardless of the method of depreciation in use, but such changes should be made only when they are substantial.

SALVAGE VALUE:

Salvage value is the value of property at the end of its useful life. It represents that amount which might be obtained from the sale or disposition of the asset after it can no longer be used productively in one's trade or business.

Since the regulations state that an asset may not be depreciated below its reasonable salvage value, an estimate of the salvage value for a piece of depreciable property must be made when it is first acquired. As with the determination of useful life, the estimate of salvage value will depend on variables such as: (1) the job conditions in which the asset is used, (2) the length of time the asset is used, (3) maintenance schedules, (4) frequency of asset use, and (5) local resale or marketing potential. Again, personal experiences in conjunction with an assessment of the local used machinery market are the best sources of information for estimating salvage values.

For depreciation calculations the taxpayer may choose to use either gross or net salvage value estimates. Net salvage value is the value of a piece of property minus the cost of removal upon disposal. If this cost is more than the estimated salvage value, the net salvage is considered to be zero, because the salvage value of an asset can never be less than zero.

If the property acquired is personal property and has a useful life of three or more years, the taxpayer may use a salvage value that is less than the actual salvage value estimate by an amount equal to 10% of the basis in the property. This 10% reduction is based on the basis in the property at the time when salvage value must be estimated. This 10% rule applies to both new and used property. The following simple example illustrates the 10% rule.

Example 6

Given:

Initial purchase price of asset = $20,000 (personal property; basis of account)

Estimated useful life = 5 years

Estimated salvage value = $3000

The salvage value of $3000 can be lowered by $2000 (10% × $20,000). The resulting salvage value for depreciation purposes would then be $1000. If, however, the estimated salvage value were initially $1500, then the salvage value for depreciation purposes would be zero since salvage may not be less than zero.

In addition to the normal depreciation deductions the taxpayer may also be entitled to deduct up to 20% of an asset's cost as *additional first-year depreciation*. This additional deduction may only be taken in the first year and applies only to tangible personal property acquired for use in one's business. In addition the property must have a useful life of at least six years, determined on the date of acquisition. In making the calculation, the cost of the property is multiplied by 20% and figured before determination of the regular depreciation deduction. The salvage value is not taken into account when performing the 20% determination. The taxpayer cannot claim additional first-year depreciation on more than $10,000 of the cost of qualifying investment.

Methods of Depreciation: There are several different methods for determining the depreciation deduction which are acceptable. The taxpayer may choose any reasonable method as long as it is applied consistently. The four methods which follow are the most commonly used in the minerals industry.

1) Unit of Production (UOP) Method—This method of figuring depreciation is usually associated only with the minerals industry and is determined by dividing the basis in the asset by the number of total units to be produced in the life of the asset. If the asset is estimated to have a useful life equivalent to the life of the property, the basis is divided by the total estimated recoverable reserves in the deposit yielding a unit depreciation rate in dollars per ton. The annual depreciation deduction is calculated by multiplying the unit of depreciation determined previously, by the number of units produced during the year (Table 7).

The UOP approach is very similar in concept to cost depletion accounting. With the UOP approach the estimated salvage value is deducted from the basis before the unit depreciation rate is determined. If the annual production rate (units produced) is constant, the resulting UOP deduction will be equivalent to straight-line depreciation.

2) Straight-Line (SL) Method—This is perhaps the easiest and most common way to calculate depreciation for a piece of property. The depreciation deduction is the same for each year with this method. The total amount of depreciation which may be deducted over the asset's life is equal to the basis minus estimated salvage value. This amount (adjusted basis) is divided by the number of years of useful life (Table 7). An example of the calculation procedure for the SL method is provided in Table 8.

3) Declining-Balance (DB) Method—Although the total amount of allowable depreciation a taxpayer can deduct over the useful life of an asset is the same with all depreciation methods, some methods provide for a larger deduction in earlier years and less in later years. These techniques are referred to as *accelerated depreciation* methods and attempt to recognize the fact that most assets lose most of their value early in their lives. One of the commonly used accelerated depreciation techniques is called the declining-balance method.

With the declining-balance method the rate of depreciation is related to the normal straight-line rate which is 1 divided by the number of years of useful life ($1/N$ in Table 7). Depending upon certain restrictions, the declining-balance rate may be 1¼, 1½, or 2 times the straight-line rate. If the rate is 2 times the straight-line rate it is referred to as double-declining balance (DDB). In order to qualify for DDB, the following conditions must be met: (a) the asset must be new, (b) it must have a useful life of 3 years or more, and (c) it must be tangible, personal property. The 1½ rate is the maximum which can be applied to used equipment. To use this rate the property must be tangible property with a useful life of 3 or more years. The 1½ or 1¼ rate may be used for real and personal property.

When calculating the depreciation deduction, the depreciable base is *not* reduced by subtracting salvage value; however, it is reduced by the amount of depreciation taken in the previous year (Tables 7 and 8). The undepreciated balance cannot be reduced below the estimated salvage value at any time.

Table 7. Depreciation Methods (Hrebar, 1977)

Symbols: Dn = Depreciation in year n
B = Depreciable basis
S = Salvage value
N = Life of asset
n = Year $(1 - N)$
Pn = Units produced in year n
R = Total production in units
T = Times straight line rate

1. Unit of Production Method:

$$Dn = \frac{(B - S)}{R} \times Pn$$

2. Straight-Line Method

$$Dn = \frac{1}{N} \times (B - S)$$

3. Declining-Balance Method:

$$Dn = \frac{T}{N} \times (B - \sum_{x=1}^{n-1} D_x)$$

where,

$$(B - \sum_{x=1}^{n-1} D_x) \geqslant S$$

4. Sum-of-the-Years-Digits Method—General Rule:

$$Dn = \left[\frac{(N + 1 - n)}{N(N + 1)/2} \right] \times (B - S)$$

Table 8. Examples of Depreciation Calculations*
(after Gentry and Hrebar, 1978)

Assuming: Basis for asset = $20,000
Useful life = 5 years
Salvage value = $2000

| | Straight-Line Method | | | |
Year	Depreciable base, $	Rate	Annual deduction, $	Unrecovered balance, $
1	(20,000 − 2000)	1/5 = 0.20	3600	14,400
2	18,000	0.20	3600	10,800
3	18,000	0.20	3600	7200
4	18,000	0.20	3600	3600
5	18,000	0.20	3600	-0-

(Double) Declining-Balance Method

Year	Book value, $	Rate	Annual deduction, $	Unrecovered balance, $
1	20,000	2 × 1/5 = 0.40	8000	12,000
2	12,000	0.40	4800	7200
3	7200	0.40	2880	4320
4	4320	0.40	1728	2592
5	2592	0.40	1037† (592)	1555 (2000)

Note that with the declining balance method there is always a balance at the end of the asset's useful life which has not been written off. This amount could be written off as a loss if indeed the asset is abandoned.

Alternatively, the taxpayer could elect to switch to the straight-line method and recover the remaining cost (less the appropriate salvage value) over the remaining useful life of the asset. In the example, if one switched to SL in year 4, the last three years of the depreciation deduction example would appear as follows:

Year	Basis, $	Rate	Annual deduction, $	Unrecovered balance, $
3	7200	0.40	2880	4320
4	(4320 − 2000)	1/2 = 0.50	1160	3160
5	2320	0.50	1160	2000 (salvage)

Although the declining balance deduction in year 4 is larger than the SL deduction, the SL deduction in year 5 is larger. In this case the switch would not be desirable from the time value of money standpoint. However, under conditions where salvage value is estimated to be zero, the switch to SL is definitely preferable because it maximizes depreciation deductions on a time value of money basis.

Sum-of-the-Years-Digits Method

Year	Depreciable base, $	Rate	Annual deduction, $	Unrecovered balance, $
1	(20,000 − 2000)	5/15	6000	12,000
2	18,000	4/15	4800	7200
3	18,000	3/15	3600	3600
4	18,000	2/15	2400	1200
5	18,000	1/15	1200	-0-

*For simplicity, this table ignores adjusted salvage value and additional first year depreciation discussed on p. 222.
†In practice the deduction in year 5 could only be $592 because of the $2000 salvage value limitation.

The taxpayer may change from the declining-balance to the straight-line method at any time during the useful life of the property. A change from the 200% or 150% declining-balance method to straight-line method for new property may be made without prior consent. Similarly, a change may be made from the 150% rate to the straight-line rate for used property. As illustrated in Table 8 the declining-balance method always leaves a residual unrecovered balance unless a switch is made to the straight-line method. With DDB the optimum time to switch to the straight-line method is the year equal to or greater than (Life/2) + 1 when there is no salvage value. After the switch has been made to straight-line, the taxpayer may not change back to the declining balance method or any other method of depreciation for that particular asset or group of assets for a period of ten years without permission from the IRS.

4) Sum-of-the-Years-Digits (SYD) Method—Like declining-balance, the SYD depreciation method is an accelerated technique. It allows the taxpayer to take larger depreciation deductions in earlier years and smaller ones in later years. With this method the salvage value is deducted from the cost or original basis before this basis is multiplied by a different fraction each year (Table 7).

The appropriate annual fraction depends on the life of the asset. The numerator is the number of years remaining in the useful life of the property. The denominator of the fraction does not change and is the sum of all the numbers in the numerator. For example, an asset with a five-year life would have a fractional denominator of $(5 + 4 + 3 + 2 + 1 = 15)$. Another means of determining the denominator is to incorporate the useful life, N, as follows:

$$\frac{(N + 1) \times N}{2}$$

In order to qualify for SYD, the property must meet the same requirements as established for DDB. As with the declining balance method, the taxpayer may switch from SYD to SL whenever it is to his advantage to do so. Table 8 illustrates the calculation procedure.

Fig. 10 shows the comparison between the various depreciation methods over time. The illustration is performed on a present value basis using a discount rate of 15%. Included in this illustration is the condition where a switch from DDB to SYD is permitted under the class life asset depreciation range (discussed in a later portion of this section). As shown, this alternative provides for the fastest write-off of equipment. Other than this special case, SYD seems to be the preferred method for assets with lives greater than or equal to seven years. DDB with conversion to SL is preferable for asset lives in the three to seven year range.

It should be remembered that one method of depreciation may be used for one particular asset, and a different method may be used for another asset of the same kind or in the same class.

Systems of Depreciation: There are many systems of depreciation accounting which may be employed. The taxpayer may establish as many accounts for depreciable assets as he wishes. He may list each asset separately or combine two or more assets into one account. Following are the major depreciation systems used.

Single Item Accounts—With this procedure each individual item of property is treated as a separate account. Because each asset has its own account and depreciation method, the bookkeeping efforts can be considerable at a typical mining operation.

Component Accounts—The taxpayer may account for depreciable property by treating each component or part of the property as a separate account. The components of a new building, for instance, might be the wiring, plumbing, roof, walls, and floor. If these items are accounted for separately, they may be depreciated separately. Such an accounting system becomes an overwhelming task at most operations.

Multiple Asset Accounts—With this system it is possible to combine a number of assets with the same or different useful lives into one account, using a single rate of depreciation for the entire account. Multiple asset accounts are generally broken down into group, classified, and composite accounts. The rates are determined

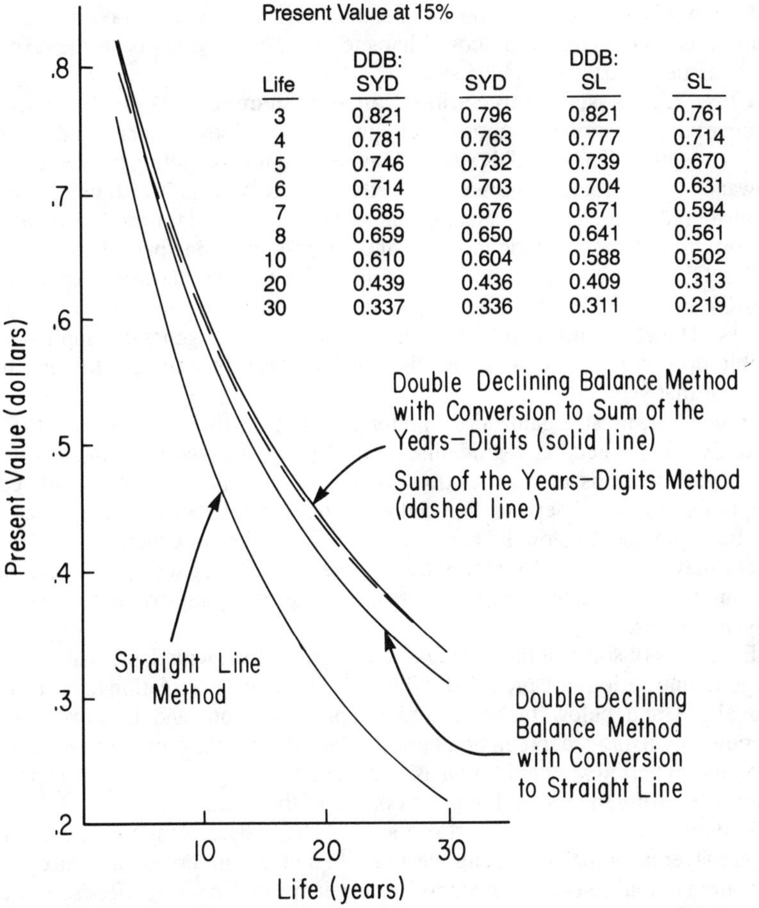

Present Value at 15%

Life	DDB: SYD	SYD	DDB: SL	SL
3	0.821	0.796	0.821	0.761
4	0.781	0.763	0.777	0.714
5	0.746	0.732	0.739	0.670
6	0.714	0.703	0.704	0.631
7	0.685	0.676	0.671	0.594
8	0.659	0.650	0.641	0.561
10	0.610	0.604	0.588	0.502
20	0.439	0.436	0.409	0.313
30	0.337	0.336	0.311	0.219

Fig. 10. Present value of $1 of depreciation vs. life of property (Gentry, Hrebar, and Martin, 1981).

differently for multiple asset accounts, depending upon the type of account established.

Group accounts. These accounts contain assets similar in kind with approximately the same useful lives. The group rate of depreciation is determined from the average of the useful lives of the assets in the account.

Classified Accounts. These consist of assets classified according to use without regard to useful life, such as machinery and equipment, furniture and fixtures, or transportation equipment.

Composite Accounts. These accounts include assets without regard to their character or useful lives. For classified and composite accounts, the applicable rate is generally determined by computing the depreciation for one year for each item or each group of similar items, and dividing the total depreciation thus obtained by the total cost or other basis of the assets. This average rate is to be used as long as later additions, retirements, or replacements do not substantially alter the relative proportions of different types of assets in the account.

Group, classified, or composite accounts may also be set up on the basis of location, dates of acquisition, cost, character, use, or any other basis the taxpayer considers necessary for his business.

Class Life Asset Depreciation Range System (CLADR): The CLADR system of depreciation accounting is unique in terms of other systems and deserves special attention. The CLADR system may be used for computing the depreciation allowance for depreciable property placed in service after 1970. Eligible property is considered to be tangible personal and real property used for producing income and for which an asset guideline class and an asset guideline period are in effect.

The CLADR system was designed to minimize disputes between taxpayers and the IRS as to the useful life, salvage value amounts, repairs, and other asset matters. The election for CLADR is made each year and generally applies to all eligible property. It may not be revoked for any year once the election has been made for that year.

In general, most organizations prefer CLADR for the simple reason that the useful lives stipulated in the guidelines period are usually shorter than those under the normal methods of depreciation. Also, salvage value is not considered in computing the annual depreciation deduction under CLADR, although no account may be depreciated below its reasonable salvage value. In general, the CLADR system may not be used for otherwise eligible property for which depreciation is computed under a method other than straight-line, declining-balance, or sum-of-the-years-digits.

Examples of some of the mining-related assets along with depreciation ranges and guideline periods are provided in Table 9. The asset depreciation range extends 20% above and below the class life. Depreciation on land improvements is computed by using a depreciation period selected from the range for the class. A depreciation period selected for an asset cannot be changed by either the taxpayer or the IRS during the remaining period of use of the asset.

Vintage Accounts—The heart of the CLADR system is the vintage account. The taxpayer must include all eligible assets first placed in service in the tax year in either item or multiple-asset accounts by the year placed in service. These accounts by the year placed in service are called *vintage* accounts. A vintage account is a closed-end depreciation account, which means that no additions to the account are permitted after the year of origin.

Each vintage account must include only assets within a single asset guideline class. However, more than one account of the same vintage (same tax year) may be established for different assets of the same class. For example, if the taxpayer purchases two drills in 1981 he may establish one in a vintage account having an 8-year life and the other in an 11-year vintage account (assuming an asset depreciation range of 8 to 12 years).

Certain kinds of assets may not be included in the same vintage account. These may be classified as follows:

1) New assets must be kept separate from used assets.

2) Property entitled to the special 10% salvage reduction must be kept separate from property not entitled to the reduction.

3) Property for which the additional first-year depreciation allowance is elected must be kept separate from property for which the additional first-year depreciation allowance is not elected.

4) Personal property must be kept separate from real property.

The depreciation methods used for calculating the annual depreciation deduc-

Table 9. Examples of CLADR System Related to Mining

Asset guideline class	Description	Asset depreciation range in years			Annual asset guideline repair allowance percentage
		Lower limit	Asset guideline period	Upper limit	
00.241	Light general purpose trucks (less than 13,000 lb)	3	4	5	16.5
00.242	Heavy general purpose trucks (13,000 lb or more)	5	6	7	10.0
00.25	Railroad cars and locomotives, except those owned by railroad transportation companies	12	15	18	8.0
00.3	Land improvements		20		
10.0	Mining	8	10	12	6.5

tions for the vintage accounts are either straight-line, declining-balance, or sum-of-the-years-digits. The rules and procedures associated with the use of these methods in the CLADR system are the same as previously discussed, except that the cost or other basis of the vintage account is not adjusted for salvage values. The same method of depreciation must be used for all property in a single vintage account. For instance, if the taxpayer selects a depreciation period of less than three years for a vintage account, he may only use the straight-line method for depreciation even if the property is new.

During the depreciation period for a vintage account, the taxpayer may make the following method changes only: (1) from the declining balance method of depreciation to the straight-line method or to the sum-of-the-years-digits method, or (2) from the sum-of-the-years-digits method to the straight-line method. Generally, no other change in method is permitted.

Percentage Repair Allowance—Under CLADR a taxpayer may continue to handle expenditures for repairs, maintenance, overhaul, and improvement of property used in his business as he has in the past. He can decide whether to deduct an expenditure currently or capitalize it on the basis of whether it appreciably prolongs the life of the asset, materially increases its value, or adapts it to a different use. However, difficulties have arisen where these expenditures have characteristics of both deductible expenses and capital expenditures. To minimize controversy between the taxpayer and the IRS, the taxpayer electing CLADR *may* elect the asset guideline class repair allowance which was designed to reduce the repair-capital expenditure question to a mechanical computation. If this election is made, a stipulated percentage (Table 9) of expenditures for repairs, etc., for each guideline class may be deducted as current expenses. If the election is not made, these expenditures are subject to the usual tax rules for determining whether an item is a repair or a capital expenditure. If the percentage repair allowance is elected, the calculation procedure is as follows:

1) Determine the total actual expenditure for repair, maintenance, rehabilitation, or improvement of eligible property in an asset guideline class.

2) Reduce the total in 1) by expenditures which are obviously considered to be capital in nature and must be recovered through depreciation.

3) The balance is deductible as repairs if it equals or is less than the amount computed as the *repair allowance*.

4) To the extent that the balance exceeds the repair allowance it is treated as property improvements and must be capitalized and recovered through depreciation.

The *repair allowance* for a class is obtained by multiplying the repair allowance percentage in effect in that class by the average unadjusted basis of repair allowance property in that class. The IRS provides some definite procedures to follow when determining the repair allowance (IRS, Publication 534). Some pertinent repair allowance percentages are provided in Table 9.

Conventions—The depreciation allowance for a vintage account (whether an item or multiple-asset account) must be determined by using one of two first-year conventions: (1) the modified half-year convention, or (2) the half-year convention. The same convention must be adopted for all vintage accounts established in a given tax year, but not for those of another tax year.

The *modified half-year convention* requires depreciation to be determined as: (1) all vintage account property placed in service during the first half of the tax year is considered placed in service on the first day of the tax year; and (2) all vintage account property placed in service during the second half of the tax year is considered placed in service on the first day of the succeeding tax year.

The *half-year convention* treats all property placed in service during the tax year as if it had been placed in service on the first day of the second half of the tax year. Example 7 illustrates this procedure.

Example 7

Assume: Initial purchase price = $10,000
(Basis in depreciation account)

Estimated useful life = 5 years
Half-year convention
Sum-of-the-years-digits depreciation method
No salvage value

Year	Basis	Rate	Annual Deduction
1	$10,000	5/15 × 0.50	$ 1,667
2	10,000	(5+4)/15 × 0.50	3,000
3	10,000	(4+3)/15 × 0.50	2,333
4	10,000	(3+2)/15 × 0.50	1,667
5	10,000	(2+1)/15 × 0.50	1,000
6	10,000	1/15 × 0.50	333
			$10,000

Economic Recovery Tax Act of 1981: In August 1981, the Economic Recovery Tax Act of 1981 was signed into law. Perhaps the most significant tax changes resulting from this law are associated with depreciation accounting. This law established the accelerated cost recovery system (ACRS) of depreciation which replaces the CLADR system for property placed in service after 1980. However, post-1980 CLADR depreciation may continue to be claimed on pre-1981 assets for which the CLADR election was made.

ACRS is an electable system in the same manner as CLADR. As such ACRS is not intended to replace other systems of depreciation such as single item accounts, component accounts, and multiple asset accounts. These systems of depreciation accounting may continue to incorporate straight-line, declining-balance, and sum-of-the-years-digits methods of depreciation. The same flexibility in choosing depreciation methods is not allowed, however, in ACRS. Under ACRS, recovery of capital costs for most tangible depreciable property placed in service after 1980 is made using accelerated methods of cost recovery over statutory recovery periods that are unrelated to and shorter than the latest CLADR lives (Commerce Clearing House, 1981a).

Some of the major characteristics of ACRS associated with depreciation tax accounting are summarized:

1) The cost recovery methods and periods are the same for both new and used property.

2) Salvage value is disregarded in computing ACRS allowances.

3) The cost of eligible property is recovered over a 3-, 5-, 10-, or 15-year period, depending upon the type of property.

4) Eligible property includes depreciable property other than (a) property for which an election to amortize was made, and (b) most property for which an elec-

tion was made to use a depreciation method not expressed in terms of years (e.g., unit-of-production).

5) The allowable percentages under ACRS depend on whether the asset is real or personal property and on the year the asset is placed in service.

6) The option which existed under the CLADR system of choosing between the half-year and the modified half-year convention does not exist under ACRS. Although a half-year convention exists for ACRS, it is different from the CLADR system. For instance, with the CLADR system an asset with a depreciation period of five years is actually depreciated over six tax years. Half a year's depreciation occurs in the first year and half a year's depreciation occurs in the sixth year. However, under ACRS half a year's depreciation occurs in the first year while the remaining depreciation is received over the next four tax years. This allows depreciation deductions over the same time period as the asset's life.

7) The unadjusted basis of real property is to be recovered over a period of 15 years.

8) The ACRS half-year convention is not used for real property. The first year percentages are determined by multiplying the allowable percentage by the fraction of the year that the realty is in service. The last year percentages are determined the same way. In most cases real property will be depreciated over 16 years.

9) The allowable percentage for real property is determined by using 175% declining balance and switching to the straight-line method at such time as the deduction is maximized.

10) In ACRS the basis can be depreciated below the salvage value, but salvage value income is considered ordinary income for tax purposes in the year it is received.

11) The taxpayer may elect to claim straight-line ACRS deductions over the regular recovery period or optional longer recovery periods. The rules vary depending on whether the asset is real or personal property. The following summarizes the options:

a) Personal Property:

Class of Property	Taxpayer may elect a recovery period of
3-year property	3, 4, or 12 years
5-year property	5, 12, or 25 years
10-year property	10, 25, or 35 years
15-year public utility property	15, 35, or 45 years

Under this election, the half-year convention must be used by claiming a half year of cost recovery in the year the personal property is placed in service and in the year following the end of the recovery period. If disposed of before the end of the recovery period, no cost recovery is allowed for that property for the year of disposition.

b) Real Property:

The optional recovery periods under the straight-line ACRS for real property are 15, 35, 45 years. The half-year convention is not used for real property. The procedure described in item 8 is utilized.

12) An election can be made to expense certain depreciable business assets. The total amount which can be expensed is limited to:

$$\$ \quad 0 \text{ in } 1981$$
$$\$ 5,000 \text{ in } 1982 \text{ and } 1983$$

$ 7,500 in 1984 and 1985
$10,000 in 1986 and beyond

Only personal property is allowed to be expensed in this manner and investment tax credit is not allowed on the expensed amount.

The allowable recovery percentages for personal property under the ACRS for 1981-84, 1985, and after 1985 are shown in Tables 10 through 12. These tables reflect the half-year convention and (1) 150% declining-balance method with a switch to straight-line method at such time as will maximize such deduction for years 1981 through 1984; (2) the 175% declining-balance method with a switch to the sum-of-the-years-digits method for 1985; and (3) the 200% declining-balance method with a switch to the sum-of-the-years-digits method for years after 1985, respectively.

For ACRS deductions recovery property means depreciable tangible property either used in a trade or business or held for the production of income. The classes of recovery property are defined in terms of Sec. 1245 class property, Sec. 1250 class property, and the CLADR class lives of assets as of Jan. 1, 1981. Each item of recovery property is assigned to one of the following classes:

1) 3-Year Property:

a) Sec. 1245 class property with a CLADR class life of four years or less (e.g., automobiles, light-duty trucks).

b) Sec. 1245 class property used in connection with research and experimentation.

2) 5-year Property:

a) Sec. 1245 class property with a CLADR class life of five or more years. This includes most mining equipment since it would be included in the 8- to 12-year CLADR classification.

Table 10. For Property Placed in Service, 1981-1984

	The applicable percentage for the class of property is:			
If the recovery year is:	3-year	5-year	10-year	15-year public utility
1	25	15	8	5
2	38	22	14	10
3	37	21	12	9
4		21	10	8
5		21	10	7
6			10	7
7			9	6
8			9	6
9			9	6
10			9	6
11				6
12				6
13				6
14				6
15				6

Source: Commerce Clearing House, 1981a.

3) 10-Year Property:

a) Sec. 1250 class property (real property) with a CLADR class life of 12.5 years or less.

Table 11. For Property Placed in Service in 1985

If the recovery year is:	The applicable percentage for the class of property is:			
	3-year	5-year	10-year	15-year public utility
1	29	18	9	6
2	47	33	19	12
3	24	25	16	12
4		16	14	11
5		8	12	10
6			10	9
7			8	8
8			6	7
9			4	6
10			2	5
11				4
12				4
13				3
14				2
15				1

Source: Commerce Clearing House, 1981a.

Table 12. For Property Placed in Service After 1985

If the recovery year is:	The applicable percentage for the class of property is:			
	3-year	5-year	10-year	15-year public utility
1	33	20	10	7
2	45	32	18	12
3	22	24	16	12
4		16	14	11
5		8	12	10
6			10	9
7			8	8
8			6	7
9			4	6
10			2	5
11				4
12				3
13				3
14				2
15				1

Source: Commerce Clearing House, 1981a.

4) 15-Year Property:

a) Sec. 1250 class property that does not have a CLADR class life of 12.5 years or less. Other real property, such as living quarters for miners, machine shops, office buildings, and warehouses which did not have CLADR guideline periods and which could only be depreciated over periods of 45 and 60 years, can now be depreciated in 15 years.

The obvious benefits of ACRS are that assets can be depreciated over shorter time periods, at faster rates, and, therefore, tax savings are realized to the firm much sooner.

Tax Equity and Fiscal Responsibility Act of 1982 (TEFRA): The large federal deficits which resulted from the tax cuts in 1981 created much concern. Resulting public and political pressures forced a reassessment of tax legislation leading to (TEFRA), which changed depreciation in two ways. First, ACRS depreciation tables for the year 1985 and thereafter were dropped. Therefore, Table 10 now applies for personal property depreciation rates and no future acceleration of depreciation allowance is recognized. Second, for years after 1982, the depreciation basis must be reduced by 50% of the ITC claimed for the asset or property. Table 13 shows the applicable depreciation rates for personal and real property when the basis is reduced. It should also be noted that the taxpayer has the option of lowering the investment tax credit (ITC) from 10 to 8% instead of reducing the depreciable basis of the asset. It can be demonstrated that the best option is to claim the full ITC and use the reduced depreciation percentages. These conditions apply to personal property which qualifies as Section 38 and Section 1245 property.

It is unclear at this writing if the basis for capitalized exploration and development costs must also be reduced by 50% of the ITC claimed. The Code is not specific on this point, and most analysts assume that such an adjustment is not mandatory—at least at present.

Table 13. Tax Equity and Fiscal Responsibility Act of 1982—ACRS Reduced Depreciation Percentages for Mining Related Personal Property (Nilsen, 1983)

Recovery year	3-year	5-year	10-year	15-year public utility	15-year real property
1	24.25	14.25	7.60	4.75	12
2	36.86	20.90	13.30	9.50	10
3	35.89	19.95	11.40	8.55	9
4		19.95	9.50	7.60	8
5		19.95	9.50	6.65	7
6			9.50	6.65	6
7			8.55	5.70	6
8			8.55	5.70	6
9			8.55	5.70	6
10			8.55	5.70	5
11				5.70	5
12				5.70	5
13				5.70	5
14				5.70	5
15				5.70	5

Gain on Disposition of Depreciable Property: In the past if a taxpayer depreciated an asset to nearly zero during its useful life and then sold the asset for more than its depreciated book value, a capital gain on the sale of the asset would result which was taxable at the relatively low capital gains rate. These possibilities are now limited with respect to so-called Section 1245 assets. Section 1245 assets generally include depreciable tangible or intangible personal property and most other depreciable tangible property except buildings or their structural components.

At the present time, gains realized on the sale or disposition of Section 1245 property is taxed as ordinary income to the extent of the depreciation taken. The impact of this procedure is to discourage the use of accelerated depreciation methods if Section 1245 property is likely to be sold.

The Code expressly provides that to the extent gain is realized from the sale of Section 1245 property, the cost of mining for purposes of the 50% net income limitation on statutory depletion is reduced by the amount of the gain. The net effect is to increase the net income from the property affected.

Investment Tax Credit (ITC)

The provision for invesment tax credit in the federal income tax law provides Congress with a mechanism for regulating, to some degree, investment in the business community at specific times. Investment can be encouraged or discouraged fairly easily by simply changing factors in an ITC calculation such as rates, qualifying percentages, and definitions of qualifying investments. Because ITC terms change frequently some financial analysts working with new mineral properties choose not to incorporate ITC into their analyses. Although the ITC has been in effect for some time now on a constant, or slightly increasing basis, there are no guarantees that it will continue in its present form by the time a new mining property is ready to come on-stream.

ITC is really an investment incentive which allows for a reduction in federal income taxes during the year in which qualifying investment is placed in service. Being a tax credit rather than a tax deduction, ITC is a particularly strong incentive. To qualify for ITC, property must:

1) Be depreciable.

2) Have a useful life of at least three years.

3) Be tangible personal property or other tangible property, excluding buildings or their structural components, used as an integral part of production or extraction.

4) Be placed in service during the year in a trade or business or for the production of income. Property is considered placed in service in the earlier of the tax year in which the period for depreciation of the property begins, or the tax year in which the property is placed in a condition of readiness and availability for service.

The amount of investment in qualifying property that is eligible for ITC depends on the useful life of the property and whether it is new or used. The useful life of the property is determined by the taxpayer at the time it is placed in service. The taxpayer must, however, use the same useful life that he used to calculate depreciation on the qualifying property to determine the amount of investment eligible for ITC. The amount of investment in qualifying property which is eligible for ITC both prior and subsequent to the Economic Recovery Tax Act of 1981 is:

Prior to the Economic Recovery Tax Act of 1981		Subsequent to the Economic Recovery Tax Act of 1981	
Life	Fraction of qualifying property eligible for ITC	Life	Fraction of qualifying property eligible for ITC
0-3 years	0	3-year property	60%
3-4 years	1/3	5-year property	100%
5-6 years	2/3		
7 or more years	100%		

Although the entire cost of new property may be considered when determining ITC for any particular year, the taxpayer may count no more than $100,000 of the cost for qualifying used property in determining ITC for any one year. This allowance was increased to $125,000 for used property for years 1981 to 1984.

The useful (depreciable) life mentioned previously presented an interesting set of options to the taxpayer. For instance, if the taxpayer elects to use the CLADR system of depreciation, he has the opportunity to place short-lived (three-six years) assets into eight-year vintage accounts for depreciation purposes and thus qualify these assets for the entire amount of ITC. However, by qualifying for this additional ITC, he would be spreading out depreciation deductions over a longer time period. The analyst should carefully explore the impacts of these trade-offs on an after-tax basis and choose the alternative which maximizes annual cash flows on a time-value-of-money basis.

After the amount of qualifying investment which is eligible for ITC has been determined, the actual ITC may be calculated. The regular ITC allowable at the present time is 10% of the *eligible* investment. The actual amount of ITC which can be declared in a particular year, however, is limited to the lesser of the tax liability, *or* $25,000 + $X\%$ (tax liability over $25,000), where

$$X = 70\% \text{ in } 1980$$
$$X = 80\% \text{ in } 1981$$
$$X = 90\% \text{ in } 1982 \text{ or thereafter}$$

If the ITC which is calculated is more than the limit described, the taxpayer may carry the unused credit back to the prior three tax years and forward over the following 15 tax years.

In 1982 TEFRA altered the conditions of investment tax credit (ITC) slightly. The limit for ITC is now reduced to $25,000 plus 85% of the total tax liability above $25,000. Also, any unused tax credits after 15 years can be deducted in the 16th year, but only for 50% of the unclaimed credit.

When performing economic analyses on new mining properties, analysts commonly assume that in the corporate case at least there is a tax liability of sufficient magnitude to allow for the declaration of the full ITC earned in a given tax year to be deducted from income taxes in that year. Under this assumption the limiting condition for ITC does not enter into the analysis. In reality this may not be a very accurate assumption for many mining companies. Evidence suggests that the tax liabilities for many mining companies, particularly in the mid-to-late 1970s, have been such that they have imposed limitations on the use of full investment tax credits generated in some years. Indeed, this same evidence suggests that the earnings and tax liabilities in some mining companies have been so low that investment tax credits have gone unused because of the limitation imposed.

The following example illustrates the calculation procedure for ITC.

Example 8

Assume the purchase of a new haulage truck in January 1982, which costs $800,000 and is expected to last five years. Calculate the ITC for the after-tax net annual cash flow assuming the firm has the following income and deductions:

    Revenue   = $5,000,000
    Operating costs = $4,000,000 (including state and local taxes)
    Depreciation   = $ 500,000 (includes new truck)
    Depletion   = $ 400,000

The ITC calculation procedure is as follows (under the Economic Recovery Tax Act of 1981):

    Qualifying investment = $800,000
    Eligible investment for ITC = $800,000 × 100% = $800,000
    ITC for 1981 = $800,000 × 10% = $80,000

The annual after-tax cash flow calculation is as follows:

Revenue:	$5,000,000
Operating costs:	4,000,000
Net after costs:	1,000,000
Depreciation:	500,000
Net after depreciation:	500,000
Depletion:	400,000
Taxable income:	100,000
Income tax liability (46%):	46,000
ITC*:	43,900
Net profit:	97,900
+ Depreciation:	500,000
+ Depletion:	400,000
− Capital expenditures (truck):	800,000
Net annual cash flow:	$ 197,900

*ITC is limited to a lesser of the tax liability ($46,000) *or* $25,000 + 90% ($46,000 − $25,000) = $25,000 + 18,900 = $43,900.

Since only $43,900 can be declared for ITC in this particular year, the remaining $36,100 ($80,000 − $43,900) would be carried back to see if it could be applied in one or more of the previous three tax years; if not, it is carried forward for use in one or more of the 15 future years.

Federal Income Tax

Federal income taxes are based on "taxable income" which may be defined in general terms for mining as gross revenue, less all costs of production, state and local taxes, depreciation, and depletion. Fig. 11 illustrates in more detail the derivation of taxable income for corporations in general. The following tax rates apply to taxable income for corporations:

	1981	1982	1983 and thereafter
On taxable income through $25,000	17%	16%	15%
On taxable income over $25,000 through $50,000	20%	19%	18%
On taxable income over $50,000 through $75,000	30%	30%	30%
On taxable income over $75,000 through $100,000	40%	40%	40%
On taxable income over $100,000	46%	46%	46%

The calculation procedure associated with determining the Federal tax liability is illustrated in the following example.

Example 9

If a corporation has determined its federal taxable income for 1981 to be $500,000, calculate the federal tax liability.

17% of the first $25,000 of taxable income	$ 4,250
20% of the next $25,000 of taxable income	$ 5,000
30% of the next $25,000 of taxable income	$ 7,500
40% of the next $25,000 of taxable income	$ 10,000
46% of $400,000 (taxable income over $100,000)	$184,000
Total federal tax liability	$210,750

When performing economic feasibility studies on mining properties, one normally assumes a marginal federal income tax rate of 46% on the premise that other divisions within the corporation generate the first $100,000 of taxable income. Such an assumption is reasonable and greatly simplifies the calculation of annual cash flows.

Minimum Tax: A part of federal income tax law which many analysts neglect is the so-called "minimum tax." This tax is applied to certain kinds of income that are given preferential treatment for tax purposes.

Corporations are subject to what is referred to as the add-on minimum tax. This tax is levied at 15% on tax preference items. The minimum tax base is the total of the taxpayer's tax preferences reduced by the greater of $10,000 or the full amount of the taxpayer's income tax liability. In other words, the minimum tax is levied on tax preference items in excess of 100% of the corporate tax liability.

The predominant tax preference items relating to mining corporations are:

1) Accelerated depreciation on real property. This is the excess of the depreciation taken during the year on real property over the depreciation that would have been allowed if the straight-line method had been used.

2) Excess depletion. This is the excess of the depletion deduction in a tax year over the unadjusted basis of the property at the beginning of the year (figured before deducting depletion for the year).

3) Exploration and development. TEFRA (1982) specified a new tax preference item equal to the difference between the deductions claimed for exploration and development and the deductions for a ten-year, straight-line amortization of these costs.

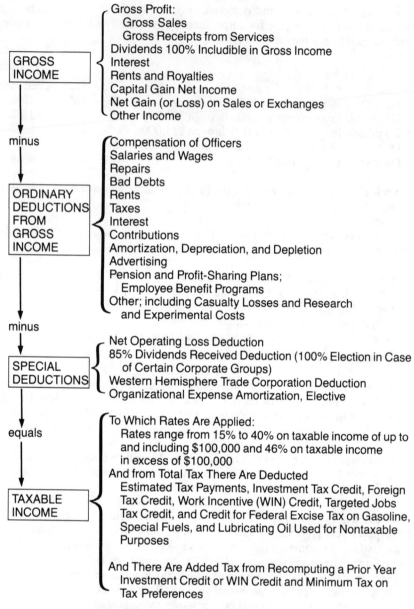

Fig. 11. Determination of corporate taxable income.

It is apparent that the major impact of these preference items on the mining industry is related to excess depletion. As the basis in the cost depletion account goes to zero, the entire depletion deduction becomes a tax preference item and subject to the 15% tax rate, depending on the corporate tax liability.

There is a slightly complicating feature which should be mentioned in regard to minimum tax on excess depletion. As discussed in the section on mine development expenditures, these expenditures may be either expensed or deferred. It was pointed out that although deferred development expenditures are not capitalized as mineral property costs for depletion purposes (not included in the cost depletion basis), they are taken into account in determining the adjusted basis of the mine or deposit for all other purposes. Consequently, in computing the amount of excess depletion which constitutes a tax preference item for minimum tax purposes, it appears that deferred development expenditures may be taken into account in order to reduce the excess depletion. As such, there are some added considerations for the taxpayer in deciding whether to expense or capitalize development expenditures.

It is also important to remember the recent changes in the statutory depletion allowance for iron ore, coal, and lignite as a result of TEFRA. When depletion is reduced by 15%, as specified previously, the minimum tax for depletion may also be reduced. The tax preference item for depletion is reduced to 71.6% of its original value, and thus may decrease the minimum tax if tax preference items exceed the federal tax liability.

Minimum tax only applies to tax preference items in excess of the corporation's tax liability. However, when calculating the corporate tax liability for minimum tax purposes, it is important to remember that ITC (if any) must be subtracted from the tax liability. The following example illustrates calculation procedure for minimum tax.

Example 10

Given the following values for depreciation, depletion, ITC, and tax liabilities for Hi-Grade Mining Co. in the tax years 1981, 1982.

	1981	1982
Accelerated depreciation (real property)	1,800,000	1,500,000
Straight line depreciation (real property)	1,100,000	900,000
Depletion claimed	5,000,000	6,000,000
Adjusted cost depletion basis	1,000,000	0
Tax liability	3,000,000	3,500,000
ITC	700,000	0

Calculate the minimum tax applicable to this situation.

	1981	1982
Accelerated depreciation on real property:	$1,800,000	$1,500,000
Straight-line depreciation calculated on real property:	1,100,000	900,000
Tax preference: depreciation on real property:	$ 700,000	$ 600,000
Depletion claimed:	$5,000,000	$6,000,000
Adjusted basis (cost depletion):	1,000,000	0
Tax preference: depletion:	$4,000,000	$6,000,000
Tax preference items (total):	$4,700,000	$6,600,000
Tax liability:	$3,000,000	$3,500,000
Less: ITC:	700,000	0
Adjusted tax liability:	$2,300,000	$3,500,000
Excess:	$2,400,000	$3,100,000
Tax at 15% of excess:	$ 360,000	$ 465,000

The positioning of minimum tax in the pro forma income statement for cash flow purposes usually follows federal income tax and precedes ITC as follows:

Taxable income
− Federal income
− Minimum tax
+ Investment tax credits
_____
Net profits

EFFECTIVE TAX RATES

During the discussion on state taxes the point was made that because of the magnitude of state taxes imposed on mining operations, the analyst was advised not to use an all-encompassing effective tax rate for state and federal taxes when evaluating mineral properties. In general this position is justified because of the complexity, interrelationships, variability, and magnitude of state taxes on the minerals industry. However, when performing order-of-magnitude or preliminary feasibility studies on mineral properties the analyst *may* choose to work in terms of an "effective" tax rate which incorporates both federal and state taxes. Certainly the use of such a rate would simplify the calculation procedure considerably, as well as reduce the time required to produce the calculation.

If the decision is made to work, at least initially, with an effective tax rate, the evaluator must decide to either construct an effective state tax rate (representing property, severance, and income taxes) or simply work in terms of an effective income tax rate which reflects only state and federal regulations. In general, most evaluators choose the latter alternative, for to do otherwise would defeat the purpose of the initial decision for simplicity.

The following material (after Stevens, 1979) illustrates the development of composite (state plus federal) income tax rates which may be used in the early stages of mineral property valuation. An attempt has been made to incorporate the more important interrelationships among mutual credits, deductions, etc., as well as the impact of the depletion allowance. The relationships which follow are believed to represent composite tax rates for the major mineral producing states in addition to federal income taxes.

The assumptions inherent in the development of the following composite tax rates are:

1) The property in question is a part of a large, profitable corporation with taxable income in excess of \$100,000. Thus, the 46% marginal tax rate applies to taxable income.

2) The basic premise used is to determine the composite tax rate on the basis of

$$t = \frac{Tf + Ts}{B}$$

where B is net income before tax, S is basis for state income tax, F is basis for federal income tax, s is state income tax rate, f is federal income tax rate, Ts is state income tax, Tf is federal income tax, sd is effective state tax rate after 50% of net depletion, D is depletion deduction (50% of net income) and t is effective composite tax rate.

Case I: State and Federal Tax Not Deductible in Computing State Tax.

A. Depletion not deductible for state income tax or 50% of net income depletion not used.

Composite effective tax rate is given by:

$$t = \frac{Ts + Tf}{B} \tag{1}$$

Since state income tax is deductible when determining federal income tax, federal income tax basis may be expressed as:

$$F = B - Ts \tag{2}$$

Also the following relationships are implied when federal income tax is *not* deductible when computing state income tax:

$$s = B \tag{3}$$
$$Ts = sB \tag{4}$$
$$Tf = fF \tag{5}$$

Substituting Eq. 2 into Eq. 5 gives:

$$Tf = f(B - Ts) \tag{6}$$

Substituting Eq. 4 into Eq. 6 gives

$$Tf = f(B - sB) \tag{7}$$

Expanding the terms in Eq. 7 yields:

$$Tf = fB - fsB \tag{8}$$

Substituting Eqs. 4 and 8 in Eq. 7 yields:

$$t = \frac{sB + fB - fsB}{B} \tag{9}$$

Dividing by B yields:

$$t = s + f - fs \tag{10}$$

B. Depletion deductible when computing state income tax and statutory depletion is limited by 50% net income.

State income tax is deductible when computing depletion, and depletion is deductible when computing state income tax; therefore:

$$D = (S - Ts) \times 0.50 \tag{11}$$

$$Ts = s(S - D) \tag{12}$$

Substituting Eq. 11 into Eq. 12 gives:

$$Ts = s\{S - [(S - Ts) \times 0.50]\} \tag{13}$$

Multiplying and collecting terms yields:

$$Ts = \frac{sS}{2(1 - s/2)} \tag{14}$$

By definition,

$$s = Ts/S \tag{15}$$

Therefore, substituting Eq. 15 into Eq. 14 yields:

$$sd = \frac{s}{2(1 - s/2)} \tag{16}$$

Substituting Eq. 16 into Eq. 10 yields:

$$t = fB + (1 - f)\left[\frac{s}{2(1 - s/2)} \right] \tag{17}$$

Case II: Federal Income Tax is Deductible in Computing State Income Tax.
A. Depletion not deductible when computing state income tax or 50% of net income depletion not in effect.
The following relationships are implied when federal income tax is deductible in computing state income tax:

$$S = B - Tf \tag{18}$$
$$Ts = s(B - Tf) \tag{19}$$
$$Tf = fF \tag{20}$$
$$F = B - Ts \tag{21}$$

Substituting Eq. 19 into Eq. 21 yields:

$$F = B - s(B - Tf) \tag{22}$$

Multiplying terms and substituting Eq. 22 into Eq. 18 yields:

$$Tf = fB - sfB + sfTf \tag{23}$$

Rearranging terms yields:

$$Tf = \frac{fB - sfB}{(1 - sf)} \tag{24}$$

Substituting Eq. 21 into Eq. 18 gives:

$$S = B - f(B - Ts) \tag{25}$$

Multiplying terms and substituting Eq. 18 into Eq. 25 yields:

$$Ts = sB - sfB + sfTf \tag{26}$$

Rearranging terms yields:

$$Ts = \frac{sB - sfB}{(1 - sf)} \tag{27}$$

Substituting Eqs. 24 and 27 into Eq. 1 yields:

$$t = \frac{\dfrac{fB - sfB}{(1 - sf)} + \dfrac{sB - sfB}{(1 - sf)}}{B} \tag{28}$$

Cancelling the B term and rearranging yields:

$$t = \frac{f + s - 2sf}{(1 - sf)} \tag{29}$$

B. Depletion deductible when computing state income tax and statutory depletion is limited by 50% of net income.
Substituting Eq. 16 into Eq. 29 yields;

$$t = \frac{f + sd - 2sdf}{(1 - sdf)} \tag{30}$$

*Case III. State Income Tax Based on Federal Income Tax Base (State Income
Tax is Deductible; Federal Income Tax is Not).*
A. Depletion not deductible when computing state income tax or 50% of net
income depletion not in effect.
The following relationships are implied when state income tax is based on the
federal income tax base:

$$S = B - sB \tag{31}$$
$$F = B - sB \tag{32}$$
$$S = F \tag{33}$$
$$Ts = sS \tag{34}$$
$$Tf = fF \tag{35}$$

Substituting Eq. 31 into Eq. 34 yields:

$$Ts = s(B - sB) \tag{36}$$

Substituting Eq. 32 into Eq. 35 yields:

$$Tf = f(B - sB) \tag{37}$$

Substituting Eqs. 36 and 37 into Eq. 1 and rearranging terms yields:

$$t = \frac{s + f}{(1 + s)} \tag{38}$$

B. Depletion deductible when calculating state income tax and statutory depletion is limited by 50% of net income.

Substituting Eq. 16 into Eq. 38 yields:

$$t = \frac{sd + f}{(1 - sd)} \tag{39}$$

Case IV: Both Federal and State Income Tax are Deductible in Computing State Tax.

A. Depletion not deductible in computing state tax or 50% of net income depletion not in effect.

The following relationships are implied under these conditions:

$$S = D - Tf - Ts \tag{40}$$
$$F = B - Ts \tag{41}$$
$$Ts = sS = s(B - Tf - Ts) \tag{42}$$
$$Tf = fF = f(B - Ts) \tag{43}$$

Substituting Eq. 43 into Eq. 42 and regrouping yields:

$$Ts = \frac{sB - sfB}{(1 + s - sf)} \tag{44}$$

Substituting Eq. 44 into Eq. 43 and rearranging yields:

$$Tf = \frac{fB}{(1 + s + sf)} \tag{45}$$

Substituting Eqs. 44 and 45 into Eq. 1 and rearranging yields:

$$t = \frac{s + f - sf}{1 + s - sf} \tag{46}$$

B. Depletion deductible when calculating state tax and statutory depletion is limited by 50% of net income.

Substituting Eq. 16 into Eq. 46 yields:

$$t = \frac{sd + f + sdf}{1 + sd - sdf} \tag{47}$$

FINANCIAL ACCOUNTING VS. TAX ACCOUNTING

A discussion on financial accounting vs. tax accounting is difficult to segregate from the related topic of standard accounting vs. cash flow concepts. As pointed out in a previous chapter, the cash flow approach to project evaluation attempts to represent the actual cash position of the project at any point in time. As such it records cash inflows and cash outflows at the time disbursements and revenues are actually made or received. In contrast the standard accounting convention attempts to match expenditures with the resulting revenue as it is produced—regardless of the time lag between the events.

Because of the procedures associated with standard accounting practices, expenditures for exploration, development, and capital equipment are recorded against subsequent revenues as they are produced. In essence, such capital expenditures are considered to be prepaid operating expenses which are written off against income as the capital item is consumed. Under this scenario exploration expenditures are often capitalized and amortized through the cost depletion account; development expenditures may be deferred and amortized on a unit-of-production basis; expenditures for capital equipment are sometimes depreciated on a unit-of-production or straight-line basis; and so on. The intent of financial accounting is to represent the on-going financial health of the firm which might otherwise be distorted by the very "lumpy" nature of capital investment. Because of this virtue, financial accounting is employed for stockholder and other public reporting. However, this approach does not accurately represent the actual timing of receipts and disbursements and, therefore, is of limited value in project evaluation.

With the cash flow approach to project evaluation, emphasis is placed on accurate representation of cash outflows and inflows during the year in which they actually occur. Such a procedure enables the evaluator to compare capital projects directly and consistently because the actual cash position of the projects is represented. Because taxes are a major cash cost of production, tax accounting is an essential part of investment analysis. Tax accounting may not be good financial accounting, but taxes do have a major effect on project profitability.

The reader will remember that in a cash flow analysis each investment receives credit for income taxes saved. Therefore, for profitable organizations, the normal objective is to maximize pretax, noncash deductions, thereby minimizing taxable income and, consequently, taxes paid. These savings in tax dollars represent a net savings in tax outflows and an earlier return of money to the firm for future use. In order to maximize noncash deductions, the firm will incorporate tax accounting procedures which expense expenditures whenever possible, and utilize the most accelerated recovery method in all other cases. Although there may be financial or other considerations which inhibit an organization's ability to maximize tax deductions as soon as possible, most firms, where possible, attempt to generate tax savings at the earliest possible time.

The importance of a basic understanding of tax regulations, as well as their interrelationships and impacts, is perhaps best illustrated in the case of trying to determine pretax deductions such that net annual cash flows are maximized. For instance, a taxpayer may be better off not to maximize the depreciation allowance if it adversely impacts the depletion deduction (50% of net limit) such that the total deduction (depreciation + depletion) is reduced. Because of the interrelationships between depreciation, depletion, investment tax credits, treatment of exploration and development expenditures, income tax, state and local taxes and minimum tax, the analyst must balance the total tax accounting procedure such that annual cash flows are maximized. However, this may mean that individual tax deductions within the entire package may not be maximized.

Table 14 illustrates the typical difference in tax treatment employed by corporations and individual mine entities. It also approximates the difference in concepts between tax and financial accounting procedures because a one-mine company cannot take advantage of the early tax deductions available due to the limited income from which to make the deductions. It should be stressed, however, that with respect to depreciation and amortization the total allowable tax deductions received in both cases are, over the long run, exactly the same. The only real

Table 14. Comparison of Federal Tax Treatment for a Typical Corporation vs. Individual Entity

Item	Corporate	Entity
Preproduction period revenue	Assume corporation is profitable with large net income	None—credits or tax savings carried forward
Exploration expenditures	Expensed during preproduction period (subject to recapture from depletion)	Capitalized into depletion account (recovered through cost depletion during production period)
Development expenditures	Expensed during preproduction period	Capitalized and recovered through amortization on unit of production basis during production period
Investment tax credits	Entire credit used immediately during preproduction period by corporation. Used by project during production with excess used by corporation	Preproduction amount carried forward to production period. During production, limited by $25,000 + 70% (tax liability—$25,000) (80% in 1981, 90% in 1982)
Federal income tax	46% of net income	Graduated rate: (see below)
"Net" investment	Usually lower	Usually higher
Production cash flow	Usually lower	Usually higher (due to deferred tax carry-forwards, etc.)

Graduated rate:

	1981	1982	1983
1st $25,000 @	17%	16%	15%
2nd $25,000 @	20%	19%	18%
3rd $25,000 @	30%	30%	30%
4th $25,000 @	40%	40%	40%
4th $25,000 @	46%	46%	46%

difference between the two tax accounting procedures is in the timing of the deductions.

Finally, the reader should recognize that the use of straight-line rather than accelerated recovery methods, while increasing income taxes, will also increase accounting profits (although cash flow will be reduced). Small mine promoters will, therefore, occasionally use the slowest recovery method—sinking fund depreciation—to "demonstrate" high profits in the early years of production.

SUMMARY

This chapter attempted to bring to light some of the more important tax regulations and considerations affecting the valuation of mineral properties. Although all possible situations could not be addressed or explored in depth, the important impacts of state and federal taxes on mineral property valuations should be rather obvious. An understanding of the various tax calculation procedures is a virtual necessity for those analysts performing feasibility studies. Simple manipulation of tax accounting procedures can significantly alter the viability of a mining venture. The interrelationships between tax regulations, mutual credits, allowable deductions, etc., are very important and can significantly affect project attractiveness.

Another area in which tax implications become important is in property lease negotiations. The negotiator who understands the various tax advantages and disadvantages associated with different leasing and financing arrangements (both from the standpoint of the lessor and the lessee) is apt to be more successful in acquiring property at reasonable costs.

Taxes represent a substantial cost of doing business in the minerals industry and, as such, should be treated completely and thoroughly when performing mining property valuations.

SELECTED BIBLIOGRAPHY

Anderson, R.C., 1977, "Taxes, Exploration and Reserves," *Non-Renewable Resource Taxation in the Western States* Lincoln Institute of Land Policy and College of Mines, University of Arizona, Tucson, Monograph No. 77-2.

Bingaman, A.K., 1970, "New Mexico's Effort at Rational Taxation of Hard-Minerals Extraction," *Natural Resources Journal*, Vol. 10, No. 3, July.

Brightwell, T.P., 1969, "Ad Valorem Taxation of Mining Properties," *Rocky Mountain Mineral Law Institute*, Vol. 15, pp. 281-304.

Burke, F.M., and Bowhay, R.W., 1978, *Income Taxation of Natural Resources*, Prentice-Hall, Inc., Englewood Cliffs, NJ.

Byrne, W.J., Jr., 1968, "Tax Climate Needed to Encourage New Mineral Industry Investment," *Missouri Mining Tax Symposium*, October, pp. 60-72.

Commerce Clearing House, 1981, *Economic Recovery Tax Act of 1981, Law and Explanation*, Commerce Clearing House, Inc., Chicago, IL.

Commerce Clearing House, 1981a, *Internal Revenue Code (as of June 18, 1981)*, Commerce Clearing House, Inc., Chicago, IL.

Commerce Clearing House, 1980, *1981 Master Tax Guide*, Commerce Clearing House, Inc., Chicago, IL.

Commerce Clearing House, 1981b, *State Tax Guide*, 2nd ed., Vols. 1, 2, Commerce Clearing House Inc., Chicago, IL.

Comptroller General, 1981, "Assessing the Impact of Federal and State Taxes on the Domestic Minerals Industry," Report to the Congress of United States, US General Accounting Office, EMD-81-13, June.

Gentry, D.W., 1977, "Financing Modeling of Mining Ventures—The Effects of State Mine Taxation," *Non-Renewable Resource Taxation in the Western States*, Lincoln Institute of Land Policy and College of Mines, University of Arizona, Tuscon, Monograph No. 77-2.

Gentry, D.W., and O'Neil, T.J., 1974, "A Short Course on Financial Modeling and Evaluation of New Mine Properties," 12th International Symposium on the Applications of Computers in the Mineral Industry (APCOM), Golden, CO, April, 149 pp.

Gentry, D.W., and Hrebar, M.J., 1978 *Economic Principles for Property Valuation of Industrial Minerals*, at SME/AIME Meeting and Exhibit, Nassau, Bahamas, September.

Gentry, D.W., and Hrebar, M.J., 1983, *Economic Principles for Industrial Mineral Property Valuations*, Society of Mining Engineers of AIME, Atlanta, GA, March, 172 pp.

Gentry, D.W., Hrebar, M.J., and Martin, J.W., 1980, "Planning and Economic Aspects—Surface Coal Mining," short course notes, Colorado School of Mines, Golden, CO.

Gentry, D.W., Hrebar, M.J., and Martin, J.W., 1981, "Planning and Economic Aspects—Surface Coal Mining," short course notes, Colorado School of Mines, Golden, CO.

Gillis, M. et al., 1978, *Taxation and Mining*, Ballinger Publication Co., MA, 350 pp.

Hansen, C.J., 1969, "An Evaluation of Mine Taxation Proposals," *Proceedings, Symposium on Mine Taxation,* College of Mines, University of Arizona, Tucson, March, pp. 11-1 to 11-8.

Hrebar, M.J., 1977, "Financial Analysis," *Mineral Industry Costs,* J.R. Hoskins and W.R. Green, eds., Northwest Mining Association, Spokane, WA, pp. 205-224.

Internal Revenue Service, 1980, *Depreciation,* Publication 534, (Rev. November).

Internal Revenue Service, 1980a, *Tax Guide for Small Business,* Publication 334, (Rev. November).

Janson, E., MacLean, J., and Wright, D., 1977, *Financial Reporting and Tax Practices in Nonferrous Mining,* 6th ed., Coopers and Lybrand.

Kearns, D.P., 1970, "Property Taxation of the Mining Industry in Arizona," *Arizona Law Review,* pp. 763-802.

Lacy, W.C., 1969, "Taxation, Assessments, and Ore Deposits," *Symposium On Mine Taxation,* College of Mines, University of Arizona, Tucson, March, pp. 2-1 to 2-25.

Laing, G.J.S., 1977, "Effects of State Taxation on Mining Industry In Rocky Mountain States," *Quarterly of Colorado School of Mines,* Vol. 72, No. 2, April.

Leaming, G.F., 1975, "The Effects of Changing Arizona Copper Industry Taxes," Arizona Economic Information Center, Marana, AZ.

Lockner, A.D., 1965, "The Economic Effect of the Severance Tax on Decisions of the Mining Firm," *Natural Resources Journal,* Vol. 4, No. 1, January, pp. 468-485.

Long, S.C.M., 1976, "Coal Taxation in the Western States: The Need for a Regional Tax Policy," *Natural Resources Journal,* Vol. 16, No. 2, April, pp. 415-442.

Mackenzie, B.W., and Bilodeau, M.L., 1979, *Effects of Taxation on Base Metal Mining in Canada,* Centre for Resource Studies, Queen's University, Kingston, Ontario, 188 pp.

Maxfield, P., 1975, *The Income Taxation of Mining Operations,* 2nd ed., Rocky Mountain Mineral Law Foundation, Boulder, CO.

McGeorge, R.L., 1970, "Approaches to State Taxation of the Mining Industry," *Natural Resources Journal,* Vol. 10, No. 1, January, pp. 156-170.

Montana Joint Conference Committee on Coal Taxation, 1975, Statement, April 16, 24 pp.

Natural Resources Bulletin, 1970, "Approaches to State Taxation of the Mining Industry," *Natural Resources Bulletin,* Vol. 10, No. 1, January.

Nilsen, M.J., 1983, "Economic Evaluation Computer Program for Mining Projects Using Sensitivity and Risk Analysis," Colorado School of Mines, Thesis, No. T-2555, April, 311 pp.

O'Neil, T.J., 1977, "Proposed Federal and Western States Mineral Tax Revisions," *Non-Renewable Resource Taxation in the Western States,* Lincoln Institute of Land Policy and College of Mines, University of Arizona, Tucson, Monograph No. 77-2.

O'Neil, T.J., Gentry, D.W., 1981, "Uranium Royalties on Federal and Indian Lands," Final Report for US Geological Survey, Reston, VA, 176 pp.

Petrick, A., Jr., 1976, "The Mine Property Tax Valuation Controversy in Arizona," *Proceedings,* AIME Council of Economics, pp. 167-179.

Paschall, R.H., 1977, "What are Our Real Goals in Minerals Taxation," *Non-Renewable Resource Taxation in the Western States,* Lincoln Institute of Land Policy and College of Mines, University of Arizona, Tucson, Monograph, No. 77-2.

Sammons, D., 1981, *Coal Industry Taxes, A State-by-State Guide,* McGraw-Hill, New York.

Smith, A., 1904, *An Inquiry into the Nature and Causes of the Wealth of Nations,* Canon ed., Methuen and Co., Ltd., London.

Sommerfield, R.M., Anderson, H.M., and Brock, H.R., 1973, *An Introduction to Taxation,* Harcourt, Brace and World, Inc., New York, 495 pp.

Spangler, E.H., 1940, "The Property Tax as a Benefit Tax," *Property Taxes,* Tax Policy League, New York, pp. 165-173.

Spence, H.M., 1976, "Mineral Severance Taxes—Causes and Effects," National Western Mining Conference, Denver, CO, January 30, 23 pp.

Starch, K.E., 1979, *Taxation, Mining, and the Severance Tax,* Information Circular 8788, US Bureau of Mines, 65 pp.

State of Texas, 1975, "Mining Lease Form 6-75," General Land Office, Austin, Texas, 6 pp.

Steel, H., 1967, "National Resource Taxation: Allocation and Distribution Implications," *Extractive Resources and Taxation,* Mason Gaffney, ed., University of Wisconsin Press, Madison.

Stevens, G.T., Jr., 1979, *Economic and Financial Analysis of Capital Investments*, John Wiley & Sons, New York.

Weaton, G.F., 1969, "The History of Minnesota Mining as Influenced by Taxation," *Proceedings*, Symposium on Mine Taxation, College of Mines, University of Arizona, Tucson, March, pp. 7-1 to 7-30.

Yasnowsky, P.N., and Graham, A.P., 1976, "State Severance Taxes on Mineral Production," *Proceedings*, AIME Council of Economics, Las Vegas, NV, 4 pp.

9

Project Evaluation Criteria

*"We can easily represent things as we wish
them to be..."*
—Aesop

INTRODUCTION

The first eight chapters of this book have primarily addressed the concepts and fundamentals associated with project valuation, particularly as they apply to mining-related projects. Specific emphasis was placed on the engineering and economic parameters and estimates necessary for quantifying line items in pro forma income statements and the subsequent calculation of net annual cash flows. Given these estimates of relative benefits, costs, and annual cash flows for a project, it then becomes necessary to convert these estimates into measures of relative desirability or attractiveness. The criteria and techniques typically utilized to determine project acceptability or desirability are, then, the topics of discussion in this chapter.

Before proceeding it may be advantageous to put the problem of project evaluation into proper perspective by reiterating some of the fundamental concepts developed in Chapter 1. The basic premise set forth in this book is that the objective of the firm should be to maximize its value to its owners (stockholders' wealth). This value or wealth is represented by the market price of the firm's common stock; consequently, the primary objective of the firm can be restated as one of maximizing the value (price) of the firm's common stock in the marketplace over the long term. This price, in effect, is an indicator or index of the market's perception of the company's progress and a reflection of the investment, financing, and dividend decisions made by corporate management. These three major decisions are interrelated and jointly impact the market price of the firm's stock. Therefore, if the objective of maximizing the value of the firm to its stockholders is to be achieved, an optimal combination of the three decisions will be necessary.

It is important to emphasize that maximizing stockholder wealth is not the same as profit maximization or even the maximization of earnings per share. Corporate managers making decisions based on the objectives of maximizing profits and/or earnings per share must recognize that these objectives do not consider: (1) the timing or duration of expected returns, (2) the business and financial risks of the promised earnings, or (3) the dividend policy of the firm. Obviously all of these factors are of concern to the company's stockholders and will have an effect on the market value or price of the firm's common stock. Maximizing market price per share, or wealth maximization, is therefore the appropriate corporate objective and any management decision which does not contribute to this objective is, in effect, a decision to redistribute part of the company's wealth to parties other than the legal owners.

This chapter addresses what is typically considered to be the most important of the three major decisions associated with financial management—the investment decision. The investment decision is directly concerned with the evaluation of proposed capital investments and the allocation of capital among alternative or competing investments. Ultimately, then, a company's total assets, the mix of these assets, and the resulting business risk associated with the firm (as perceived in the marketplace) are a result of the firm's investment decisions.

To better explain and illustrate a major aspect of the investment decision—capital budgeting—this chapter is divided into two parts. Part I deals with capital budgeting from the standpoint of evaluating the attractiveness of various investment proposals under consideration. It specifically discusses the evaluation criteria typically used for project selection. Part II of the chapter addresses the problem of selecting among alternative projects for optimum allocation of capital. It is important to point out that the capital budgeting discussions throughout this chapter are based on the assumption that the risk or quality of all investment proposals under review does not differ from the risk of existing investment projects within the firm, and that the acceptance of any proposal or group of proposals does not change the relative business risk of the firm.

PART I—BASES FOR PROJECT COMPARISON

When a company is confronted with several investment proposals, it becomes necessary to evaluate the attractiveness of each proposal. Any evaluation criterion utilized should provide company management with a means of distinguishing between acceptable and unacceptable projects in a consistent manner. In other words, the criterion should help answer the question, "Is Project A and/or Project B good enough to justify capital investment by the company?" In addition it is also desirable, as discussed in Part II, for the evaluation criterion to provide a ranking of the proposals under consideration in the order of their desirability. To provide this necessary information for investment decision making, any satisfactory evaluation criterion must respect two basic principles (Quirin, 1967):

1) Bigger benefits are preferable to smaller benefits.
2) Early benefits are preferable to later benefits.

The project evaluation criteria presented in this section are not intended to represent an exhaustive list available to the analyst. Rather, those discussed represent the major evaluation criteria utilized for evaluating investment proposals within the minerals industry. It should be recognized however that many variations of these basic techniques exist within the industry. Typically these variations have evolved as the result of companies trying to assess, prioritize, and quantify what they perceive to be the most critical parameters affecting an investment decision.

An extremely important point to remember is that project evaluation criteria do not, by themselves, provide investment *decisions*. They only provide *guidelines* for making decisions. Ultimately, managers must make the actual investment decision after consideration of all the engineering/economic analyses, the large amount of relevant qualitative information that impacts any major decision, and the unique risk and uncertainty possessed by each investment alternative. Investment decision making is a complex process in which quantitative economic studies are of considerable assistance. However, in a world where future values of critical variables are subject to large estimating errors, there is no substitute for sound managerial judgment.

Degree of Necessity

There are times when company management is faced with making investment decisions based on only limited quantitative data. These types of investments may be referred to as "degree of necessity" investments and are characterized by the fact that the decisions are either obvious or can be quantified only to a limited degree. To illustrate the concept, suppose the main production hoist at a large, profitable underground mine suddenly failed. Under these conditions it is conceivable that some comparative analyses could be performed in order to help decide what kind, model, etc., of hoist should be purchased. However, the investment decision to actually purchase a new hoist need not be predicated on rigorous economic analyses resulting in calculated rates of return or some other yardstick of profitability, unless the operation is already marginal. Indeed, performing a formal benefit: cost analysis for an investment proposal indicated in this example is not a simple task, and, in this case, the likely results are often extraneous to the actual decision required.

Other examples of some degree of necessity investment decisions in the manufacturing industries can be illustrated by advertising and marketing expenditures. How does one justify the millions of dollars spent promoting toothpaste, razor blades, and laundry detergents? Obviously most of these organizations attempt to invest enough in these activities to maintain a competitive position in the marketplace, and detailed economic rates of return are not crucial to the decision.

Examples more closely related to the minerals industry are expenditures in the areas of research and development (R&D) and exploration. How much, if any, capital should be allocated to R&D activities? What quantity of capital should be allocated to exploration activities? Should this be a fixed percentage of the annual corporate budget, some amount commensurate with industry average, or what? Certainly the successes resulting from these kinds of activities are extremely difficult to anticipate or predict. Nonetheless it is apparent that if the mining business is to endure over the long run it must support some level of exploration activity. Furthermore, those firms which have maintained significant, well-designed exploration programs have generally secured mining rights to the best ore deposits around the world.

Answers to the foregoing questions require careful consideration of factors such as the availability and cost of alternative technology, the company's desire to maintain or secure a particular competitive position within an industry or commodity market, the attraction and retention of key personnel, the availability of capital, and so on. Clearly these kinds of investment decisions fall within the realm of management judgment and are predicated more on corporate strategies than on any specific economic criterion.

Accounting Rate of Return

One of the more common versions of the accounting rate of return calculation is often referred to as the average rate of return. The average rate of return is calculated by dividing average annual profits after taxes by the average investment in the project (average book value after deducting depreciation).

Example 1: Table 1 illustrates the calculation procedure for the average rate of return on a project requiring an investment of $10,000 with an estimated salvage

Table 1. Accounting (Average) Rate of Return

	Year 1	Year 2	Year 3	Year 4	Average
Net operating income	3000	4000	5000	6000	4500
Depreciation	2000	2000	2000	2000	2000
Taxable income	1000	2000	3000	4000	2500
Taxes @ 50%	500	1000	1500	2000	1250
Net profit	500	1000	1500	2000	1250
Book value:					
January 1	10,000	8000	6000	4000	
December 31	8000	6000	4000	2000	
Average	9000	7000	5000	3000	6000

$$\text{Average rate of return} = \frac{1250}{6000} \times 100 = 20.8\%$$

value of $2000 at the end of year 4. Estimated annual profits are given. Depreciation is on a straight-line basis.

Another version of the accounting rate of return uses original investment for the denominator rather than average book value. In the example presented, the calculation would be represented as: $1,250/10,000 \times 100 = 12.5\%$. This version is somewhat less informative since income is averaged but investment is not. However, as an approximation to the internal rate of return (see Internal Rate of Return) it usually provides better results than using the average investment.

The primary advantages of the accounting rate of return criterion are: (1) it is simple to calculate, (2) it makes use of readily available accounting information, and (3) it provides a "rate of return" number to which most managers seem to relate. Once the calculation has been performed for a project, the rate can be compared with the company's required, or cutoff, rate to determine whether the project should be accepted or rejected.

The principal disadvantages of the method are that: (1) it is based on accounting profits rather than actual cash flows, and (2) it does not take into account the timing of these profits. These are very serious disadvantages, as they violate some basic concepts and requirements set forth in this chapter and earlier in the book. In reality it takes little additional effort to work with actual annual cash flows and incorporate the concept of time value of money into the analysis.

Payback (Payout) Period

One of the most common evaluation criteria used by mining companies is the payback or payout period. Whereas it was once used as a primary investment criterion, the payback period today is generally used in conjunction with other, more informative methods. The payback period is simply the number of years required for the cash income from a project to return the initial cash investment in the project. Although benefits resulting from the investment can be measured in terms of net profits for calculation purposes, modern practice has resulted in the use of annual net cash flows for the denominator.

The calculation procedure for determining payback period is illustrated in the following example.

Example 2: Table 2 illustrates five investment proposals having identical cap-

Table 2. Payback Period

	Proposal A	Proposal B	Proposal C	Proposal D	Proposal E
	Annual net cash flows				
Initial investment	$10,000	$10,000	$10,000	$10,000	$10,000
Project year 1	2,000	7,000	1,000	6,000	6,000
2	2,000	2,000	2,000	2,000	2,000
3	2,000	1,000	7,000	2,000	2,000
4	2,000	2,000	2,000	-0-	3,000
5	2,000			-0-	4,000
6	2,000			-0-	1,000
7	2,000			-0-	1,000
8				-0-	500
Payback period (years)	5.0	3.0	3.0	3.0	3.0

ital investment requirements but differing expected annual cash flows and lives. The payback period is calculated for each.

The investment decision criteria for this technique suggests that if the calculated payback period for an investment proposal is less than some maximum value acceptable to the company, the proposal is accepted; if not, it is rejected. In other words an investment proposal having a payback period of three years is acceptable to a company having a hurdle value of five years and is preferable to a second project having a payback period of four years.

An interesting situation arises when calculating the payback period for a typical new mining venture where several years of negative cash flows (investment) are anticipated prior to project start up. Fig. 1 illustrates the situation, in a very simplified sense, for a new mining property having a preproduction period of five years followed by positive annual cash flows.

From Fig. 1 the total investment in the project is expected to be $40. The payback period for this example, calculated from the start of the project, is then determined to be nine years. However, some will argue that the payback period is four years because it only takes four years from the time production begins to recapture the investment in the project. Which of these payback period calculations is correct? From an economic and financial standpoint, the nine-year payback period is theoretically more correct because it represents the commitment of investment throughout the preproduction period, particularly the opportunity cost associated with the investment during this period. However, both payback period calculations provide useful information to the decision-maker, and consistency in application is the most important feature. It is often important to know how long it will take to recoup the investment in a project after actual start-up. When such information can aid in helping management make the correct investment decision, it should be used even though it may not be theoretically as correct as another calculation procedure.

When analyzing the payback period more closely it becomes apparent that the criterion has some significant disadvantages. First, the payback period fails to consider cash flows after the payback period; and, therefore, it cannot be regarded as a suitable measure of profitability. For example, proposals D and E in Table 2 have identical payback periods and would be rated equally by the criterion. However,

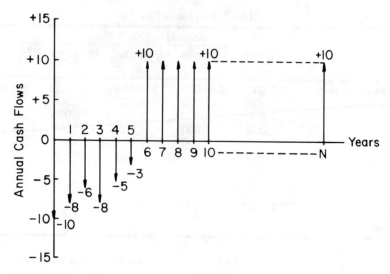

Fig. 1. Annual cash flows for a new mining project, illustrating effect of preproduction period on payback period calculation.

proposal D never earns a profit while proposal E continues to generate income after its payout period. Obviously proposal E is better than proposal D, but the payback period does not recognize this distinction.

An additional disadvantage of the method is that it does not consider the magnitude or timing of cash flows during the payback period. Rather, it only considers the recovery or payback interval as a whole. For example, proposals B and C in Table 2 have identical payback periods. However, proposal B is clearly preferable to proposal C because of the large return of funds early in proposal life which could then be reinvested by the firm. Therefore, when time value of money concepts are taken into consideration, one proposal is clearly preferable to the other even though the payback period does not indicate this difference. This troublesome point may be alleviated by calculating a *discounted payback period* for investment proposals. The discounted payback period calculation for proposal A in Table 2 using a 7% interest rate would take the following form.

Proposal A		Present worth factor (P/F) at 7%	Present value of cash flows
Initial investment: $-$ $10,000		1.0	$-$ $10,000
Cash flows:			
Project year 1	$ 2,000	0.9346	1,869
2	$ 2,000	0.8734	1,747
3	$ 2,000	0.8163	1,633
4	$ 2,000	0.7629	1,526
5	$ 2,000	0.7130	1,426
6	$ 2,000	0.6663	1,333
7	$ 2,000	0.6228	1,246

Undiscounted payback period:	5.0 years	Discounted payback period:	6.4 years

As expected, the discounted payback period will always result in a longer payback interval than the undiscounted calculation procedure.

The last major disadvantage with the payback period is the problem of establishing the appropriate hurdle rate or maximum acceptable value. Someone must decide whether or not a proposal having a four-year payback period is acceptable. Establishing this rate, or hurdle, is often the result of subjective and arbitrary decisions and falls in the domain of management judgment. Should one hurdle rate be applied to all investment proposals even though some may be considerably more risky than others? What should the appropriate hurdle rate be and how is it determined for investments having varying degrees of risk? Purely subjective determinations are often unsatisfactory when utilizing analytical techniques to assist with investment proposal decision-making.

In view of these very serious disadvantages of the payback period criterion discussed previously, why does the technique continue to be so widely utilized throughout the minerals, as well as manufacturing, industries? Following are some reasons typically given by managers in support of the payback period. Where appropriate, some additional abbreviated discussion is provided on the validity of these reasons. A more complete discussion on the strengths and weaknesses of the payback period and its relationship to risk and uncertainty is provided by Gentry (1971).

1) Payback period is simple and easy to calculate and also provides a single number which can be used as an index of proposal profitability.

There is little question that the simplicity and ease of calculation contribute significantly to the technique's widespread use. However, as previously demonstrated, it is a rather poor or inadequate measure of project profitability.

2) Payback period prevents management from exposure to excessive risk (protection against risk).

The argument is that risky investments should have shorter payback periods due to the risks involved. Although this may be true, someone still has to decide what an acceptable payback period should be. These are often subjective decisions having no theoretical basis. Also, Chapter 13 presents alternatives for analyzing risk in mining ventures which are inherently superior to simply adjusting the payback period.

The payback period does not protect against unforeseen poor outcomes, or the dispersion of possible outcomes, in the initial payback years. Rather, it protects against bad results subsequent to the payback period by simply ignoring them. Thus, a decision to opt for projects promising the shortest payback period can cause the firm to accept high-risk, short-lived projects and reject projects having longer lives but with little or no risk. The payback period then in reality acts as a constraint on the *time distribution* of expected cash flows, but has no effect on their variability. It is really a *measure of the rate* at which the uncertainty of the project is *expected* to be resolved and nothing more.

3) Payback period can minimize "lost opportunity risk" to the firm.

The concept is that projects with short payback periods will minimize lost opportunity risks since cash inflows will be returned to the firm within a short span of time, thus allowing the firm to seize unexpected investment opportunities which may become available. This is really an argument for liquidity and is probably valid—particularly for the cash poor company which must look to early recovery of invested funds to finance normal company operations. However, most firms plan for the generation of operating funds from the investment program as a whole or from operations generally. Under these conditions, the liquidity argument is really not the same as the liquidity constraint for the firm as a whole and therefore becomes less important.

4) Payback period represents a break-even point.

The argument is that projects with lives greater than the payback period will contribute profit to the firm while those with lives shorter than the payback period will result in a loss to the firm. Therefore, the *probability distribution* of the project's life in effect becomes a measure of the *distribution of project profitability*.

This rationale suggests that the longer the payback period for a project, the greater are the chances of incurring a loss in the project; whereas a project having a shorter payback period would suggest a higher probability of profit and hence would be more attractive. The reasoning is that short lives and early cash flows can be predicted with more certainty than those occurring in the distant future. In other words the shorter the payback period, the sooner the investment's profitability will become known.

An objective appraisal of the payback period indicates that it can provide some useful information to the decision-maker when considering investment proposals. However, the technique has too many drawbacks to be used alone. It should *not* be used as the sole quantitative tool for making investment decisions, but rather in a supplementary role to other more sophisticated methods. Many firms use the payback period criterion as a hurdle which investment proposals must clear before progressing to more rigorous and sophisticated forms of analyses. The payback period should appropriately be regarded as a *constraint* on the acceptability of an investment proposal and *not* as a criterion to be optimized.

Present, Future and Annual Value

The *present value*, or present worth, method of measuring investment proposal desirability is a widely used technique. The term present value (PV) simply represents an amount of money at the present time ($t = 0$) which is equivalent to some sequence of future cash flows discounted at a specified interest rate. In other words this technique recognizes the time value of money and provides for the calculation of an amount at the present time which is equivalent in value to a series of future cash flows.

Present value calculations are most frequently performed to determine the present worth of income-producing property, such as an existing mining operation. If the future annual cash flows can be estimated, then by selecting an appropriate interest rate, the present value of the property can be calculated. This value should provide a reasonable estimate of the price at which the property could be bought or sold.

In the more general case of investment proposal evaluation, one is interested in determining the difference between cash outflows and cash inflows associated with the proposal on a present-value basis. This calculation procedure is referred to as the *net present value (NPV)* method and is simply the difference between the sum of the present value of all cash inflows and the sum of the present value of all cash outflows. NPV can be expressed as follows:

Net present value = Σ present value of cash benefits − Σ present value of cash costs

If the NPV of the proposal is a positive value (NPV>0), then the project should be accepted. A positive NPV indicates that the investment proposal will provide for the recovery of invested capital, a return on the unrecovered capital each year throughout the project life at the stipulated interest rate utilized in the calculation, as well as some surplus amount. In other words, the project promises to yield a return in excess of that rate used in the calculation procedure. If the rate used in the calculation is the rate of return investors expect the firm to earn on investments, then proposals having a positive NPV should increase the wealth of the firm. Similarly, proposals yielding a negative NPV at the required discount rate should be rejected.

The following example illustrates the calculation procedure for net present value determinations.

Example 3:

Initial investment:	$100,000
Project life:	10 years
Salvage value	$ 20,000
Annual receipts:	$ 40,000
Annual disbursements:	$ 22,000
Minimum acceptable rate of return (discount rate):	12%

Solution:

	Present value
1. Annual receipts = $40,000 (P/A,12%,10)	$226,000
2. Salvage value = $20,000 (P/F,12%,10)	6,440
Total PV of cash inflows:	$232,440
3. Annual disbursements = $22,000 (P/A,12%,10)	124,300
4. Initial investment =	100,000
Total PV of cash outflows:	$224,300
Net present value (NPV): (PV inflow-PV outflow):	$ 8,140

Since the NPV>0 in example 3, the project should be accepted according to this criterion. Note that annual *net* benefits (receipts—disbursements) could have been discounted as a single stream, and the NPV would be the same. This is *not* the case for the benefit/cost ratio criterion (see Benefit/Cost Ratio) where separate discounting of receipts and disbursements is essential.

Under certain conditions the present value concept can also be used for evaluating projects on a cost basis. For example, it is often desirable to determine what it will cost, in today's equivalent, to operate alternative pieces of equipment over some future period. When comparing investment proposals the decision criteria would suggest that the proposal promising the largest PV of inflows should be selected, or if working with costs, the proposal promising the lowest PV of outflows should be selected. However, the analyst must use caution in any study based upon cost minimization. Often the lowest cost alternative is to do nothing; the objective function really is to minimize cost *subject to the completion of some particular task*. Thus, selection of the alternative based on minimum PV of costs assumes that alternative can perform the required job.

The present value technique has a number of characteristics which make it suitable as an accept/reject criterion for proposal evaluation. First, it takes into account the time value of money by utilizing a specified interest rate in the calculation. Second, it provides a single number, or cash equivalent, which can be used as an index for comparison at a specific point in time ($t = 0$). Third, the value of the present value amount is always a unique quantity for a given interest rate.

This last point can be illustrated by calculating the net present value associated with different interest rates for an investment proposal. Fig. 2 represents the present value profile for the problem represented in example 3. Clearly, the present value of an investment proposal is a function of the interest rate selected. The connection between the interest rate selected for calculation purposes and the firm's required rate of return is obvious in that only those projects promising a positive NPV at the required rate of return should be accepted.

Investment proposals can also be evaluated on the basis of how much money they will accumulate at some future point in time, usually the end of project life. Value determinations calculated in this manner are referred to as *future values* and represent the future worth amount for a specific proposal at some point in time at a given interest rate.

Recognizing the time-value-of-money concept, it is apparent that this method is just the reverse of the present value concept. In fact the future value amount can be calculated by first determining the present value amount of the cash flows in the following manner:

$$F = P\,(F/P, i, n)$$

likewise,

$$P = F\,(P/F, i, n)$$

In other words, for given values of interest rate (i) and years (n) the future value amount is simply the present value times a constant. Consequently, if an investment proposal has a present value amount two times as large as an alternative proposal's present value, it will also have a future value amount two times as large as the alternative proposal.

In view of the foregoing discussion, it makes no theoretical difference if projects are evaluated on the basis of future values or present values. However, most project evaluators prefer to work in terms of present value amounts, probably because they are considering amounts in equivalent dollars at the time the accept/reject decision is under consideration ($t = 0$). In general most investment decisions in the minerals industry are based on present value determinations rather than future value determinations.

There are times when it may be more convenient to evaluate investment proposals in terms of *annual value* or *annual cost* as opposed to present value or future value amounts in the aggregate. For example, many analysts prefer to compare pieces of operating equipment in terms of annual costs simply because associated benefits are often difficult to quantify and cost records are readily available.

Annual value refers to a uniform annual series of money (annuity) for a given period of time which is equivalent in amount to a specified sequence of annual cash flows under consideration. The concept of equivalency suggests that the equivalent annual value for a series of cash flows can be determined by first calculating the present value amount of the actual cash flow series and then multiplying this amount by the capital recovery factor which is, in effect, an annualizing factor (Chapter 3). Therefore, the annual value amount (sometimes referred to as the *equivalent uniform annual value (EUAV)*) may be represented as follows:

Annual value (EUAV) = (Σ PV of cash flows) $\times$ capital recovery factor

The following example is intended to demonstrate the calculation procedure for determining the annual value of a project and illustrate the close relationship with the present value method.

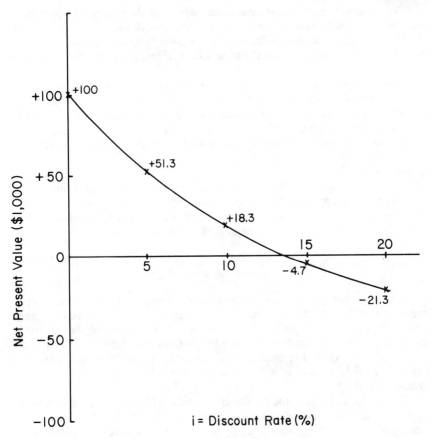

Fig. 2. Present value profile for example 3.

Example 4: Suppose a new piece of equipment is being considered for purchase which promises to generate annual benefits in the amount of $10,000, annual costs of $5,000, and a life of 10 years. If the initial cost of the machine is $40,000 and the expected salvage value at the end of 10 years is $2000, what is the net annual worth if interest on invested capital is 10%?

Solution:

	Annual worth
Benefits: $10,000 per year	$ 10,000
Salvage: $2000 (P/F,10%,10) (A/P,10%,10)	125
Costs: $5000 per year	− 5,000
Investment: $40,000 (A/P,10%,10)	− 6,508
Net annual value:	− 1,383

Since the net annual value is less than 0, the calculation shows that the project is expected to earn less than the 10% interest rate used. If the equipment is purchased it will result in an average net annual cost to the firm of $1,383 per year at the interest rate specified. Therefore the project should be rejected.

Comments: Present, future, and annual values are all measures of equivalency and differ only in the times at which they are determined. Therefore, for fixed values of i and n the techniques provide consistent bases for comparison of investment proposals. They will all provide the same accept/reject decision for the proposal in question.

An important point to make at this time is that an interest rate must be provided in order for these criteria to be determined. This interest rate is an important factor and must be equated to the firm's required rate of return if these criteria are to be used to make investment decisions which generate wealth to the organization. More detail is provided on the determination of this interest rate in Chapter 11.

Benefit/Cost Ratio

The benefit/cost ratio (B/C ratio), often referred to as the profitability index (PI), is generally defined as the ratio of the sum of present value of future benefits to the sum of the present value of present and future investment outlays and other costs (Quirin, 1967). This ratio is expressed as follows:

$$\text{B/C ratio (PI)} = \frac{\Sigma \text{ PV of net cash inflows}}{\Sigma \text{ PV of net cash outflows}}$$

In order to perform this calculation an interest rate must be specified prior to present value determination. If the calculation results in a PI⩾1.0, the investment proposal should be accepted; if not, it should be rejected. This is the same as saying the project should be accepted if it has a NPV⩾0. Indeed, the only difference between the NPV calculation and the PI calculation is that the NPV is the difference between the present value of inflows and outflows, whereas the PI is the ratio between the two.

For any given project, the NPV method and the PI will provide the same accept/reject decision, assuming the calculation is performed at the same interest rate. However if a choice must be made between two investment proposals, these methods may provide inconsistent project rankings. This aspect is illustrated in the following example.

Example 5:

	Project A	Project B
Present value cash inflows:	$500,000	$100,000
Present value cash outflows:	300,000	50,000
Net present value:	$200,000	$ 50,000
Benefit/cost ratio:	1.67	2.0

In this example the NPV criterion suggests that project A is preferable to project B, whereas the B/C ratio clearly indicates project B is superior to project A. Which of these present value-based techniques is correct and why do they provide different rankings?

There are really two answers to these questions. In absolute terms the expected economic contribution of project A to the firm is greater than that promised by project B. This aspect is correctly represented by the NPV technique. However, the B/C ratio reflects the *relative* profitability of the two projects. In this case the B/C ratio indicates that project B promises a greater return per dollar of outflow. Stated in other terms, the relative gain from the capital resources committed to project B is greater than from those committed to project A.

If A and B are mutually exclusive projects, an important question to ask is, "What happens to the extra $400,000 if project B is selected?" If this is invested in a third project, X, then the correct comparison is the B/C ratio for A with the composite B/C ratio for projects B and X. Selecting an optimum portfolio of projects is covered more thoroughly later in this chapter under Capital Budgeting.

Early in this book rationale was provided which advocated the use of annual cash flows instead of net profits as the appropriate means of measuring project performance on an annual basis. At that time the statement was made that it is really more correct to compare *total annual benefits* with *total annual costs* resulting from an investment where risk is present. The point was also made that it is generally acceptable to use *net* annual cash flows in either a risk-free environment or where risk is explicitly handled elsewhere in the analysis. Because this book addresses methods by which business risk may be incorporated into investment proposal evaluations, it was decided to work in terms of net annual cash flows throughout the book.

The PI, however, is the one investment criterion where it makes a difference whether one treats separate streams of benefits and costs rather than net benefits. This is illustrated in the following example.

Example 6: Given the following investment proposals, select the best alternative based on the PI criterion assuming a 10% interest rate.

	Project A	Project B
Initial investment:	$300,000	$300,000
Annual benefits (after-tax):	55,000	250,000
Annual costs (after-tax):	5,000	200,000
Life:	10 yrs.	10 yrs.

Solution: Table 3 shows the PI calculations for the cases where *net* annual values and *total* annual values are used.

Example 6 shows that there is no difference (both have a PI = 1.02) when evaluated using *net* annual benefits. When *total* benefits and costs are used however, project B is less attractive than project A. Characteristic of the PI it shows that the relative gain from the capital resources committed to project A is greater than the relative gain from those resources committed to project B. From another standpoint, the risk might be considered greater in project B because of the higher commitment of funds. Unless they are mutually exclusive, both projects should be accepted because they both promise a PI⩾1.0. However this example does illustrate that the PI is an important ranking technique where risks are involved and have not specifically been handled elsewhere in the analysis. Under these conditions the use of total benefits and costs associated with an investment proposal is preferred.

In summary, the NPV method is preferred for determining the absolute expected economic contribution of a project. However, it is often relative profitability of a project that is of interest, particularly in capital-rationing situations, and it is here that the project-ranking capability of the B/C ratio is most appropriate. Further discussion on capital rationing is contained in Part II of this chapter.

Internal Rate of Return

When evaluators in the minerals industry speak of a rate of return on an investment proposal they are almost always referring to the so-called discounted cash

Table 3. PI Calculations Using Net and Total Annual Values

	Project A		Project B	
	Net	Total	Net	Total
Investment	300,000	300,000	300,000	300,000
Total benefits		55,000(6.145) = 337,975		250,000(6.145) = 1,536,250
Total costs		5,000(6.145) = 30,725		200,000(6.145) = 1,229,000
Net benefits	50,000(6.145) = 307,250		50,000(6.145) = 307,250	
PI	$\dfrac{307,250}{300,000} = 1.02$	$\dfrac{337,975}{330,725} = 1.02$	$\dfrac{307,250}{300,000} = 1.02$	$\dfrac{1,536,250}{1,529,000} = 1.005$

flow return on investment (DCF-ROI) or the discounted cash flow rate of return (DCF-ROR). These terms are special versions of the more generic term, *internal rate of return (IRR)*, or marginal efficiency of capital. This criterion is employed more in the minerals industry for investment proposal evaluation than perhaps any other technique.

The internal rate of return is defined as that interest rate which equates the sum of the present value of cash inflows with the sum of the present value of cash outflows for a project. This is the same as defining the IRR as that rate which satisfies each of the following expressions:

Σ PV cash inflows $-\ \Sigma$ PV cash outflows $= 0$

$$NPV = 0$$
$$PI = 1.0$$

Σ PV cash inflows $=\ \Sigma$ PV cash outflows

In general, the calculation procedure involves a trial-and-error solution unless the annual cash flows subsequent to the investment take the form of an annuity. The following examples illustrate the calculation procedures for determining the internal rate of return.

Example 7: Given an investment project having the following annual cash flows; find the IRR.

Year	Cash flow
0	$ – 30,000
1	– 1,000
2	5,000
3	5,500
4	4,000
5	17,000
6	20,000
7	20,000
8	– 2,000
9	10,000

Solution: Step 1. Pick an interest rate and solve for the NPV. Try 15%.

NPV $= -30,000(1.0) - 1000(P/F,1,15\%) + 5,000(P/F,2,15) + 5500(P/F,3,15) + 4000(P/F,4,15) + 17,000(P/F,5,15) + 20,000(P/F,6,15) + 20,000(P/F,7,15) - 2000(P/F,8,15) + 10,000(P/F,9,15)$

$\quad = \$ + 5619$

Since the NPV>0, 15% is not the IRR. It now becomes necessary to select a higher interest rate in order to reduce the NPV value. If $r = 20\%$ is used, the NPV $= \$ - 1664$ and therefore this rate is too high. By interpolation the correct value for the IRR is determined to be 18.7%

Example 8: Given an investment which promises the following uniform annual cash flows; find the IRR.

Year	Cash flow
0	$ – 20,000
1	6,000
2	6,000
3	6,000
4	6,000
5	6,000

Solution: Since the annual cash flows subsequent to the investment are in the form of an annuity, the solution is simpler and takes the following form:

$$\Sigma \text{ PV outflows} = \Sigma \text{ PV inflows}$$
$$\$20,000 = 6,000(P/A,5,\ X\%)$$
$$\frac{20,000}{6,000} = (P/A,5,\ X\%)$$
$$3.33 = (P/A,5,\ X\%)$$

The solution is found by looking for the value 3.33 under the P/A column for n = 5 years in the appropriate interest table. The value 3.33 lies between 15 and 16%. By interpolation the correct IRR is 15.2%.

The acceptance or rejection of a project based on the IRR criterion is made by comparing the calculated rate with the required rate of return, or cutoff rate, established by the firm. If the IRR exceeds the required rate the project should be accepted; if not, it should be rejected. If the required rate of return is the return investors expect the organization to earn on new projects, then accepting a project with an IRR greater than the required rate should result in an increase in the price of common stock (i.e., shareholders' wealth) in the marketplace.

There are several reasons for the widespread popularity of the IRR as an evaluation criterion. Perhaps the primary advantage offered by the technique is that it provides a single figure which can be used as a measure of project value. Further, this figure is expressed as a percentage value. Most managers and engineers prefer to think of economic decisions in terms of percentages as compared with absolute values provided by present, future, and annual value calculations.

Another advantage offered by the IRR method is related to the calculation procedure itself. As its name suggests, the IRR is determined internally for each project and is a function of the magnitude and timing of the cash flows. Some evaluators find this superior to selecting a rate prior to calculation of the criterion, such as in the profitability index and the present, future, and annual value determinations. In other words, the IRR eliminates the need to have an external interest rate supplied for calculation purposes.

Because the internal rate of return is such a popular evaluation criterion throughout the minerals industry and others, it deserves careful scrutiny to ensure that it is truly worthy of this popularity. The following points are offered for consideration.

1) Even though the IRR provides for the determination of an internal percentage rate, it must still be compared with the hurdle, cutoff, or required rate of return established by the firm before the accept/reject decision can be made. Presumably this stipulated required rate of return established by the firm is related to the firm's cost of capital (see Chapter 11) or required cutoff rate and carries with it the implicit borrowing and reinvestment assumptions of any discounting process. If this is not the case and the required rate of return is a subjectively determined value, then similar criticism to that offered in the payback period discussion is warranted.

2) Perhaps the most serious problem associated with the IRR lies in what engineers and managers perceive it to mean. When people speak of a project's projected rate of return they are typically thinking in terms of the project's "rate of return on investment." This implication brings forth some interesting questions.

First, what is the real return generated by a project? Should this return be measured in terms of profits or cash flows, since cash flows represent a return *on* and a return *of* investment?

Second, what is the investment? Since part of the investment is returned annually, the amount of investment remaining must be continually declining. Therefore, *all* of the investment is not working on an annual basis. Consequently, what should be used for "investment" in the calculation procedure?

Third, what is an acceptable rate of return? Should a new mining project promising a rate of return of 12% be accepted? Would a project promising a rate of 20% always be acceptable?

Clearly these are complex questions and anyone who has worked in the area of project evaluation appreciates the difficulty involved in arriving at answers. However, these questions do illustrate some misconceptions many individuals have with respect to the meaning of "rate of return." The following discussion is offered in hopes of clarifying any possible confusion.

Rate of return is defined as the percentage or rate of interest earned on the *unrecovered* portion of the investment such that the payment schedule makes the unrecovered investment equal to zero at the end of the investment's life. The following example illustrates this calculation procedure.

Example 9: Given an investment proposal promising the following cash flows after an initial investment of \$10,947 (a) find the rate of return, and (b) determine the repayment schedule for the investment.

Year	Annual cash flow
1	\$4500
2	3500
3	2500
4	1500
5	1000
6	750
7	500

Solution: (a) By a trial-and-error process as illustrated in example 7, the rate of return is found to be 11%. (b) This part of the problem asks for the repayment schedule on the investment at the rate of return (11%).

Year	Cash flow	Unrecovered investment at beginning of year	11% return on unrecovered investment during year	Investment repayment at end of year	Unrecovered investment at end of year
0	−10,947				10,947
1	4,500	10,947	1204	3,296	7,651
2	3,500	7,651	842	2,658	4,993
3	2,500	4,993	549	1,951	3,042
4	1,500	3,042	335	1,165	1,877
5	1,000	1,877	206	794	1,083
6	750	1,083	119	631	452
7	500	452	50	450	2*
			3305	10,945*	

*Rounding error; the account goes to zero if all figures are carried.

This example illustrates that the 11% rate of return is a true rate in the sense of the foregoing definition. It clearly shows (other than rounding error) that the investment was recovered at the end of the project's life *and* that an 11% return was earned annually on the *unrecovered* investment.

The Implied Reinvestment Question: Many people hold the misconception that the 11% rate of return determined in example 9 represents the rate of return on the initial investment ($10,947) in the project over its life (7 years). This is clearly *not* the case as demonstrated in example 9. The project actually earns net returns in the amounts of $1204, $842, $549, $335, $206, $119, and $50 from years 1 through 7, respectively, on an investment that decreases annually from an initial amount of $10,947 to $0 at the end of project life.

The only way the initial investment could earn an 11% rate of return on the total investment ($10,947) over the project's life would be if each annual cash flow were reinvested at an 11% interest rate from the time of receipt to the end of the project's life. To illustrate, if the $10,947 were invested in an account which paid 11% it would yield:

$10,947(F/P,11%,7) = 10,947(2.076) = $22,726 at the end of 7 years

However, the only way the project cash flows can generate $22,726 at the end of 7 years is if they are invested at 11% from the time of receipt to the end of year 7.

Year	Cash flow	Interest factor at 11%		Amount generated at end of project
1	4500	(F/P,11%,6) = 1.870	=	$ 8,415
2	3500	(F/P,11%,5) = 1.685	=	5,898
3	2500	(F/P,11%,4) = 1.518	=	3,795
4	1500	(F/P,11%,3) = 1.368	=	2,052
5	1000	(F/P,11%,2) = 1.232	=	1,232
6	750	(F/P,11%,1) = 1.11	=	833
7	500	(F/P,11%,0) = 1.0	=	500
			Total	22,724*

*Rounding error.

This example shows that the rate of return on total investment over a project's life can only be achieved if the annual cash flows are reinvested at the calculated rate of return from the time of receipt until project termination. When people rank projects using the IRR they often fail to realize this *reinvestment rate* assumption. By definition the IRR is that interest rate which equates the sum of the present value of expected cash outflows with the sum of the present value of expected cash inflows. This interest rate carries with it the fundamental premise of any discounting process; namely, that money can be borrowed and reinvested at this rate. As a result, the internal rate of return method, when used to compare projects, implies that funds released from the project are reinvested at the IRR over the remaining life of the project.

In terms of investment proposal evaluation this means that if the funds released or generated by a project are reinvested at some rate less than the calculated IRR (e.g., the required rate of return) then the IRR may overstate the value of the project in relation to some alternate investment. For instance, if a new mining project promised an IRR of 20%, but an actual reinvestment rate of only 15% could be achieved on cash flows generated by the project, then obviously the project would yield a composite rate of return less than 20% over the full life of the investment.

There is nothing inherent in the calculation procedure for the IRR that assumes reinvestment of cash inflows. However, when the IRR is used to *compare projects*, the analyst makes the implicit assumption that inflows from each project can be reinvested at the IRR of that project over its remaining life. This is clearly illustrated in example 9.

The Multiple Roots Question: One of the disconcerting aspects associated with the internal rate of return is that more than one interest rate may satisfy the calculation. The solution procedure for IRR is essentially the solution for an nth-degree polynomial of the form:

$$\text{NPV} = 0 = A_0 + A_1 X + A_2 X^2 + A_3 X^3 + \dots + A_n X^n$$

where $X = 1/(1 + r)$.

For a polynomial of this type there may be n different real roots, or values of r, which satisfy the equation. Multiple positive rates of return may occur when the annual cash flows have more than one change in sign.

Descartes' Rule of Signs is helpful in identifying the possiblity of multiple rates of return for an nth-degree polynomial of the type which is of interest in the IRR calculation. Essentially the rule states that the number of real positive roots of an nth-degree polynomial is never greater than the number of sign changes in the sequence of its coefficients, and if less, always by an even number. It must be understood that this rule of signs only provides an indication as to the *maximum* number of possible rates of return that may occur. The *actual* number of rates that do occur will also depend upon the *magnitudes* of the cash flows involved.

The following example illustrates the possibility of multiple rates which satisfy the definition of IRR.

Example 10: Suppose a mining operation has a remaining life of eight years, but an investment is considered to increase the production rate. This will result in depleting the deposit in five years. Assuming the following cash flows, is the investment justified?

Year	Annual cash flows ($1000's)		
	Existing	Proposed	Δ
0		-15	-15
1	$+10$	$+16$	$+6$
2	10	16	$+6$
3	10	16	$+6$
4	10	16	$+6$
5	10	15	$+5$
6	10	15	$+5$
7	10	0	-10
8	10	0	-10

Solution: Because there are two sign reversals in the cash flows, Descartes' Rule of Signs indicates there are a maximum of two real roots to the IRR polynomial. Solving for these roots by trial and error yields the following:

Discount rate (r):	0%	4%	5%	10%	12%	15%
Net present value:	-1	-0.07	$+0.06$	$+0.05$	$+0.03$	-0.25

Graphically this appears as shown in Fig. 3. The rates at which NPV $= 0$ are, by definition, the internal rates of return. By interpolation, the two solving rates of

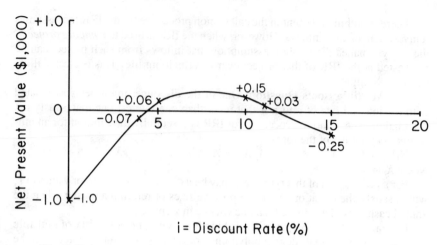

i = Discount Rate (%)

Fig. 3. Net present value profile for example 10, illustrating multiple rates.

return for this example are approximately 4.5 and 12.3%. Should the firm invest in the project or not? If both rates were above the firm's required rate of return there would be no problem and the firm would accept the project. However, what if the required rate of return is 10%? Which of the calculated IRR values is correct? The answers to these questions are that they are both *mathematically correct*, but they are *meaningless* from an *economic* standpoint. Neither of these rates can be considered an adequate measure of the project's rate of return because a project can not earn more than one rate of return over its life. Therefore, the calculation of an IRR value(s) does not always enable the decision-maker to make accept/reject decisions on investment proposals.

Some might wonder how often this problem of multiple rates actually occurs. As one might expect the possibility of multiple-rate occurrences is perhaps more prevalent in the case of new mining ventures than in most other industries. The authors are aware of several examples where multiple rates did occur as the result of anticipated negative cash flows scattered throughout the project's life. These negative cash flows are typically the result of anticipated periods of reduced market prices, major capital expenditures for equipment replacement, expansion programs, and/or major environmental expenditures—particularly at the end of project life.

Because of the possibility of multiple rates and the reinvestment assumption when using the IRR to rank projects, the evaluator must carefully consider the exclusive use of this technique for decision-making.

External Rate of Return

In order to overcome many of the disadvantages associated with the evaluation criteria presented thus far—and particularly those associated with the internal rate of return—two financial measurement criteria have been proposed which have considerable promise. These two criteria are similar in concept, if not in exact calculation procedure, and are presented jointly in this section. They are grouped under the heading external rate of return because an external interest rate must be provided prior to calculation of the criterion itself. These techniques may be regarded as methods for calculating the worth of an investment proposal plus reinvestment opportunities, rather than the worth of an investment proposal alone. This is an important concept because the decision-maker is really presented with alternative courses of action and not simply the narrow choice between investment projects.

Wealth Growth Rate (WGR): The wealth growth rate (WGR) is defined by Berry (1972) as that interest rate which equates the *future* value of the capital investment with the *future* value of the cash flows resulting from the project. The time horizon for both future values is the termination date of the project. The positive net annual cash flows subsequent to the investment are assumed to be reinvested at the firm's reinvestment rate to the termination date of the project. If investment occurs over several years (i.e., preproduction development), these negative cash flows are discounted to time 0 (initial investment) using the same reinvestment rate. This recognizes that the discounting process implies that the borrowing and reinvesting rates are the same. The WGR is then the compound rate at which the cumulative discounted capital investment must grow in order to equal the future wealth generated by the project. The reinvestment rate is specified by the firm (external to the calculation); and, therefore, the WGR determination is a rather simple process.

Although a number of variations on calculation procedures have been developed over the years, the following simple example illustrates the normal procedure.

Example 11: A firm is considering investing in a new mining property which is anticipated to have the following annual cash flows. If the firm's reinvestment rate is 12%, find the WGR.

Year	Annual cash flow (1000's)
0	− 1000
1	− 900
2	− 800
3	1000
4	500
5	1000
6	2000
7	1000

Solution:

Step 1. First the annual positive cash flows must be reinvested at 12% to termination of the project in year 7.

Year	Annual cash flow (1000's)	Interest factor at 12%		Amount at end of year 7
3	$1000	(F/P,12%,4) = 1.574	=	1574
4	500	(F/P,12%,3) = 1.405	=	702
5	1000	(F/P,12%,2) = 1.254	=	1254
6	2000	(F/P,12%,1) = 1.12	=	2240
7	1000	(F/P,12%,0) = 1.0	=	1000
			Total	6770

Graphically the problem now has the following form:

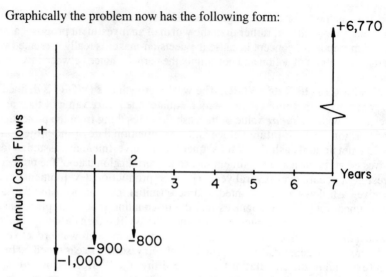

Step 2. The negative cash flows occurring in years 1 and 2 are discounted to time 0 to determine the equivalent capital investment in the project.

Year	Annual cash flows	Interest factor at 12% (P/F)	Amount at time 0
0	− 1000	1.0	− 1000
1	− 900	0.8929	− 804
2	− 800	0.7972	− 638
		Total	− 2442

The problem now has the following graphical representation:

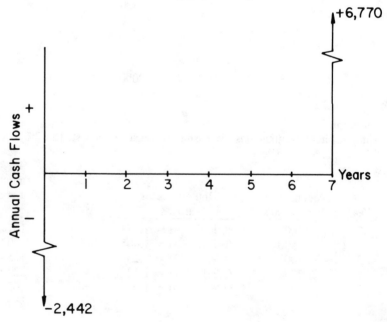

Step 3. The WGR is determined by finding the rate at which an investment of $2442 must grow in order to equal $6770 in 7 years. Therefore, the appropriate interest factor is

$$F = P (F/P,i,n)$$
$$\$6770 = \$2442 (F/P,i,7)$$
$$(F/P,i,7) = 2.7723$$

By scanning the interest tables in the F/P column for $n = 7$, the WGR is determined to be between 15% and 16%. The actual WGR is 15.7%.

The accept/reject decision is determined by comparing the calculated wealth growth rate with the firm's required target rate. This target or threshold rate may be a rate stipulated by the firm or may be the reinvestment rate used in the calculation procedure. If the WGR is equal to or exceeds the required or target rate the project should be accepted; if not, it should be rejected.

There are several unique properties of the WGR which provide this technique with some distinct advantages over some other criteria. First, the WGR uses annual cash flows as opposed to profits and it recognizes the time value of money. The technique enables the firm to specify the actual or anticipated reinvestment rate which it can reasonably expect during a project's life. Thus, when used to rank projects the WGR provides a uniform and consistent reinvestment rate for *all* projects rather than a different rate (the IRR) for each project. Another advantage is that the WGR provides a unique solution which quantifies the rate of wealth growth and expresses this solution in terms of an annual rate which can be directly compared with the firm's reinvestment rate. In other words, this criterion determines the average rate of growth of the firm's accumulated wealth resulting from a capital project.

The last major advantage associated with the WGR is that the criterion allows for the comparison of project alternatives having different finite lives, which is a difficult exercise with other criteria discussed thus far. Because a project's life is explicitly treated in the calculation procedure, the technique provides an excellent means for ranking investment alternatives.

An important point in this regard, however, is the inherent assumption associated with the reinvestment rate at project termination when comparing alternatives with different lives. The WGR method assumes reinvestment rates for projects under consideration up to the termination date of the longest-lived project. As previously stated, the rate assumed during the life of each project is the firm's reinvestment rate which is externally supplied. However, this technique then uses the calculated WGR rate to compound the cumulative value at the end of an individual project's life to the termination life of the longest-lived project. This assumption is defended by Berry (1972) on the grounds that near the conclusion of a project, management, because of its experience and learning with the project in question, should be able to search for and implement a replacement project with an equivalent or superior WGR. Some analysts take exception to this assumption and view it as a disadvantage of the WGR.

Growth Rate of Return (GRR): The growth rate of return (GRR) is identical to the wealth growth rate except for the common terminal date assumption. As developed by Capen, et al. (1976), the GRR is calculated by first compounding all the positive cash flows forward to some time horizon, t years in the future. Any cash flows occurring after time t are discounted back to time t. The rate at which these cash flows are compounded or discounted is the reinvestment rate or opportunity rate of the firm and is externally supplied for calculation purposes. This total amount of money determined for time horizon t then represents the expected revenue from the project plus the earnings or interest generated by the reinvestment rate (reinvestment in future projects). The negative cash flows resulting from the investment decision are discounted to a present value ($t=0$) in order to obtain an equivalent investment at this point in time. The discount rate is again the same as the reinvestment (opportunity) rate supplied by the firm. By definition the GRR is that interest rate at which the investment would have to grow in order to equal the total amount of money accumulated by the project at time t.

The calculation procedure for GRR is illustrated in the following example.

Example 12: Calculate the GRR for a project having the following anticipated cash outflows and inflows. The firm's reinvestment rate is given to be 12% and the project should be evaluated at time horizon $t = 7$.

Year	Annual cash flows (1000's)
0	− 1000
1	− 900
2	− 800
3	1000
4	500
5	1000
6	2000
7	1000
8	1000

Solution: The problem can be illustrated graphically as follows:

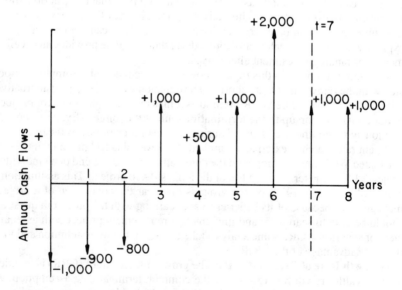

Step 1. The first step is to bring the positive cash flows occurring prior to time $t = 7$ forward to that point at the stipulated reinvestment rate.

Year	Annual cash flows	Interest factor at 12% (F/P)	Amount at time t = 7
3	1000	1.574	1574
4	500	1.405	702
5	1000	1.254	1254
6	2000	1.120	2240
7	1000	1.000	1000
		Total	6770

The problem is now graphically represented as follows:

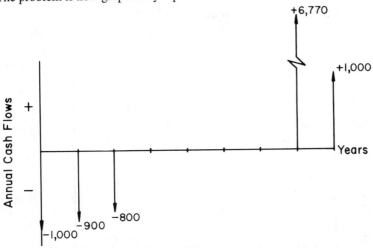

Step 2. Next, the positive cash flow occurring after $t = 7$ must be discounted back to time $t = 7$ at the reinvestment rate.

Year	Annual cash flow	Interest factor at 12% (P/F)	Amount at time t = 7
8	1000	0.8929	893

The graphical representation of the problem at this point is

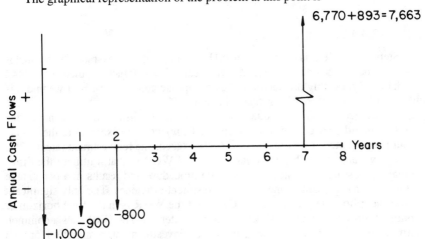

Step 3. The annual investment (negative cash flows) must be discounted to an equivalent investment at $t = 0$.

Year	Annual cash flow	Interest factor at 12% (P/F)	Amount at time t = 0
0	− 1000	1.00	− 1000
1	− 900	0.8929	− 804
2	− 800	0.7972	− 638
		Total	− $2442

The problem has now been simplified to the following:

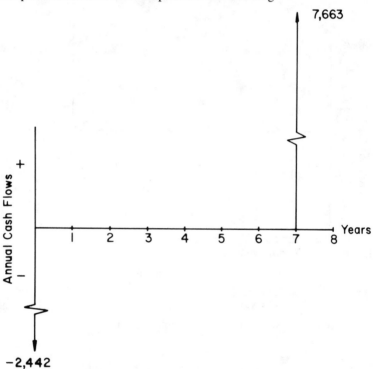

Step 4. The GRR is now determined by finding that interest rate which equates an investment of $2442 to $7663 in seven years. The F/P factor is then $7663/2442 = 3.138$. By locating this factor in the appropriate column for $n = 7$ we find it is between 17% and 18%. The actual value is 17.6%.

If the calculated GRR is equal to or greater than the firm's reinvestment rate, the project should be accepted; if not it should be rejected. In example 12 the project would be accepted since the GRR of 16.7% exceeds the reinvestment rate of 12%.

The virtues of GRR are similar to those of WGR in that it utilizes the firm's actual reinvestment rate in the calculation procedure and results in a percentage return which represents a measure of investment efficiency. The only significant difference, however, between the GRR and the WGR is in the time horizon or common terminal date for calculation of the criterion. The GRR uses a common terminal date for all projects so that the same reinvestment rate assumption for cash flows is made for all projects. This eliminates the potential problem with WGR where the longest-lived project is taken as the base for comparative purposes and projects with shorter lives are assumed to have two reinvestment rates (the stipulated rate to project termination and the WGR from project termination to the end of the longest-lived project).

Some people prefer not to use GRR because the calculated rate for a project depends on the time horizon, t, chosen. This is absolutely true. However, when this technique is used by a firm to evaluate investment proposals, all projects are compared using the same values of t and r, the reinvestment rate. Under these conditions the GRR will provide the same accept/reject decisions as present value determinations and the same project ranking as the profitability index. Also, the relative rankings of projects do not change as the time horizon changes, nor does the accept/reject decision change.

Hoskold's Method

Perhaps the first mining-related evaluation technique which incorporated the concept of present value was developed by Henry Hoskold in 1877. The basic premise underlying this technique was that the property owner would require a certain return on his invested capital in a mining property, and also the owner would establish a sinking fund such that at the end of the property's life the owner would have accumulated an amount equivalent to his initial investment to enable the owner to purchase another property to replace the exhausted mine.

The Hoskold method has historically been applied to uniform annual net *profits* resulting from an investment. The idea is that some portion of the annual net profits will be placed in a sinking fund which earns interest at a so-called "safe" rate (e.g., government securities) and the remainder of the annual profits provides the return on invested capital yielding the so-called "speculative" rate of return. This higher speculative rate would be equivalent to the reinvestment or hurdle rates used previously in this chapter.

The Hoskold formula for present worth is provided in Parks (1957, p. 193) and is represented as:

$$Vp = \frac{A}{\dfrac{r}{(1+r)^n - 1} + r'}$$

where Vp is present value, r is safe rate on redemption of capital, r' is speculative rate on invested capital, and A is uniform annual net profit.

This method for determing the present worth of a mining property is illustrated in the following example.

Example 13: An investor can purchase a mining property for an initial investment of $30,000,000. If the mine has an estimated remaining life of 15 years and promises to yield annual net profits in the amount of $3,000,000, what is the net present value of the property assuming a safe rate of 5% and a speculative rate of 15%?

Solution: The present worth of the future annual profits ($3,000,000) can be determined from the foregoing equation or by finding the appropriate interest factor in Parks (Table 7, p. 401):

$Vp = \$3,000,000 \ (5.5481) = \$16,644,300$

Therefore, NPV $= \$16,644,300 - \$30,000,000 = -\$13,356,000$

The decision criterion for this technique is to reject projects having a negative present value. This is essentially the same concept as with the present value techniques discussed earlier except that two separate interest rates are involved in the Hoskold Formula.

It is interesting to note that if a sinking fund and the associated safe rate of interest were not a part of the calculation the present value of example 13 would be:

$$PV = \$3,000,000 \ (P/A,15\%,20)$$
$$= \$3,000,000 \ (6.259) = \$18,777,000$$

This indicates that the Hoskold approach, utilizing a sinking fund, will result in a lower estimate of value than the net present value method using the speculative rate for discounting. This results from the fact that a sinking fund reduces the net benefits available to the investor by the amount required for the sinking fund. Obviously the lower the safe rate of interest, the more cash must be placed annually in the sinking fund.

Modern financial management practices rarely include the establishment of sinking funds, unless absolutely necessary. Companies simply prefer not to establish sinking funds for recouping investments, just as they do not methodically put aside depreciation allowances into a special fund for future equipment replacement. The opportunity costs associated with these funds are too high.

Another major problem with this technique is that it utilizes annual net profits rather than cash flows. This is excusable because at the time this technique was developed, there was no income tax to create a difference between profit and cash flow. This is not the situation today, however, and cash flow is the appropriate means of measuring the performance of a project.

Relationship of Methods

To this point in the chapter some of the more important project evaluation criteria have been briefly discussed, and calculation procedures demonstrated. In view of the various advantages, disadvantages, and unique characteristics of these criteria, a logical question at this point is, "How do they compare with each other for mining project evaluation?"

A one-to-one comparison of each of the criteria presented in this chapter, particularly in quantitative terms, would require considerable time and space. However, it is useful to discuss some of the relationships which exist between major criteria which are often utilized jointly in project evaluation.

Table 4 represents a comparison of the evaluation criteria discussed. Although general in nature, this comparison does illustrate some of the major differences which exist between criteria and leads the analyst to some tentative conclusions regarding the appropriateness of specific criteria in mining project evaluation work.

The following discussion relates to some specific relationships between evaluation criteria which are generally of interest to investment proposal analysts.

IRR vs. Payback Period: The internal rate of return and the payback period continue to be two of the more popular evaluation techniques utilized in the minerals industry. Although some relationships do exist between these criteria, they occur under some rather unique circumstances. For instance, where projects have long lives, substantially in excess of the payback period, and where income streams are uniform each year, the payback period is a good approximation to the

Table 4. Comparison of Financial Measurements*
(after Berry, 1972)

Item*	Accounting rate of return	Payback period	Discounted payback period	NPV	PI	IRR	WGR	GRR	Hoskold
1.	Profit	Either	Either	CF	CF	CF	CF	CF	Profit
2.	No	No	Yes	Yes	Yes	Yes	Yes	Yes	Yes
3.	No	No	No	No	Yes	No	Yes	Yes	Yes
4.	No	No	No	No	No	Yes	No	No	Yes
5.	Yes	No	No	No	No	Yes	Yes	No	No
6.	No	No	Maybe	No	No	Yes	No	No	No
7.	Yes	Maybe	No	Yes	Yes	Yes	Yes	Yes	No
8.	Yes	No	No	Yes	Yes	No	Yes	Yes	Yes
9.	No	No	No	Yes	No	No	Yes	Yes	Yes
10.	No	No	Yes	Yes	Yes	No	Yes	Yes	Yes
11.	No	No	No	No	No	No	Yes	Yes	No

*Explanation of items for financial measurement:
1. Use profit or cash flow?
2. Recognizes time value of money?
3. Requires reinvestment rate in calculation?
4. Assumes a sinking fund?
5. Results in form of a rate of return?
6. Can yield multiple solutions?
7. Compares different investment requirements?
8. Accounts for benefits after payback period?
9. Appraises market value of project?
10. Ranking may vary with different reinvestment rates?
11. Explicitly recognizes life of the project?

reciprocal of the internal rate of return. In the unique case where $n \rightarrow \infty$, this relationship exists precisely.

Gentry (1971) presented an expression for the reciprocal of the payback period in terms of IRR and project life, n, under the assumptions of a project having equal annual cash flows (an annuity) and a zero terminal value. Percentage point deviations were calculated between the payback reciprocal and the IRR. He was able to show that, over a rather broad range of economically significant combinations of IRR and n, the payback method can result in a substantial upward bias in profitability estimates.

Although attempts have been made to develop relationships between the payback period, IRR, and project life, little success has been achieved except where some rather demanding and simplifying assumptions have been involved. In the general case, where all types of projects are under consideration, little direct correlation exists between IRR and the payback period.

IRR vs. NPV: The internal rate of return and the present value methods provide identical accept/reject answers for a single investment proposal. However, it is important to recognize that these two discounted cash flow techniques can give contradictory results when comparing mutually exclusive projects (only one can be selected). Consider two mutually exclusive investment proposals, A and B, which are expected to generate annual cash flows as follows:

Year	Annual cash flows (1000's)	
	Project A	Project B
0	$ − 10	$ − 10
1	1	7
2	5	5
3	6	3
4	7	1

Fig. 4 illustrates the net present value profiles for each of these projects. Notice that the IRR for projects A and B are 24.7% and 30.5%, respectively. Therefore, project B is considered superior to project A. However, if the required rate of return were 15%, the NPV of proposals A and B are 2.60 and 2.41, respectively, and project A is preferred to project B because it contributes more wealth to the firm. At a required rate of approximately 16.5% there is virtually no difference between the two projects.

The conflict between these two criteria is the result of different implied assumptions for reinvestment rates for funds generated by the project. The internal rate of return method assumes that funds are reinvested at the IRR over the remaining life of the project. The present value method assumes reinvestment at a rate equal to the required rate of return used as the discount rate.

In view of this conflict, which method is best for evaluating investment proposals? The answer to this question depends upon what is considered to be the appropriate reinvestment rate for the intermediate cash flows generated by the project. If one must choose between these two techniques, most believe the present value method is superior—at least from a theoretical standpoint.

When using the internal rate of return to rank projects, a high reinvestment rate

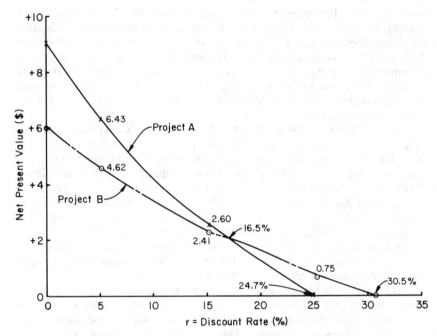

Fig. 4. Present value profiles for two projects, showing NPV and IRR values.

is assumed for a proposal having a high IRR whereas a proposal having a low IRR carries with it a low reinvestment rate assumption. Only by coincidence will the IRR calculated be the same as the reinvestment rate actually available to the firm for intermediate cash flows. Also, there is seldom any sound reason to assume that reinvestment opportunities following one project would be more favorable than those following another project.

In contrast the present value method assumes the required rate of return is the reinvestment rate which remains the same for *all* proposals. This rate is intended to represent the minimum return on opportunities available to the firm. This may introduce some conservatism into the calculation since actual reinvestment rates may exceed these minimum rates. Nonetheless it has the advantage of being applied consistently to all investment proposals. For these reasons the present value method is preferred over the IRR method for ranking projects.

WGR vs. Other Criteria: The wealth growth rate method is unique because it (1) enables the firm's reinvestment rate to be employed explicitly in the calculation procedure, (2) allows for the comparison of projects with different finite lives, and (3) expresses the financial advantage of each alternative as an annual rate which can be directly compared with the firm's reinvestment or required rate. These unique features enable the WGR to assist in investment decisions where other criteria are unsatisfactory.

It can be shown that the WGR equals the required, or reinvestment rate, when NPV = 0 or PI = 1. Therefore, it will provide the same accept/reject decision as these present value methods. However, because it provides a means of comparing projects with different lives (remembering the reinvestment rate assumption at the end of a shorter-lived project), the WGR is a powerful ranking technique for investment proposals. For instance, projects with the same net present value or profitability index, but with different lives, can be differentiated with the WGR in that the shorter-lived project would be selected (Berry, 1972). In the normal case where NPV and PI are different for various proposals, the WGR ranking might also be different due to the required investment being different or the lives being different.

The WGR cannot be directly correlated with the payback period, the accounting rate of return, or the Hoskold method because of the major differences in the basic concepts involved.

GRR vs. Other Criteria: The growth rate of return (GRR) is essentially the same as the wealth growth rate (WGR) except for the common terminal date assumption. The GRR evaluates investment proposals at some time horizon, t, as compared with WGR which evaluates projects at the end of their lives. With the WGR, project comparisons were made on the basis of the longest-lived project. It was assumed that the accumulated wealth at the end of the shorter-lived project could be reinvested at the project's WGR to the terminal date of the longer-lived project. The GRR method eliminates this assumption by comparing all projects on the basis of constant values for the time horizon, t, and the reinvestment rate, r. In essence, the selection between the WGR and the GRR is a matter of accepting the more realistic reinvestment rate assumption.

Although the calculated value of GRR for a project depends on the time horizon chosen, as long as t and r are the same for all projects the GRR will yield the same accept/reject decisions as present value methods and the same ranking of projects as the PI evaluation criterion. This stems from the nature of the calculation procedure and the reinvestment assumptions.

Comments

The preceding discussions, examples and illustrations pertaining to the evaluation criteria presented in Part I of this chapter were intended to provide some insight into the advantages, disadvantages, unique characteristics, and assumptions inherent with each evaluation criterion. At this point it should be rather obvious that anyone who believes there is *one* universal evaluation criterion clearly does not understand the problem.

As stated very early in this chapter, there are two basic principles which any satisfactory evaluation technique must respect. The methods which most closely adhere to these principles are those utilizing discounted cash flow (DCF) concepts. In general, these DCF methods will provide the same accept/reject decision for investment proposals; however, it has also been clearly demonstrated that they may provide different rankings of projects. As a result it is important to point out that no single evaluation technique will always provide the *correct* answer for investment decisions and, therefore, no one criterion should be used for the purpose of making investment decisions. Rather, a combination of several techniques should be utilized in evaluations before a decision on any investment proposal is reached.

PART II—EVALUATING ALTERNATIVES

Part I of this chapter dealt primarily with the problem of evaluating individual investment proposals from the standpoint of measuring proposal acceptability to the firm. The evaluation criteria discussed are simply methods for helping the firm make the appropriate accept/reject decision for a given investment proposal. However, the final investment decision is not based solely on the outcome of the accept/reject decision resulting from evaluation of the investment proposal. Rather, the investment decision must also be based on determination of which of the acceptable investment proposals are best in terms of meeting the objectives of the firm. Consequently the problem is not one of simply asking "Are projects A and B acceptable to the firm," but also, "Is project A better or worse than project B?" It is this problem of ranking or evaluating investment alternatives that is discussed briefly in Part II.

Since these issues are essential to the problem of capital budgeting, it may be beneficial to place the problem in proper perspective. The following discussion is intended to provide a brief background with respect to the concept of capital budgeting.

Capital Budgeting

Briefly, capital budgeting is the process of allocating available capital in an optimal manner to investment proposals, the benefits from which are to be realized in the future. In a general sense capital budgeting encompasses the (1) generation of investment proposals, (2) estimation of annual cash flows for the proposals, (3) evaluation of the cash flows, (4) selection of projects based on an acceptance criterion, and (5) continuous reevaluation of the proposals after acceptance. This chapter deals with that aspect of capital budgeting relating to the selection of projects for investment. This final selection of projects for investment is based on determination of which projects are acceptable to the firm, and, ultimately, the relative desirability of these investment proposals. This latter aspect of evaluating investment alternatives is critical to the capital budgeting process.

Most financial managers view the capital budgeting problem as one concerned with the choice of a group of investment proposals from a larger suite of *acceptable* proposals. This problem of *budgeting* capital stems from one or more constraints which prohibit the funding of all acceptable projects available to the firm. If there were no capital constraints imposed, the firm would simply invest in all *acceptable* proposals and life would indeed be easy. Unfortunately few firms have this luxury, and the general case finds constraints of one form or another imposed on the organization which affect the ultimate investment decision. Typically these constraints result from a shortage of available capital for new investment proposals, although restrictions of materials and supplies, limited labor availability, and the mutual exclusiveness of investment proposals may also exist.

If the financial objective of the firm is to maximize stockholder wealth (as discussed earlier), then capital budgeting decisions should be based on the following basic principles (Stevens, 1979, p. 157):

1) Every increment of capital expenditure must justify itself.

2) An acceptable investment proposal today is better than the speculation that a better proposal will become available in the future.

Some of the problems associated with project ranking were introduced in Part I of this chapter when project evaluation criteria were discussed and compared. Some of these examples will be referred to subsequently in this portion of the chapter. In addition some of the more typical problems associated with evaluating investment alternatives are presented in the following sections.

Dependent vs. Independent Projects: An investment proposal is said to be *independent* when the acceptance of this proposal from a suite of proposals has no effect on the acceptance of any of the other proposals contained in the suite. It is doubtful if very many proposals in a firm are truly independent, but they are generally considered to be independent if they are functionally different. For example, proposals concerning the purchase of a new rotary drill rig, air-conditioning the corporate office building and undertaking a new marketing campaign would generally be considered to be independent proposals.

The capital budgeting problem associated with choosing between independent investment proposal alternatives is generally quite easy. Under these conditions an appropriate evaluation criterion is used to make the accept/reject decision. These criterion values can then be ranked and the relative proposal desirability determined.

A potential problem with ranking independent investment proposals arises when the IRR and present value methods are used. As pointed out in Part I of this chapter the IRR method will give consistent results with the present value method for accept/reject decisions. However, these criteria can provide inconsistent ranking of independent investment proposals as previously discussed and illustrated in Fig. 4.

The more general case involving investment proposals is where they are not independent but are related to one another in such a way that the acceptance of one proposal from within the suite of proposals will influence the acceptance of others. One type of relationship which often exists between proposals is the result of making an investment in some project which spawns a number of other possible auxiliary investment proposals. These auxiliary proposals are referred to as *contingent* proposals since their acceptance is contingent on the acceptance of a prerequisite proposal. However, the acceptance of the initial or prerequisite proposal is independent of the contingent proposals.

The most common case of investment proposal interdependencies is that of *mutually exclusive* proposals. Mutually exclusive proposals refer to the situation where a group of proposals are related to one another in such a manner that the acceptance of one proposal from the suite of proposals precludes the acceptance of any of the other proposals in the suite. An example of mutually exclusive investment proposals might be the situation where a mine requires an additional piece of primary loading equipment. The investment proposals to be considered might include an electric shovel, hydraulic shovel, and front-end loader. The selection of any one of these proposals would preclude any of the other options from further consideration since only one alternative is necessary to perform the job. Mutually exclusive investment proposals are a special case of the capital budgeting problem. The appropriate method for analyzing and ranking mutually exclusive proposals is covered later in this chapter.

It is perhaps important to note that whenever there is capital rationing—constraints on the amount of capital available for investments—and the aggregate investment cost of all acceptable proposals exceeds the capital available for investment, financial interdependencies are introduced between investment proposals. This may occur, for example, when one project that ranks lower than another is accepted so that a higher return on the entire capital budget is achieved. These interdependencies can occur whether the proposals are independent, contingent, or mutually exclusive. Although these interdependencies are often not obvious, they are nonetheless introduced whenever budget constraints are imposed and amplify the importance of the capital budgeting problem.

Projects Having Unequal Lives

Up to now the examples presented have been used to demonstrate the application of various evaluation criteria on investment proposals having equal lives. Only the wealth growth rate and growth rate of return methods attempted to address the problem of comparing investment proposals with unequal lives. Both of these techniques, however, employed some very specific assumptions regarding reinvestment rates and the common terminal date for comparing proposals. The comparison of mutually exclusive investment proposal alternatives having different lives is a common one and represents a special case in the capital budgeting exercise. The purpose of this section is to develop a basis for comparison of mutually exclusive alternatives having unequal lives.

It should be noted at this point that the capital budgeting decision in this case concerns investing in two courses of action, not just investing in two projects. The fundamental consideration centers on the total economic impact generated by each of the two courses of action at a given point in the future. Also, these opportunities need not be mutually exclusive. For instance, under the constraints of capital rationing one may be faced with the classic problem of investing in a large long-life mine promising a modest rate of return vs. small, short-life mines promising high rates of return. An important question here is, "Can the firm continue to generate small mine reinvestment opportunities (i.e., what are the reinvestment assumptions)?"

When comparing investment proposal alternatives with unequal lives, the basic principle that all alternatives under consideration must be compared over the same time span is fundamental to sound decision-making. Equal time spans for investment alternatives must be assumed if the effect of undertaking one alternative is to be compared directly with the effect of undertaking any other alternative.

The basis for comparing mutually exclusive investment proposal alternatives with unequal lives is generally based on one of the following three common assumptions regarding future alternatives (Stevens, 1979, p. 161):

1) Assume that money generated (cash flows) by each alternative will be invested by the firm in other assets that will earn the minimum or required rate of return for a period of time equal to the life of the longest alternative.

2) Assume that each investment alternative will recycle for a period of time equal to the least common multiple of the alternative lives. When alternatives are recycled under this assumption the initial investmemt, life, salvage value, and annual disbursements are assumed to be identical to the estimates used for the first life cycle.

3) Make specific assumptions (estimates) about future investment opportunities for a period of time equal to the life of the longest alternative.

The appropriate assumption to use will depend upon the type of problem and the assumption which is believed to be the most accurate representation of the future. The following example illustrates assumption 1.

Example 14: Suppose the following cash flows represent two mutually exclusive investment proposals which have to deal with expanding a mine's production. If the required rate of return is 12%, which alternative should be selected?

Solution:

	Annual cash flows (1000's)	
Year	Project A	Project B
0	− 150	− 200
1	55	60
2	55	60
3	55	60
4	55	60
5	55	60
6		60
7		60
8		60
IRR	24.3%	24.9%

The IRR indicates that project B is slightly better than project A.

Calculating the NPV of both projects using a required rate of return of 12% yields the following:

$$NPVa = 55(F/A,12\%,5)(F/P,12\%,3)(P/F,12\%,8) - 150$$
$$= 55(6.353)(1.405)(0.4039) - 150 = \underline{48.29}$$

$$NPVb = (F/A,12\%,8)(P/F,12\%,8) - 200$$
$$60(12.300)(0.4039) - 200 = \underline{98.06}$$

The NPV analysis suggests that project B will maximize the value to the firm, and therefore project B should be accepted under assumption 1.

Assumption 2 is often applied to situations where mutually exclusive investment proposal alternatives are measured in terms of negative cash flows. These are typically investment proposals common to most operating divisions and pertain to cost comparisons or equipment replacement alternatives. The following example illustrates this procedure.

Example 15: Assume the following two machines can perform a given job equally well. If the initial investment and annual disbursements are as given and the required rate of return is 10%, which machine should be selected?

	Machine X	Machine Y
Initial investment (1000's)	150	200
Annual operating disbursements (1000's):	18	10
Life:	5 yrs.	10 yrs.
Salvage value (1000's):	2	0

Solution: Under assumption 2, comparison of the two machine alternatives may be represented as follows:

Year	Annual cash flows (1000's) Machine X	Machine Y
0	− 150	− 200
1	− 18	− 10
2	− 18	− 10
3	− 18	− 10
4	− 18	− 10
5	− 18, + 2, − 150	− 10
6	− 18	− 10
7	− 18	− 10
8	− 18	− 10
9	− 18	− 10
10	− 18, + 2	− 10

The NPV calculations are as follows:

$$NPV_x = -18(P/A,10\%,10) + 2(P/F,10\%,5) - 150(P/F,10\%,5) - 150 + 2(P/F,10\%,10)$$
$$= -18(6.1446) + 2(0.6209) - 150(0.6209) - 150 + 2(0.3856)$$
$$= \$-351.73$$

$$NPV_y = -10(P/A,10\%,10) - 200$$
$$= -10(6.1446) - 200$$
$$= \$-261.45$$

Expressed in equivalent uniform annual costs (EUAC) the machines would have the following costs:

$$EUAC_x = (-351.73)(0.1628) = \$-57.26$$
$$EUAC_y = (-261.45)(0.1628) = \$-42.56$$

Based on this analysis machine Y promises the firm an annual cost savings of $14,700 (57,260 − 42,560) per year over the 10-year interval if it is selected over machine X. Therefore machine Y is the preferred alternative.

In the situation where the machine would only be required for use during some time period which is shorter than the least common multiple of lives, then this time frame would serve as a suitable analysis period. For instance, if the machines described in example 15 were only required through year 8, it would be acceptable to base the cash flow analysis over this time frame and compare the machines at the termination of year 8. Estimates of machine salvage values at that time would be required for these situations.

As indicated previously assumption 2 is perhaps most often utilized because it is quite logical for problems involving cost comparisons and equipment replacement analyses. However, the assumption of replacing a machine in x years with an exact replica is troublesome. Obviously if this does indeed happen, then the foregoing anlaysis is sound. However, it is rare that a sequence of alternatives will replace itself exactly because technological progress will likely lead to improved alternatives in the future. In general this approach to the unequal lives problem tends to overstate the differences between alternatives when it assumes that the differences will occur over a time span that exceeds the service lives of the current proposal alternatives.

Certainly if assumptions 1 and 2 are independently applied to the same problem, they *may* give conflicting decisions. In essence the analyst must determine the degree to which future investment alternatives for the shorter-lived project can be predicted. If these future investment alternatives can be predicted with reasonable accuracy, they should be utilized and analyses performed in conjunction with assumption 2. This condition is perhaps most commonly satisfied when analyzing activities or processes of a continuous nature. It happens that these types of activities are most often measured in terms of negative cash flows. In mining, perhaps the classic examples of investment alternatives suitable for analysis utilizing assumption 2 pertain to the comparison of alternative production systems and specific equipment replacement options as illustrated in example 15. For instance, the replacement alternative for a specific piece of short-lived mining equipment can be predicted with reasonable accuracy since the analyst can often assume that the equipment can be replaced with essentially an exact duplicate. However, when future investment alternatives cannot be acceptably predicted, the analyst is advised to incorporate assumption 1 into the analysis. These types of investment alternatives often are associated with projects which do not necessarily reflect ongoing activities and are normally measured in terms of positive cash flows. Therefore, it is very important that the type of problem be carefully considered and the appropriate assumption utilized in the analysis.

Perhaps the most appealing of the three assumptions is assumption 3. Although this assumption should provide the most realistic approach to the problem—at least from a theoretical viewpoint—it is difficult to employ in practice. Estimating investment opportunities which may occur in the future is not an easy task and introduces considerable subjectivity into the analysis. In practice most analysts prefer to work with assumption 2 since it represents a base position throughout the analysis. For example, one is reasonably certain that an existing machine can be replaced with one exactly like it at some future date, but future technological advances in machine production capabilities, design changes, etc., are extremely difficult to predict with any degree of certainty. Therefore, although assumption 2 may be a conservative assumption, it does provide a consistent basis of comparison for all alternatives.

Projects Having Unequal Investment

When comparing mutually exclusive investment proposal alternatives there are two main principles which should apply. These are as follows (Canada, 1971, p. 62):

1) Each increment of investment capital must justify itself.

2) Compare a higher investment project against a lower investment only if the lower investment project is justified.

Based on these principles the criterion for choosing between investment proposal alternatives then becomes, "select the proposal which requires the highest investment for which each increment of invested capital is justified."

It is this concept of "bigger is better" that is discussed in this section. Obviously if two proposals have the same indicated rate of return but different initial investment requirements, the project requiring the larger investment will generate the largest magnitude of total benefits or wealth to the firm. In essence, the problem is one of maximizing use of the investment dollar.

Optimizing use of the investment dollar is really not a troublesome issue in the situation where a firm has adequate investment capital available to undertake all investment proposals which promise returns in excess of the firm's required rate of return. Under these conditions the wealth of the firm will, in theory, be maximized by simply investing in all projects which surpass the cutoff rate. However, where capital rationing does exist, the problem of optimum utilization is an important one.

Under the capital-rationing constraint, all investment proposals which exceed the firm's required or cutoff rate may not be chosen for investment. Additionally, the firm may generate more wealth by selecting several smaller, less profitable proposals that fully utilize the capital budget than to accept one large investment proposal that results in only partial utilization of the budget. The following example illustrates this concept.

Example 16: Suppose that the following investment proposals were available to a firm. If the capital budget constraint is $500,000 for the period, select the optimal investment portfolio.

Proposal	Profitability index	Initial capital investment
7	1.14	$400,000
3	1.13	200,000
5	1.11	300,000
4	1.05	250,000

Solution: The objective is to find that combination of investment proposals which provides the highest net present value to the firm. There are three primary combinations.

Alternative No. 1
  Proposal 7: $400,000(1.14 − 1.0) = $56,000 = NPV

Alternative No. 2
  Proposal 3: $200,000(1.13 − 1.0) = $26,000
  Proposal 5: $300,000(1.11 − 1.0) = $33,000
                              NPV = $59,000

Alternative No. 3
  Proposal 3: $200,000(1.13 − 1.0) = $26,000
  Proposal 4: $250,000(1.05 − 1.0) = $12,500
                              NPV = $38,500

This solution shows that alternative No. 2 (proposal 3 and 5) should be chosen since the NPV to the firm is maximized with the selection of these proposals. The reason is that more of the available budget is utilized with this combination of proposals, even though a more profitable individual proposal was available to the firm.

Example 16 also illustrates the importance of initial capital outlays when functioning under the constraints of capital budgeting. Implied in the foregoing example is the assumption that uninvested capital has a NPV = 0. This is the same as assuming it is placed in an investment which has a yield equal to the required rate of return. If it cannot be reinvested such that NPV is equal to 0, then full utilization of the investment capital available is even more important.

In the example, it is apparent that all proposals promising a positive net present value are not necessarily accepted under capital-rationing. Certainly the required rate of return (used to calculate PI) establishes the lower limit for proposal consideration and proposals yielding less than this required rate would not be considered even if the budget were not exhausted. However, when capital rationing results in rejection of investment proposals which promise a positive NPV, the firm is incurring an opportunity cost. This opportunity cost may be considered as the yield foregone on the most profitable investment proposal rejected. Because of this aspect, capital rationing usually results in an investment program that is less than optimal.

The cost of capital is discussed in detail in Chapter 11.

Incremental (Marginal) Analysis

In the preceding section it was noted that one of the main principles which should apply when comparing mutually exclusive investment proposal alternatives is that each increment of investment capital must justify itself. This is an aspect which is often overlooked in many analyses, but one which is fundamental to the capital budgeting problem—particularly under capital-rationing constraints.

Incremental or marginal analysis is a technique which can help the evaluator choose between mutually exclusive projects having unequal investments. The concept is to (1) calculate the differential investments and annual cash flows between the projects, and (2) compare the calculated rate of return on the differential cash flows with the required rate of return. If this rate exceeds the required rate, the additional incremental investment is justified. The following example illustrates the procedure.

Example 17: Suppose the following cash flow estimates represent four mutually exclusive investment proposals. If the firm's required rate of return is 15% which proposal should be chosen?

		Cash flows		
Year	Proposal A	Proposal B	Proposal C	Proposal D
0	$ – 12,000	$ – 15,000	$ – 19,000	$ – 21,000
1	3,000	3,700	4,200	4,600
2	3,000	3,700	4,200	4,600
2	3,000	3,700	4,200	4,600
2	3,000	3,700	4,200	4,600
2	3,000	3,700	4,200	4,600
10	3,000	3,700	4,200	4,600
IRR (%)	21.4	21.0	17.8	17.5
NPV (at 15%)	3,056	3,569	2,079	2,086

The IRR for all four proposals exceeds the 15% required rate of return and therefore each would be acceptable to the firm. However if only one proposal is required, the IRR criterion suggests that proposal A is superior.

At this point it is necessary to perform a rate of return calculation on each increment of investment to determine if the incremental proposal investments can be justified. A comparison between proposals A and B shows:

$$B/A: (P/A, r\%, 10) = \frac{15,000 - 12,000}{3,700 - 3,000} = 4.2857$$

$$r = 19.36\%$$

This indicates that proposal B is preferred to proposal A at this point because the return on the incremental investment of $3000 exceeds 15%. The next comparison is between proposals C and B

$$C/B: (P/A, r\%, 10) = \frac{19,000 - 15,000}{4,200 - 3,700} = 8.00$$

$$r = 4.28\%$$

This comparison indicates that proposal C should be eliminated because the return on the incremental investment of $4000 is less than the required rate of 15%. The last comparison is between proposals D and B

$$D/B: (P/A, r\%, 10) = \frac{21,000 - 15,000}{4,600 - 3,700} = 6.6667$$

$$r = 8.14\%$$

The final comparison indicates that the rate of return on this incremental investment is also less than the required rate, and therefore proposal B is the final choice.

In Part I of this chapter a comparison was made between the IRR and NPV criteria with respect to inconsistent rankings of investment proposals. The simple example used to demonstrate this feature was as follows:

	Annual cash flows (1000's)	
Year	Project A	Project B
0	$ − 10	$ − 10
1	1	7
2	5	5
3	6	3
4	7	1
IRR (%)	24.7	30.5
NPV (15%)	2.60	2.41
Required rate of return:	15%	

If an incremental analysis were performed on this example, the following would result:

Year	Incremental cash flows: A-B ($1000's)
0	0
1	−6
2	0
3	3
4	6

The IRR which equates -6 at the end of year 1 with $3 and $6 at the end of years 3 and 4, respectively, is 16.54%. Because this rate exceeds the required rate of return of 15%, project A should be selected, even though project B has the larger IRR.

It is interesting and important to note that both of these examples illustrate situations where the proposal with the largest IRR is not necessarily the best proposal when mutually exclusive proposals are being considered. Proposal choice in the mutually exclusive case is, of course, dependent upon the required rate of return and the associated reinvestment rate assumption discussed earlier in this chapter. The incremental analysis illustrated in both examples resulted in choosing investment proposals with the highest net present values. These proposals could have been selected simply by comparing NPV values initially. Therefore, it is possible to generalize and state that the internal rate of return (IRR) and net present value methods give the same results in capital budgeting problems *if incremental analysis* is used on mutually exclusive projects.

Other Factors

In addition to the economic parameters which affect investment decisions, there is typically a host of other factors which either directly or indirectly influence corporate decisions. Although attempts have been made to quantify many of these factors for direct inclusion into the decision-making process, it is debatable whether or not the end justifies the means.

The relative risk associated with investment proposals is a good example of a factor which is not directly economic in nature, but one which certainly affects investment decisions. Early in this chapter the assumption was made that the acceptance of any investment proposal, or group of proposals, would not alter the business-risk profile of the firm as perceived by those in the marketplace. Yet, it is fairly obvious that different investment proposals do have different degrees of risk—particularly in the mining business. Therefore, consideration must be given to the potential change in the perceived business-risk complexion of the firm as a result of investing in any given proposal or group of proposals. Investors in the marketplace may not view the impact of recent investment decisions in the same light as the firm's management. When this occurs compensation for perceived risk changes generally takes the form of stock price adjustments. In essence this suggests that individual investment proposal risk may not be as important as its impact on the marginal change in the overall risk associated with the firm's investment portfolio. The concept of portfolio theory applied to financial management is an important one, but beyond the scope of this book. However, Chapter 13 addresses the problem of incorporating project risk analysis into the decision-making process.

Another factor which often affects investment decisions is the firm's philosophy on the importance of diversification. In the case of mining companies, diversification may take the form of investments in business or manufacturing activities which are not directly mining-related such as processing and fabrication. More typically, however, the diversification issue is associated with investments in various mineral commodities. For example some corporations attempt to diversify by investing in some combination of base metal, precious metal, industrial mineral, and/or energy commodity mining ventures. Commodity diversification is often advocated as a means of smoothing out earnings which fluctuate considerably with the cyclic performance of most mineral commodity prices. The argument is that a more stable earnings and investment performance record is preferred by the shareholders, reduces the overall risk associated with the investment program, and enhances future financing at reasonable costs.

On the other hand, some corporations prefer to concentrate their efforts on only one or two commodities. In fact, some firms even prefer to work in a very limited number of commodities which are typically mined by only one method. The rationale in this instance is usually based on doing a few things very well and hoping to maintain or achieve a dominant position within a very competitive, and often specialized, commodity market. The firms following this philosophy often argue that it is too expensive to maintain overhead associated with acquiring and maintaining high quality geological and engineering staffs which have the expertise necessary to find, develop, and extract a wide variety of mineral occurrences in diverse environments. They take the position that in the long-term it is preferable to develop and maintain a high level of expertise in a fairly narrow spectrum of mining activities. Certainly, the corporate philosophy with respect to diversification will not only influence the firm's investment decision, but the financing and dividend decisions as well.

The experience factor cannot be ignored when making corporate investment decisions. The term "experience" may relate to different considerations at different levels within the firm. For example, in the case of a mining investment proposal, it might relate to the experience a firm has in the development, extraction, and marketing of mineral commodities. Another experience factor might be associated with the mining and marketing of a specific mineral commodity which is significantly unlike those normally mined by the organization. Experience of technical staffs is also a major consideration. For example, engineering staffs which have specialized in cut-and-fill mining operations over the last 20 years would probably find it difficult to effectively and efficiently design a large open pit mining complex. Should the firm choose to undertake such a project, it would surely find it necessary to procure the necessary technical capabilities from outside the organization. The firm's ability and success in securing this external expertise would probably be a reflection of managerial experience.

Perhaps one of the most important and most difficult factors to consider with respect to a firm's investment decisions is the competence of its managerial personnel. In a general sense engineers are required to carefully analyze investment proposals and subsequently make a recommendation as to proposal acceptability on the basis of engineering/economic analyses. Corporate management, on the other hand, is required to make the actual investment decision on a given proposal recommendation after consideration of all the direct and indirect factors and constraints which may impact, or be impacted by, the investment decision. The effectiveness of these investment decisions is difficult to evaluate *a priori*. However, the marketplace will perform its evaluation of the investment program and issue its report card through adjustments in stock prices. These adjustments, over the long-term, can be compared with marketplace evaluations of other competitors in the same investment risk class. The dispersion of stock values may well be related to differences in managerial competency. Probably the biggest problem facing most financial managers is the trap associated with investment decisions predicated on short-term risk aversion, or some artificial goal achievement, as opposed to long-term wealth maximization. In the long-term the marketplace will segregate the competent financial managers from those less adept.

APPENDIX A. WHAT CONSTITUTES "INVESTMENT"

Part I of this chapter dealt with the major evaluation criteria normally used to determine the attractiveness of projects for investment. Throughout the discussion, reference was made to performing some rather simple mathematical manipulations on cash "outflows" and cash "inflows," on either a discounted or undiscounted basis. The terms "benefits" and "costs" were also used and referred to cash inflows and outflows respectively.

Under normal circumstances an investment opportunity will require some amount of investment (negative cash flows) from which future benefits (positive cash flows) will accrue. In this regard, and in accordance with the discussion in Part I of this chapter, all negative cash flows are considered to be "investment" in the project—no matter when they occur during the life of a project. For example, when performing a cash flow determination on a new mining property it is not uncommon to have six or eight years of anticipated negative cash flows during the preproduction period followed by the project coming on-stream with anticipated positive cash flows. It is also not uncommon in new mining ventures to actually anticipate certain years during the operating life of a property when the annual cash flows will be negative. These negative cash flows may be anticipated as the result of planned expansion programs, replacement of major segments of the equipment fleet, soft markets, environmental expenditures, or any number of other reasons.

In the classical sense these intermittent negative cash flows are considered "investment" in the project and are treated exactly the same as the negative cash flows occurring during the preproduction period. But are these intermittent negative cash flows really an investment in the project in the sense that they represent a drain on the corporate treasury? One logical distinction is that the real "investment" in the project is the total amount of negative cash flows necessary to bring the project on-stream (when positive cash flows begin), and any future negative cash flows subsequent to project start-up could be covered from the earnings generated by the project itself and would not constitute a commitment from the corporate treasury. If, however, the cumulative project earnings *were not* sufficient to cover the intermittent negative cash flow(s), then a commitment from the corporate treasury would be necessary, and *this amount would presumably be treated as "investment" in the project in the normal sense.*

The appropriate treatment of this issue can be important when evaluating investment opportunities using the wealth growth rate (WGR) and the growth rate of return (GRR) criteria. To illustrate the difference between approaches, consider the following two projects and associated cash flows:

Year	Project A	Project B
0	$-10,000	$-10,000
1	4,000	4,000
2	6,000	6,000
3	4,000	2,000
4	-2,000	0
5	3,000	3,000

These two projects are identical except for the cash flows in years 3 and 4. The choice between these two projects seems to be a simple one. Would the investor rather have $2000 this year and nothing the next year as in project B; or would he rather have $4000 this year with the obligation of returning $2000 next year? Our understanding of the concepts of time value of money make the second option the obvious choice. Therefore, project A is preferred.

But will this obvious choice in projects be confirmed through the evaluation criteria? As indicated previously, the method of treating the -2000 cash flow in year 4 (project A) will affect the WGR and GRR criteria due to their reinvestment rate assumptions. To illustrate, suppose the GRR is calculated at the end of the project ($t = 5$ years) at a 10% reinvestment rate. In this case the GRR and WGR calculations would be identical since GRR is calculated at the termination of the project.

In the classical or traditional approach to handling investment in this calculation, the -2000 in year 4 is discounted to time zero and included as part of the investment. The calculation procedure is as follows:

Year	Project A cash flows	Present value of outflows @ 10%	Future value of inflows @ 10% ($t = 5$ years)
0	$-10,000$	$-10,000$	
1	4,000		$5,856
2	6,000		7,986
3	4,000		4,840
4	$-2,000$	$-1,366$	
5	3,000		3,000
	Totals:	$-11,366$	$21,682
		GRR $= 13.79\%$	

In the case where we assume that the -2000 in year 4 is covered by cumulative earnings from the project, the calculation procedure would appear as follows:

Year	Project A cash flows	Present value of investment @ 10%	Cumulative project earnings by year @ 10%
0		$-10,000$	
1	4000		4,000
2	6000	$4,000(1.1)+6,000 =$	10,400
3	4000	$10,400(1.1)+4,000 =$	15,440
4	-2000	$15,440(1.1)-2,000 =$	14,984
5	3000	$14,984(1.1)+3,000 =$	19,482
	Totals:	$-10,000$	$19,482

$$\text{GRR} = 14.27\%$$

A comparison between these two calculation procedures for project A and the GRR for project B yields the following:

	Project A (Year 4 cash flow treated as part of primary investment)	Project A (Year 4 cash flow carried by project earnings)	Project B (Calculated by the the normal procedure)
GRR @ 10%	13.79%	14.27%	14.01%

In view of our earlier determination that project A is intrinsically "better" than project B, it therefore follows that the evaluation criteria (in this case GRR) should be higher for project A. This example clearly illustrates that the GRR (and in this case WGR) for project A is superior to that of project B only when the intermittent negative cash flow was assumed to be covered from project earnings. In the case where the intermittent negative cash flow was taken to be part of the investment, project B's GRR was superior to that of project A. This, then, appears to be an incorrect approach when calculating evaluation criteria such as WGR and GRR that deal with future values and associated reinvestment rates.

A comparison of the GRR for the foregoing example with NPV and IRR is as follows:

	Project A (all negative cash flows treated as investment)	Project A (negative cash flow in year 4 carried by project earnings)	Project B (calculated in the normal procedure)
GRR @ 10%	13.79%	14.27%	14.01%
NPV @ 10%	20.97		19.60
IRR	20.32%		19.23%

The results indicate that if GRR is to provide the same ranking of projects for investment decision-making as NPV and IRR, the negative cash flows occurring after project start-up should be considered to be covered from project earnings when calculating external rate criteria. Although this conclusion is not based on exhaustive analytical testing, it would appear to be a logical and practical approach based on heuristic reasoning.

Similar logic was developed by Lin (1976) in an attempt to modify the internal rate of return calculation so that the intrinsic problems associated with multiple rates and the reinvestment rate assumption could be alleviated. The net result is essentially the same as that proposed previously for GRR and WGR.

SELECTED BIBLIOGRAPHY

Berry, C.W., 1972, "A Wealth Growth Rate Measurement for Capital Investment Planning," Ph.D. dissertation, The Pennsylvania State University, University Park.

Canada, J.R., 1971, *Intermediate Economic Analysis for Management and Engineering*, Prentice-Hall, Inc., Englewood Cliffs, N.J.

Capen, E.C., Clapp, R.V., and Phelps, W.W., 1976, "Growth Rate—A Rate-of-Return Measure of Investment Efficiency," *Journal of Petroleum Technology*, May, pp. 531-534.

Fabrycky, W.J., and Mize, J.H., 1977, *Engineering Economy*, 5th ed., Prentice-Hall, Inc., Englewood Cliffs, N.J.

Gentry, D.W., 1971, "Two Decision Tools for Mining Investment and How to Make the Most of Them," *Mining Engineering*, November, pp. 55-58.

Gentry, D.W., 1979, "Mine Valuation: Technical Overview," *Computer Methods for the 80's in the Mineral Industry*, edited by A. Weiss, ed., AIME, New York, pp. 520-535.

Grant, E.L., Ireson, W.G., and Leavenworth, R.S., 1976, *Principles of Engineering Economy*, 6th ed., The Ronald Press Co., New York.

Lin, S.A.Y., 1976, "The Modified Internal Rate of Return and Investment Criterion," *The Engineering Economist*, Vol. 21, No. 4, Fall, pp, 237-247.

Newnan, D.G., 1980, *Engineering Economic Analysis*, revised ed., Engineering Press, Inc., CA.

Parks, R.D., 1957, *Examination and Valuation of Mineral Property*, 4th ed., Addison-Wesley Publishing Co., Inc., Boston, MA.

Quirin, D.G., 1967, *The Capital Expenditure Decision*, Richard D. Irwin, Inc., IL.

Smith, G.W., 1973, *Engineering Economy: Analysis of Capital Expenditures*, 2nd ed., Iowa State University Press, Ames, IO.

Solomon, E., 1956, "The Arithmetic of Capital-Budgeting Decisions," *Journal of Business*, Vol. 29, No. 2, April, pp. 124-129.

Stermole, F.J., 1974, *Economic Evaluation and Investment Decision Methods*, 2nd ed., Investment Evaluations Corp., Golden, CO.

Stevens, G.T., Jr., 1979, *Economic Financial Analysis of Capital Investments*, John Wiley & Sons, New York.

Van Horne, J.C., 1980, *Financial Management and Policy*, 5th ed., Prentice-Hall, Inc., Englewood Cliffs, NJ.

Young, D., 1975, "Expected Present Worths of Cash Flows Under Uncertain Timing," *The Engineering Economist*, Vol. 20, No. 4, Fall, pp. 259-268.

10

Inflation in the Mine Investment Decision

"We should be concerned about the future because we will have to spend the rest of our lives there."

—Charles Kettering

INTRODUCTION

Since the early 1970s, there has been no economic phenomenon that has been more widely discussed or experienced in the United States than inflation. Although its precise definition and measurement are elusive, inflation has had profound and extensive impacts on the manner in which nearly everyone lives. It is painfully obvious to consumers that inflation is characterized by a continuing upward spiral of prices. In a capital-intensive industry like mining, those increases can be positively breathtaking. An interesting example is shown in Table 1.

Table 1 shows that on a unit basis, the equivalent 1980 capital cost of Cuajone was over six times the cost of the adjacent and quite similar Toquepala operation completed in 1959. Such sobering statistics weigh heavily on the minds of executives who are contemplating major investments in new mining facilities.

In spite of its monumental presence, inflation has usually been either inadequately covered or totally ignored in textbooks on engineering economy. The reasons for this are not entirely clear, but a contributing factor is the inherent difficulty in accurately predicting inflation. However, under certain conditions, inflation may become the most important factor in a mining investment, and it can rarely be safely ignored in capital investment analyses.

DEFINING AND MEASURING INFLATION

Defining inflation and related terms is relatively easy. Its measurement, however, is the source of almost endless debate in economic and political circles. However, much of that debate is not relevant to the scope of this book, so that only those issues which impact directly on the capital investment decision are covered in this section.

Inflation is the general decline in the value of money as measured by what it will buy. It is clear, then, that the rate of inflation varies with the particular currency under consideration as well as the type of goods and services being purchased. One would find, for example, that the general rate of inflation relative to German marks

Table 1. Impact of Inflation on Capital Costs in the Copper Industry

Project	Cost estimate			Cost per annual ton of copper produced
	Date	Type	Amount, $ millions	
Toquepala, Peru	1959	Actual	$ 237	$ 1,700
Cuajone, Peru	1973	Estimate	550	3,055
Cuajone, Peru	1977	Actual	726	4,033
Cuajone, Peru	1977	Actual (1977 $s)	1200	6,700
Cuajone, Peru	1980	Actual (1980 $s)	1950	10,800

Sources: Barber, C.F. (1980) ASARCO (1973).

is different than for US dollars, and different for petroleum products as compared to computer components.

Although a general rate of inflation exists, it is obvious that price trends often vary considerably from one product or service to the next. For example, the price of refined petroleum products rose over 500% during the 1970s, whereas the prices of certain electronic goods actually declined. This situation where prices are changing for different products at different rates is called *escalation* in this book to distinguish it from the overall general rate of inflation.

Productivity and Conservation: When discussing inflation, it is important to recognize that price increases alone do not signify inflation. If the price increase is accompanied by a commensurate increase in productivity of the product, the final economic result will not be a reduction in the purchasing power of the dollar. For example, the increased price of large haulage trucks may not be inflationary if design improvements offer a corresponding increase in productivity. In this case, a price series which is heavily influenced by haulage trucks would not be an appropriate measure of inflation.

A second problem also arises when historic price series for production inputs are extrapolated into the future in financial analyses. Firms are often able to escape some of the burden of rapidly rising price by conserving, consuming less, or using substitutes whose prices are rising less rapidly. In most cases, for example, industry has been able to partially offset the spiraling cost of electric power by initiating energy conservation programs. Custom smelting contracts that carry escalators based upon the unit costs of inputs also tend to overstate the impact of rising prices because fewer units are generally consumed.

The impact of productivity improvements and conservation efforts are extremely difficult to quantify for investment studies. Often, these factors are omitted, but they should not be ignored. It is worthwhile noting that when escalating prices for goods and services are factored into the analyses, their impact will be softened, however modestly, by improved technology and increased conservation.

Measuring Inflation: Because of differential price movements of various products, inflation is usually measured by a dimensionless index which measures average price trends over a wide variety of goods and services. The indexes most widely used to measure the general rate of inflation in the United States are described in the following and are plotted in Fig. 1.

Consumer Price Index (CPI)—The CPI is the most widely used cost of living

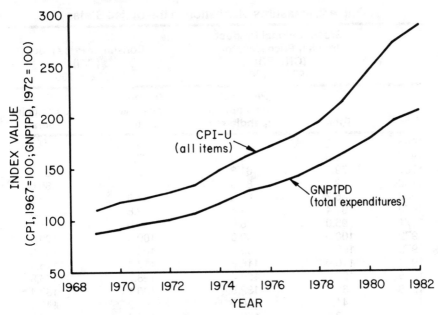

Fig.1 Indexes for the general rate of inflation.

indicator in the country. Many labor contracts contain a provision for periodic cost of living adjustments, and most use the CPI for this purpose. The hourly wage rate for most Arizona copper miners, for example, moves 1¢ per hr for every 0.3 point movement in the CPI.

The CPI measures the average change in price over time for a fixed-market basket of goods and services. Data are collected monthly by the Bureau of Labor Statistics of the US Department of Labor from 85 different urban areas through actual observations in the marketplace and through mail questionnaires. Included in the data are costs to the consumer for food, clothing, shelter (including property taxes), fuels, transportation fares, doctor's fees, drugs, and many more items.

The raw data are aggregated by region and by commodity so that, in reality, there is a large number of Consumer Price Indexes. The national average CPI for all items is most widely quoted, but often a regional index is a better indicator of cost of living trends in a particular application. Table 2 illustrates the difference between the CPI national average and the average for the Los Angeles metropolitan area.

An important revision in the CPI was made in January 1978 when two separate indexes were formed:

CPI-U: This index covers all urban consumers, roughly 80% of the population

CPI-W: Covered here are urban wage earners and clerical workers, which includes roughly 50% of the individuals in the CPI-U index.

This change was made primarily to reflect the different mix of goods and services which constitutes the cost of living to each group. Thus, the contents of the market basket are somewhat different for these two indexes.

The CPI has the advantages of a broad data base, a long history, and a rigorous collection methodology, although when used as a measure of the cost of living, the

Table 2. Measures of Inflation in the United States

	Gross National Product Implicit Price Deflator (GNPIPD) 1972 = 100		Consumer Price Index (CPI) 1972 = 100	
Year	Total	Personal consumption expenditures	National average all cities	Los Angeles area
1965	74.3	77.1	75.4	78.3
1966	76.8	79.3	77.6	79.7
1967	79.0	81.3	79.8	81.8
1968	82.6	84.6	83.1	85.0
1969	86.7	88.5	87.6	89.0
1970	91.4	92.5	92.8	93.5
1971	96.0	96.6	96.8	96.9
1972	100.0	100.0	100.0	100.0
1973	105.8	105.5	106.2	105.6
1974	116.0	116.9	117.9	116.5
1975	127.2	126.4	128.7	128.9
1976	133.7	132.8	136.1	137.4
1977	141.1	140.0	144.9	146.9
1978	152.1	150.0	155.9	157.7
1979	165.5	163.3	173.5	174.7
1980	178.6	179.2	224.2	228.1
1981	195.5	194.5	252.3	255.5
1982	207.1	206.0	267.7	271.2

fixed mix of goods and services becomes troublesome. Differential price movements and changing consumer tastes affect purchasing patterns. For example, the fourfold increase in gasoline prices in the 1970s has not generally been accompanied by similar increases in individual expenditures for gasoline because consumers are driving less and benefitting by better gasoline mileage in their vehicles. In spite of this defect, the CPI remains one of the best measures of the general rate of inflation.

Many publications carry CPI data, but the most comprehensive coverage is provided by the Department of Labor's monthly, *CPI Detailed Report*, available from the US Government Printing Office.

Producer Price Index (PPI)—The PPI (formerly the Wholesale Price Index, WPI) is analogous to the CPI, except that the index is organized to be compatible with producer rather than consumer expenditure patterns. As a consequence, PPI data are categorized on (1) an industry/commodity basis and (2) a stage of production basis. Thus, whereas the CPI measures prices to the final consumer, the PPI catalogues prices for crude, intermediate, and semi-finished as well as finished products.

Like the CPI, the PPI is maintained by the Bureau of Labor Statistics of the US Department of Labor who collect some 10,000 monthly quotations, primarily by questionnaire. As noted previously, these are aggregated by commodity and by stage of processing and finally into composite series for all items.

Because the PPI measures trends in prices for products other than finished goods, it is a less accurate measure than the CPI of the cost of living, and therefore, the PPI is rarely used as a measure of the general rate of inflation. Clearly, however, the PPI is highly correlated with the CPI because producer price changes are ultimately passed through to the consumer.

Although the PPI is not commonly used to measure the general rate of inflation, it is very useful in tracking the changing cost of doing business in various sectors. Thus, later in this chapter, various PPI's will be used to project future production costs for a hypothetical mining operation.

The primary reference for PPI data is the monthly publication, *Producer Prices and Price Indexes,* available from the US Government Printing Office.

Gross National Product Implicit Price Deflator (GNPIPD)—The GNPIPD has recently become a popular measure of the general rate of inflation, rivalling the CPI. The GNPIPD has the advantage of measuring shifts in the composition of purchases, as opposed to the fixed-market basket approach of the CPI; but on the other hand, the GNPIPD is a derived statistic, not being based on actual market observations (as is the CPI). Fig. 1 contains a plot of the GNPIPD total expenditures index, and Table 2 gives comparative data in tabular form for both the CPI and GNPIPD series. The two major indexes are clearly related and a reconcilliation between the CPI and the GNPIPD (personal consumption) is provided in monthly issues of the *Survey of Current Business,* prepared by the US Department of Commerce, and available from the US Government Printing Office. This publication is also the primary source for GNPIPD information.

In summary, the CPI and the GNPIPD are both widely used measures of the general rate of inflation, with the latter series becoming more popular. This is partly due to the fact that the CPI has risen more sharply since 1978 because of spiralling housing and home-financing costs which are weighted more heavily in the CPI. The GNPIPD was able to simultaneously adjust for the precipitous decline in housing purchases over the same period.

Measuring Escalation: As discussed extensively in Chapter 6, there are a wide variety of index series available which measure various cost/price trends. These may be placed in two categories: (1) factor cost indexes and (2) project cost indexes.

Factor cost indexes measure the cost trends for a specific type of product, e.g., explosives, industrial chemicals, etc. A project cost index gives the overall relative cost for an entire project which typically involves several individual factor inputs. Good examples are the *Engineering News-Record* indexes which include a fixed combination of structural steel, portland cement, lumber, and labor costs.

The best known factor cost indexes are the *Producer Price Indexes* (PPI), published by the US Department of Labor, which apply to the output of specific industries. These are particularly useful in studying historic trends and projecting future levels of operating costs. A number of these indexes relevant to the cost of mining are listed in Chapter 6 (Table 10).

Project cost indexes are most applicable to capital cost analyses and projections. The series most useful in mining are:

1) *Engineering News-Record* (ENR) Building and Construction Cost Indexes.
2) Marshall and Swift (formerly Marshall and Stevens) Cost Indexes.
3) *Chemical Engineering* (CE) Plant Construction Cost Index.
4) Nelson Refinery Construction Cost Index

The ENR and Nelson indexes, both having large labor cost components, generally move in parallel. Until very recently they have increased at a more rapid rate than the M&S and CE indexes. These latter two indexes also track closely but are influenced less by labor. Each of these indexes is described in the following, and the ENR and M&S are plotted in Fig. 2 and tabulated in Table 3.

1) *Engineering News-Record Cost Indexes (ENR)*—The ENR indexes are based on the costs of labor and building materials in the following proportions: 25 cwt of structural steel, 6 bbl of portland cement, 1088 bd ft of 2x4 lumber and 200 hr of labor (*common* labor for the Construction Cost Index, *skilled* labor for the Building Cost Index).

2) *Marshall and Swift Cost Index (M&S)*—The M&S index has several values. The value most often used and referred to is the all-industry equipment index. This index is the average of the indexes calculated for 47 different industries. Other M&S indices are calculated individually for eight process industries (cement, chemical, clay products, glass, paint, paper, petroleum products, rubber) and four related industries (electrical, power equipment, *mining and milling,* refrigerating and steam power). The M&S index for mining and milling is important to studies of mining costs.

M&S indices are based on equipment appraisals, installation labor, modifying factors, and judgments concerning current economic conditions. The M&S index reflects changes in installed equipment costs. The index is published in *Chemical Engineering* magazine, as is the CE index below.

3) *Chemical Engineering Plant Construction Cost Index (CE)*—The CE index is based on equipment erection and installation, material, labor, engineering, and

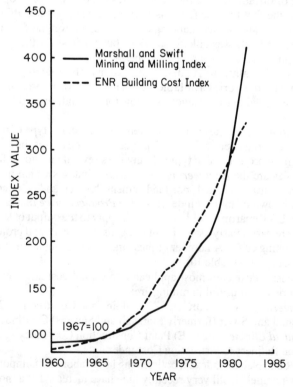

Fig. 2. Project cost indexes (1967 = 100).

Table 3. Project Cost Indexes (1967 = 100)

Year	Marshall and Swift Mining and Milling Equipment Cost	ENR Construction Cost	ENR Building Cost
1960	91.3	76.7	82.8
1961	90.8	78.9	84.1
1962	91.0	81.2	85.9
1963	91.1	83.9	87.9
1964	92.1	87.2	90.6
1965	93.1	90.4	92.7
1966	96.0	95.1	96.6
1967	100.0	100.0	100.0
1968	103.7	107.5	106.7
1969	108.0	118.3	117.0
1970	114.8	128.4	123.6
1971	121.9	146.2	139.7
1972	125.9	161.2	154.6
1973	130.1	176.6	168.6
1974	149.6	188.0	178.3
1975	171.2	205.7	193.3
1976	183.3	223.4	210.9
1977	197.7	240.0	228.6
1978	209.2	258.4	247.7
1979	235.0	279.5	269.3
1980	289.4	301.4	287.7
1981	349.6	332.4	312.7
1982	411.3	353.4	329.3

supervision costs in the following percentages: 61% equipment, machinery, and support; 22% erection and installation labor; 7% buildings, materials, and labor; and 10% engineering supervision and manpower.

4) *Nelson Refinery Construction Cost Index (NR)*—The NR index is used primarily for estimations in the petroleum industry. The total index is based on material and labor in the following percentages: 20% iron and steel, 8% building material, 12% miscellaneous equipment, and 60% labor (65% of which is skilled and 35% common).

EFFECT OF INFLATION ON INVESTMENT ANALYSIS

If the prices for all goods and services were rising at the same rate, and if there were no income taxes, the quantitative impact of inflation on investment decisions would be small. However, escalation and income taxes do exist and are not likely to disappear. Under these circumstances, improper handling of inflation in capital investments analysis can lead to serious errors. Four such errors are discussed in the following.

Depreciation: A Declining Tax Shelter

Chapter 2 showed that, as a noncash expense, the sole impact that the depreciation allowance has on the economics of mining ventures is to reduce the income tax liability. However, most tax codes limit the amount of depreciation to the original purchase price of the asset, ignoring any change in replacement cost which usually is higher due to inflation. Therefore, while other costs and prices continue to rise (again due to inflation), depreciation remains tied to the original purchase price and becomes a less effective tax shelter over time. The following simple example illustrates this phenomenon for inflation of 10% per year and straight-line depreciation.

	Year 1	Year 5	Difference, %
Revenues	100	146	46
− Cash expenses	60	88	46
− Depreciation	10	10	0
Taxable income	30	48	60
− Income taxes @ 50%	15	24	60
Net profit	15	24	60
+ Depreciation	10	10	0
Cash flow	25	34	36

Note that a 46% increase in revenue and cash costs resulted in a 60% increase in taxes and only a 36% increase in cash flow. Clearly, the tax saving afforded by the depreciation allowance declines in proportion to the general rate of inflation. For this reason, ignoring inflation in a capital investment analysis will overvalue the project, perhaps encouraging investment where it is not justified.

Working Capital Drain

Another subtle way in which inflation can impair a capital project such as a new mining venture is through ever-increasing needs for working capital. It is normal to provide for working capital only at the beginning of a new venture, the assumption being that, once in operation, sales will generate sufficient revenue to pay for the inputs to the production process. Under conditions of high inflation, however, the spiralling prices of inputs may not be covered from sales which would require further infusions of working capital.

This problem has been particularly onerous during the long preproduction development periods characteristic of mining. Returning to the board of directors for supplemental appropriations before the project even begins is generally not a pleasant experience. To guard against this situation, a number of mining companies factor inflation into their projections during the preproduction period, but then work in constant dollars after start-up. Although this procedure may solve the working capital problem, it creates other serious problems and is generally not advisable. This is further discussed in the following and in Heath, et al. (1974).

Capital Gains Taxation

The acquisition cost of some assets—most notably land—is not deductible for tax purposes. The rationale for this is that land is not consumed in the production process. It, therefore, presumably does not lose value and can be sold at any time to recover the initial investment. However, if the price of the land has increased in current dollars, even though its value in constant dollars remains the same or even declines, the seller must pay taxes on the amount of the apparent increase, usually at the capital gains rate. This phenomenon is analogous to the ratcheting effect of the progressive rate structure for personal income taxes during periods of high inflation.

Discount Rate Adjustment

A very serious problem exists with many firms today that evaluate capital investments in constant-dollar terms and then use a market-determined cost of capital as the minimum acceptable rate of return. It is vitally important that consistency be maintained here. If a market-determined discount rate (e.g., the cost of capital) is used, the rate will contain a component for inflation and should, therefore, *only* be used when revenues and costs are also adjusted for inflation. Similarly, if the analysis is performed in constant dollar revenues and costs, the discount rate should not contain a component for inflation.

Using an inappropriate discount rate is one reason that large new mining ventures look so unattractive to some companies in an inflationary economy. If constant dollar analyses are used, it becomes immediately obvious that only the rare bonanza discoveries can possibly show an acceptable rate of return when a discount rate of 20% or higher and a preproduction period of 5 to 8 years are used. A simple example will vividly illustrate this:

Given: Preproduction period: 8 years
           Present value (PV) of investment: $100 million
           Discount rate: 20%
           Mine life: 10 years
Find: Required minimum annual cash flow (CF)

$$NPV = PV_{CF_{1\text{-}8}} - PV_{CF_{9\text{-}18}}$$

$$0 = 100 - \frac{1}{(1 + 0.2)^8} \left(\sum_{n=1}^{10} \frac{CF_n}{(1 + 0.2)^n} \right)$$

$$0 = 100 - \frac{CF}{(1 + 0.2)^8} \left(\sum_{n=1}^{10} \frac{1}{(1 + 0.2)^n} \right)$$

$$100 = \frac{CF}{(1 + 0.2)^8} \left(\sum_{n=1}^{10} \frac{1}{(1 + 0.2)^n} \right)$$

$$CF = \$102.6 \text{ million}$$

In other words, the project must yield more than its *total* investment *every year* of operation in order to be acceptable. Using constant-dollar cash flows with inflation-adjusted discount rates will discourage even the most enthusiastic investor.

If the analysis is modified to correctly delete the inflation component in the discount rate (assume roughly 12% here), the new required minimum annual cash flow is much less:

$$100 = \frac{1}{(1 + 0.08)^8} \left(\sum_{n=1}^{10} \frac{CF_n}{(1 + 0.08)^n} \right)$$

$$CF = \$27.6 \text{ million}$$

In a specific application, the difference may not be quite as dramatic, but it is clear that the annual cash flows and the discount rate must be determined on a consistent basis in regard to inflation.

An Inflation-Adjusted Required Rate of Return (RRR)

A common error that occurs is when the rate of inflation is simply added to the constant-dollar minimum required rate of return (RRR) to derive an inflation-adjusted hurdle rate. To illustrate, let i_c = constant-dollar RRR, i_i = rate of inflation, and i_e = effective interest rate, or inflation-adjusted RRR.

Then, if the cash flows are expressed in constant dollars

$$\text{NPV} = \text{CF}_0 = \frac{\text{CF}_1}{(1 + i_c)} + \frac{\text{CF}_2}{(1 + i_c)^2} + \dots + \frac{\text{CF}_n}{(1 + i_c)^n}$$

However, if the cash flows are in inflated dollars,

$$\text{NPV} = \text{CF}_0 + \frac{\text{CF}_1}{(1 + i_i)(1 + i_c)} + \frac{\text{CF}_2}{(1 + i_i)^2(1 + i_c)^2}$$

$$+ \dots + \frac{\text{CF}_n}{(1 + i_i)^n(1 + i_c)^n}$$

Therefore

$$(1 + i_e) = (1 + i_i)(1 + i_c)$$
$$i_e = i_i + i_c + i_i i_c$$

At low rates of inflation, simply adding i_i and i_c usually does not result in a significant error. However, at higher rates, this may not be the case as the correct rate may be more than a whole percentage point higher than the rate obtained by simply adding i_c and i_i. This phenomenon is illustrated with an example later in this chapter.

Each of the foregoing potential problem areas often has special significance in mining due to some of the unique characteristics of the industry. The distortions caused by the depreciation allowance and working capital drain are magnified by the *capital intensity and long lead time* inherent to mining projects. Escalation is also particularly pronounced in mining where *mineral prices,* often established on world markets, fluctuate widely and, in the short run, are rarely in phase with production cost trends. Finally, the inherent *high risk* of mining ventures is further aggravated by the tremendous uncertainty of inflation.

Thus, mining is particularly vulnerable to the ravages of inflation and, therefore, special care is warranted at the evaluation stage to quantify the impacts of inflation.

HANDLING INFLATION IN INVESTMENT STUDIES

It should now be clear that inflation must be integrated into all investment analysis studies to avoid a potentially serious error. There are several options available to the analyst for this purpose. One such option which will result in consistently correct answers is to observe the following fundamental rule:

Convert all net annual cash flows into constant dollars before applying any investment criterion.

If current, or inflated dollars are used to calculate, say, a payout period, the result is meaningless. If the currency value changes from year to year, it is the same as stating one year's cash flow in German marks, the next in British pounds, and so forth. The annual cash flows *only* have relative meaning if they are expressed in units of constant value. Otherwise, an elastic yardstick is being used to measure investment value. As a consequence, all cash flows should first be converted to the same currency, usually constant present-day dollars, before investment criteria are applied.

Impact on Sales Contracts

Inflation can have a variety of impacts upon capital projects based upon the nature of anticipated future cost and price increases. Basically, the degree of impact is dependent upon how closely the individual cost and revenue items respond to general inflationary pressures. Two extremes in this respect are shown in the following example.

Example: Mole Coal Co. supplies two power plants under different sales contracts.

Power plant #1—Fixed price contract of $9.50 per ton over the seven-year mine life.

Power plant #2—Initial price of $8.50 per ton, with escalation allowed at the general inflation rate.

Assume that inflation of 6% compounded annually is anticipated over the project life. Which is the better contract to Mole if the firm's required rate of return is 10% (no inflation component)?

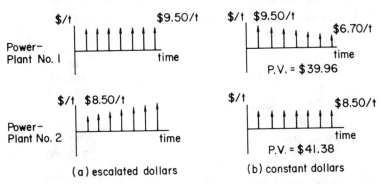

(a) escalated dollars   (b) constant dollars

In the first case, there is no response by price to inflation, so that real revenues to the firm decline steadily over the entire mine life. A number of mining companies in the coal and uranium business have been saddled with fixed-price contracts during times of high inflation which seriously eroded their profits. Virtually no long-term sales contracts are signed today without a provision for adjusting prices for inflation.

Annual revenues from power plant #2, however, respond perfectly to inflation by remaining unchanged in constant dollars over the life of the project. As indicated, this contract has a higher present value to the mining company than the fixed-price arrangement.

General Approaches to the Problem

Three general cases are defined to illustrate the impact of inflation on an investment decision. To aid in this illustration, consider the following mining investment opportunity.

Example: Donnaker Crushed Stone is available for purchase for $3.8 million. The current relevant production and cost data have been estimated as follows:
1) Production rate: 650,000 tpy; 5-year life
2) Sales price: $6.00 per ton
3) Mining cost: $1.20 per ton
4) Crushing and screening cost: $0.75 per ton
5) General plant expense: $1.05 per ton
6) Transportation to market (contracted): $1.35 per ton

    Cost items 3, 4, and 5, are further broken down as follows:
        Wages and salaries-40%
        Payroll overhead-20%
        Power-4%
        Parts and supplies-18%
        Fuel-10%
        Taxes, insurance, misc.-6%
    Percentage depletion for crushed stone: 5%
    Anticipated rate of inflation: 9% per year

If inflation is ignored in the analysis of this proposed investment, annual net cash flow and the internal rate of return for the investment would be calculated as follows:

Revenue:	$3,900,000
Transportation	877,500
Gross income from mining	3,022,500
Operating costs	
Labor	819,000
Power	78,000
Parts, supplies	351,000
Fuel	195,000
Taxes, insurance, misc.	117,000
Fringes	390,000
Net operating income	1,072,500
Depreciation	760,000
Depletion	151,125
Pretax net income	161,375
Income tax @ 50%	80,688
Net profit	80,687
Cash flow	991,813

$$\text{DCFROI} = i, \text{ where } \left(\sum_{n=1}^{5} \frac{\$991,813}{(1+i)^n} \right) - \$3,800,000 = 0$$

$$i = 9.6\%$$

Now, evaluate this investment opportunity based on each of the following three assumptions: (1) uniform inflation, no income taxes, (2) uniform inflation, with income taxation, (3) differential inflation (escalation), with income taxation.

Uniform Inflation, No Income Taxes: This case assumes that all prices will increase at the same rate as the general rate of inflation. Furthermore, the absence

Table 4. Donnaker Crushed Stone Uniform Inflation—
No Income Taxes

	Inflation = 9% per yr. Amount in thousands of $				
	Year				
	1	2	3	4	5
Revenue	4251	4634	5051	5505	6000
Costs					
Transportation	957	1043	1137	1239	1351
Labor	893	973	1060	1156	1260
Power	85	93	101	110	120
Parts, supplies	383	417	455	495	540
Fuel	213	232	253	275	300
Taxes, insurance, misc.	128	139	152	165	180
Fringes	425	463	505	551	601
Net cash flow (current $)	1167	1274	1388	1514	1648
	$\div (1.09)^x$				
Net cash flow (constant $)	1071	1072	1072	1073	1071

Purchase price—year 0 = $3.8 million
DCFROI—current $ = 22.8%
DCFROI—constant $ = 12.7%

of income taxes is assumed. The results for Donnaker Crushed Stone can be examined in Table 4. Note that throughout this example end-of-year cash flows are assumed.

It may appear trivial to first escalate costs and prices at a fixed rate and then deflate the residual cash flow at the same rate, but the procedure does clearly show that inflation could have been ignored entirely in this case, and the same rate of return would have resulted.

It is interesting to note that if a rate of return had been computed using escalated, or current year, cash flows, the result would have been a DCFROI of 22.8%, rather than the correct value of 12.7%. The current-dollar rate of return could have been derived directly by using the relationship developed in the earlier section, Effect of Inflation on Investment Analysis.

$$i_e = i_i + i_c + i_i i_c$$
$$= 0.09 + 0.127 + 0.011$$
$$= 0.228 = 22.8\%$$

The DCFROI of 22.8%, having been adjusted for the impact of inflation, should only be used to compare with a hurdle rate which also contains an inflation component, such as the acquiring firm's weighted average cost of capital. In general, however, fewer errors will arise if the procedure advocated at the beginning of this section is followed, that is, first convert all cash flows to constant dollars. Then, use constant-dollar values of investment criteria.

Occasionally, mine promoters will produce "inflation-adjusted" income statements of the type shown previously in an attempt to attract investors. In such cases, it is obvious that if costs and prices are rising at the same rate, the margin will also always increase, in current escalated dollars. In purchasing power, however, returns from the project would not improve over time.

Uniform Inflation, With Income Taxes: Having explored the hypothetical situation depicted in the first example, now return to the very real world of income taxation. Allowable income tax deductions are almost always based upon cost, and where such deductions may extend over several years, no adjustment for inflation is permitted. The most important example of this concept is depreciation, which is really capital cost allocated over the life of the acquired asset. The amount of depreciation is limited by the original purchase price of the assets involved so that depreciation deductions in future years will be in lower-valued, inflated dollars. Thus, firms are really not able to deduct the total cost of their capital assets, let alone generate sufficient funds through depreciation to purchase higher-priced replacement units. It is clear, then, that depreciation charges do not respond to inflationary pressures and, therefore, alter the evaluation even when all other revenues and costs respond uniformly to the general rate of inflation.

The preceding comments are illustrated in Table 5. In this case, all items above depreciation respond uniformly to the inflation rate, rising at an annual compound

Table 5. Donnaker Crushed Stone Uniform Inflation—Income Taxes

| | Amount in thousands of $ | | | | |
| | Year | | | | |
	1	2	3	4	5
Revenue	4251	4634	5051	5505	6000
Transportation	957	1043	1137	1239	1351
Costs					
Gross income from mining	3294	3591	3914	4266	4649
Operating costs					
Labor	893	973	1060	1156	1260
Power	85	93	101	110	120
Parts, supplies	383	417	455	495	540
Fuel	213	232	253	275	300
Taxes, insurance, misc.	128	139	152	165	180
Fringes	425	463	505	551	601
	1167	1274	1388	1514	1648
Depreciation straight-line)	760	760	760	760	760
Depletion	165	180	196	213	232
Income tax @ 50%	121	167	216	270	328
Cash flow (current $)	1046	1107	1172	1244	1320
	$\div (1.09)^x$				
Cash flow (constant $)	960	932	905	881	858

DCFROI—current $ = 15.9%
DCFROI—constant $ = 6.3%

rate of 9%. Depreciation, being limited to costs incurred at the beginning of the project, is constant over the life of the project even though $760,000 in year 5 certainly does not have the same value as $760,000 in year 1. With depreciation remaining constant and all other pretax items rising at 9% per year, taxable income and, therefore, taxes rise at a much greater rate. In this particular example, it turns out that annual increases in income taxes ranged from 21% to 38%. It will come as no surprise, then, that the after-tax DCFROI for the project declines from 9.6% (inflation ignored) to 6.3% (with inflation).

The important conclusion here is this: when costs and prices are assumed to rise together uniformly at the general rate of inflation, an evaluation which compensates for inflation will *always* yield a lower after-tax internal rate of return than an evaluation which ignores inflation. This may prove to be an unsettling conclusion to firms that have assumed that the effects of costs and prices rising at the same rate nullify each other, so that inflation can be safely ignored in analyzing long-term investments. Although the difference may or may not be significant, ignoring inflation always leads to overvaluing an investment on an after-tax basis.

It should be noted here that some authors categorize interest with depreciation as a cost item which remains constant in escalated dollars in an inflationary economy. While interest expense does, indeed, resist the tides of inflation, correct capital investment analysis does not permit interest to be levied against a specific project. Interest is one component of the cost of capital which, in turn, is used as either the minimum acceptable rate of return or the discount rate, depending on which investment criterion is employed. Therefore, charging interest to a specific project would result in double-counting the cost of capital.

The fixed-cost nature of interest has encouraged the notion that investors should vigorously pursue debt financing in inflationary times. This certainly is not a dictum which should be followed blindly, for (1) lenders generally understand the workings of inflation rather well and modify their rates accordingly, and (2) negative leverage on a floundering firm in a period of high inflation can provide a cruel coup de grace.

Finally, one might reasonably ask what impact the depreciation method has on the foregoing results. As with any other case where taxable income exists, accelerated depreciation will produce a higher internal rate of return than straight-line methods. Beyond this, however, it is very difficult to generalize the amount of difference which can be anticipated without considering project-specific information.

Differential Inflation (Escalation), With Income Taxation: This is the most complex and realistic case involving the treatment of inflation in investment analysis, but few sweeping generalizations can be made. In essence, this case recognizes that prices of individual inputs rarely—if ever—rise at the same rate. This is, of course, particularly true in the short run where market forces often amplify or dampen price swings caused by general monetary inflation. Thus, in Table 6, fuel costs were assumed to rise at an annual compound rate of 12%; power, at 8%; labor and fringes, at 10%; parts and supplies, at 9%; and other costs, at 11%. Product price increases were estimated to rise at 7% annually and contract transportation costs were estimated to be level for three years before rising by 10%. The values used here are quite arbitrary and are intended only to be illustrative.

In the example shown in Table 6, the DCFROI for the project declines even further to 4.4% in comparison to 6.3% return when uniform inflation was assumed. No particular sigificance should be attached to these values, however, as they are simply a product of the escalation rates selected.

Table 6. Donnaker Crushed Stone Case 3—Escalation

| | General rate of inflation = 9% Amount of thousands of $ | | | | | |
| | Year | | | | | Escalation rate, % |
	1	2	3	4	5	
Revenue	4173	4465	4778	5112	5470	7
Transportation	878	878	878	966	966	varies
Gross income from mining	3295	3587	3900	4146	4504	
Operating costs						
Labor	901	991	1090	1199	1319	10
Power	84	91	98	106	114	8
Parts, supplies	383	417	455	495	540	9
Fuel	218	245	274	307	344	12
Taxes, insurance, misc.	130	144	160	178	198	11
Fringes	429	472	519	571	628	10
	1150	1227	1304	1290	1361	
Depreciation (straight-line)	760	760	760	760	760	
Depletion	165	179	195	207	225	
Income tax @ 50%	113	144	175	162	188	
Cash flow (current $)	1037	1083	1129	1128	1173	
	$\div (1.09)^x$					
Cash flow (constant $)	951	911	872	799	762	

DCFROI—current $ = 13.8%
DCFROI—constant $ = 4.4%

The foregoing analyses quantify only the impact of the depreciation allowance on investment rates of return in an inflationary economy. The other impacts (working capital drain and capital gains taxation) require more detailed assumptions than had been provided in the Donnaker Crushed Stone example. Allen (1976) argues that these considerations will further reduce the DCFROI from a case where inflation is ignored, although the degree of impact is usually less than for depreciation. Thus, although the preceding section illustrates the need to consider inflation in investment analyses, the quantitive impact may well be *greater* than that shown with the oversimplified Donnaker Crushed Stone model.

SELECTING INFLATION OR ESCALATION RATES

It is obvious that the results in the previous examples could be changed greatly by using different rates of inflation or escalation. The selection of such rates for an actual evaluation exercise is, in fact, extremely difficult.

The earlier discussion on cost index series implied that historic trends could be used to estimate future costs and prices. For the immediate future, this usually is a reasonable assumption, but for the more distant future, exogenous factors quickly overwhelm relationships which are in operation today. Because of this, it is often useful to divide the future into separate periods for the purpose of projecting costs and prices. A reasonable rule, for example, is to employ escalation in the near term, but adopt uniform inflation beyond, say, four years. Certainly, it becomes more difficult to justify differentiated rates of increase in the distant future when other uncertainties are so large.

A Note on Exponential Growth

Inflation is routinely expressed as an annual percentage rate of increase. Although this is a useful way to characterize inflation in the short run, compound growth rates over a longer term often produce questionable results due to the dynamics of exponential growth. For example, production costs at one mining company rose at an annual rate of 17% in the early 1980s. However, if this factor were used to project costs 20 years in the future, a 2300% rise would result! Even though it is impossible to accurately predict the future, this outcome would certainly be unprecedented and quite unlikely to occur. A more likely scenario is that inflation rates will rise and fall over the 20-year period, and 17% is probably above the average rate which will prevail over that time.

Thus, the analyst should be aware that high compound growth rates are usually an accurate portrayal of reality over only relatively short periods. Some other assumption (e.g., constant *amount* of increase, estimate individual years) may generate more plausible projections over a longer period.

IMPLEMENTING INFLATION ANALYSIS

Because capital budgeting methodology has been developed only in the last three decades, the formal integration of inflation into that methodology was of little concern until the 1970s. Since 1974, however, a high rate of inflation has been the investment analyst's constant companion. Nonetheless, no specific procedure for investment decision-making under conditions of inflation has become widely accepted in practice. Hayes (1977) found that most major corporations had not adjusted their capital budgeting methods for increased inflation, a situation which appears to persist today.

The Role of Constant Dollar Analysis

Mining is a highly technical business, and mining project analyses are, therefore, usually performed—and often directed—by engineers and scientists, some of whom possess financial analysis skills. To require these individuals to adjust their analyses for inflation could result in two potential problems.

1) Not having expertise in analyzing economic trends, these analysts might generate results of insufficient credibility to the decision-makers and, furthermore, the process might divert attention away from the crucial technical analysis required.

2) Technical analysts need to constantly examine their projections from the standpoint of common sense. At the evaluation stage, it is important to be able to distinguish between future cost changes created by technical factors and those created by inflation assumptions. Inflation-adjusted future cost projections tend to obscure the underlying technical relationships.

As a consequence, the need to consider inflation in investment studies does not normally eliminate the need to conduct constant-dollar analyses. Rather, project evaluation tends to become a closely coordinated, two-stage iterative process where technical specialists provide development alternatives and economics specialists then factor in the impact of inflation. Inflation is now of sufficient importance that nearly every major firm can justify staff economics counsel to perform this function.

The form in which final evaluation results are presented is guided by the needs of the decision maker. For internal presentation to senior management, however, both constant-dollar and inflation-adjusted data and results are helpful. The constant-dollar analysis (1) generates greater technical understanding of the project and (2) permits management to study separately the sensitivity of project profitability to technical risk and to economic risk.

SUMMARY

Many firms, when faced with the vagaries of estimating future inflation rates, adopt the position that, over the long run, the rate of increase for production costs will be matched by the rate of increase of product sales prices. Assuming that this compensates for inflation, the firms then often use some market-determined cost of capital for the discount rate, or required rate of return. This tends to generate two significant errors in the results.

1) Due mainly to the depreciation deduction, ignoring inflation (i.e., assuming cost and price rises are perfectly offsetting), results in *overvaluing* an investment project. This was illustrated in the section on Handling Inflation in Investment Studies, and the size of the error can be significant.

2) A market-determined cost of capital includes a component which is based on investor's perceptions of future inflation. If this rate is applied to project cash flows which are not adjusted for inflation, the project will be seriously *undervalued*.

Thus, this commonly used procedure for handling inflation is usually unacceptable. To be sure, there are financial analyses of widely varying precision carried on almost continuously in the development of a mine. Some of the most preliminary studies will obviously not warrant the time and expense needed to include inflation. Nonetheless, inflation must be included in any analysis whenever major capital requests are submitted. The importance of doing so is directly related to the rate of inflation, and it is likely that this rate will continue at recent, higher levels rather than return permanently to the modest rates of the 1960s.

In the capital-intensive mining industry where construction periods for new projects are exceptionally long, inflation is a factor of major importance in investment analyses. If inflation is ignored, mining projects tend to be overvalued—often seriously so. Therefore, inflation must be included in the evaluation process in a quantitative manner. The best method is to separate cost and revenue items into categories which track well with published cost index series. Individual escalation rates can then be applied to these categories for the near term (up to four years), followed by uniform inflation thereafter.

The need for inflation-adjusted financial analyses does not eliminate a continuing need for constant-dollar studies. The latter permit greater technical insight into the project and allow management to better analyze the sources of risk to the project. In the final presentation, both types of analyses are important.

REFERENCES

Allen, B., 1976, "Evaluating Capital Expenditures Under Inflation: A Primer," *Business Horizons*, December, pp. 30-39.

ASARCO, 1973, "Annual Report to Shareholders," ASARCO, Inc., New York.

Barber, C.F., 1980, "Mineral Investment in an Anxious World," Paper 32, Conference on National and

International Management of Mineral Resources, Institution of Mining and Metallurgy, London, England, 7 pp.

Hayes, S.L., III, 1977, "Capital Commitments and the High Cost of Money," *Harvard Business Review*, May-June, pp. 155-161.

Heath, K.C.G., Kalcov, G.D., and Inns, G.S., 1974, "Treatment of Inflation in Mine Evaluation," *Proceedings*, Institution of Mining and Metallurgy, London, England, pp. A20-A32.

11

Selecting A Discount Rate

*"There is nothing so disastrous as a rational
investment policy in an irrational world."*
John Maynard Keynes

INTRODUCTION

The principles of time value of money concepts were discussed and illustrated in Chapter 3. The importance of time value of money and the role it plays in the discipline of engineering economy have been well established. Chapter 9 further illustrates that project evaluation criteria that fail to consider this aspect are seriously deficient in providing appropriate capital budgeting decisions. Obviously future project receipts and expenditures must be discounted to permit valid comparisons with current cash flows.

Although the concept of discounting is widely accepted, the selection of the appropriate discount rate has been the source of considerable debate and much disagreement. This chapter, then, provides an overview of the theoretical and practical considerations for determining or calculating the appropriate rate for discounting, the corporate cost of capital. Note specifically that the assumption of constant business risk among the various investment alternatives is maintained throughout this chapter.

THE INTUITIVE APPROACH

The fact that interest exists suggests that all money has a cost associated with its use. The cost of this money may be the result of either explicit or implicit charges or some combination of the two. Any time an individual or a firm consumes less than the total earnings generated in a given time period that individual or corporation has either consciously or unconsciously decided to "invest" these excess funds in some type of activity. For instance, the individual or corporation may decide to invest the excess funds in tangible property, place the funds in a money market account, or simply keep these funds in cash. In any case, these funds have been invested in some type of activity which promises the investor some return on the invested capital.

Remembering that the firm's primary objective should be the maximization of shareholder wealth, then intuitively it seems reasonable to suggest that the firm should not invest in any project where the anticipated return does not exceed the cost of funds (capital) committed to the project. Indeed, if the firm always invests in projects having returns in excess of the cost of capital committed to them, then the wealth of the firm (as measured by the price of common stock) should be in-

321

creased to the stockholders. This logic suggests that the firm's cost of capital for investment opportunities is very important in the capital budgeting process in that it represents the hurdle or base against which project acceptability may be measured.

Following our intuition further, it seems reasonable to accept the fact that no money is free. Certainly there is a cost associated with raising investment capital. In the simplest case, for instance, there is a cost associated with going to the debt markets and borrowing money for investment purposes. This cost of borrowing results from the fact that interest exists and is in reality an explicit cost. The level of interest rates will vary, depending upon, among other factors, the supply of and demand for such borrowings. It is important to note that an investor who purchases common stock has the same expectations of making a return on his investment as a banker who loans money. Although there is no legal obligation to pay dividends as there is to pay interest, there is, nonetheless, clearly a cost for equity funds for an ongoing firm to meet the expectations of its shareholder owners.

Additionally it seems important to recognize that an individual or corporation can allocate investment capital to alternative investment opportunities which may be available. For instance, a firm may decide to invest its capital in certificates of deposit (CDs), money market funds, commercial paper, or any number of other relatively safe short-term opportunities. Consequently, any competing investment alternative should offer a return in excess of the return promised by the low-risk investment avenues available to the firm. This concept of alternative investment opportunities suggests that there are also opportunity costs associated with investment capital. Opportunity cost refers to the return foregone by the firm as a result of not investing in the next most desirable project available to the firm. These opportunity costs are implicit in nature and, along with any explicit costs, should be a consideration in selecting a discount rate.

As discussed elsewhere in this book, one of the principal characteristics of the mining business is the long lead times associated with bringing a new property on-stream. These long lead times in conjunction with high inflation rates can significantly impact the evaluation and viability of mining-related investment opportunities. It seems reasonable to suggest that this aspect should also be incorporated into the determination of the appropriate discount rate since inflation seems to be at least one component in the cost of obtaining or utilizing investment capital. This correlation has been well tested in the US—particularly during the 1970s and 1980s.

From the previous preliminary discussion it seems clear that there is definitely a cost associated with procuring and utilizing investment capital. This cost consists of explicit as well as implicit costs and is influenced, at least to some extent, by inflationary trends and expectations. At this point it seems reasonable to suggest that the determination of the overall cost of investment capital to a corporation may be a rather complex exercise—especially when considering the basic variables involved. Nonetheless investment capital does have a cost and this cost must be related in some manner to the determination of the appropriate discount rate to be utilized in discounted cash flow analyses.

COMPONENTS OF DISCOUNT RATE

There are four major components to the discount rate. Some of these components stem from sound theoretical and practical foundations, while others are judgmental. The following discussion is intended to illustrate the major components of discount rate selection for DCF analyses.

Rarely is the combined discount rate calculated by summing these separate components, as the capital markets do a good job of determining the overall cost of funds to various uses. Nonetheless, the combined rate must be sufficient to cover these four elements.

Base Opportunity Cost

As discussed in the preceding section, there is some base opportunity cost associated with procuring and utilizing investment capital. This cost is the return foregone by diverting funds from the next most attractive project and exists whether the funds are obtained externally or internally. In either case, funds appropriated for a particular use will carry a cost related to the return those funds could have achieved elsewhere. Note that opportunity cost is the clearest illustration that there is, indeed, a cost to investment funds obtained from retained earnings.

The other components of the discount rate are typically added to this base opportunity cost for investment capital.

Transaction Cost

Whether corporate investment capital is procured from the debt market or the equity market, the firm will experience transactions costs regardless of whether or not a new security issue is involved. These costs include broker and investment banker fees, costs of prospectuses and various filings, sales discounts, and other flotation costs. Transaction costs are typically much higher for equity financing than for debt financing. Flotation costs for new securities as a percentage of gross proceeds decline with increasing size of the issue. In the range of $20 to $50 million issues, flotation costs run about 1% to 5% for bonds and around 5% for common stock. Although the aggregate transaction costs are generally much less than the base opportunity cost, they are significant and should be incorporated into the determination of the appropriate discount rate.

Increment for Risk

The cost of procuring funds, either debt or equity, includes a component for the investor's perception of risk. Naturally, the cost will rise with higher perceived risk.

Many organizations add some *additional* percentage increment for risk to the cost of raising investment capital before applying this rate to discounted cash flow analyses. The rationale here is that the funding source is looking at the risk of the entire firm whereas the firm is concerned with the risk in the particular project under consideration. Thus, many firms believe that high-risk projects should be discounted at some higher rate to compensate for their relative riskiness. Obviously the higher the discount rate, the lower will be the calculated present value of the project and the rationale is that the project's perceived risk will be negated in some manner.

One approach utilized by some organizations is to establish risk classes for projects having increasing levels of risk and then assign additional increments to the cost of capital which will be used to evaluate projects in these various risk classes. For example, suppose a firm can raise investment capital at a cost of 12%. Under these conditions, it might separate its investment opportunities into risk classes with the associated discount rates shown in Table 1. The increment above

Table 1. Risk Adjusted Discount Rates

Investment opportunities	Risk class and associated discount rates			
	I, 12%	II, 14%	III, 18%	IV, 25%
1. Replacement of equipment at operating properties	X			
2. Expansion program at operating property		X		
3. Develop a new property, same commodity (domestic)			X	
4. Develop a new property, new commodity (foreign)				X

12% used in the respective discount rates represents the firm's perceived risk of projects in that class and is intended to compensate for this risk.

There are two fundamental problems associated with this approach to accounting for project risk. First, the increments for risk assigned to the discount rate are subjective by nature and cannot be quantitatively equated to project risk in any systematic manner. Indeed, it is unlikely that very many individuals would categorically agree with the discount rates suggested in the foregoing examples for the types of projects described. One of the problems then with this approach to handling project risk is determination of the appropriate discount rate—remembering that a large upward adjustment of the discount rate can make even the best project appear unattractive.

The second problem with this approach is the manner in which it addresses project risk. As referred to here project risk is synonymous with the business risk of the project. Business risk refers to the risks inherent with normal operations of the project on a routine basis and is measured by the variability in annual project cash flows. It should be noted that the discount rate only adjusts for the present value of money; it does not by itself adjust for project business risk. The discount rate provides no information to the firm on how the annual project cash flows might vary nor does it provide any insight into why these cash flows might fluctuate from year to year. Consequently the discount rate does not provide a measure of project risk and, in general, should not be used to do so. Superior approaches to accounting for project risk are discussed in some detail in Chapter 13.

Increment for Inflation

Spiraling inflationary pressures throughout the world over the last decade have introduced a new dimension to the science of capital budgeting and the proper approach to evaluating mining ventures in particular. As discussed in Chapter 10, the need to consider inflation in investment studies seems obvious but does not eliminate the need to conduct constant-dollar analyses. However, the question arises as to "What is an appropriate discount rate to use in conjunction with each of these approaches to financial analysis?" Certainly it seems reasonable to assume that investment capital raised in the marketplace would include some cost or premium for investor expectations of future inflation. Therefore, presumably a market-de-

termined cost of capital includes a component which is based on investors' perceptions of future inflation. As a result, this market-determined cost of capital should
be used as the appropriate discount rate when analyzing inflation-adjusted project
cash flows. However, the use of this same rate when analyzing constant-dollar cash
flows (not adjusted for inflation) will result in seriously *undervaluing* the project.
The amount of downward adjustment in the market-derived cost of capital for use
with constant-dollar analyses is roughly equal to the general rate of inflation as
shown in Chapter 10. Although the preparation of constant dollar *pro forma* cash
flows is highly recommended to promote better technical understanding of the project, discounting should always be performed on inflated dollars to avoid mis-
calculation of income taxes as also pointed out in Chapter 10.

THE COST OF CAPITAL

As previously stated, the firm's objective should be the maximization of shareholder wealth (i.e., maximization of its common stock prices over time). While in
practice linkages between the firm's activities and its stock prices are, at best, im-
precise, a logical and intuitively appealing theory has been developed directly related to earnings per share, because earnings—again over the long run—are significantly affected by the firm's investment policy. This leads directly to the important
conclusion that the returns on investments must be greater than the costs of the
funds utilized if share prices are to ultimately rise. This cost of funds, known as the
cost of capital, is the direct linkage between investment policy and the firm's
objective.

The cost of capital also relates the financing decision to the investment decision. It should be emphasized that the cost of capital is the *only* link between these
two decisions. Otherwise, in the general case the financing and the investment decisions are largely separate considerations. The relationship between the financing
decision and the investment decision in corporate finance via the cost of capital
may be illustrated as follows:

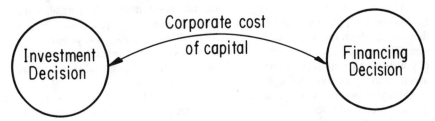

The minimum acceptable rate of return (MARR) may be defined as the minimum
rate of return a firm must earn on its investments in order to leave unchanged the
value of existing shares of its common stock. Investing at the cost of capital will
achieve this objective and, therefore, the MARR is quantitatively equal to the cost
of capital. The logical conclusion from this discussion is that this cost of capital
represents the "hurdle rate" or the appropriate discount rate to be used in conjunction with discounted cash flow (DCF) analyses of investment opportunities available to the firm. It should be noted that it is incorrect in most cases to explicitly
include financing costs in the evaluation of a project when the firm's cost of capital
is utilized as the discount rate in DCF analyses because this "double counts" the
cost of capital. This aspect is illustrated in detail in Chapter 12.

Most analysts agree that the manner in which a firm chooses to finance itself will largely determine the cost of capital to the organization. Thus the *capital structure* of the firm, defined as the mix of long-term sources of funds used to finance assets, becomes important to corporate managers in that the resulting cost of capital to the firm directly influences the investment decision. Since the firm's investment decisions influence common stock prices and therefore the value of the firm, the capital structure of the firm becomes a key parameter in corporate finance. Consequently, management wishes to know whether the value of the firm is increased by using more or less debt financing relative to equity financing. The capital structure of the firm refers to long-term sources of funds and traditionally excludes short-term liabilities.

At this point it might be beneficial to distinguish between the capital structure of the firm, as defined previously, and the *financial structure* of the firm which includes all sources of funds to the firm, short-term, as well as long-term. Financial structure characteristically refers to the use of a fixed cost source of funds, such as debt, in the capital structure of the firm. The use of debt financing results in financial leverage and may impact the firm in either a positive or negative sense. Because debt represents a fixed obligation to the firm, financial leverage will benefit the firm during good times because as earnings increase, the residual amount available to shareholders after fixed-debt serving costs rises at a much faster rate. However, financial leverage also amplifies declines in earnings available to shareholders for the same reason. As earnings available to shareholders drop, so do stock prices and the value of the firm—a position many mining companies have painfully experienced.

As stated previously, debt creates financial leverage and increased risks to the shareholders. This increased risk which must be borne by the stockholders is known as financial risk as opposed to business risk which was mentioned previously and is discussed in detail in Chapter 13.

Cost of Specific Types of Capital—The General Case

While the selection of a proper discount rate is essential in the pursuit of the firm's objective, other factors have a bearing on common stock prices also. Two of these over which the firm exercises control are dividend policy and capital structure. The relationships between these factors and stock prices are difficult to quantify with empirical data. Therefore, dividend policy and capital structure are assumed to be constant in the following discussion.

Many corporations choose to develop a capital structure by utilizing a mixture of debt, preferred stock, and equity financing. Each of the various financial instruments available within these general categories has its own specific cost to the firm.

The cost of each specific type of financing is that rate which equates the funds received with all discounted future outlays relevant to that source.

$$I_0 = \frac{C_1}{(1 + k)} + \frac{C_2}{(1 + k)^2} + \cdots + \frac{C_i}{(1 + k)^n} \tag{1}$$

where I_0 is net proceeds received at time zero, C_i is cash outflow in the ith period, and k is cost of this specific source of financing.

The reader will note that this is directly analogous to determining the internal rate of return for a capital investment project. I_0 is net of underwriting and other issuing costs. The C_i's are fairly easily determined in the case of debt and preferred stock financing. Equity financing costs, however, have generated a great deal of debate, and, while Eq. 1 applies in theory, one of several proposed simplifying approximations must be adopted to determine the costs of equity financing in practice.

Cost of Debt

In keeping with the preceding discussion, the cost of debt is that discount rate which equates the net proceeds from the debt issue to the future cash outlays. Further, these future payments are net of taxes. This is very significant with debt financing since interest payments, unlike dividends, are tax deductible. If k_d is the cost of debt and k is the *yield to maturity* determined by the application of Eq. 1,

$$k_d = (1 - t)k$$

where t is the marginal tax rate.

For example, assume Zesty Zinc Co. sold 30-year, $7\frac{7}{8}\%$ debentures in 1981. Assuming that they *netted* $1000 face value per bond, and that the marginal tax rate is 46%, the cost of debt here would be

$$k_d = (1 - 0.46)\,0.07875 = 4.25\%$$

Costs of debt instruments that sell at a discount are slightly more difficult to determine. In the previous example, suppose that Zesty Zinc netted only $940 per bond. Discounts of this magnitude are rare, but are useful to demonstrate the concept. Using the generalized formula

$$940 = \frac{76.75^*}{(1+k)} + \frac{76.75}{(1+k)^2} + \dots + \frac{76.75}{(1+k)^{30}} + \frac{1000}{(1+k)^{30}}$$

$$= \sum_{i=1}^{30} \frac{76.75}{(1+k)^i} + \frac{1000}{(1+k)^{30}}$$

*net of $2 per year amortization of the discount.

Solving for k, the yield to maturity of the instrument, gives k $= 8.37\%$; and k_d then equals 4.52%.

In many cases the accounting rate of return method (average annual return divided by average investment) gives a good approximation to the correct cost of debt. The approximation becomes better with more uniform annual payments. In the second case, this method would produce the following results, where P is the bond's face value

$$k = \frac{C_i + \frac{1}{n}(P - I_0)}{\frac{P + I_0}{2}} = \frac{78.75 + \frac{1}{30}(1000 - 940)}{\frac{1000 + 940}{2}} \tag{2}$$

$$= \frac{80.75}{970} = 8.31\%$$

and

$$k_d = (1 - t)k = 4.49\%.$$

Finally, with perpetuities, the cost is simply the annual payment divided by the net proceeds from the issue. Again, an adjustment for tax deductibility of interest must also be made.

Cost of Preferred Stock

The popularity of financing with conventional preferred stock has declined considerably over the years, and new issues are uncommon today. In theory, because preferred stock dividends do not represent a legal obligation on the part of the firm, preferred stocks offer a distinct advantage over debt financing. In practice, however, the omission of a preferred stock dividend is viewed by the financial community as an indication of serious financial difficulty in the firm. This can result in damage to the company's credit standing and a decline in its common stock prices. Also, preferred stocks often have additional cumulative or participating features, and some issues give voting power to holders of preferred if dividends are omitted. Because of these reasons, firms usually view preferred stock dividend payments as de facto obligations.

The cost of preferred stock is usually computed in a manner similar to debt perpetuities since both instruments have fixed annual costs and no maturity. There is one important difference, however. Preferred stock dividend payments are not tax deductible to the firm.

If the annual dividend paid is D dollars, and I_0 represents the net proceeds per share for a new stock issue, the cost of preferred, k_p, is

$$k_p = \frac{D}{I_o} \tag{3}$$

Suppose Zesty Zinc has a straight preferred stock issue on the market which pays \$8.00 dividends annually. If it originally sold for its \$100 par value, its cost was 8%. However, its current market price (1983 range = 66 to 74½) is a better indicator of what a similar new issue would sell for today. The present applicable cost of preferred to Zesty is then around:

$$k_p = \frac{D}{I_o} = \frac{8.00}{70.00} = 11.4\%$$

Based on the high level of interest rates today, this is a far more realistic cost than the 8% coupon rate.

From an investment viewpoint, preferred stocks offer one advantage over bond investments to a corporation. Eighty-five percent of dividend income to the corporation is nontaxable, while interest income from bonds is taxable at the full applicable rate. Since the individual investor does not have this tax advantage, he will frequently find preferred stock prices to be relatively inflated in comparison to the yield offered.

Cost of Equity Capital

Until the 1970s, the mining industry historically relied on equity financing for most of its investment capital—usually internally generated funds. While this has been cited as excessively conservative financial policy, wide fluctuations in earnings of mining companies—particularly the nonferrous miners—certainly cautions against a high degree of leverage. There are examples where large interest payment obligations coupled with a sharp downturn in business fortunes have resulted in liquidation or external acquisitions of firms that might otherwise have been able to weather the storm.

Equity financing comes from two sources, new common stock issues and retained earnings. It is clear that there is a cost of funds obtained through the issuance of new common stock. Proceeds from such a transaction must be invested to return sufficient earnings to maintain the firm's earnings per share. Because after the stock offering there would be a larger number of common shares outstanding, the newly financed investment must obviously "carry its own weight," or common stock prices will decline.

That there is also a cost of retained earnings (net profits minus dividends) is less obvious. However, if the firm retains, rather than distributes, earnings, and it is dedicated to the maximization of shareholder wealth, it clearly must invest these funds above some minimum rate of return. Futhermore, the minimum rate in general is the cost of common stock financing, for this is the rate the investor implicitly agreed to by purchasing the stock in the first place. If no investment opportunities are available to the firm above this minimum rate, all earnings should be returned to the owners (shareholders) to allow them to reinvest the funds themselves.

Thus, there is no fundamental difference between the various types of equity financing. Shareholder expectations are the same whether the funds come from new issues or from retained earnings.

While all analysts agree that there are real costs involved in equity financing, no such agreement exists on calculating these costs in practice. Two of the most popular methods are discussed in the following.

Dividend Valuation Model: The basis for this model is that common stocks only have a value because of an expected future stream of dividends. It is true, of course, that some stocks command high prices with low, or in some cases, no dividend payments. However, these firms usually have attractive earnings growth histories that imply handsome dividends *sometime* in the future.

In theory, this model is consistent with the generalized cost of capital model discussed early in this chapter.

$$P_0 = \frac{D_1}{(1 + k_e)} + \frac{D_2}{(1 + k_e)^2} + \cdots + \frac{D_i}{(1 + k_e)^i} = \sum_{i=1}^{\infty} \frac{D_i}{(1 + k_e)^i} \qquad (4)$$

where P_0 is the value of stock at time 0, D_i is the dividend payment in the ith year, and k_e is the cost of equity applicable to firms in this risk class.

Obviously, dividend payments in future periods are unknown at the present time. Investors, however, have subjective estimates of what these payments will be, and these estimates are generally based on the current dividend and the historical long-run growth of dividends. An intuitive expression of these relationships might be:

$$k_e = \frac{D_0}{P_0} + g \qquad (5)$$

where D_0 is the current dividend and g is the annual growth rate of dividends.

In fact this intuitive model can be derived from the preceding dividend valuation model (Eq. 4) by assuming that dividends will grow perpetually at a rate g and by using continuous compounding.

Eq. 5 for determining the cost of common stock financing was first proposed by Gordon and Shapiro (1956). While being conceptually simple, it is often difficult to apply in practice.

To illustrate the procedure, suppose Zesty Zinc's financial results over the past ten years are as follows:

	1972	1973	1974	1975	1976	1977	1978	1979	1980	1981	1982
Earnings per share	$1.65	1.78	1.94	2.10	2.31	2.52	2.72	2.93	3.14	3.39	3.62
Dividends	$0.40	0.44	0.49	0.54	0.60	0.67	0.76	0.85	0.96	1.10	1.23

Zesty shows a steady upward trend in both earnings and dividends. Over the past ten years dividends have increased an average of 11.9% per year and earnings by 8.2% per year. Assuming the future for Zesty continues to look bright, it would seem reasonable that, as a first approximation, g might equal about 10% for Zesty. With a current market price of $48¼, Zesty's cost of common stock financing is approximately

$$k_e = \frac{D_0}{P_0} + g = \frac{1.23}{48.25} + 10\% = 2.5\% + 10\% = 12.5\%$$

While considerable judgment pertaining to Zesty's future outlook, etc., is required in computing the firm's cost of common, companies with greater deviation of annual earnings and dividends pose an even greater problem.

E/P Ratio: The second method for computing the cost of common stock financing is advocated by Solomon (1955). He argued that the investor really "purchases" earnings when he buys common stocks and that future earnings per share determine the value of a share of common stock. A logical extension of this model would be for all firms to eliminate dividends, for this would provide maximum funds for reinvestment and would, therefore, maximize subsequent earnings. However, there is much empirical evidence that indicates that dividend policy, in many cases, does affect common stock prices. While this casts doubt upon general usage of the E/P model, we have previously assumed a constant dividend policy for this discussion and have, therefore, netted out this source of disturbance.

The E/P model obviously is more suitable than the dividend valuation model for firms that pay low dividends or none at all. For firms that have strong earnings growth records, a growth term is sometimes appropriate in the E/P model to reflect shareholder expectations that such growth will continue.

Let's now consider our previous example, Zesty Zinc Co., and compute its cost of common stock with the E/P model. With current earnings of $3.62 per share and recent earnings growth of 8% per year, Zesty's cost of common is:

$$E/P_0 = \frac{3.62}{48.25} = 7.5\%$$

$$E/P_0 + g = 7.5\% + 8\% = 15.5\%$$

The costs of common computed with the E/P model differ from the cost computed with the dividend valuation model. The next logical question is, "which method is correct?" Obviously, there is no one "correct" value, and considerable judgment is required in selecting an appropriate cost of common. Shareholders no doubt expect that Zesty's strong record of earnings and dividend growth will continue, so that the models containing growth terms are more reliable here. Furthermore, the stability and predictability of dividends give added support to the dividend valuation model. From the preceding discussion a good estimate of Zesty's cost of common would be about 13%.

Market prices for common stock were used throughout the foregoing discussion. However, if a new common issue is planned, P_f, net proceeds per share, should be used rather than P_0, the current market price. P_f will be lower than P_0 by the amount of discount required to sell the new issue, plus the underwriting costs.

Another procedure for estimating the cost of equity that has recently gained in popularity is the *capital asset pricing model*. This method is based upon the premise that R, the required rate of return on common stock (i.e. the cost of equity), is determined as follows:

$$R = i + (R_m - i)\beta \tag{6}$$

where i is a risk-free interest rate, R_m is the average return on a broad market portfolio, and β is the beta coefficient measuring the correlation of the systematic risk of the stock in question to that of the market in general.

The beta coefficient is the slope of the line derived by plotting historical data sets for the incremental return above the risk-free rate for (1) the stock in question and (2) a broadly diversified portfolio of common stocks. Therefore, prices of stocks with betas above 1.0 tend to be more volatile than the market in general, and that volatility raises the cost of equity capital to such firms. The opposite is, of course, true for stocks with betas below 1.0.

This procedure introduces greater mathematical rigor into estimating the cost of equity and provides a deterministic result. However, a significant data time series is needed to compute a reliable beta, which again raises the question of the suitability of using historical data to project investors' expectations about the future.

The capital asset pricing approach to estimating the cost of equity is described thoroughly in Van Horne (1980).

As was demonstrated earlier, the *cost of retained earnings* is correlated to the cost of common stock financing. Retained earnings are really stockholders' assets that are being reinvested in the firm, so the subsequent return must at least match that from common stock funds (i.e., the cost of common). There are two basic differences between common stock and retained earnings financing, however.

1). There are no flotation costs with retained earnings. Therefore the cost of retained earnings, if based on a new common stock issue, should be reduced by the amount of flotation costs.

2). If earnings are distributed in dividends rather than retained, investors would not have the full amount of the dividend payments for reinvestment. The reason is that dividend income is subject to personal income taxation. Therefore, an argument is often advanced that the cost of retained earnings is less than the cost of common by an amount determined by the average marginal tax bracket of all shareholders.

Aside from the practical problem of determining this tax rate, this complication can be disputed by considering the external yield criterion. This criterion notes that a third alternative exists for disposing of earnings—purchasing shares of another firm having a similar degree of total risk. Assuming that equilibrium exists in the marketplace, the firm can thus earn a return on its investment equal to the return offered by its own common stock. In fact, the firm could actually repurchase its own stock. The cost of retained earnings is, therefore, the opportunity costs of such external investments which avoids the personal income tax complication discussed above. Further, this opportunity cost is the return afforded by common stocks of firms with similar risk (i.e., the firm's cost of common stock funds).

In summary, the cost of retained earnings is the cost of common adjusted, where applicable, for flotation costs.

Cost of Convertible Securities

Securities that are convertible into a fixed number of shares of common stock are usually considered to be simply delayed equity financing. Therefore, one approach to determining the cost of convertibles is to use Eq. 4 and substitute the conversion price, P_c, for the current market price, P_0, and solve for k_c.

$$P_c = \sum_{t=0}^{\infty} \frac{D_t}{(1 + k_c)^t} \tag{7}$$

Since P_c is generally 10% to 20% above the current market price, k_c will be less than k_e, which is usually one of the prime objectives in the issuance of convertibles. Subject to the same assumptions used in deriving Eq. 5 from Eq. 4, Eq. 7 will yield:

$$k_c = \frac{D_0}{P_c} + g \tag{8}$$

Unfortunately, the foregoing procedure does not consider an "overhanging" convertible, (i.e., a convertible issue that is not converted). In this case the issue becomes a fixed obligation and does not supply the desired additional equity cushion for future debt financing. This undesirable outcome represents a "cost" also, so that overall, the cost of convertible securities is somewhat between k_c, computed previously, and k_e, the cost of equity.

MARGINAL WEIGHTED AVERAGE COST OF CAPITAL

In a previous section we made the point that the cost of capital maintains a very prominent position in corporate finance since it represents the minimum rate of return that a firm must earn on new investments to maintain the value of existing shares of common stock. As such, the cost of capital becomes the hurdle rate which future investment opportunities must surpass. We also made the point that the capital structures of firms (i.e., the proportion of debt, preferred stock, and equity financing) vary as do the costs associated with specific types of capital. Consequently it seems reasonable to calculate a firm's cost of capital as a weighted average of the costs of each source of financing in the firm's target capital structure. That is, the weighted average cost of capital to be used to evaluate prospective investments should refelect the mix of debt, equity, and preferred which the firm intends to use in the future. However, for many firms the target capital structure is the same as the current capital structure. The weighted average cost of capital is expressed on an after-tax basis, which conforms to analyses utilizing after-tax annual cash flows.

Archer, Choate, and Racette (1979) appropriately point out that in practice the cost of funds will vary with the scale of the proposed investment outlay and must be computed at each scale of outlay under consideration by the firm. Consequently it is the marginal cost of each increment of capital obtained, and not the average cost of all sources of financing, that is relevant for deciding which investment opportunities will be accepted or rejected. Van Horne (1968) notes that in theory a marginal-sequential costing-of-funds approach allows the firm to combine the investment decision directly into the financing decision. However in practice this approach is extremely difficult to employ directly and most organizations revert to the weighted average cost of capital approach with the assumption that the firm will finance itself in the future in the same relative proportions as in the past. Under these conditions and from an investment standpoint, we need only ensure that the promised rate of return from a project surpasses the marginal weighted average cost of capital to the firm. In this context the marginal weighted average cost of capital refers to the increment of additional financing required to support a new investment opportunity or project, given the firm's existing capital structure.

Before providing an example illustrating the procedure utilized to calculate a marginal weighted average cost of capital, it might be beneficial to clarify some of the issues which often create concern with the cost of capital concept.

Future vs. Historic Costs: Future investments will be financed with funds raised in the future, so that the relevant costs are those that will be incurred in raising new funds and not what was paid in the past for similar funds. Furthermore, the pertinent future costs are really the *marginal* costs, the cost of the next increment of capital, given the firm's current capital structure.

Spot vs. Normalized Costs: The financing decision is very much influenced by spot or current market capital costs, but normalized values are more appropriate in the investment decision because fianancing costs are known only approximately at the project evaluation stage. These normalized costs are intended to smooth out short-run random fluctuations in the capital markets. Moving averages over time are commonly used to normalize common stock prices, but this is invalid if an abrupt change in the company's earnings prospects has occurred recently. The cost of debt is usually less volatile, so that spot costs are generally more reliable.

Specific vs. Composite Costs: Each financing source has its specific cost, but the cost of capital is an average of the costs of all financing sources to be used in the future by the firm, weighted by the expected proportionate use of each source. The cost of the specific type of financing used for a given project is *not*, in general, the cost of capital used to evaluate the project. This is frequently a difficult point to grasp, and an example is presented later in this chapter to demonstrate this point.

Market Weights vs. Book Value Weights: Book value weights are traditionally used in determining the weighted average cost of capital. Book values are simply the amounts recorded on the firm's books, generally at cost. Market values, particularly for common stock, often differ drastically from book values, and since financing costs are measured by market costs, market weights would appear to be more desirable. Major differences between book and market values are not unusual and can affect the cost of capital significantly.

However, as significant as these differences may be, the use of book value weights is favored by tradition. Investors, creditors, and analysts usually judge a firm's financial position from its balance sheet, and these judgments affect financing costs to the firm. In addition to ease of computation, book value weights are not subject to the daily fluctuations of market weights. In summary, although market weights may be more precise, the financial community pays more attention to book values.

Example Calculation—Marginal Weighted Average Cost of Capital

To illustrate the calculation procedure for determining the marginal weighted average cost of capital for a mining corporation, the following example is provided. Although the numbers utilized in this example are patterned after an actual mining company, this exercise is not intended to calculate the cost of capital for a specific company. Rather, it is intended to illustrate the calculation procedure normally utilized and the problems associated therewith.

When performing a cost of capital calculation the primary sources of information are (1) the company's annual stockholder reports, (2) market prices of securities in such publications as the *Wall Street Journal*, and (3) as much knowledge as possible about the company's operations, management, and future prospects.

Although the company in question is basically a base-metal producer, it is a diversified company having operating mines producing a wide variety of metals, coal, and nonmetallics. Over the years the company (let's call it Minty Mining Co.) has managed to maintain a relatively low cost of capital by making good use of its debt capacity and diversification.

The annual report for 1982 showed the capital structure for Minty to be as follows:

	$ in thousands
Long-term debt (notes and debentures)	958,300
Preferred stock (nonconvertible)	191,500
Common stock, including capital surplus	822,400
Retained earnings	1,147,700
Total	3,114,900

The long-term debt consists mostly of notes, intermediate-term loans, and sinking-fund debentures. The last several debenture offerings have been either privately placed or have a foreign listing and have been issued at coupon rates of 10% to 11%. However, one of the older 8% sinking fund debentures having a 1986 maturity date was recently traded at 79 ⅜. This results in a yield-to-maturity of 10% before taxes, or [10% x (1 - 0.46)] = 5.4% after taxes. It seems reasonable to assume from the information available that the cost to Minty of new sinking fund debentures of the type previously offered would approximate 5.4% after taxes. It should be noted, however, that should Minty choose to take on new term loans the cost would be something in excess of the prime rate. Minty would probably have to pay a premium of about 50 basis points (½%) above prime for debt of this type.

The preferred stock issue in Minty's capital structure is small compared to the magnitude of the capital structure and is not very significant. However, it is relatively easy to calculate a cost for a new preferred stock issue for Minty should it choose to make such an offering. The last issue of Minty's preferred stock has a cumulative annual dividend of $7.25 and is currently selling in the market for 38¼. The after-tax cost of preferred stock to Minty is calculated to be:

$$k_p = \frac{D}{I_o} = \frac{\$\ 7.25}{\$38.25} = 18.9\%$$

As stated previously, estimating the cost of equity for Minty, or any other company, is much more judgmental than for other portions of the capital structure. Using the dividend valuation model

$$k_e = \frac{D_o}{P_o} + g = \frac{\$\ 1.85}{\$19.25} + g = 9.6\% + g$$

The foregoing calculation results from the fact that Minty declared a dividend in 1982 of $1.85 per share while the average selling price in the market during that period was 19¼.

Again, g is difficult to determine. To get a better feel for what might be an appropriate value, the dividend and earnings records over the last ten years are listed for Minty as follows:

Year	Earnings per share, $	Percentage change, %	Dividend per share, $	Percentage change, %
1973	1.30	—	0.73	—
1974	2.00	+54	0.78	+6.8
1975	2.53	+27	0.85	+9.0
1976	2.05	−19	0.92	+8.2
1977	2.12	+3	0.92	0
1978	0.77	−64	0.92	0
1979	2.04	+165	0.98	+6.5
1980	4.78	+134	1.34	+37.0
1981	5.73	+20	1.85	+38.1
1982	2.52	−56	1.85	0

A review of these numbers reveals some interesting aspects about the minerals industry in general and Minty in particular. First, Minty has experienced extreme fluctuations in earnings over this ten-year period. This is rather typical for companies within the minerals industry and reflects the relative riskiness of the industry—particularly for companies producing commodities traded in international markets. The wide fluctuations in earnings per share (EPS) values suggests it would be extremely difficult to arrive at a reliable value for g from these numbers. This is a common phenomenon with mining companies.

The second interesting aspect about Minty's record is that Minty has maintained a positive dividend policy. That is, the dividends have not been allowed to fluctuate with earnings; rather, dividends have increased during periods of high earnings and have remained constant during periods of stagnant or reduced corporate earnings. This action by Minty suggests that the company believes a positive growth record for dividends is very important to its shareholders and is extremely reluctant to reduce dividends even when earnings are poor.

Unfortunately Minty's dividend policy does not shed much light on what might be an appropriate value for g. The average growth in dividends over this ten-year period is 10.5%. However this value does not necessarily have any real meaning since the investment community is interested in what Minty will do in the immediate future. Looking ahead the future does not appear particularly bright for Minty—at least over the short-term. During the first two quarters of 1983 commodity markets associated with Minty's primary production deteriorated to the point where Minty was producing at 40% capacity at its operations. This has resulted in a need to terminate significant portions of operating and technical staffs. These problems were anticipated by the investment community and explain in part the low P/E ratio (7 to 8) for Minty stock in 1982. Remember, g is the anticipated *future* growth in dividends. In the absence of other relevant information the historic growth record is sometimes used, but in Minty's case the dark clouds on the horizon had already been identified.

In view of the foregoing information what is likely to be the impact on Minty's dividends for 1983? If Minty is able to maintain its current policy, one would expect that there would be no net increase in dividends and they would remain at $1.85 per share. Even if conditions turned around in the last half of 1983, it is hard to imagine that Minty's dividends could grow more than about 5% in view of the relatively poor earnings performance in the minerals industry as a whole.

Before settling on a value of g for Minty, it might be advisable to check the cost of equity via the earnings model. The E/P value for Minty's cost of equity is:

$$\frac{\$\ 2.52}{\$19.25} = 13\%$$

The growth component was omitted because earnings growth projections are as questionable as they are for dividends. This calculation suggests that a value of 3.4% might be reasonable for g in the dividend valuation model. If g is assumed to be 3%, the dividend valuation model yields an after-tax cost of common stock to Minty of 9.5% + 3% = 12.5%. However, one must remember that if Minty decides to raise capital through the sale of common stock there will be a flotation charge associated with this offering. If this charge is assumed to be 3%, the net after-tax cost of new common stock to Minty is:

$$k_e = \frac{12.5\%}{1 - f(\%)} = \frac{12.5\%}{1 - 0.03} = \frac{12.5\%}{0.97} = 12.9\%$$

Since there is no flotation charge associated with retained earnings, the after-tax cost of this capital to Minty is simply 12.5%.

Using the cost of individual financing sources calculated previously, the marginal weighted average cost of capital for Minty is as follows:

Type	Amount ($1000's) ×	After-tax cost =	Product
Long-term debt	958,300	0.054	51,748
Preferred stock	191,500	0.189	36,194
Common stock	822,400	0.129	106,090
Retained earnings	1,142,700	0.125	142,838
	$3,114,900		336,870

$$\text{Marginal weighted average cost of capital} = \frac{336,870}{3,114,900} = 10.8\%$$

Minty's weighted average cost of capital is then about 11%—remarkably low for a mining company as will be discussed in the next section. Under normal circumstances then, this would be an appropriate discount rate for Minty to use in evaluating investment projects, remembering that no adjustment for risk has yet been made and that Minty will finance itself in the future in the same relative proportions as it has in the past.

Practical Problems with the Cost of Capital Calculation

From an academic standpoint the preceding calculation procedure for the weighted average cost of capital is reasonably straightforward. However, let's look at the Minty Mining Co. example from a practical aspect to ascertain whether or not the cost of capital determined is reasonable.

First, what about the 10% before-tax cost of long-term debt? A closer look at Minty shows that the debt-to-equity (D/E) ratio is 49% and that long-term debt constitutes 31% of the firm's capital structure. Certainly this is a substantial amount of debt financing for a mining company. It appears that Minty is aware of the attitude of the marketplace to new debt issues because they have raised all new debt capital within the last five years either through foreign listings or private placement. As indicated in the example, Minty will probably incur before-tax costs in excess of prime rates if they choose to acquire term loans. Indeed, when they borrowed long-term in 1981 the before-tax cost was 18⅛%. Furthermore, Minty realizes that taking on more long-term debt with its associated fixed interests costs in view of its 1983 prospects could have disastrous effects on its shareholders. All of these factors place Minty in a very difficult position, in that the real cost of obtaining new debt will probably be substantially higher than the 5.4% after-tax cost calculated in the example. After considering Minty's future prospects for the next year or so and the amount of debt in the present capital structure, one must question whether or not there would be many buyers for new bonds from Minty at anything less than a high discount which would result in a very high cost to the firm.

The second item of interest in the cost of capital relates to the cost of preferred stock. In the example it seems unrealistic in view of Minty's future prospects and present status to have preferred stock at a cost in excess of that of common stock. Since preferred stock has a fixed dividend and is, therefore, generally considered less risky than common stock during bad times, investors should be willing to pay more for preferred stock, thereby reducing the cost of financing to the firm. Clearly the 18.9% cost for preferred stock is unrealistically high. One explanation could be that investors perceive Minty's weakened financial condition as a threat to preferred dividends and, therefore, the preferred stock is selling at a low multiple.

The last item in the cost of capital example which needs attention is the cost of equity. As stated previously, determination of the value for g in the dividend valuation model is perhaps the most difficult variable to quantify in practice. After reviewing the earnings and dividends history for Minty, as well as the prospects for the immediate future, it seems reasonable that the investment community would expect Minty's dividends to remain constant (0% growth) or at best increase only slightly for 1983; therefore, a value for g of between 0-5% seems reasonable. However, how many individuals or financial institutions would be willing to buy

Minty's common stock for an overall return of 12 to 13%, considering that the ability of Minty to sustain this level is highly suspect? Certainly few investors would find this opportunity attractive when they can invest in much less risky money market accounts having yields in excess of 10%. The overall return must rise, which means g must increase, the price must decline, or both.

It is interesting to note that many mining companies occasionally find themselves in exactly the same financial situation as Minty. Their capital structures and earnings prospects become such that the firms are essentially unable to obtain additional long-term debt financing at any reasonable cost and the cost of floating new common stock issues approaches infinity under then-current earnings and mineral market forecasts. How then are companies like Minty to raise new investment capital under these circumstances? The answer to this question is certainly not simple; however, most companies in this situation are resigned to looking to internally generated funds (retained earnings) in order to improve their battered balance sheets. Unfortunately, this also is not an easy or rapid solution to the problem due to the poor earnings records of mining companies over the last decade. Some mining companies have managed to obtain external investment capital over the last few years, albeit at the ultimate cost of being acquired by other corporations.

In summary, it seems reasonable to conclude that the weighted average cost of capital model is sound in theory and gives reasonable results when applied to firms having fairly stable earnings and dividend histories. However, mining companies often fail this stability test. Therefore, considerable judgment is involved when dealing with a specific mining company, and considerable attention must be given to the overall economy, the industry's performance within that economy, and the specific company's performance within its industry. Inflationary trends, along with the overall poor performance of mining companies over the last decade and the growing importance of project financing (see Chapter 12), suggest that considerable thought and additional research should be performed on the concept of weighted average cost of capital for mining corporations.

SEPARATING THE INVESTMENT AND FINANCING DECISIONS

One of the fundamental canons of corporate finance is that the investment decision should be distinct and separate from the financing decision. In other words, the first step is to decide whether or not to invest in a project. This consideration was addressed in Chapter 9. Once the decision has been made to invest in a project, the second step is to procure the required financing for the project. The general forms of financing are discussed briefly in this chapter as well as Chapter 12. These sections clearly illustrate how the investment and financing decisions are related only by the marginal weighted average cost of capital. That cost was expressed as an interest rate, such rate then being used as the discount rate for future cash flows in evaluating prospective investments by the firm. Finally it was demonstrated that if the estimates of project performance were accurate, this procedure would guarantee that the financial return generated by the project would be greater than the cost of funds employed. This is, of course, the fundamental objective of capital investment.

Financing and investment decisions become much more interrelated when lenders look to the specific project for security rather than to the overall corporation. This aspect is discussed further under Project Financing in Chapter 12.

One of the most common errors made in discounted cash flow financial analysis involves using the cost of specific financing in evaluating projects rather than the average cost of all capital used by the firm. When a capital project involves an expenditure of, say, $5 million and a financial decision is made to raise the entire $5 million through a bank loan carrying an interest rate of 12%, it often is difficult to understand that (1) the interest charges from the loan should *not* be levied against the project for evaluation purposes, and (2) the appropriate discount rate to use is *not* 12% (or its approximate after-tax equivalent of 6½%), but rather the marginal weighted average cost of *all* capital used by the firm. These errors are sufficiently common that a more detailed discussion of each follows.

Error No. 1. Charging Specific Capital Costs to the Project

Being the link between investment and financing, the cost of capital occupies a unique place in the evaluation procedure. All other project costs such as operating costs, plant overhead, and depreciation are handled explicity in the *pro forma* income statement. The cost of capital, however, *is* the discount rate and does not receive a separate line in our statements. Such a procedure might be applicable in other types of economy studies (discussed later in this section) but in discounted cash flow analyses, it is incorrect. This has led many practitioners to believe that the cost of funds has been omitted in the analysis and should be listed separately in a manner similar to the other cost items. This misconception occurs most frequently when debt capital is raised and the accompanying interest payments are not charged directly to the project in the DCFROI calculations. However, it is important to recognize that to do so would clearly double-count the cost of capital—once through the discount rate and a second time through the explicit interest charges. Consider the following situation which illustrates this point.

Example 1: Suppose Hikki Mining Co. contemplates an investment opportunity having the following anticipated revenues and costs.

Investment = $100
Net annual cash flows (year 1-5) = $27.74
Cost of capital = 12%
NPV = $27.74 (P/A, 12, 5) − 100 = 0

In other words the project should just break even afer returning the investment plus the 12% cost of capital. This is verified below.

	\multicolumn{6}{c}{Year}		
	1	2	3	4	5	Total	
Cost of capital @ 12%	12.00	10.11	8.00	5.63	2.97	38.71	
Total money owed before year-end payment	112.00	94.37	74.63	52.49	27.72		
General payment	27.74	27.74	27.74	27.74	27.74	138.70	Total payments
Money owed after year-end payment	84.26	66.63	46.86	24.75	—		
						100.00	Principal

Thus, we have assured that all project costs are covered—including return *of* and return *on* capital. To have levied financing charges explicity against the project prior to discounting would clearly have forced the project to pay these costs twice. It bears repeating here that the discount rate provides for the return of costs of capital and does not exist to compensate for any risk specific to the project. Protection against risk should normally be handled separately as discussed in Chapter 13.

It nonetheless continues to bother many that when the firm has acquired additional debt obligations which must be met from the earning power of its investment, the resulting interest payments are not charged directly to the project. This has given rise to procedures used by some firms where projects are evaluated on (1) a full-equity basis and (2) various degrees of debt leveraging. Example 2 will serve to illustrate the types of calculations resulting from this misunderstanding. Please note that, in our opinion, these procedures are improper and are presented here for illustrative purposes only.

Example 2: An underground conveyor system is proposed to reduce haulage costs at a zinc mine. Benefits to be derived from this project are estimated to be $200,000 annual savings in operating costs. Depreciation is straight-line and the income tax rate is 50%. Depletion is constrained by 22% of gross income from mining and can, therefore, be ignored as being constant in all alternatives. If the total investment is estimated to be $1,146,000, and the service life is 12 years with no salvage value, calculate the rate of return on equity capital for: (A) all equity case; (B) 50/50, debt/equity financing; and (C) 95/5, debt/equity financing.

Assume the before-tax cost of debt is 8% compounded annually, and for simplicity assume that the entire principal amount will be repaid at the termination of the project.

Solution

1) Equity investment
   A) $1,146,000
   B) $ 573,000
   C) $ 57,300

2) Debt investment to be repaid at the end of year 12
   A) 0
   B) $ 573,000
   C) $1,088,700

3) Annual cash flows

	Case A	Case B	Case C
Net operating savings	$200,000	$200,000	$200,000
Less: depreciation	− 95,500	− 95,500	− 95,500
: interest	0	− 45,840	− 87,096
Pretax net income	104,500	58,660	17,404
Less: income tax	52,250	29,330	8,702
Incremental net profit	52,250	29,330	8,702
Add: depreciation	95,500	95,500	95,500
Incremental net cash flow	147,750	124,830	104,202

4) DCF return on equity

Case A: $1,146,000 = 147,750 (P/A, i, 12)

$$i = 7.5\%$$

Case B: $573,000 = 124,830 (P/A, i, 12) − 573,000 (P/F, i, 12)

$$i = 14.8\%$$

Case C: $57,300 = 104,202 (P/A, i, 12) − 1,088,700 (P/F, i, 12)

$$i = 182.5\%$$

The foregoing analysis seems to say that by using debt financing one can turn a very marginal 7.5% ROI project into one yielding a phenomenal 182.5% ROI project. But is this really a meaningful comparison?

It is important to note that the DCFROI on the equity portion of the investment is always increased with leveraged financing if the after-tax cost of debt is less than the DCFROI of the project figured for the all-equity case. Thus, the most marginal of investments can often be made to appear positively superb, as illustrated previously, when a high percentage of debt capital is used. Furthermore, the comparison of investment alternatives becomes very obscured using the foregoing procedure. Does one compare projects using the same debt/equity *ratios*, or using the same *amount* of debt financing? Is a project having a DCFROI of 25% with 40% debt financing "better" or "worse" than one having a return of 20% but with only 25% debt capital? The frame of reference is not constant, and it is impossible to tell, without further analysis, which project is more consistent with the objective of the firm—i.e. maximization of shareholder wealth. A basic principle of investment analysis is that a project must stand on its own merits and must compete for funds on an equal basis with other investment opportunities. Each such opportunity will share fully in the benefits of debt leveraging if the same weighted marginal average cost of capital is used for all projects as previously prescribed.

At this point, it should be noted that leveraging cuts both ways. If the investment falls short of expectations, debt leveraging serves to amplify losses just as it amplifed earnings under a favorable outcome. The same fixed interest charge which allows high earnings under favorable conditions becomes a serious drain on financial resources when earnings decline. In essence, leveraging makes good projects appear better, and bad projects, worse.

In summary, the following should always be observed in project investment analysis.

1) To avoid counting capital costs twice, do not charge costs of specific capital sources against the project. These charges are embedded in the discount rate.

2) Calculate ROI's on total capital invested (i.e., all-equity case) to insure that all projects are compared on an equal basis.

Error No. 2. Using Specific Capital Costs as the Discount rate

Here we are not concerned with the assessment of interest charges to specific projects in the discounted cash flow evaluation process, as this error was treated in the preceding discussion. Debt financing, however, is the source of another common error in capital project evaluations. There is a tendency on the part of some analysts to use the cost of a specific source of financing—rather than a weighted average of all capital sources—as the discount rate in measuring the attractiveness of a project. The flaw in this reasoning can best be illustrated with an example modified from Quirin (1967).

Example 3: Global Mineral Ventures, Inc. was presented with a similar set of investment opportunities in three successive years as shown:

	Year 1		Year 2		Year 3	
Project	Amount of investment, $	Rate of return, %	Amount of investment, $	Rate of return, %	Amount of investment, $	Rate of return, %
A	100,000	20	100,000	20	100,000	20
B	200,000	15	200,000	15	200,000	15
C	200,000	11	200,000	11	200,000	11
D	200,000	8	200,000	8	200,000	8
E	200,000	6	200,000	6	200,000	6

In year one, Global had no long-term debt and the corporate treasurer found that the full $900,000 could be raised by selling debentures bearing an annual interest rate of 5 ½%. He convinced Global's board that, since every project returned more than 5½%, they should all be accepted.

In year two, the treasurer was able to borrow a further $700,000 at 7 ½%, and using the same logic as the previous year, accepted all projects except E, which offered a return below the 7 ½% marginal cost of debt.

In the third year, however, Global's treasurer found only a limited amount of additional debt available to him—$100,000 at 18% from a finance company. Since this was still below the estimated 19% for a new equity issue, the treasurer used the debt to accept only project A.

Thus, over the three-year period by using the cost of a specific source of debt as his investment criterion, the treasurer had invested $1.7 million at a weighted average return of 12%. It is clear, however, that the treasurer's eagerness to accept projects in year 1 precluded the acceptance of better projects in year 3. In fact the same $1.7 million could have been invested to yield an average rate of 13.3% by accepting the following projects:

Year 1   A, B, C, D
Year 2   A, B, C
Year 3   A, B, C

What went wrong? Where was the treasurer's concept of marginal cost of capital in error? In essence, the treasurer failed to recognize that debt financing is only possible if an adequate equity base exits. If expansion of the equity base does not keep pace with borrowings, the firm will reach a point—as Global did in year three—where the firm's financial risk is too high for further credit at reasonable rates. Clearly the marginal cost of capital in this case, then, was not simply the cost of debt but also the cost of equity which will be required in the future to support the added debt. As Quirin notes, "When the capital structure is changed by an issue of debt, the relevant cost is not only the out-of-pocket cost of the debt itself, but includes the increase in the cost of equity resulting from the higher risk premium attached to the shares as a result of the debt." Thus, every investment must carry its proportionate share of the necessary—but higher cost—equity funds. In the above example, using the marginal weighted average cost of *all* capital sources would have resulted in rejecting the lower-value projects in year one, thereby permitting the acceptance of higher-value projects in year three.

As noted in the next section, corporate finance is "lumpy" insofar as marginal capital acquisitions are seldom in the same proportions of debt and equity as is the target capital structure. During one year new capital might come mainly from a new debenture whereas in the next it might be exclusively equity. The applicable marginal weighted cost of capital will, however, continue to reflect the overall relative usage of each type of financing in the firm's target capital structure, and, thus, will normally change only by a modest amount from year to year. Indeed, it would not necessarily be erroneous to use the proceeds from an equity issue costing 14% to invest in a project returning only 12%—providing enough debt occurred in the target capital structure to reduce the overall marginal cost of capital to below 12%.

The main point in the preceding discussion, then, is that it is incorrect to use the cost of a specific source of capital as the cash flow discount rate or, equivalently, as the minimum acceptable rate of return for capital projects.

OPTIMUM CAPITAL STRUCTURES

Does it matter how investment proposals are financed; and, if so, what is the optimal capital structure for the firm? Can the firm affect the market price of its stock as a result of its financing decision? Can the firm affect its total valuation and its costs of capital by changing its financing mix? How should a firm be financed if the value of the firm and the wealth of its stockholders is to be maximized?

To answer these questions it is necessary to consider the concept of whether or not there is an optimal capital structure for a coporation. There is considerable debate on this issue and universal agreement does not exist. However, most practitioners adhere to the traditional position that an optimal capital structure does exist for a given organization such that the overall cost of capital can be minimized. Assuming we accept the premise that a corporation can achieve an optimal capital structure, then how is it determined and how do we know when it is reached?

Given the preceding discussions on the specific costs associated with the various types of financing available to a corporation, it seems reasonable to conclude that it should be possible to combine these various forms of financing into a corporate capital structure which would result in a minimum weighted average cost of capital to the firm. In other words, the implication is that there apparently is an optimum capital structure which will minimize the firm's weighted average cost of capital. This concept developed by Solomon (1955) is referred to as the "traditional" approach to optimal capital structure. Fig. 1, adopted from Van Horne (1968), illustrates this traditional view of capital structure. It is intuitively appealing and has received much empirical support.

Fig. 1 illustrates that successive additions of debt to the firm's capital structure will first reduce k_0 and then increase k_0 as the firm's financial risk becomes excessive. In other words, low cost debt will only be issued by lenders if there is a sufficiently large equity cushion to absorb most of the business risk of the firm. It becomes clear that a firm cannot continue to use low-cost debt financing exclusively because ultimately a point will be reached whereby the firm can borrow no more— no matter what the cost—until the equity base is increased. Furthermore, at high debt/equity ratios, the cost of equity will also rise rapidly because the high level of

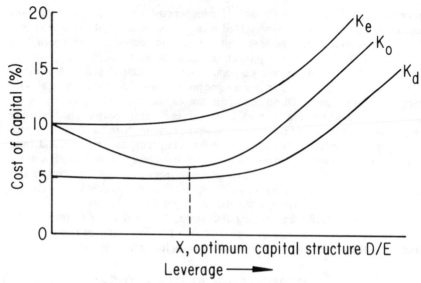

Fig. 1. Traditional view of optimum capital structures. k_0 is marginal weighted average cost of capital, k_e is cost of equity, and k_d is cost of debt.

interest payments jeopardizes earnings available to shareholders. In summary, judicious use of debt will reduce the firm's weighted average cost of capital to a minimum. Beyond that point additional debt will cause the costs of both debt and equity to rise rapidly.

Due to notoriously fluctuating earnings, mining companies find their optimum capital structures at relatively low debt/equity ratios. On the other end of the scale are industries such as public utilities where earnings have historically been very predictable.

The traditional approach to an optimum capital structure suggests that an optimum structure can be obtained which will result in a minimum weighted average cost of capital to the firm. Therefore, if a firm finances itself in this optimum manner, it should generate the lowest possible weighted average cost of capital. If this cost of capital is then used as the appropriate discount rate when evaluating new investment opportunities, the firm should find that more investments become acceptable. This should subsequently result in an optimum investment program and generate more wealth to the shareholders. In summary, then, the traditional approach suggests that it does make a difference how the firm is financed and this optimum financial structure will tend to maximize the value of the firm to the shareholders.

The logical extension of the traditional approach to optimal capital structures is to conclude that each firm should select its long-term financing in a manner to provide a minimum overall cost of capital. Unfortunately, it is not easy to determine a firm's optimum capital structure, for there is no reliable way to anticipate the market's reaction to changes in leverage. If investors behaved rationally, many thorny finance problems could be solved, but the stock market's history is not encouraging on this point.

If a company has an estimate of its optimum leverage, it presumably would finance in that manner in the future, and this target capital structure would be used for the weights in the cost of capital calculation. Since most firms either have not attempted to determine their optimum structure or view their current capital structures as optimum, future financing most likely will be in the same proportion as present financing. The appropriate weights in this case are those in the current balance sheet.

Finally, in practice individual fianancing packages rarely are composed of the various sources of funds in their optimum ratios. Financing is lumpy, with borrowings used for this project and perhaps equity for the next. As a consequence the firm's actual capital structure may deviate from its target capital structure at any given point in time. The optimum structure is generally considered to be fairly robust so that modest excursions in either direction will not create large penalties.

SUMMARY

Determining with precision the appropriate discount rate to use for financial analysis studies is a difficult, much-debated topic. It is clear that the cost of acquiring investment funds must cover opportunity costs, transaction costs, compensate for risk, and must cover anticipated inflation. Although this places some bounds on the problem, each of these items is also difficult to quantify with precision.

Happily, a high degree of precision is normally not essential, and a good answer can usually be obtained by intelligent use of the cost of capital models adjusted by knowledge of the current status and future prospects for the firm in question. In virtually every case, the most difficult question is, "What is the cost of equity?" The answer is determined by the expectations of investors. For a firm with a fairly stable history of earnings and dividends, the analyst can comfortably assume that investors expect the future to be largely an extension of historic trends. In such cases, the cost of capital models give reliable estimates of the firm's cost of procuring investment funds.

In industries such as mining, however, where strong business cycles are evident, estimates of future earnings and dividends are subject to large errors. As a consequence, greater judgment is needed to produce a reliable answer from the cost of capital models. This solution creates uneasiness in those searching for scientific precision, but engineers are comfortable with the fuzzy nature of real problems where the optimum is rarely attainable. The logic behind using the marginal weighted cost of capital as the discount rate in evaluating capital investments is clear and compelling. The application of this theory is often difficult, "Nevertheless" as Van Horne (1968) notes (p. 120), "estimates must be made."

SELECTED BIBLIOGRAPHY

Archer, S.H., Choate, G. M., and Racette, G., 1979, *Financial Management—An Introduction*, John Wiley and Sons, New York, 724 pp.

Bussey, L.E., 1978, *The Economic Analysis of Industrial Projects*, Prentice-Hall, Englewood Cliffs, NJ, 491 pp.

Gentry, D.W., and O'Neil, T.J., 1974, "Principles of Project Evaluation in the Minerals Industry," Short Course, 12th International Symposium on Application of Computers in the Minerals Industry, Golden, CO, May, 199 pp.

Gordon, M.J., and Shirpiro, E., 1956, "Capital Equipment Analysis: The Required Rate of Profit," *Management Science*, October, pp. 102-110.

Quirin, G.D., 1967, *The Capital Expenditure Decision*, Richard D. Irwin Inc., Homewood, IL, 258 pp.

Solomon, E., 1955, "Measuring a Company's Cost of Capital," *Journal of Business*, October, pp. 128-140.

Van Horne, J.C., 1968, *Financial Management and Policy*, Prentice-Hall, Englewood Cliffs, NJ, 583 pp.

Van Horne, J.C., 1980, *Financial Mangement and Policy*, 5th ed., Prentice-Hall, Englewood Cliffs, NJ, 808 pp.

12

Special Evaluation Problems in Mining

"If you can hold a board of cross directors
In happiness against their gauzy schemes;
If you can dodge the wrath of the electors
Till dividends will flow as in their dreams;
If you can make a mine pay from the grass roots
No matter what the time or place or year;
Then on my soul until the final blast shoots
We'll add the title "MINING" when we call
you "ENGINEER."

—From Wayne Darlington,
"The Mining Engineer,"
The Engineering and Mining Journal,
Vol. 98, Sept. 26, 1914, p. 587.

REASONS FOR THIS CHAPTER

Up to this point in the book the material presented has addressed the fundamental parameters associated with the evaluation of mining ventures and other investment alternatives. The procedures and techniques illustrated represent the basic approach to project evaluation, particularly when considering the normal engineering economic analysis of investment proposals. The evaluation procedure is complicated, however, when dealing specifically with mining-related investment proposals. These complicating factors are invariably directly related to the many special characteristics of the mining industry. Many of these special features such as tax policies, revenue and cost estimation, and ore reserves have been addressed in previous chapters.

In addition to the many special features of mining which directly impact, in a quantitative manner, the various line items in the pro forma income statements of a cash flow analysis, other features of mining affect evaluation procedures in a more general manner. The main purpose of this chapter, then, is to provide an exposition of some of these more qualitative factors. We hope this will enable the reader to recognize these factors in mine evaluation practice and to make suitable adjustments in his or her analysis.

One of the special features which can significantly complicate the evaluation process for mining ventures is that of ownership. The complexity of ownership associated with many mineral lands or deposits can significantly increase the difficulty in evaluating property.

To illustrate the problem, consider a common example which often confronts mining companies on a routine basis. Suppose an entrepreneur decides that an old mining district has important mineral potential in view of recent changes in metal prices, increased technology which can significantly reduce operating costs or increase recoveries, or perhaps there is a new theory of ore genesis which seems applicable to this particular district. For whatever reason, the entrepreneur decides to establish a land position in the district in order to test his hypothesis and sets about trying to tie up a large, contiguous block of land in the most favorable part of the district.

Any land person who has undertaken such a task appreciates the painful ordeal ahead for this entrepreneur. This individual will typically be dealing with second or third-generation heirs of the individuals who originally staked or subsequently owned the claims. By this time ownership of the claim, or block of claims, is often split among many descendants, each having different leasing objectives. Few, if any, understand the mining business or processess, and all are convinced that great grandad's claim contains a bonanza deposit worth millions. Consequently, the entrepreneur finds himself negotiating widely differing lease agreements with property owners, depending on their individual leasing objectives. Various types of rental agreements, minimum payments, buy-out arrangements, and annual production or payment schedules will result. In addition, agreements must be consummated on royalty provisions (type, rate, base) and/or profit-sharing arrangements with each property owner. Ownership complications are vastly escalated when surface ownership does not include ownership of mineral rights.

After the land position has been secured, the entrepreneur often finds the need for considerable additional capital in order to test the exploration target. At this point he generally approaches one or more mining companies in hopes of structuring a deal whereby the company performs the needed exploration work in exchange for an interest in the property. The financial arrangements in this deal must recognize the prior lease commitments as well as mutually acceptable economic considerations for the entrepreneur and the mining company. It is vitally important that the evaluator of these deals ascertain the actual ownership position in any future mining operations which might result from these exploration activities. This is important from both legal and tax standpoints because it will have an impact on project viability.

The potential owner/operator must also be aware of any previous lease agreements which may impact mine planning or operating practices. For instance, some leases may stipulate that a given amount of production must be produced from the lease in a certain time period, or the lease must be producing by a certain date. Also it may be necessary for the operator to segregate ore produced from different leases in order to determine royalty obligations to different leasors within the general mine unit. The magnitude and timing of minimum royalties, advance royalties, and rentals must be analyzed with respect to buy-out provisions and overall project economics. Certainly the complexity of ownership for most mineral-bearing lands presents additional problems for mining property evaluators from the standpoint of determining who has economic interest in the property at any point in time and who incurs the associated tax liabilities or benefits.

Another special feature associated with mining which complicates the evaluation procedure for mining ventures is the ever-increasing complexity of financing. Traditionally, new mining ventures were financed through internally generated funds or through the sale of new equity issues. If external financing was required, it was obtained on the strength of the mining company's balance sheet which usually contained little debt. Since the 1960s, however, there have been some profound changes in the area of financing new mining ventures. These changes were hastened by governmental intervention in the areas of taxation, price controls, and environmental regulations. In addition, inflationary pressures have significantly escalated capital requirements for new ventures at the same time the industry has been experiencing substandard financial results. The net result has been to force mining companies to seek and rely on external sources of funding for new projects and ventures. As a consequence, mining companies have experienced increasing financial risk as a result of the unprecedented debt levels now contained in many corporate financial structures.

In order to maintain and expand domestic and worldwide mineral production, mining companies have found it necessary to obtain considerable external funding from various groups under various types of organizational umbrellas. This external financing can take many forms such as production payments, carried interests, term loans, and many more. These funds may be obtained from other mining companies, suppliers, consumers, export creditors, international development institutions, commercial banks, and others. Because of the increased use of so-called multilayer financing, most mining companies are establishing new mining ventures under various organizational structures. These structures may take the form of joint ventures, carried interests, partnerships, unusual subsidiary arrangements, or complete new operating companies owning only one mine or mineral deposit. In general the organizational structures established are a result of the external financing mix, the tax considerations for the parties involved, and the unique set of political factors which surrounds every project. From an individual mining company's standpoint it is important to accurately ascertain the income, obligations, and liabilities of each participant in these complex financial arrangements. The priority and repayment schedule of senior debt, subordinated debt, loans, etc.—as well as each sponsor's other obligations and liabilities—are critical items when evaluating the relative corporate exposure and project viability resulting from these evolving organizational forms and financing mechanisms.

Perhaps the most difficult special feature of mining to adequately address in evaluation procedures is that associated with the international nature of mineral projects. Clearly, domestic mining companies must consider the impacts of major new mineral discoveries in foreign countries on domestic operations and the international markets for mineral commodities. For example, the discovery of Saskatchewan potash had a strong impact on US producers. The potential impact of the tremendous uranium discoveries in Canada and Australia are as yet not fully understood, but, nonetheless, are of great concern to US producers experiencing a sagging market.

Major new discoveries of some mineral commodities can also have a dramatic effect on international mineral markets. US producers must compete in these markets and are directly affected by domestic and foreign government actions. For instance, US producers point to decreasing productivity levels coupled with increasing production costs resulting from runaway inflation rates and the costs associated with environmental regulations as some of the reasons for their inability to maintain a competitive position in some commodity markets. Also, in many countries governments subsidize production either directly or indirectly for any number of reasons and thus expand mineral supply and depress price levels. Such actions by foreign governments, in conjunction with uneven environmental commitments, can significantly influence the relative competitiveness of producers in international markets.

The international nature of mineral projects is very pronounced in the area of project financing. It is not uncommon to find mineral producers, suppliers, and purchasers from several different countries involved in the joint financing of mineral ventures. In recent years various governments have taken an active role in the financing and development of mineral properties as have commercial banks, international development institutions, and export credit agencies. These types of financing arrangements are necessary in order to raise the vast amount of capital needed to bring new mines into production—particularly in lesser developed countries (LDCs) where infrastructure must also be developed. Most countries recognize the necessity of international trading in mineral commodities. Almost all of the major industrialized nations of the world have become net importers of mineral commodities and look to developing nations for a large portion of these vital materials. Joint venturing of mineral developments in these countries not only secures a supply of minerals, but also minimizes a company's exposure to political risk.

Certainly the changing economic and political climates in the world have significantly affected mineral development in the past 10-15 years. Recent approaches to mineral project financing involving multilayer package arrangements between mining companies, world investment institutions, banking houses, mineral consumers, and foreign governments typify the international nature of minerals projects that has evolved over this period. The involvement of these various financing sponsors and the complexity of the resulting marketing and pricing structures, loan guarantees, and respective liabilities between these entities often cloud the basic principles of project valuation. Nonetheless, project evaluators must carefully ascertain the relevant inflows and outflows which are attributed to each participant along with the pertinent liabilities and risks associated with potential project returns.

FORMS OF ORGANIZATION

The organizational structure under which a mineral deposit is developed and exploited will depend upon the size of the organization(s) involved, the method of financing the project, and, perhaps more importantly, on legal and tax considerations. Indeed, the tax consequences of each type of organizational entity play an important role in making a final decision on the appropriate organizational structure.

In the case of a domestic mining company planning to bring a new mine into production, the decision would probably be made to leave the property under the corporate umbrella, generally through the establishment of a wholly owned subsidiary to operate the property. Subsidiaries or operating divisions are typically established for liability, tax, and accounting reasons. Most significant operations are separately incorporated to provide greater operating flexibility.

All too often situations arise where the cost of bringing a mining property into production exceeds the capital resources available to a given corporation, small mining company, or group of individuals. Under these conditions, alternate forms of financing must be obtained in addition to establishing some organizational structure through which the property can be operated. This organizational entity should facilitate the financing arrangements and optimize after-tax returns to the participants.

Following are some examples of organizational entities, other than the corporation, which have been used for mineral property development. In many cases the organizational structure is, in effect, a reflection of the type of financing employed on the project. As a result, some discussion on financing is also included with the various types of organizational structures.

Partnerships

A partnership is the formalized relationship between two or more persons or corporations who join together to carry on a trade or business. Each partner contributes money, property, labor, or skills to the partnership and expects to receive in return a share of the profits and losses of the business. Generally, a partner's share of income, gain, loss, deductions, or credits is determined through the partnership agreement. This partnership agreement stipulates specifics regarding each partner's interest in the partnership, distribution of income, liabilities, capital requirements, salaries, and other rights and obligations.

For federal income tax purposes, the term partnership includes a syndicate, group, pool, joint venture, or other unincorporated organization which is carrying on a business and which may not be classified as a trust, estate, or corporation. A joint undertaking merely to share expenses is not a partnership (*1981 Master Tax Guide*, Commerce Clearing House).

A partnership is not a taxable entity even though it must file an informational return. A partnership determines its annual taxable income in much the same way as an individual determines net income. The annual partnership income or loss is then allocated to the partners based on each partner's distributive share. Each partner's distributive share of any income or loss is typically stipulated in the partnership agreement. This share of partnership gain or loss is included in each partner's separate income tax determination for the year. Therefore, only the individual partners are taxed on their individual distributive shares of the partnership taxable income, whether it is distributed to them or not. Partnership losses are also allocated on the basis of each partner's distributive share, although sharing ratios for profits and losses need not be the same. The partnership cannot protect one partner from liability incurred by another partner in the name of the firm. In general, for large capital projects, partnerships suffer some serious disadvantages in comparison to corporations:

1) Each partner is individually liable for the acts of every other partner made in the name of the firm. That is, each partner can act as an agent of the firm without specific authorization of the remaining partners.

2) Each partner has unlimited financial liability for the firm's business obligations. In practice, corporations often form thinly capitalized subsidiaries to hold partnership interests in an attempt to protect the parent from partnership liabilities.

3) Upon the death or withdrawal of any partner, the firm is immediately dissolved. This instability of life is a major shortcoming of the traditional partnership in raising debt capital.

Because of these serious liabilities in the partnership form, a number of related business organizations have evolved which retain many of the benefits of partnerships. One such variation of the partnership which became popular in the late

1960s for financing capital acquisitions is the *limited partnership*. Limited partnerships include one or more general partners as well as inactive limited partners who have no operating responsibility or authority. Under these agreements a limited partner is one whose personal liability for partnership debts is limited to the amounts which he or she contributed, or is required to contribute, to the partnership. Obviously these types of agreements are popular in more speculative, high risk ventures because the downside risk is limited while the sharing of profits is unconstrained. Limited partnerships have been used frequently for high-risk mineral exploration ventures as an incentive to investors who can classify their contributions as mineral exploration expenses for tax purposes. Limited partnerships have also been used in the financing and development of small-to-medium sized mining ventures with considerable success.

The *mining partnership* is another organization type that has existed for many years and is really a form of joint stock company. This organization provides limited liability to the partners and permits members to join or withdraw without dissolution of the firm. Finally, the venturers may elect to operate the property simply as *co-owners* (tenants in common) through an operating agreement as opposed to establishing a formal partnership agreement. Tax consequences for co-owners vis-á-vis the partnership are similar although some added flexibility may be gained through the partnership form (see Burke and Bowhay, 1982). Co-ownerships are described more fully in the next section.

One of the prime objectives in mining joint ventures is usually to avoid corporate tax status for the venture so that part of the project earnings will not be taxed three times—first at the project level, second as corporate dividend income (15% of which is taxable) to the parent, and third as dividend income to the individual shareholders of the parent corporation. The IRS has defined several tests which partnerships and joint ventures must meet to avoid being taxed as a corporation. These are described in the following section.

Joint Ventures (Co-Ownerships)

In mining, the term "joint venture" has come to mean the pooling of resources by two or more firms for the purpose of exploring for new ore bodies, developing or mining ore bodies, or expanding existing operations. Usually a single mineral property or producing region is involved, and the organization normally dissolves at the termination of the project. Limited scope and duration are usually desirable and are often keys to avoiding corporate tax status for the organization.

As will be discussed, joint ventures can operate through any one of a variety of business forms, but overall managerial control usually resides in a board of directors, or management committee composed of representatives from each venturer in proportion to that partner's ownership share in the organization. In most cases, one of the partners will act as the operator of the project.

For many reasons, joint ventures and partnerships have become increasingly common in the minerals industry. Some of the more important of these are as follows.

Capital Intensity: Mining has always been one of the most capital intensive industries, and capital costs in recent years have soared to breathtaking heights. Copper investment costs per annual ton of refined capacity have, for example, risen from about $3000 in 1972 to over $10,000 in 1982. New copper projects costing less than $100 million are rarely encountered today. In surface coal mining the capital cost has now reached the $20-$30 per ton of annual capacity range.

These costs have risen far more rapidly than has the earning power of the major producing firms. Consequently, the ability of many individual firms to take on major new mining projects unilaterally has been seriously impaired.

Technical Characteristics: As labor costs continue to spiral upward and ore grades sink gently downward, often the only path to profitable operation is to adopt (1) more sophisticated technology and (2) larger projects to benefit from increased economies of scale. Both of these alternatives add greatly to the industry's capital requirements, compounding the cost explosion cited previously.

Risk: Joint ventures are particularly attractive for large capital projects in high risk environments. New mineral ventures today often face serious risks of two types. One with which the mining community has considerable experience is geologic risk. Ore deposits being found today are often completely concealed and are deeper and lower grade than discoveries made in the past. Problems of grade continuity, ground control, and other geology-related factors are more difficult to analyze at depth. Even with the many technological advances in exploration and mining, the risks have never been greater, nor the finding costs higher.

The second risk category—political risk—has become more intense in recent years. Virtually everywhere, local governments are increasing their control and regulation of mining ventures. The ultimate risk—expropriation—is the most obvious threat, but uncertainty surrounding flourishing environmental regulations, mineral leasing constraints, and tax environments, has significantly raised the risk profile of mining projects even in the United States.

Joint ventures offer a very appealing hedge against risk. Because of the enormous capital investments involved in mining, even the largest firms become potential victims of gambler's ruin unless they share the associated risks with partners. The Alaskan North Slope oil development is an excellent case study of risk sharing through joint ventures for an extremely costly mineral development. The first efforts in ocean mining—a costly, new, high-risk business—also seem to be selecting the joint venture mode.

Acquisition of Expertise: Upon occasion firms having limited expertise in mining acquire control of a major mineral deposit and then take on a more experienced partner to develop and operate the mine. A good example of this was the Lakeshore mine, a copper operation in Arizona. El Paso Natural Gas, owner of the mining rights to the deposit, invited Hecla Mining Co., a firm with considerably more mining experience than El Paso, into a joint venture to develop and manage the property. Such arrangements are becoming common in coal and uranium production also.

Marketing Factors: A firm may form a joint venture to acquire a market for the final product particularly if demand for the product is fairly weak. When mineral supplies are tight, a similar phenomenon occurs, although now the initiating party is the consumer who may have experienced difficulty in obtaining adequate supplies at reasonable cost. During the tight copper market of the early 1970s General Electric and Essex Wire and Cable eagerly sought out possible joint ventures for just this reason.

Joint ventures may include many types of participants such as corporations, buyers of mineral commodities, suppliers, individuals, trading companies, and others. One of the most common type of joint venture in mining relates to exploration activities. Most of these arrangements are established such that one of the participants (venturers) serves as the operator and conducts the entire scope of exploration activities on the project. This type of joint venture is generally referred to as a horizontal arrangement and is typical of exploration joint ventures.

The classical example of exploration joint ventures is where an individual or group has acquired a land position which has promising mineral potential. This group often seeks to joint venture with a mining company for one or more of the following reasons:

1) To minimize risk by transferring part of it to the joint venturer(s).

2) To reduce the drain on financial resources and allow for other exploration activities.

3) To maximize exploration activities on a number of different properties.

4) To secure a market for the commodity through the joint venture.

5) To acquire needed financial resources.

6) To acquire needed technical expertise for the exploration program.

In general the joint venture (mining company) receives an operating interest in the property in exchange for performing some needed exploration efforts.

Obviously, if the exploration program is successful, the original property owner will be forced to share future rewards with the venturer who assumed part or all of the exploration risk. This brings up an important point in regard to exploration joint ventures. The venture agreement should always include specific arrangements regarding the development and operation of the project should the exploration program be successful. Potential problem areas which should be addressed in the agreement, in addition to specific economic provisions, are (Erdahl, 1975):

1) What criteria are to be used in determining when the property will be developed and put into production?

2) Who will be the operator and how will he be compensated for management services and home office costs?

3) What are the limitations of the operator's liability and authority?

4) When does the venture terminate and how may a venturer withdraw?

5) What act or failure to act constitutes a default under the agreement?

Joint ventures related to mining property development or operation may be either horizontal or vertical in nature. In a vertical agreement, more than one venturer contributes to the scope of the project. For example, one venturer may mine the ore, another may mill the ore, and yet another may smelt or refine the output. As might be imagined, vertical joint venture agreements can become extremely complex where multiple venturers are involved. The establishment of transfer prices, allowable costs, and priorities of processing must be addressed to the satisfaction of all venturers. Failure to do so will inevitably result in disputes and adversary positions among venturers.

Although joint ventures are popular in the mining industry at the present time, the project evaluator assessing these arrangements must first determine that there is a definite reason for such agreements. He must ascertain if, and how, any of these factors may affect project viability. In addition to the motives stated previously for joint ventures, there are some others which should be considered with respect to project economics. First, a joint venture may be the only way a company can generate the financial resources necessary to develop the property. Second, on the other end of the financial scale a property may be too small to support the significant overhead costs of a large company, whereas a joint venture with a small mining company acting as the operator could enhance project economics to both venturers. Third, some properties should be developed and mined in conjunction with a neighboring producing property in order to enhance project economics. An interesting example of such a property where the affected parties have yet to come to a satisfactory agreement is the open pit copper deposit near Tucson, AZ, separate parts of which are owned by Asarco, Cyprus Pima, and Anamax Mining Co. Any or all of these considerations should be carefully analyzed when formulating and structuring joint venture agreements and associated economic parameters.

From an evaluation standpoint, the tax effects of joint ventures can have a significant impact on project economics. In a sense, joint ventures come under the general classification of partnerships for federal income tax purposes. Under federal income tax law there is one primary distinction between partnerships and unincorporated joint ventures. The basic test is whether a separate entity has been formed to act as principal in the operation of the mine, or the operator is simply acting as an agent for the owners through a joint operating agreement and no separate entity is involved. In the case of a joint operating agreement between co-owners, the venture may elect to be excluded from all or part of the IRS provisions relating to partners and partnerships and elect not to file the partnership return. In order to make this election, the following requirements must be met (Burke and Bowhay, 1982):

1) The venture must be organized for the joint production, extraction, or use of property.

2) The participants retain ownership of the property as co-owners, in fee or under a lease or other form of contract granting exclusive operating rights.

3) Each participant reserves the right to take his share of production in kind, although he may delegate authority to sell his share of production for the time being for his account, but not for a period of time in excess of the minimum needs of the industry, and in no event for more than one year.

4) The organization does not have as one of its principal purposes cycling, manufacturing, or processing operations for others.

For as long as this election remains in effect, the joint venture would file information returns for income payments for each venturer. Under these conditions each venturer reports its share of income and expenses on its own tax return. In general most joint operations elect not to become subject to the partnership provisions because coordinated tax planning between co-owners is less rigorous, and less administrative expense is incurred.

If the joint venture chooses not to elect to be excluded from the IRS partnership provisions, then the venture follows the provisions established for partnerships. As such the partners may introduce some flexibility in tax planning by electing to allocate income and expenses in proportions which are different from their respective ownership interests. One point of caution is offered, however, with respect to taxation of the organization in question. It is possible for partnerships (also limited partnerships) to be treated as an association taxable as a corporation. Unincorporated organizations (partnerships and joint ventures) will be considered associations and taxed as corporations if (1) these organizations have associates and (2) they are organized to carry on business for profit and divide any gains. In addition, they must have a *majority* of the following corporate characteristics: (3) continuity of life, (4) centralization of management, (5) limited liability, and (6) free transferability of interests (Burke and Bowhay, 1982).

From the foregoing discussion, it is obvious that the business form adopted by the joint venturers is usually greatly influenced by two factors: taxation and financing. As joint ventures are particularly attractive for foreign investments, local conditions are also very important. Investment planners must carefully analyze such items as local tax codes, monetary exchange rates, and profit repatriation rules. Each of these may affect the type of joint venture established and the project's impact on the parents of the partners. Income taxation in the United States, as discussed previously, treats corporations quite differently from other business forms. The corporation possesses certain major advantages over other business forms in raising investment capital. The corporate form imposes definite tax disadvantages on most joint ventures but is adopted often by the venturers to limit their liability in high risk situations or to accommodate local requirements in foreign operations. Nevertheless, partnerships remain a very common type of joint venture in mining.

Working (Operating) and Carried Interests

Whenever two or more individuals or companies become involved in a mutual effort to explore, develop, or exploit a mineral property, certain stipulations are required between the participants regarding responsibilities, liabilities, income sharing, who will serve as operator and much more. Consequently the agreement should specify the terms of the operating interest in the property, any carried interest considerations, royalties, overriding royalties, and so forth. Depending upon how these agreements are structured, the financial and economic impacts on the project and the participants can be significant. As a result, it is important for the evaluator to understand the various terms included in these agreements and their tax implications.

The following discussion abstracts heavily from Burke and Bowhay (1982), an excellent reference on the subject. A working or operating interest in a mineral property is, in effect, the mineral interest minus any nonoperating interest. For federal tax purposes, an operating interest is an interest in minerals in place which is burdened with the cost of development and operation of the property. Because the operating interest must bear the burden of development and operating costs, this interest generally receives a disproportionately large share of the total benefits from the property. A typical example of such an arrangement is when a land owner sells an operating interest in his claims to a mining company for perhaps a fixed fee in addition to some royalty provision. In exchange the mining company agrees to explore the mineral property to some mutually agreeable level and subsequently to bring the property into production should the exploration program be successful. Thus, the working or operating interest is created, or carved out of the mineral interest, in most cases, by the granting of a lease.

In most cases the lease does not require the lessee (operating interest) to complete specific tasks such as core drilling, development of the property, etc. However, the lessee usually agrees that the lease will expire if a certain amount is not spent on exploration, assessment work, or property development, in a given time period. Ordinarily the lessee may abandon the property without incurring any penalty.

The lessee, as owner of the operating interest, may choose to leave the interest intact throughout the period of the lease, or he may decide to carve it up into other property interests. He may decide to choose the latter alternative if he needs additional funds to finance the project or if he wants to spread the risk inherent in developing the property. The lessee may carve up his operating interest in a number of ways. One method is to assign a production payment to another firm or organization in return for financing or for some specified work on the property. Another means of carving up his operating interest is through the sale of an *overriding royalty* for cash and then use these funds to develop the property. An overriding royalty is similar to a royalty for tax purposes in that each represents a right to minerals in place that entitles its owner to a specified fraction of production (in kind or value), and neither is burdened with the costs of development or operation. The primary difference between a standard royalty and an overriding royalty is that the latter is created from the operating interest in the property.

Whether the lessee chooses to carve up his operating interest through production payments or with overriding royalties, he has in effect retained the operating interest in the property and has the primary responsibility for development and subsequent operation. Also, since both of these interests (production payment and overriding royalty) are created from the working interest, they can not exist beyond the life of the working interest from which they were created.

It is possible for the lessee to assign his working interest in the property to another party and retain an overriding royalty or some other form of interest in the property. When the lessee assigns the working interest to another party and retains a fractional share of production free of development and operating costs, the overriding royalty is said to be retained. The shifting of development and operating responsibilities from the operating interest owner to an assignee is referred to as a *farm-out* in the extractive industries.

A *carried interest* is different from an operating interest in a mineral property in that a carried interest is an arrangement between two or more co-owners of a working interest, whereby one agrees to advance all or some part of the development costs on behalf of the others and to recover such advances from future production accruing to the other owners' share of the working interest. A carried interest may extend for full or partial development of the property.

Although a carried interest may come about through an agreement between the various co-owners of a working interest, it is generally created in connection with an assignment of a portion of the working interest agreements:

1) The owner of the operating interest may assign the entire working interest, or some fraction thereof, for a period of time measured by the recoupment of cost by the assignee, at which time a portion of the working interest assigned reverts to the assignor.

2) The owner of the operating interest may assign an undivided fraction thereof, together with a production payment, measured in terms of dollars of expenditure incurred by the assignee for the benefit of the assignor, and payable out of the assignor's retained fractional interest.

3) The owner of a fraction of the working interest may mortgage his interest to the other co-owners as security for a loan for which he is not personally liable.

Although each of these arrangements produces the same economic results with respect to division of income and expense, they each have a different tax consequence. In general the IRS considers any carried interest arrangement to amount to an assignment of the entire operating interest for a period of time with a reversionary interest in the assignor. Also, if the agreement provides that one party is to be carried for the entire life of the property, the carried interest is considered to resemble a net profits interest in the property. Differing tax treatments of these carried interest arrangements can be extremely important when evaluating mining properties involving several participants.

INTERNATIONAL VENTURES

The economic evaluation of international mining ventures is more complex than the evaluation of domestic projects. Not only must the analyst worry about the normal engineering and economic project variables, but he must also consider the unique financing arrangements which are often involved, the special arrangements which may have been made with the host country with respect to economic factors, and the various laws and regulations pertaining to mineral development in that country. Perhaps the most important factors affecting overall project economics of foreign ventures are the policies and regulations pertaining to currency exchange, taxation, and royalties.

Because of problems with balance of payments and for other economic reasons, some foreign governments place rigid constraints on money leaving the country. In general most developing countries prefer to retain as many US dollars in the country as possible. As a result, some foreign governments place limits on the amount of dollars that may leave the country. This not only creates a problem for corporations, but often for their employees as well. Some countries are now requiring that portions, often significant, of annual profits derived from mining activities be reinvested in the host country. This was a major source of funds for the Cuajone Copper project in Peru.

An important financial consideration with international mining ventures is fluctuating currency exchange rates. When sales transactions are based on one currency and production costs are incurred primarily in a second currency, severe economic dislocations can occur if there is a signficant variation in the relative values of the two currencies. For example, a Chilean copper producer may sell in terms of US dollars but pay for labor and materials largely in Chilean pesos. If, as occurred from 1982 to 1983, the peso is devalued by 29% relative to the US dollar *and there are no unit cost increases*, the mining firm benefits substantially from changing foreign exchange rates.

In practice, of course, the impact of exchange rate fluctuations is more difficult to determine. Generally, if a currency is depreciating, local unit costs will rise, tending to reduce the overall impact. Furthermore, a major part of the miner's purchases may be made in US dollars. Thus, the net impact of exchange rate movements is determined by (1) the proportion of production costs incurred in the local currency and (2) the ability of the mining company to resist unit cost increases in these items in the face of the depreciating currency.

Predicting exchange rates is often a futile exercise, but the host country's history of fiscal stability and its current financial condition may be important indicators. If a devaluation appears likely, a sensitivity analysis on this variable should be performed.

One of the prime considerations when evaluating international mining ventures is the impact of foreign taxes and their interrelationships with those of the operating company's home country. Many foreign taxes have different bases, rates, allowable credits, and deductions from those in the US. As a result it is difficult to directly compare the impact of foreign taxes with US taxes on project economics. However as a generalization, it appears that the taxes imposed by most mineral-producing foreign countries on mining operations exceed those imposed by the US. On the other hand, foreign governments—particularly in lesser developed countries—often offer individually negotiated tax concessions and are more amenable to providing broad tax incentives to mineral producers. These items have the effect of lowering the overall tax obligation on the venture.

These tax incentives are generally designed to achieve one or more of the following objectives:

1) Reduce unemployment.

2) Establish a revenue generating base.
3) Improve balance of payments.
4) Increase labor skill levels.
5) Develop the country's infrastructure.

Designed to attract investment capital, these tax incentives may take the form of tax holidays for some specified time period, liberalized depreciation and depletion rates, special tax rate reductions for preferential hiring agreements, and so forth. Some foreign governments have chosen to negotiate sophisticated royalty agreements, most of which have a profit-sharing provision of one type or another, in addition to some fixed financial commitment by the operating company. These complex agreements generally seek to maximize the recovery of the resource and provide for periodic renegotiation. Although foreign ownership of mines is still permitted in some countries, the trend is definitely in the other direction. Indigenous ownership, at least for a minority position, and negotiated tax and royalty terms are now the rule rather than the exception around the world, and there is hope that such partnership arrangements with the host country will enhance security and operating stability. Cynics will note that tax incentives have occasionally disappeared or been greatly diluted after the mining investment has been made. Nonetheless, if there is a community of interest between the mine and host government, there is reason to believe that greater stability will result.

The interrelationships between foreign and US taxes is an important one to consider when evaluating foreign ventures. In general, US mining companies can engage in foreign activities in one of several structures: (1) directly, either as a corporation or as an individual; (2) US subsidiary company; (3) US or foreign partnership; (4) nominee, cost, or farm-out company, or (5) foreign subsidiary company. Selecting the form of organization through which mining or exploration activities will be conducted is influenced by the tax laws of the host country. For instance, some foreign governments have a requirement that exploration or mining permits may be issued only to legal entities organized within that country. Where possible, however, most companies find it to their advantage to utilize a US entity such that foreign tax deductions may be claimed against US source income.

The applicability of various tax concepts peculiar to the natural resource industry and the impact of foreign taxes can have a significant effect on foreign project viability. Consideration must be given to the treatment of exploration, development, operating expenses, depletion allowances, foreign tax credits, limitations on credits, determination of foreign source income, and other factors during the evaluation procedure. The exploration and development of foreign mineral deposits can generate substantial tax deductions such as losses resulting from abandonment of properties, mine exploration expenditures up to $400,000, and mine development expenditures for US entities. These deductions can be offset against US income that would otherwise be taxable. However the Tax Reduction Act of 1975 has had the effect of reducing these deductions and the relative benefits therefrom.

A US operator is entitled to claim foreign income taxes paid or accrued as a credit when computing US taxable income. In order to claim the credit or deduction, the operator must show that the foreign taxes were imposed on the operating company, that it bears the economic burden of these taxes, and that the taxes are based on income. Legally a foreign tax qualifies as a legitimate US tax deduction only if it is within the US concept of an income tax, or is in lieu of an income tax. The amount of credit which can be claimed against US income tax is limited to the amount of US income tax liability attributable to foreign source income. The over-

Table 1. Reduction of Foreign Taxes on Mineral Income Under Section 901(e) of the Tax Reform Act of 1969

	Country			
	A	B	C	D
Foreign tax rate	37.4%	40%	48%	53%
1. Foreign mineral income ($1000 less depletion of $220)	780	780	780	780
2. US tax (46% of $780)	359	359	359	359
3. Foreign tax paid	374	400	480	550
4. Hypothetical US tax (46% of $1000)	460	460	460	460
5. Difference between foreign tax paid and US tax (line 3 less line 2)	15	41	121	191
6. Difference between hypothetical US tax and US tax (line 4 less line 2)	101	101	101	101
7. Reduction of foreign tax (smaller of line 5 or 6)	15	41	101	101

After Burke and Bowhay, 1982.

all limitation is computed by determining the ratio of total foreign source taxable income to the entire taxable income and applying it to the US income tax before credits (Burke and Bowhay, 1982). In other words, foreign tax credits cannot be used to reduce income taxes on US source income.

For purposes of foreign tax credit, the Tax Reform Act of 1969 reduces the amount of foreign tax eligible for US tax credit to the extent attributable to US percentage depletion. Foreign mineral income is considered to include income from the extraction, processing, transportation, distribution or sale of minerals, or of the primary products derived from minerals. The amount of foreign tax disallowed is the smaller of the difference, determined separately for each country, between the foreign tax on mineral income from sources in that country and the US tax with respect to its income, or the difference between the hypothetical US tax and the actual US tax (Burke and Bowhay, 1982). The hypothetical US tax on the foreign mineral income must be computed without percentage depletion; cost depletion, computed on the adjusted basis of the property, is to be used instead (US Treasury Regulation Section 1.901-3(a)(2)(iii)). Table 1 illustrates the calculation procedure associated with this reduction of foreign taxes on mineral income. This reduction comes into effect before applying the limitation mentioned previously with respect to US income tax attributable to foreign source income.

Another tax problem encountered in foreign operations is that associated with foreign government imposed export and import taxes. In many cases governments will impose extensive taxes on imports such as machinery, equipment, parts, and supplies in an attempt to develop domestic industries. Although this action is understandable where a country wants to provide a stimulus to its own manufacturing industries, it nevertheless generally has the effect of increasing operating costs because of higher prices, reduced availability, and in some cases inferior quality.

Also, many countries impose severe export taxes on mineral commodities leaving the country—particularly those mineral products not in final refined form. The objective of the tax is to encourage further processing of the raw material within the host country prior to export and sales. Again, this is an attempt by the country to build its processing and manufacturing industries and capture value-added revenues.

The tax concepts and factors involved in foreign mineral production are complex and can have a significant impact on project viability when viewed in light of US corporate tax laws. Also the form of organization under which foreign operations are conducted can be critical from a tax standpoint. These impacts, as well as problems which may be associated with currency exchange, must be carefully considered when foreign ventures are analyzed.

PROJECT FINANCING

As discussed in the first section of this chapter, the financing of mineral projects has progressed over time from total equity financing using internally generated cash flows to more sophisticated and complex multilayered financing agreements. These changes in financing have occurred for a number of reasons. Among these are:

1) Rapidly increasing capital requirements per unit of production.

2) Deteriorating grades of ore with associated requirements for increased production to maintain a given level of profitability.

3) Environmental restrictions and high-cost control mechanisms.

4) Tax policies.

5) Worldwide competition for scarce capital resources.

In addition, the rampant inflation of recent years has inflicted a heavy toll on the minerals industry. This fact, in conjunction with the relative stagnation of earnings in mining, has resulted in the inability of industry to generate sufficient funds to meet its spending needs. The resulting massive drain on capital has forced most mining companies to borrow heavily in the last decade. Although mining was once a relatively debt-free industry, the long-term debt of many mining companies today is approaching maximum prudent levels. As a result the liquidity and earning power of many such companies has deteriorated to the point where traditional financing techniques are no longer possible for new mining ventures.

The more popular forms of financing new mineral projects in recent years have been in association with commercial banks. One mechanism by which commercial banks have helped finance new mine development is through the technique of *production payments*. Production payment financing remains a popular form of financing in the US even though the 1969 Tax Reform Act requires carve-outs to be treated as loans for tax purposes (see Chapter 8). With production payment financing a mining company typically sells a stated amount of future production from new or existing mines to an arms-length third party (the buyer). The buyer pays the miner for this advance sale with the proceeds of a bank loan. The loan is usually provided in a series of installments which coincide with mine development schedules. Repayment of the loan plus interest is from a stipulated share of future mine income, usually known as the "dedicated percentage." The loan is fully repaid when the production payment is liquidated. Fig. 1 from Bispham (1982) illustrates the transactions for the production payment financing of a coal project.

With production payment financing the bank often secures itself with a mortgage on the production payment in addition to an assignment of the pledged income from production. This share of production income is also secured by a commitment from the mining company to develop and operate the mine on a continuous basis. Except for this commitment to continuously develop and operate the property, the mine owner often has no financial obligation for the debt incurred. Because the company has not directly assumed the debt but has simply sold future production, mining companies often carry production payment financing as de-

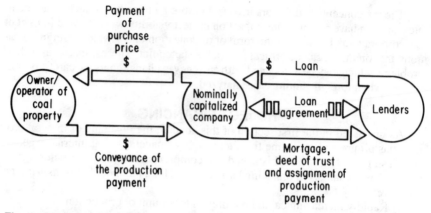

Fig. 1. Carved out production payment financing involves the conveyance of the production payment—a share of coal in place from the coal property—terminating when a sum from the sale of such coal has been realized and made payable out of a percentage of coal produced. The nominally capitalized company is formed solely to buy the production payment and usually borrows 100% of the purchase price to buy the production payment.

ferred income instead of long-term debt on their corporate books. The advantage, of course, in recognizing production payments only as deferred income on the corporate books is that it may not adversely affect the company's additional borrowing capacity. For tax purposes, the production payment is not taxable income to the miner, as it is viewed as a capital investment in the property. Similarly, expenditures for mine development funded from production payments are not deductible by the miner. However, revenue received for the dedicated percentage is excluded from taxable income to the miner, although costs incurred in producing these minerals are currently deductible. An example from Janson, et al. (1980), illustrates these transactions.

Example 1:

A, the owner of mineral property X, sells a $1 million production payment to B. A agrees to use the proceeds from the sale to develop mineral property X, and in fact does so, thereby creating a development carve-out. Under the terms of the agreement, B receives a percentage of the gross income from property X until he receives $1 million plus an interest equivalent. B has no recourse for payment other than mineral property X.

For tax purposes, A does not recognize taxable income when the production payment is sold to B, and B is deemed to have invested in a mineral interest (i.e., the production payment). Neither A nor B enjoys a tax deduction for the development financed with the proceeds from the sale of the production payment.

When production commences on property X, B receives his share of the gross income from mining and A excludes that amount from his taxable income. However, the cost incurred by A for B's share of production is currently deductible by A. B recognizes gross income from mining for his share of the payout from the property and deducts an allocable amount of his investment in the production payment, which constitutes cost depletion. (B is entitled to claim either cost or percentage depletion for the gross income from the property, but percentage depletion rarely exceeds cost depletion in these instances.) Thus, *B* realizes ordinary income for the interest equivalent of the production payment. To illustrate:

Assume	Total	A	B
Gross income from property *X*	$10,000,000	$8,600,000	$1,400,000*
Cost of mining	8,000,000	8,000,000	—
Taxable income before depletion	$ 2,000,000	$ 600,000	1,400,000
Depletion		300,000†	1,000,000‡
Taxable income		$ 300,000	$ 400,000

*Represents full payout of the production payment, including interest equivalent.
†Percentage depletion limited to 50% of net income.
‡Cost depletion on production payment.

In recent years commercial banks have been actively engaged in a form of financing referred to as *project financing*. Project financing has evolved—particularly at the international level—as the result of the previously mentioned changing economic and political climates throughout the world.

Pike and Thibodeau (1981) define project financing as:
"The provision of funds for a specific economic unit where the financing is secured principally by the assets of the unit with repayment derived from the unit's cash flow rather than the earnings of a parent company or group of sponsors."

Thus, the project cash flows and economics are segregated from the sponsors' other operations and evaluated on a stand-alone basis. Financing is generally not guaranteed by the sponsors or owners, once the design performance of the unit has been demonstrated. Under these conditions the credit worthiness of the sponsors becomes less important than the ultimate viability of the venture itself. Consequently, lenders associated with the project financing have found it necessary to develop skills needed to assess the technical merits of large mining projects. It should be noted that even though the sponsors of a project being financed via this mechanism would prefer the borrowing on a nonrecourse basis such that their balance sheets or credit standing is not affected, the lender will require some credit support from either a direct or indirect source. This support is usually obtained from the sponsors directly in the form of completion guarantees, but once the project has met its technical design criteria, the lenders generally have no recourse to the sponsors.

Because of the importance placed on the future viability of the particular project being considered for financing, it is obvious that the project must be supported with a complete feasibility study accompanied by detailed engineering and economic analyses. This comprehensive study, usually prepared by a highly regarded independent consultant, is generally known as the "bankable" document. The primary reason for these detailed studies is to provide the lenders an opportunity to assess the various engineering and economic alternatives, the risk of project development completion within the assumed time and cost forecasts, the reliability of projected annual cash flows, and the overall project viability. Creditors want to ensure that funding for the total project (including contingency) is available and committed at the start of the project, and that the projected annual cash flows are adequate to service the debt obligations associated with the loan.

A check of the debt or repayment capabilities of a project is one of the first steps normally undertaken by creditors. This so-called debt coverage ratio is simply the relationship between projected annual cash flows and the proposed debt servicing requirements (interest plus repayment of principal). The debt coverage ratio (DCR) is generally calculated on an annual basis as follows:

$$DCR = \frac{\text{Earnings before interest and taxes}}{\text{Interest} + \text{principal payments}/(i - t)}$$

The denominator reflects that interest is tax deductible, whereas principal payments are not.

The actual ratio required by the lending institution will depend upon a number of factors such as sales agreements, perceived accuracy of the feasibility study estimates, and presence of specific guarantees from sponsors. In general, however, because lending institutions are not willing to accept much risk, most commercial banks require DCRs for any given year to be in the range of 1.2 to 1.6.

One of the most essential ingredients in the structuring of any project financing arrangement is the debt to equity ratio for the project. From the lender's viewpoint, the larger the contribution of equity funds from the sponsors, the greater the loan security. Obviously the reverse is true from the viewpoint of the sponsors. The extent to which equity funds are provided is perceived to be one measure of the borrower's commitment and dedication to the project. From the lender's viewpoint, a minimal contribution in terms of equity funds provided by the project sponsors represents a potentially serious economic problem. Low equity inputs by sponsors in projects using nonrecourse financing have resulted in the lender being subjected to considerably more risk due to the absence of a sufficient equity cushion. The net result has been that bankers have been asked to provide investment capital to risky ventures for low returns. Because these returns are not considered to be commensurate with the risks involved, lenders have either required sponsors to support the project with some guarantees or they have placed their capital in industries and investments other than mining.

There is a direct correlation between the debt coverage ratio, the amount of leverage, and the lender's risk in a project. In general the amount of leverage a project can withstand depends upon (1) the ability of the project to meet its scheduled debt payments in a satisfactory manner and (2) the ability of the project to maintain the level of structural credit strength required by the lender (Ulatowski, 1977). If the lender perceives the degree of leverage to be extreme, the sponsors must structure the project such that the lender's perceived risk exposure is reduced. Perhaps the best way to accomplish this is to increase the equity base in the project. This base may be increased by the sponsors themselves or by interested third parties such as product buyers, users, or suppliers to the project. The equity base may be increased directly through sale of common stock or even through the issuance of subordinated debt, which has the effect of increasing support to senior debt and, therefore, reducing the relative risk level. However, it is obvious that subordinated debt cannot be used entirely in lieu of equity. In short, most lenders insist on a financial structure whereby the sponsors have a financial, as well as legal, responsibility to the project. Therefore some level of equity must be provided to the project. An excessive level of debt financing produces a high level of financial risk and the venture becomes unattractive to the lending institution. In general, the amount of financing available from lenders will vary between 100-400% of the equity contributions of the sponsors.

Under certain circumstances no equity may be required in a project-financed venture. Usually a necessary condition for such an arrangement is a "hell or high water" marketing arrangement such as a production payment facility or a "take or pay" contract (see p. 366).

As might be imagined, the drawdown and repayment schedules associated with international venture financing can be complicated indeed. Much depends on the host country's economic, social, and political track record; the mining companies involved; management responsibilities and capabilities; the organizational structure; the availability of local funds; location; and the host country's laws on foreign ownership and transfer of mining rights—to mention only a few considerations. The participation of other parties (e.g., mineral buyers, consumers, project suppliers, international financing institutions, governments) in these multilayered structures complicates the issues even more from an evaluator's standpoint.

Because project loans are not secured by the sponsors, the lender must give serious consideration to his exposure to a number of mining-related risks. Perhaps the more important risks facing the lenders are as follows (Rau, 1980):

Completion Risk: Timely completion of mine construction within the capital expenditure budget.

Reserve Risk: Sufficiency of the quality and quantity of mineral reserves necessary to produce the projected tonnage of minerals over the life of the project.

Production Risk: Ability to produce the projected tonnage of minerals annually, while maintaining projected levels of operating profit.

Marketing Risk: Ability to sell output at prices that will maintain profit margins. If sold under long-term contracts, the strength and geographical diversity of purchasers, while contracts contain appropriate force majeure provisions.

Political and Exchange Risks: The possibility of expropriation, nationalization, business interruptions, tax changes, and insurrection. Also the possible inconvertibility of local "soft" currency into "hard" currency necessary to repay the financing.

When project financing is obtained for a venture, the loan is typically drawn down throughout the preproduction period in accordance with some predetermined development schedule. The normal time period required for this preproduction activity is in the range of 3-5 years but can vary considerably. Once the project comes on-stream, the loan is to be repaid from annual cash flows over some time period. Herein lies another problem for the lending institution. Commercial banks prefer not to provide loans with tenors in excess of 8 to 12 years from the time of commitment. Furthermore, the economic life of the unit should be at least the length of the term of the loan. Although in some cases the terms can be slightly longer, the 8 to 12 year time is consistent with the short- to intermediate-term nature of bank deposits. Given the normal range of preproduction periods associated with new mine developments, there is only limited time available for debt amortization, and the resulting debt repayment schedules can become quite burdensome on the project. As a result, the sponsors must carefully consider when project financing should be acquired to ensure that the project can fulfill the obligation of the loan.

Depending on the capital structure of the venture and the level of risk perceived by the lending institution, the sponsors may be required to provide some additional support to the venture before funding can be obtained. This support may take many forms, but typically it is related to specific commitments or guarantees on the part of the sponsors. The following are examples of the types of support a lending institution may require.

Completion Guarantee

One of the major areas of concern for a lending institution financing a new venture is the risk of capital cost overruns beyond the amount of capital available to complete the project. With capital cost overruns the lender is faced with either the prospect of project noncompletion due to insufficient capital funding, or the strong probability of deteriorating project economics overall due to the escalated capital requirements. This, in turn, jeopardizes the anticipated debt coverage ratio, and the overall project risk increases.

To protect against capital cost overruns and/or unsatisfactory operating performance, the lender often requires the sponsors to undertake a completion guarantee whereby the sponsors agree to fund all capital cost overruns either through equity or subordinated debt and to ensure completion of the project. The sponsors are released from this guarantee only upon satisfaction of some predetermined completion criteria. Typically the completion stipulations require that the project produce or operate continuously for some time period (six months to one year), produce a specified quantity of output (e.g., 80-90% of design capacity) at or below expected cost and efficiency levels; output must meet quality requirements and possibly cash flow requirements or constraints must be satisfied.

Operating Deficiency Agreement

After the completion agreement is satisfied, the lender's recourse is limited to the assets of the venture itself and its future performance. If there appear to be considerable operating risks, the lender may require the sponsors to enter into an operating deficiency agreement. Under such an agreement, the sponsors are obligated to maintain a specified level of production. Even though such an arrangement does not guarantee repayment to the lender, it does ensure that the sponsors will continue with the project and output will result.

Sales Contracts

Lenders often require long-term sales contracts. Lenders like to have sales prices set at a level which allows for the project to cover operating and debt servicing costs even under the worst circumstances. They prefer the sales agreement to run for some time beyond final debt repayment and to provide payment for the product as soon as possible (f.o.b. mine). Lenders prefer to have sales contracts with strong and diverse purchasers who will agree to escalation provisions or periodic renegotiation of prices.

"Take or pay" sales contracts are most popular with lenders. In these agreements, the purchaser has a noncancellable obligation to pay for a fixed amount of product irrespective of whether or not the purchaser takes delivery of this entire quantity.

Marketing Guarantee:

Lenders may require the sponsors to purchase sufficient unsold output such that annual tonnage or unit sales reach some minimum level. Even though these agreements normally set the sales or purchase price at prevailing market rates, the lender is still responsible for commodity price risk. These agreements are often associated with long-term sales contracts.

Working Capital Maintenance

Lenders may require sponsors to maintain a specified level of working capital. Should working capital fall below this level, the sponsors are required to contribute funds to the project to bring working capital to the required level. By definition, working capital includes all debt falling due within 12 months. Consequently this type of agreement essentially guarantees the lender payment of project debt for the duration of the working capital maintenance agreement.

Mortgages and Other Security

Creditors may obtain mortgages or liens on readily salable assets of the venture, including improvements, land, mineral titles or rights, ore reserves, and mining rights. Most lenders realize, however, that these forms of security are of questionable value because the real value of a mineral venture lies in the future cash flows generated from mining. Salvage values for mining ventures are notoriously low. As a result lenders often secure their position by placing controls on project cash flows. For example, they may require long-term sales contracts as discussed previously, or they may reserve the right to control sales proceeds in order to preserve the overall security of their loan. Controls on cash flows are much more effective than mortgages on ore reserves or mining rights, particularly in foreign ventures. Many foreign governments consider ore reserves, mineral titles, and mining rights to be the property of the state. Under these conditions the lender has virtually no loan security other than that associated with project assets and improvements.

Project financing offers a number of advantages over conventional methods. Perhaps the most significant virtue is the fact that project financing can be custom-tailored to individual projects. There are few set rules or regulations regarding specific structuring. Loan disbursements and repayment schedules can be directly correlated with each specific project. Disbursement and repayment periods associated with project financing are generally longer than those with conventional financing, particularly when additional project support through the previously mentioned agreements is provided by the sponsors. Disbursements normally range from one to four years during the preproduction or development period and repayment periods may be as much as 15 years, although the average is closer to 10. Repayments may be structured in the form of an annuity, in increasing amount over time, or tailored to the project's anticipated cash flow profile.

As mentioned earlier in this chapter mining companies have recently experienced significant increases in debt in their capital structures. In some cases the magnitude of debt acquired has resulted in certain limitations and restrictions being associated with additional borrowings. These limitations generally relate to the acquisition of additional debt by the company and reflect concern about the financial structure and future of the firm as perceived by existing creditors. Under these conditions, mining projects—particularly joint ventures—may be financed through project financing even though the individual company(s) might not qualify for additional debt financing. Furthermore, because project loans are generally not financial obligations of the sponsors, the balance sheets of the sponsors are not encumbered by new debt. However, a footnoted obligation will probably be required, depending on the level of financial risk borne indirectly by the sponsors through various covenants.

In general, project financing provides a reduced burden on project sponsors through reduced equity contributions and reduced liability. Indeed, the liability of mining project sponsors may, in some cases, be virtually eliminated with the use of project financing. Project financing may be structured so that there is some support to the lender over the life of the financing or, at the other extreme, the arrangement may be entirely nonrecourse in nature. However, bankers are not risk takers and even nonrecourse loans generally require at least completion guarantees so that the sponsors bear the risk until the project is operating as designed.

Although project financing offers some important advantages, there is no free lunch in mining project finance. The primary risks associated with marketing, long lead times, and cost overruns do not disappear just because the project is financed differently. The unique aspect of project financing is that these risks can be shared to a certain extent, and the project can be financed on its own merits. As with most any financial venture, expected returns are generally commensurate with perceived risk levels. The sharing of mining project risks through financing mechanisms also implies the sharing of project returns. Lending institutions fully expect to share in project returns in proportion to the risks they share or acquire. If these returns cannot be achieved, investment capital will be placed elsewhere and mineral development throughout the world will stagnate.

OTHER TYPES OF ECONOMIC STUDIES

The focus of this book is on the financial analysis of projects for the purpose of capital investment decision-making. We have advocated methods which compare the cash derived from a project with cash invested in the project, suitably adjusted for timing by the use of an appropriate discount rate. There are, however, other procedures that are sometimes used to study the economic viability of a project. Some are valid under limited circumstances, but they generally do not stem from a sound theoretical base which relates the investment decision to the objective of the firm. In general, these procedures arise from accounting concepts and are not compatible with discounted cash flow investment decision-making. Some of these procedures involve the assignment of the cost of specific capital sources to specific projects and, therefore, result in unending confusion in discounted cash flow investment analyses where—as demonstrated in Chapter 11—such assignments are incorrect. A few situations where the analyst might encounter these accounting-based analyses are listed in the following.

The Representative Year

One procedure that is often used early in the evaluation process is to prepare a pro forma income statement for a "typical" year to get a general idea of the project's chances for success. Frequently, the statement is based upon cost per unit of product (e.g., cents per pound of copper or dollars per ton of coal, etc.). Revenues and costs are estimated, and the statement is prepared in a conventional manner with two important differences. First, because average values are sought, straight-line depreciation and average costs are used, and detailed income tax algorithms are avoided. Second, as the resulting average net annual benefit is not discounted, it is appropriate to charge costs of capital directly to the project. Unlike the discounted cash flow situation described in Chapter 11, double-counting of capital costs is not a problem here.

Unfortunately, in representative year statements only out-of-pocket costs are usually considered, and opportunity costs are ignored. That is, costs of debt are included but costs of equity are not, so that it is impossible to conclude whether or not the residual earnings are satisfactory. Furthermore, it is important to remember that, regardless of how profitable or unprofitable a project is in a given year, generally no conclusions can be drawn with respect to its DCFROI until results from several years are available. Single-year income statements and rate of return on investment are different concepts and are not directly comparable.

Cash Budgets for Lenders

Any major creditor will likely demand a pro forma cash budget for the project so that it can be determined whether or not the risk associated with the loan is within acceptable limits. The recent rapid growth in project financing has expanded the need for such analyses. The lender may be particularly interested in the debt servicing ability—the debt coverage ratio—of the project. That is, what is the magnitude of the project's cash flow before interest and income taxes in relation to annual interest and principal obligations?

Cash budgets, of course, include all cash transactions and, therefore, interest payments are once again a legitimate deduction. As a consequence, cash budgets are not directly usable for discounted cash flow calculations.

Joint Ventures

Joint ventures may have debt obligations which are *not* secured by any of the individual partners. In such a case, the net benefits to each partner are determined after the deduction of all costs—including interest and principal payments if any. That is, the venture must dispose of internal obligations before distributing any remaining cash to its owners. In any event, every partner should compute its rate of return upon its total investment, regardless of the source of funds.

It cannot be emphasized too strongly that, although the preceding types of economy calculations may be interesting and useful in specific situations, they are not compatible with investment decision-making procedures described in earlier chapters. This book consistently evaluates capital projects in terms of increasing shareholder wealth.

EVALUATING SPECIAL MINING PROBLEMS

Some of the special problems of mining highlighted in this chapter have centered on organizational forms and associated taxation implications, foreign and international ventures, and the financing of mineral ventures. Although the complexities inherent in these various arrangements often confuse the issue considerably, the principles involved in evaluating these unique situations are basically the same as for any other type of capital investment decision. The analyst must simply remember to determine the magnitude and timing of the actual investment, operating costs, and revenues expected from the project for each particular firm. The actual procedure for evaluating the investment decision remains unchanged. For example, when evaluating joint ventures each partner must determine separately whether its share of the investment will provide a satisfactory rate of return. More specifically, each partner must define its investment and annual net benefits over the life of the project. The internal rate of return, NPV, or wealth growth rate calculation can then be performed routinely.

If the joint venture is taxed as a partnership, the annual net benefits will be determined from each partner's distributive share of the project's revenues and costs and his specific tax circumstances. No income taxation is imposed at the project level, and all income and costs flow through to the individual venturers. If the joint venture is taxed as a corporation, the venturers receive their annual benefits in the form of after-tax dividends, 15% of which are taxed again as income to the parent corporation.

It is important to note that in both of the foregoing cases "investment" means total investment, debt plus equity. If one of the participants invests, say, $4 million in cash plus $20 million of bank borrowings in the project, his investment—as discussed at length earlier in this chapter—is the entire $24 million.

One potential source of confusion can arise here in regard to project financing. If debt capital used in a project is truly nonrecourse, being secured *only* by the earning power of the project itself, such capital should not appear in the investment account of any of the venturers. Furthermore, the interest and principal paid on this debt is a legitimate deduction from project earnings in arriving at net benefits distributed to participants for their individual rate of return calculations. In this case there is no double-counting of the costs of capital as project capital is accounted for at the project level, whereas each venturer's cost of capital is handled through the discounting process when each determines its rate of return on invested capital. The following example has been constructed to illustrate these concepts.

Example 2

Bung Mines, Ltd., a 50/50 joint venture of Apex Minerals and Zap Uranium has been formed to develop and operate a uranium mine in Wyoming.

Production and cost estimates for the project are as follows:

Bung Mines Production Statistics
Ore reserves: 3.5 million tons @ 0.149% U_3O_8
Mining rate: 1000 tpd ore; 10-year life
Stripping ratio: 20:1
Estimated product price: $35 per lb U_3O_8
Mill recovery: 96%
Percentage depletion rate: 22%

Capital costs	*Operating costs*
Mine, $8.0 million	Mining, $15 per t ore
Mill, $15.0 million	Milling, $13.50 per t ore
Support facilities, $13.0 million	General and admin., $10 per t ore
	Transportation to market, $3 per t ore

A) 50/50 joint venture, Apex and Zap
  Taxed as partnership

Joint Venture results:	($000)
Sales revenue	35,045
Costs:	
Transportation	1,050
Mining	5,250
Milling	4,725
General and admin.	3,500
Net oper. income	20,520
Depreciation	3,600
Other fixed charges	1,800
Depletion	7,479
Pretax net income	7,641

Apex and Zap then each include 50% of these revenues and costs in their individual corporate accounts and would calculate their after-tax returns on the $18 million which each had invested. In partnerships, tax elections (e.g., depreciation method) are made by the partnership. For simplicity, straight-line depreciation was assumed here. In joint ventures taxed neither as partnerships nor corporations, elections are made separately by the partners.

Assuming a marginal tax income of 50%, each partner would have post-tax annual cash flows of [0.50 (7641) + 3600 + 7479] ÷ 2 = 7450, yielding a DCFROI of roughly 40%.

B) 50/50 joint venture, Apex and Zap
Taxed as corporation.
Combined income tax rate = 50%
$6 million project loan, 9% interest, payable in equal annual installments

Corporation results (operating year 6)

Sales revenue	35,045
Costs:	
Transportation	1,050
Mining	5,250
Milling	4,725
General and administration	3,500
Net operating income	20,520
Depreciation	3,600
Other fixed charges	1,800
Interest charges	270
Depletion	7,425
Pretax net income	7,425
Income taxes	3,713
Net profit	3,712
Depreciation	3,600
Depletion	7,425
Net cash flow	14,737
Debt repayment	600
Distributable cash flow	14,137

In this case, net cash benefits to each parent for year 6 would be

$\frac{14,137}{2} \times [0.85 + 0.15(.5)] = \$6538$. Investment by both Apex and Zap

is now $\frac{36 - 6}{2} = \$15$ million.

If we make the simplifying assumption of constant annual cash flows, the resulting DCFROI is approximately 42%. It is interesting to note that in this example the partial double taxation of dividends is more than offset by the positive leveraging impact of the project loan, so that both venturers' returns are increased in the corporate alternative. Obviously, different assumptions might yield different results so that no generalized statement can be made here regarding the advantages of corporations or partnerships.

SUMMARY

One of the fascinations of capital investment analysis in mining is the unique problems which every project presents. Some of these unique features have been introduced and briefly described in this chapter. Many of these concepts are rich areas for further study and research, and the accompanying bibliography provides some guidance to the reader for additional study.

Perhaps the most important point emphasized in this chapter is that regardless of the complexity of the venture, the capital investment decision is still based on a comparison of the costs incurred and benefits enjoyed by the investing organization. The relevant costs include all investment—both debt and equity—but exclude obligations that are the sole responsibility of the specific venture (e.g., nonrecourse loans). The relevant benefits include the investor's share of net after-tax cash flows received from the venture, usually on a fully repatriated basis.

For large joint ventures having multinational ownership and multilayer financing the evaluation can be a lengthy and costly activity. In fact, one of the dangers of such evaluations is that the high level of effort required may divert management's attention from more fundamentally important technical questions pertaining to the venture. It is easy to become so engrossed in the financial analysis that the dubious nature of some of the data becomes obscured in a stack of computer printouts. Therefore, in capital investment analysis, as in all quantitative disciplines, there is no substitute for careful, methodical, documented progress toward a well-defined goal.

SELECTED BIBLIOGRAPHY

Bailey, C.C., and Boericke, W.F., 1953, "Financing Domestic Mining Ventures, Part I," *Mining Engineering*, Vol. 5, No. 6, June, pp. 584-588.

Bailey, C.C., and Boericke, W.F., 1953, "Financing Domestic Mining Ventures, Part II," *Mining Engineering*, Vol. 5, No. 7, July, pp. 679-684.

Bispham, T.P., 1982, "Coal Mining Project Financings: Lenders' and Borrowers' Considerations," Engineering Bulletin, Dames & Moore, April, pp. 27-32.

Burke, F.M., and Bowhay, R.W., 1982, *Income Taxation of Natural Resources*, Prentice-Hall, Inc., Englewood Cliffs, NJ.

Castle, G.R., 1975, "Project Financing—Guidelines for the Commercial Banker," *Journal of Commercial Bank Lending*, April, pp, 14-30.

Commerce Clearing House, Inc., 1980, *1981 Master Tax Guide*, Chicago, IL.

Erdahl, L.O., 1974, "Economic Aspects of Joint Ventures," *Mining Congress Journal*, Vol. 60, No. 6, June, pp. 24-26.

Erdahl, L.O., 1975, "Economic Aspects of Joint Ventures," *Mining Engineering*, Vol. 27, No. 9, September, pp. 33-35.

Frohling, E.S., and McGeorge, R.M., 1975, "How Stepwise Financing Can Turn Your Prospect into an Operating Mine," *Mining Engineering*, Vol. 27, No. 9, September, pp. 30-32.

Hardin, D.C., 1978, "Some Problem Areas and Suggestions in Financing the Next Decade of Mineral Expansion," remarks at Technical Forum, Outlook for Mineral Supplies and the US Mineral Industries, Nonfuel Minerals Policy Review, Denver, CO, September 19.

Haworth, G.R., 1977, "How Will Major New Mines Be Financed in the Future?," SME-AIME Preprint 77-K-125, AIME Annual Meeting, Atlanta, GA.

Internal Revenue Service, 1980, *Tax Guide for Small Business*, Publication 334.

Janson, E.C., Knup, S.P., and Wright, D.T., 1980, *Financial Reporting and Tax Practices in Nonferrous Mining*, 9th ed., Coopers & Lybrand, New York, 136 pp.

Lewis, M.F., and Bhappu, R.B., 1975, "Evaluating Mineral Ventures via Feasibility Studies," *Mining Engineering*, Vol. 27, No. 10, October, pp. 48-54.

Lindley, A.H., et al., 1975, "Mineral Financing," Chap. 3.11, *Economics of the Mineral Industries*, 3rd ed., W.A. Vogely, ed., AIME, New York, pp. 420-432.

Maxfield, P., 1975, *The Income Taxation of Mining Operations*, 2nd ed., Rocky Mountain Mineral Law Foundation, Boulder, CO.

Nevitt, P.K., 1978, *Project Financing*, 2nd ed., Vol. 1, AMR International, Inc., 319 pp.

Nevitt, P.K., 1978, *Project Financing*, 2nd ed., Vol. 2, AMR International, Inc.

O'Neil, T.J., 1980, "Inflation and the Capital Investment Decision in Mining," Short Course Notes, University of Arizona, Tucson, December.

Pike, R.E., and Thibodeau, J.T., 1981, "The Role of Banks in Mining Projects," *Mining Magazine*, October, pp. 285-288.

Quirin, D.G., 1967, *The Capital Expenditure Decision*, Richard D. Irwin, Inc.

Rau, W.I., 1980, "Project Financing for International Mining Ventures," *Mining Engineering*, Vol. 32, No. 8, August, pp. 1262-1264.

Ulatowski, T., and Frohling, E.S., 1977, "Delaying Debt During Early Development: A New Approach to Mine Financing," *Engineering & Mining Journal*, Vol. 178, No. 5, May, pp. 65-70.

Ulatowski, T., 1977, "Financing of Mining Ventures," *Mineral Industry Costs*, J.R. Hoskins and W.R. Green, eds., Northwest Mining Association, Spokane, WA, pp. 179-188.

Ulatowski, T., 1980, "Importance of Financing in Project Planning," *Mining Engineering*, Vol. 32, No. 6, June, pp. 688-692.

13

Accounting for Risk in Mining Investments*

"October. This is one of the peculiarly dangerous months to speculate in stocks. The others are July, January, September, April, November, May, March, June, December, August and February."

—Mark Twain

Although the preceding chapters describe the type of preliminary evaluation frequently performed in the mining industry, it has been assumed that input data are known with certainty—clearly an enormous simplification. In reality, estimates of ore grade, mining cost, metal price, etc., are subject to varying degrees of uncertainty due to the inability to predict the future with much precision. Management would probably not, for example, be equally attracted to two projects each offering a 20% DCFROI—one for a new gold venture in Nevada and the other for an undersea manganese nodule operation in the Pacific. The high risk with the latter investment tells the obvious; there is more to investment decision-making than a deterministic analysis using "best estimates" of input parameters.

The science of decision theory normally distinguishes between decisions under uncertainty, where probabilities of various outcomes are unknown; and decisions under risk, where such probabilities can be estimated. The first case is rarely encountered in practical capital investment analysis in mining, so the remainder of this chapter focuses on that class of problems where rational probability distributions for the future states of input variables can be estimated.

RISK IN MINING

Risk in the context of this chapter is the unforeseen deviation of individual cash flows from expected values for a capital project. For a mining venture, the source of this uncertainty could be any number of factors relating to such items as ore grade, ore reserve tonnage, operating costs, product prices, etc. With conventional

*Adapted from O'Neil, T.J., 1979, "Procedures for the Preliminary Financial Evaluation of Metal Mining Ventures," *Computer Methods for the 80's*, A. Weiss, ed., AIME, New York, pp. 556-573.

deterministic evaluations a point estimate of each of these factors is made. Subsequent operating results usually reveal these estimates to be in error, thus giving rise to different cash flows than expected.

Demonstrating the existence of risk is one thing; measuring it is quite another. Many people feel comfortable with point estimates of all evaluation parameters but realize that no parameter value is known with certainty. By analyzing uncertainty they see a bewildering task of evaluating an infinite number of possible outcomes using some decidedly speculative guesses about the future—a costly, time-consuming exercise which inevitably requires some glaring simplifying assumptions. Nonetheless, uncertainty in the investment decision should be handled explicitly; one should not surrender to the complexity of the problem and be satisfied with a rule of thumb "ignorance factor" as protection against risk. Simplifying assumptions, naturally, are required, but by quantifying the distribution of possible outcomes with the best available information, a much clearer picture of overall project risk results. The risk still exists, but its sources and dimensions are identified. This could indicate, among other factors, where additional expenditures in the evaluation stage can reduce the magnitude of uncertainty.

Mining ventures have always been well-known—perhaps, notorious—as risky investments. This reputation is well-deserved due not only to the inherent technologic and geologic risks encountered with mines, but to the many highly speculative mining issues traded on every stock exchange as well. Whereas the smaller investor may be more concerned with this latter risk, this chapter is directed toward methodologies available to analyze the former. Thus, the issue here is how management can best cope with the inherent uncertainty involved in capital investments in mining.

The statement that mining is risky business seems to be well-accepted by nearly everyone. Miners often emphasize the point in defending the minerals depletion allowance, and since no one can really see through rock, the argument that mining is a high-risk enterprise has considerable intuitive appeal. Examples abound with political risk (expropriation), geologic risk (ore continuity, mineralogy), hydrologic risk (mine flooding), and technologic risk (equipment and system failures). Huge financial losses can be cited for specific projects when unanticipated difficulties arose.

CHARACTERISTICS OF MINING RISK

Two basic types of risk occur for mining companies: financial risk and business risk. Financial risk refers to the uncertainty in earnings which the firm voluntarily accepts due to its capital structure. Financially risky firms employ a relatively large amount of debt financing and are, therefore, highly leveraged. Relatively small changes in revenue are amplified in net income due to the fixed charges for debt servicing. As a consequence net earnings rise (and drop) faster than gross income. Management is, of course, interested in the firm's financial risk, but this topic is treated adequately in the financial literature.

The analysis of business risk is the focus of this chapter. Business risk is defined by the type of business a firm is engaged in and is reflected in the fluctuation of net earnings before interest charges and income taxation. Examples of firms with low business risk are public utilities for which sales and costs can be projected into the future with considerable accuracy. This permits utilities to assume rather high financial risk through extensive debt financing without seriously jeopardizing expected earnings to common stockholders. Recently, however, even the electric utility industry has suffered losses due to high business risk resulting from trouble-plagued nuclear power plants.

The reverse is true with most mining companies where relatively high business risk has not encouraged highly leveraged capital structures. In fact, low levels of debt financing have historically been a good indication that mining is, indeed, characterized by a high degree of business risk. Recent large increases in debt financing by many mining companies may seem to contradict this statement. However, such borrowing has often been for mandated investment in pollution control equipment which has resulted [when compounded by poor markets (business risk)] in operating losses for many firms.

The methods available for risk analysis are not restricted to any particular industry. Nonetheless, the mining industry does bear unusually severe risks in two areas. First the industry is extremely capital intensive. Among the industrial groupings analyzed in *Fortune* magazine's annual listing of the 500 largest industrial firms, mining usually ranks near the top in assets per employee and near the bottom in terms of annual sales per dollar of invested capital. Thus, with much more at stake in the event of an unfavorable outcome, mining faces greater risk than most other industries.

The second characteristic of mining which significantly affects the industry's risk profile is the inherent long lead time of new mining investments. A five to eight year preproduction period—or even longer—is often required in mining before a positive cash flow commences. Because uncertainty is directly related to time, mining investments, where cash inflows are delayed for several years, are decidedly risky. For example if new copper mining facilities were under development in 1956 based on the then current price of only $0.418 per lb, resulting income in 1958 would have reflected a copper price of only $0.258 per lb, 40% lower than anticipated. This is not an academic example, for the early years of the huge San Manuel mine were plagued by unexpectedly low revenues brought on by depressed copper prices—a classical example of a risky parameter.

ACCOUNTING FOR RISK

Many different attitudes and philosophies have developed over time regarding uncertainty in the new mine investment decision. Most mining firms have established procedures for evaluating capital projects with uncertainty built into the system in some manner that reflects their experience in mining projects. No system (including the methods suggested in this chapter) can be substituted for good managerial judgment. Factors such as competitive strategies, available human resources, social goals of the firm, and a myriad of other nonquantitative factors must be considered in a major capital investment decision. There is no substitute for good judgment in any capital investment decision, but this does not discharge the decision maker from his responsibility to minimize uncertainty by quantifying risk whenever it is economically feasible to do so.

A firm can take uncertainty into account in many ways. Some of the more common traditional methods are discussed in the following.

Risk-Adjusted Payout Period

One method used to compensate for risk is to demand shorter payout periods for riskier projects. Aside from the shortcomings of the payout period as a capital investment criterion, the technique encounters additional problems here. Payout period does not measure risk directly; in fact, it is a rather poor measure of profitability. Risk in the early years of a project (when cash inflows are most important) is not affected by the payout period. Risk is considered only by ignoring cash flows after the payout period. By adjusting the maximum acceptable payout period, risk may be resolved sooner but will not be diminished.

Difficulty arises with risk-adjusted payout periods as to how much adjustment to make for a particular project. One might agree that the maximum acceptable payout period for a large new mine in a politically turbulent part of the world should be less than for a similar investment in the United States. The question is, of course, "how much less?" for which there is no valid, objective answer. Subjective decisions are required and considerable arbitrariness can creep into the decision. The decision-maker can select any course of action he or she pleases simply by assigning prohibitive maximum acceptable payout periods for all other alternatives.

As supplementary information, the payout period is useful in studying the uncertainty in an investment project. Risk-adjusted payout period, however, is a poor primary method for accounting for risk in capital projects.

Risk-Adjusted Discount Rate

It is very common for firms to require risky projects to have higher rates of return on investment than safer ones. For example, a firm might establish three classes of capital projects and associated rates of returns as follows:

Class 1: Replacement of equipment in an ongoing operation. Market is known, technology is proven, so that risk is fairly low. Maybe a discount rate of 10% is acceptable here.

Class 2: Expansion of present mine or plant facilities. Many technical problems have already been solved, but there may be a question in marketing. Can the additional output be sold at preexpansion prices? The added risk here may indicate that a higher discount rate is in order, say, 15%.

Class 3: Opening of a new operation, entering a new market, etc. Here the sources of uncertainty are many, justifying, perhaps, 20% as the minimum acceptable rate of return on such projects.

The magnitudes of the discount rates used in the example are not important; only the concept of risk-adjusted discount rates is of concern.

The major drawback of this method is the same as with risk-adjusted payout period—subjective establishment of hurdle values. There is no method for assigning acceptable risk-adjusted rates of return to individual projects in an objective manner, so that inconsistency is inevitable. Nonetheless, the method is easy to apply and is not likely to disappear soon.

Risk-Adjusted Input Parameters

When considering the extensive uncertainty encountered in evaluating new copper mining ventures, industry executives often compensate for this risk by using very low prices for copper. Similarly, conservative values for other input parameters such as mining costs, ore grade, etc., during the evaluation stage can screen out all but the best projects. Of course, by being excessively conservative, all projects can be rejected, and this is a real danger with risk-adjusted input parameters. When more than one variable is adjusted, compounded conservatism can easily reject nearly any project. Also, subjectivity in quantifying the risk-adjusted variables causes the same type of problems as in "Risk-Adjusted Payout Period" and "Risk-Adjusted Discount Rate."

In capital intensive mining projects where investments exceeding $100 million are common, treating uncertainty exclusively with one of the arbitrary methods previously described is, in general, unsatisfactory. More quantitative data relevant to the sources and magnitude of risk can be developed and with such large sums at stake, additional analysis is certainly justified. Rather than using "ignorance factors" to discount risk, an attempt should be made to identify the sources and estimate the magnitude of this risk. The decision maker may be a plunger, wishing to accept a high degree of risk if the reward is great enough; or, more likely, a risk-averter, selecting a project with relatively low risk even if expected profits are also low. By quantifying risk in this manner, it is possible to estimate the chance of achieving a given level of profitability with a specific project. Management then has much more information at their disposal, so that they can make an intelligent trade-off between risk and expected profits.

Stochastic Risk Analysis Models—The general method by which the level and magnitude of risk associated with capital projects in mining is determined is to establish probability distributions for the input variables rather than treating each of these parameters as being known with certainty, i.e., in the deterministic case, an ore grade of 3.4% Zn may be used, when it is clear that in reality, the ore grade may be as low as 2.5% Zn or as high as 4.7% Zn. This possible range of values for ore grade will now be recognized by establishing a probability distribution for ore grade. Furthermore, with recent advances in geostatistical methods of ore reserve estimation, analysts no longer need rely on subjective probability estimates by experts to establish these distributions. The appropriate definition of probability in this case is that of likelihood of occurrence (identical to the Weather Bureau's chance of rain being 20% for tomorrow, for example) and not long-run relative frequency. Thus, if the probability that an ore grade of 2.5% Zn is estimated at 50%, the intended interpretation is that it is just as likely that the actual ore grade will turn out to be 2.5% Zn, as not.

To demonstrate the two most common stochastic risk analysis models, a simple example problem is presented. The problem is first solved deterministically for its DCFROI, assuming that all pertinent parameters were known with certainty. These simplified assumptions are then relaxed in the probabilistic cases.

Example: Dip Slip Mines is considering the expansion of its medium-sized open-pit copper mine and mill. For a capital investment of $13.5 million, the operation can be expanded from 9,000 to 12,000 stpd ore production with a commensurate increase in waste stripping. Given the data in Fig. 1, determine the DCFROI for this expansion project.

Solution:

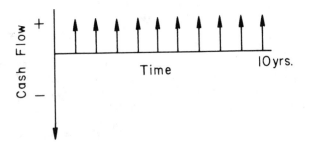

Fig. 1. Data for determining the DCFROI for the expansion project.

Ore grade = 0.8% Cu Net copper price from custom smelter = $0.50 per lb
Mill recovery = 90%   Total annual cash costs including income taxes =
                      $3,500,000
Mine life = 10 years No salvage value
Construction period = 1 year
350 operating days per year

Annual copper production: $0.008 \times 2000 \dfrac{\text{lb}}{\text{tons}} \times 3000 \dfrac{\text{tons}}{\text{day}} \times 350 \dfrac{\text{days}}{\text{year}}$

$$\times\ 0.9 = 15,1200,000 \text{ lb}$$

Gross income @ $0.50 per lb = $7,560,000
Less: total cash costs   =   3,500,000
Net post-tax cash inflow $4,060,000

$$P = A[P/A, i, n]$$
$$\$13.5 \text{ million} = \$4.00 \text{ million } [P/A, i, 10]$$
$$3.3251 = [P/A, i, 10]$$
$$i = 27.4\% = \text{DCFROI}$$

The problem has been very much simplified to demonstrate the various risk analysis techniques, so the reader should not be alarmed if the example is somewhat unrealistic.

Discrete Probability Distributions

The previous solution required estimates of several parameters for future time periods. However, since knowledge of the future is imperfect, one might decide that rather than formulating point estimates of these parameters, it would be preferable to permit these parameters to assume other values. For example, although the best estimate of mill recovery might be 90%, it could drop as low as, say, 85% with a probability of 0.4. Thus, we could state

$$\text{mill recovery} = \begin{matrix} 90\% \ (p = 0.6) \\ 85\% \ (p = 0.4). \end{matrix}$$

Since the random variable mill recovery has been permitted to assume only a finite number of values (two in this case), a discrete probability distribution results.

In a similar manner, ore grade might vary from 0.75 to 0.85% Cu, with 0.8% Cu being the best estimate. One possible probability distribution for this situation is

$$\text{ore grade} = \begin{matrix} 0.75\% \text{ Cu } (p = 0.4) \\ 0.80\% \text{ Cu } (p = 0.5) \\ 0.85\% \text{ Cu } (p = 0.1) \end{matrix}$$

Finally, experience with new plant construction might show that such expenditures can run considerably higher than estimates, but it is quite unlikely that actual costs will be lower than estimated. The following discrete probability distribution might reflect the best judgment on this parameter:

$$\text{investment} = \begin{matrix} \$13.0 \text{ million } (p = 0.05) \\ 13.5 \text{ million } (p = 0.55) \\ 16.0 \text{ million } (p = 0.40) \end{matrix}$$

Note that in all three cases the sum of the probabilities is 1.0, i.e., the parameter must assume one of the discrete values specified; no other values are permitted. Ore grade cannot equal 0.78% Cu, for example. While certainly restrictive, this situation is clearly more flexible than that deterministic case when ore grade could assume only one value—0.8% Cu. As will be seen, however, this restriction is eliminated if continuous probability distributions are assumed, as in the next section.

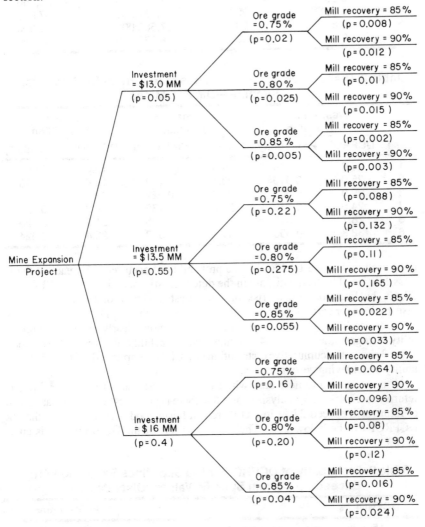

Fig. 2. Outcome possibilities in Dip Slip Mines' expansion project.

With three parameters defined by discrete probability distributions there are now 18 possible outcomes (2 mill recoveries × 3 ore grades × 3 investments). Fig. 2 is a flow diagram that illustrates each of the possible outcomes and its respective probabilities of occurrence. Thus, from one outcome in the deterministic case, 18 possible outcomes have been enumerated. Tables 1 and 2 show the calcula-

Table 1. Gross and Net Income for Various Operating Parameters—Dip Slip Mines

Case	Ore grade, % Cu	Mill recovery, %	Annual gross income, $	Annual net cost flow, $
1	0.75	85	6,693,750	3,193,750
2	0.80	85	7,140,000	3,640,000
3	0.85	85	7,586,250	4,086,250
4	0.75	90	7,087,500	3,587,500
5	0.80	90	7,560,000	4,060,000
6	0.85	90	8,032,500	4,532,500

Table 2. Discounted Cash Flow Rates of Return for Each Possible Outcome—Dip Slip Mines

Case	Investment = $13 million		Investment = $13.5 million		Investment = $16 million	
	DCFROI	Probability	DCFROI	Probability	DCFROI	Probability
1	20.9	0.008	19.8	0.088	15.0	0.064
2	25.0	0.010	23.8	0.110	18.6	0.080
3	29.0	0.002	27.6	0.022	22.1	0.016
4	24.5	0.012	23.3	0.132	18.2	0.096
5	28.7	0.015	27.4	0.165	21.9	0.120
6	32.8	0.003	31.4	0.033	25.4	0.024

tions used to derive a DCFROI and a probability of occurrence for each of the 18 possibilities. The best estimate in the deterministic case showed a DCFROI of 27.4%, but by accounting for risk through the establishment of discrete probability distributions for certain input variables, it is obvious that the DCFROI might deviate considerably from this value. This is shown more clearly in Fig. 3, a probability histogram, and in Fig. 4, a cumulative probability histogram of outcomes. Table 3 shows the cumulative distribution of DCFROI's arranged to show the probability of achieving certain values.

It is interesting to note that, whereas the best estimate of the internal rate of return was 27.4%, risk analysis shows that there is only a 24% chance that the rate of return will exceed 27%. In fact there is a fairly good chance (33.6%) that the DCFROI will be below 21%. The expected value of the DCFROI here is about 22.7%.

Table 3. Probability of DCFROI for Dip Slip Mines Expansion Project Exceeding Certain Specific Values (Discrete Case)

×, DCFROI, %	Pr (outcome ≥ ×)
15	1.000
18	0.936
21	0.664
24	0.286
27	0.240
30	0.036
33	0.000

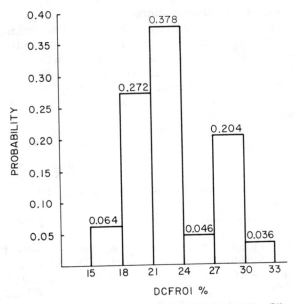

Fig. 3. Probability histogram of DCFROI for Dip Slip Mines.

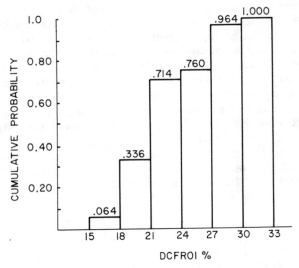

Fig. 4. Cumulative probability histogram of DCFROI for Dip Slip Mines.

Although this is a very simplified example, the value and shortcoming of the approach should be evident. By explicitly accounting for risk, Dip Slip's management should have a better feel for the magnitude of risk they must accept if they proceed with the project.

Continuous Probability Distributions

When input variables are described by discrete probability functions, there are a finite number of possible outcomes, each of which can be calculated along with its respective probability of occurrence. For example, in the previous example problem, there were three probabilistic variables which lead to 18 possible outcomes with probabilities of occurrence ranging from 0.008 to 0.165.

Now, if one permits continuous, rather than discrete, probability functions to describe these input parameters, there is no longer a finite set of possible outcomes. Each of the three random variables can assume an infinite number of values within specified ranges. Thus, it is impossible to compute the DCFROI for every possible outcome, but samples can be drawn from this population and parameters of the parent population can be inferred from this sample. For such a sample to be valid, it must be selected randomly, and it must be large enough to achieve the desired level of accuracy.

With the example problem concerning Dip Slip Mines' expansion project for a copper mining and milling project, let investment, ore grade, and mill recovery now be described by continuous probability distributions. In order to obtain one sample from the continuous distribution of resulting DCFROI's, a random sample from each of the three input parameter distributions must be taken, and the corresponding DCFROI is then computed. If this process is repeated many times, the underlying parent population of DCFROI's will begin to emerge. A point will ultimately be reached where the distribution of sample outcomes very closely approximates this population. At this point further sampling will not change the overall sample statistics significantly and should be discontinued.

Repetitive calculations such as those previously described are, of course, ideally suited to high-speed digital computers. The computer can generate such a large number of sample outcomes or simulations in a short period of time that obtaining a sample of sufficient size is seldom a problem. For most financial analysis problems, 100 simulations will usually achieve the necessary steady-state condition, and this requires very little time on most computers.

Using the same example problem for Dip Slip Mines, and assuming continuous probability distributions for the three variables in question, the problem is now analyzed using full simulation risk analysis.

Investment: The best estimate of capital investment is $13.5 million, although it is recognized that cost escalation in the construction contract could increase this amount considerably. Thus, the following probability density function might represent capital investment for the project (Fig. 5).

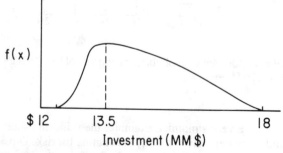

Fig. 5. Probability density function which might represent capital investment for the project.

As with any continuous probability density function, the probability that capital investment, x, will be between two values, a and b, $Pr(a \geqslant x \geqslant b) = {}_a\!\int^b f(x)dx$. The integral $\int f(x)dx = F(x)$ is the cumulative distribution function (CDF) of x (capital investment in this case). Furthermore, by definition,

$$-\!\!\!\int_{\infty}^{\infty} f(x)\, dx = 1.0$$

for any probability density function. For capital investment in the example problem this reduces to

$$_{12}\!\!\int^{18} f(x)dx = 1.0$$

as the random variable is only defined within these limits. For example, the probability of capital investment being less than or equal to \$13 million is

$$_{12}\!\!\int^{13} f(x)dx.$$

A plot of the indicated CDF would look like the example in Fig. 6.

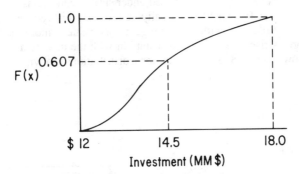

Fig. 6. A plot of the indicated CDF (cumulative distribution function).

Note that cumulative probabilities range from 0 to 1.0, and that random numbers in the same range (0 to 1.0) can be easily generated, either manually or by the computer. If a three-digit random number selected were 0.607, a plot of the CDF could be entered on the ordinate at $y = 0.607$, and the corresponding capital investment is shown on the abscissa to be about \$14.5 million. This is demonstrated in Fig. 6. An equivalent process on the computer is to find the upper limit of integration such that the cumulative probability is equal to the random number. In this example, solve for a, such that:

$$_{12}\!\!\int^a f(x)dx = 0.607$$

Therefore a uniform random number can be used to generate a random sample from a specified probability distribution either manually or with a computer.

Ore Grade: For illustrative purposes, it was further assumed that possible ore grades are normally distributed, with mean, $\mu = 0.80\%$ Cu, and standard devia-

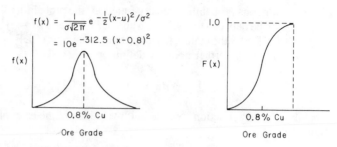

Fig. 7. Density and distribution function for ore grade.

tion, $\sigma = 0.04\%$ Cu. The density and distribution functions for ore grade would then be as shown in Fig. 7.

As the normal distribution ranges from $-\infty$ to $+\infty$, it may be desirable to exclude any randomly sampled ore grades that do not fall in the range $\mu \pm 3\sigma$. There are numerical approximations to the cumulative normal distribution which will permit the sampling of this distribution on the computer. Split-normal and triangular distributions have also been used under similar conditions.

Mill Recovery: Perhaps it is virtually certain that mill recovery will lie between 85% and 90%, but within that range no one value is more likely than any other. This gives rise to a uniform distribution with the density and distribution functions shown in Fig. 8. Sampling uniform distributions is obviously a relatively simple task.

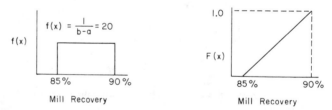

Fig. 8. A uniform distribution with density and distribution functions.

Fig. 9 demonstrates random sampling of all three distributions using uniform random numbers and shows the resulting DCFROI's for seven trials. The mean, $\bar{X}$, DCFROI for these seven samples is 27.1%, and the standard deviation, s, is 3.84%. However, it is very unlikely that only seven samples will closely represent the underlying parent population. Remote values, such as the second trial where the DCFROI = 19.5%, might weight the mean too heavily. Therefore, to obtain a more representative sample, a total of 100 trials were performed. The resulting probability histogram is shown in Fig. 10, and Table 4 shows the same data in tabular form. After 100 simulations, the average DCFROI has dropped to $\bar{X} = 23.99\%$, with the standard deviation being 3.96%. In this example problem, the three probabilistic variables are all well behaved in that they are all unimodal. Thus, one would expect the distribution of DCFROI's to be similarly well defined. Fig. 10 is, therefore, judged to be representative of the underlying population of DCFROI's, although in practice it is good to test the effect of additional simulations on the results.

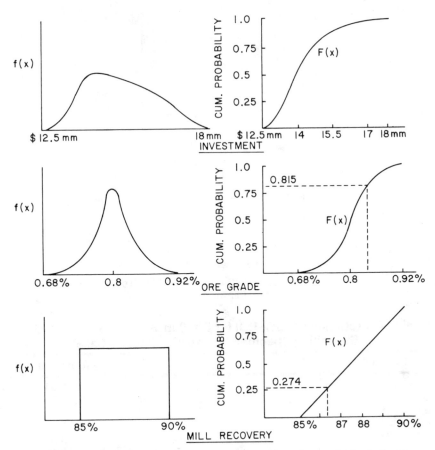

Fig. 9. Simulated sampling of continuously distributed random variables in the financial analysis of Dip Slip Mines.

Sample Values of Random Variables

Simulation No.	Random	Investment	Ore Random	grade	Random	Mill recovery	DCFROI, %
1	0.006	12.5	0.815	0.836	0.274	86.3	30.3
2	0.946	16.8	0.448	0.795	0.822	89.1	19.5
3	0.305	13.5	0.465	0.796	0.882	89.5	26.7
4	0.536	14.0	0.949	0.866	0.446	87.2	29.2
5	0.028	12.6	0.579	0.808	0.416	87.0	28.2
6	0.757	14.5	0.779	0.831	0.434	87.1	25.3
7	0.500	13.9	0.984	0.886	0.310	86.6	30.5

In summary, the Dip Slip Mines' example has been used to demonstrate deterministic analysis as well as risk analysis using both discrete and continuous probability distributions for input parameters. Each succeeding procedure provided an increased information base for the decision-maker. With full simulation, the full

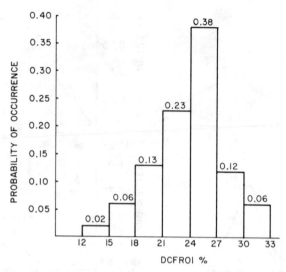

Fig. 10. Probability distribution of DCFROI's after 100 simulations, Dip Slip Mines.

Table 4. Probability of DCFROI for Dip Slip Mines Expansion Project Exceeding Specified Values (Continuous Case)

×, DCFROI, %	Pr (outcome ≥ ×)
12	1.00
15	0.98
18	0.92
21	0.79
24	0.56
27	0.18
30	0.06

range of possible outcomes is available for management's inspection. This distribution is, however, no more accurate than the data from which the distribution was derived, i.e., the input parameter distributions. This condition is true, of course, for any evaluation procedure; there is no monopoly on garbage in-garbage out. With carefully prepared estimates of evaluation parameters, full simulation risk analysis provides detailed information on the magnitude of risk that the firm accepts if it proceeds with a given project. This technique is, therefore, of considerable value in the new mine investment decision process.

LIMITING ASSUMPTIONS

Independence of Input Parameters

Implicit in the full simulation risk analysis procedure is that random variables are independent, i.e., the value of any one parameter is not affected by the value of any other. In reality, some of these variables are, of course, related. For example, in the preceding example ore grade and mill recovery are probably interrelated. Another example would be ore grade and ore reserve tonnage which are clearly interdependent for most mining ventures.

There are a number of alternative courses of action for the analyst if interdependence of variables exists. It may be that treating these variables as independent random variables creates serious bias in the result, perhaps even more serious than that caused by simplified deterministic analysis. If this is true the analyst may decide to treat the interdependent variables as constants, rather than as random variables. This will, however, seriously impair the ability of risk analysis to quantify the magnitude of risk inherent in the project.

Risk analysis models have been developed to handle interdependent variables, although in general the complexity of these models and the data requirements are staggering. Some improvements are forthcoming in this area, but the time and expense necessary for formulating such models and running the analyses will likely be prohibitive for most preliminary financial evaluations.

On balance, the analyst should always be alert for interdependent variables, but the problems created by such variables are seldom sufficiently serious to forego the many other benefits of risk analysis.

Perfect Correlation Over Time

Ore grade and mill recovery in the previous example were assumed to be constant over the mine life for any given simulation. This is clearly unrealistic, although any errors that are introduced by this assumption might be insignificant if enough simulations are run. Alternatively, each random variable could be sampled for each year of the mine's life. Such repetitive sampling vastly increases the number of required calculations, but this requires no extraordinary effort with access to a computer. If annual sampling is adopted, however, one must decide what impact, if any, previous values of a random variable has on subsequent values. Time independence may be no better than total time dependence as used in the preceding example. If a random sample of, say, mining cost is unusually high in one period (e.g., $\mu + 3\sigma$), it is exceedingly unlikely that such cost in the following period will be unusually low (e.g., $\mu - 3\sigma$). It might be appropriate to modify the distribution of permissible outcomes for future periods according to what actually occurred in previous periods.

In theory full simulation risk analysis models can be developed to include these relationships, but as with interdependent variables, the time and expense involved in developing such complex models should be examined carefully.

With financial analysis models, as with other computer systems models, there is the inevitable trade-off required between reality and generality. A model can usually be developed that is a very accurate analog of the prototype (although some simplifications are always necessary), but this model will likely be applicable only for the specific problem for which it was developed. For major projects this may be acceptable—indeed, desirable. For the majority of cases, however, construction of the model can only be justified if it can be applied to the solution of a number of problems. Including interdependent variables and time dependency routines in a financial evaluation package may be justified, but the range of applicability of the model will no doubt suffer.

Clearly, successful capital investment decision-making in mining requires many components, the most important of which is sound managerial judgment. Risk analysis is one analytical tool which helps to improve and condition such judgment by permitting greater insight into the source of risk and its impact on project financial results.

14

Case Study

There are mines that make us happy,
There are mines that make us blue,
There are mines that steal away the tear-drops
As the sunbeams steal away the dew.
There are mines that have lost the ore chute faulted,
When the ore's forever lost to view,
But the mines that fill my heart with sunshine,
Are the mines that I sold to you.
                                        —Anonymous

INTRODUCTION

The preceding chapters have discussed, in various degrees of detail, the primary considerations necessary to perform a cash flow analysis of a mining-related investment opportunity. Although the previous discussions regarding the manipulations, calculations, and interrelationships of the various components of the pro forma income statement were illustrated in the text, many individuals are unfamiliar with the format and appearance of a final spread sheet for the evaluation of a mining property. As a result, it seems appropriate to conclude this book with an abbreviated case study of an actual mine property (currently in production) showing the cash flow analysis and some back-up data. Obviously all of the supporting data and calculations for the case study presented cannot be provided; however, reference is made to the major assumptions utilized and to the appropriate procedures employed in the calculations, as described elsewhere in the text.

The problem associated with using any case study for illustrative purposes is that one tends to be overwhelmed by detail. For instance, specific costs, prices, expenditure timings, and other variable values may be challenged because they either do not conform to one's expectations or are not consistent with specific operating experiences. When this occurs the instructional value of the problem becomes obscured. The purpose of most case studies, and particularly the one presented in this chapter, is to expose the student to the methodology, procedure, and format associated with performing a cash flow analysis of a mining property. It is *not* to laboriously wade through the specific details associated with operating and capital cost development, mine plans, or equipment schedules. It should be noted, however, that the values represented in this case study result from specific mine plans, operating procedures, equipment spreads, manpower requirements, and other engineered variables associated with mining a specific property. As mentioned else-where in this text, the evaluation procedure is an iterative process requiring the evaluation of numerous alternatives to determine the maximum value for any given property.

CASE STUDY: LIMESTONE MINING PROPERTY

This case study is based on a mine property model developed by Gentry and Hrebar (1983). This particular case study is associated with a near-surface lime-stone deposit located on a hogback formation on the eastern slope of the Rocky Mountains in the state of Colorado. The property is contained within a 1/4 section and the limestone units of interest have a strike length of one-half mile and underlie the eastern 1100 ft of the property.

Although the hogback structure provides outcroppings of the limestone units and related stratigraphy, the overall thickness, grade, continuity, and uniformity of the limestone units are not substantiated. However, based on information from a nearby operation, outcrop mapping, sampling, and data from geologic publica-tions, the following preliminary information is available on the stratigraphy at the property:

Sandstone "A": Uniform, blocky, quite competent; average thickness of 15 ft although it gradually increases downdip; average dip is 20°E, strike N-S; overlain by alluvium downdip.

Limestone "A": Uniform, consistent, blocky; average thickness is 20 ft; over-burden increases downdip; average dip is 20°E, strike N-S; estimated average grade is 98.5% $CaCO_3$.

Sandstone "B": Same basic description as sandstone "A"; average thickness is 27 ft.

Limestone "B": Same basic description as limestone "A"; average thickness is 22 ft; average grade is estimated at 96% $CaCO_3$.

The owner of the property has approached Limy Mining Co. about purchasing the property for an initial payment of $500,000 plus a 10% royalty on all products (f.o.b. mine). Limy's senior management is requesting an economic feasibility study of the property to determine if acquisition is advisable.

Considerations and Assumptions

In the course of developing a feasibility study on a potential mining property, a number of assumptions must be formulated and subsequently checked for accuracy as more information becomes available. Pertinent considerations and assumptions are developed as needed throughout the feasibility study in order to perform the required analysis.

One consideration which should be noted at the very start, however, is the as-sumption that Limy Mining Co. is a division of a large corporation which consists of a number of other large, profitable divisions. As such, Limy chooses to expense whenever possible rather than capitalizing and amortizing expenditures. In other words, tax deductions are maximized and declared as soon as possible by consol-idating them into the parent corporation, thereby reducing the present value of cor-porate income taxes paid. Consequently, this property is provided the maximum benefit from tax savings resulting from the corporate umbrella.

Another fundamental assumption is that Limy Mining Co. presently does not have a producing mining property within the state of Colorado. Thus, if Limy should choose to acquire the property in question and bring it into production, the revenue generated would represent the only income derived by Limy within the state.

Finally, model cash flows were prepared in constant dollars. Although this contradicts our assumptions in Chapter 10 that inflation must be considered in discounted cash flow calculations, the case of constant dollars makes it far easier for the reader to follow the sometimes tortuous path of calculations in the model. For those who wish to study the escalated dollar version of this example, a brief description and summary tables are given later in the chapter in the section on Escalated Dollar Analysis.

Ore Reserve Assessment

Based on the preliminary information available on property dimensions and limestone thickness, the in-place reserves for both limestone units (A and B) may be estimated if a specific gravity is assumed. If a value of 2.6 is assumed to be the specific gravity for all materials, the indicated in-place limestone reserves are as follows:

42 ft combined thickness $\times$ 2640 ft long $\times$ 1100 ft wide
$\times$ 2.6 $\times$ 62.4 lb per cu ft $\div$ 2000 lb per ton = 9.9 $\times$ 10^6 tons

If mining recovery is estimated at 95%, a total of 9.4 million tons of limestone is available.

Marketing, Products, and Production Requirements

Industrial minerals are often described as "place-value" commodities. That is to say, the relative value of an industrial mineral deposit is often a function of its location relative to markets. Therefore, one of the first steps in conducting a feasibility study on these deposits is to assess the status of markets for potential products from the operation. Assume a detailed marketing study resulted in an indication that the following limestone products could be sold in the quantities and at the prices specified:

Product	Specifications	Production, tpy	Price, $ per ton f.o.b. mine
Sugar rock	5½ × 3-in. and 3 × 2-in; +95%CaCO$_3$	235,600	7.26
Cement plant feed	3/4-in. − : +88%CaCO$_3$	85,700	2.50
Cattle feed granules and poultry feed	These four products will require pulverization and screening.	avg. 26,800	22.00
Glass material		avg. 37,500	20.50
Industrial fillers		avg. 16,000	20.00
Rock dust		avg. 26,800	20.00
Total tons of product(s):		428,400	

Assuming a recovery rate of 90%, the associated mining rate necessary to produce the required salable products is

$$\frac{428,400 \text{ tpy}}{0.90} = 476,000 \text{ tpy of limestone rock}$$

If it is also assumed that a reasonable estimate of a stripping ratio for both limestone units within the property limits will average about 1:1, the associated stripping requirement is

$$476,000 \ \frac{\text{tons limestone rock}}{\text{year}} \times \frac{1 \text{ ton overburden}}{1 \text{ ton limestone rock}} = 476,000 \text{ tpy}$$

Therefore, the daily required stripping rate on a one-shift basis is:

$$476,000 \text{ tpy} \times \frac{1 \text{ year}}{238 \text{ days operation}} \times \frac{1 \text{ day}}{1 \text{ shift}} = 2000 \text{ tpd}$$

Thus, the total mining rate is 4000 tpd of material.

In conjunction with the marketing requirements, it is decided to select crushing and screening equipment to provide the following size distribution and product proportions:

Product	% total	Tpy
Sugar rock 5½ × 3-in. and 3 × 2-in.	55%	235,600
Cement plant feed, ¾-in. −	20%	85,700
Remainder, (2 × ¾-in.)		
("pulverizing" plant feed)	25%	107,100
Total	100%	428,400

The 2 × ¾-in. product is used as feed material to the pulverizing and screening plant to produce the four fine-based products. The initial anticipated production mix from the 107,100 tons feed is as follows:

Product	% total	Tpy
Cattle feed granules	25%	26,800
Glass material	35%	37,500
Rock dust	25%	26,800
Industrial fillers	15%	16,000
Total	100%	107,100

Based on the anticipated mining recovery rate of 95% of in-place geologic reserves and the production requirement of 476,000 tpy of limestone rock, the operating life of the property is calculated at 20 years as follows:

$$9,900,000 \text{ tons} \times 0.95 \div 476,000 \text{ tpy} = 20 \text{ years}$$

Using projected product prices and production rates, annual project revenues can be calculated. These are illustrated in Table 1.

Mining Method Selection

The next step associated with preparation of the feasibility study is development of a proposed mine plan for the property. Given an ore production requirement of 2000 tpd and an estimated average stripping ratio of 1:1, it appears

Table 1. Project Production Revenue Calculations

Product	Specifications	Tons per year	F.o.b. mine	Revenue, $ per year
Sugar rock	5½ × 3 in. and 3 × 2 in.: + 95% CaCO₃	235,600	7.25	1,708,000
Cement plant feed	¾ in. − : + 88% CaCO₃)	85,700	2.50	214,200
"Pulverizing" plant	A. Cattle feed granules	26,800	22.00	589,600
products	B. Glass material	37,500	20.50	768,750
	C. Rock dust	26,800	20.00	536,000
	D. Industrial fillers	16,000	20.00	320,000
Total annual project revenue				4,136,550

Table 2. Capital Equipment Summary List

	Item	Unit cost, $	Quantity	Total cost, $
Mine:	Primary jaw crusher, 30 × 42-in.	180,000	1	180,000
	Secondary cone, 20 × 4½-in. (or comparable gyratory)	172,000	1	172,000
	Feeder radial stackers, feed	120,000	1	120,000
	Portable conveyors	25,000	8	200,000
	Screening sections	62,000	3	186,000
	7—CY loader	330,000	1	330,000
	5—CY loader	240,000	1	240,000
	25—T trucks	200,000	3	600,000
	Air tracks and compressors	140,000	2	280,000
	Dozers (D-8)	275,000	2	550,000
	Pickups	14,000	2	28,000
	Shop and misc. equipment	—	—	20,000
	Total mine			2,906,000
Pulverizing plant:	5—CY loader	240,000	1	240,000
	Plant (pulverizers, screens, bucket elevators, cyclones, dryer bins, truck scales, etc.)			2,500,000
	Total pulverizing plant			2,740,000
Mine yard (real):	Shop and warehouse			100,000
	Office			50,000
	Lab equipment			35,000
	Total mine yard (real)			185,000
Total capital				5,831,000

reasonable to select a surface technique which does not require an excessive capital investment. In view of the 20° dip on the limestone units of interest, it would be difficult for mobile equipment to work directly on the contact between the limestone units and the underlying sandstones. A more reasonable approach would be to push the blasted material downdip with a bulldozer to a loading point where front-end loaders could pick up the material and load it into trucks. This approach could be used for overburden materials as well as for the limestone. It is envisioned that mining would progress downdip in a stairstep configuration until the property boundaries are reached.

Equipment Selection and Manpower Requirements

Preliminary mine design and equipment selection are essentially simultaneous considerations. Some idea of a mining method must be generated before a develop-

Table 3. Estimated Manning Table

	Total	Wages per day, $	Cost per year, $
Mine (1 shift/day)			
Loader operators	2	95.20	45,315
Crusher operator	1	80.00	19,040
Dozer operators	2	95.20	45,315
Truck drivers	3	74.00	52,836
Mechanics	2	96.00	45,696
Oiler	1	76.00	18,088
Welder-fabricator	1	78.00	18,564
Drill operators	2	95.20	45,315
Laborers	4	58.00	55,216
Subtotal	18		345,385
Fringes @ 40%			138,154
Total mine			483,539
Pulverizing plant (2 shifts per day)			
Loader operators	2	70.00	33,320
Mechanic	1	80.00	19,040
Plant operators	2	76.80	36,557
Subtotal	5		88,917
Fringes @ 40%			35,567
Total pulverizing plant			124,484
Total project, direct labor	23		608,023
Salary:			
General manager	1		50,000
Quarry superintendent	1		38,000
Quarry foreman	1		27,000
Clerk-secretary	2		27,608
Office manager	1		27,000
Subtotal			199,608
Fringes @ 40%			79,843
Total salaried	7		279,451
Total project labor	30		887,474

ment sequence can be determined. In some cases, development constraints may prohibit the use of some mining alternatives.

In this case study, a downdip surface mining technique was selected. From the generalized geology, it appears that some preproduction stripping will be required prior to actual limestone production. Stripping can presumably be accomplished in a manner similar to the mining technique previously described and utilize the same equipment. After determination of a production rate (476,000 tons of limestone rock + 467,000 tons of waste), it is possible to select the type and size of equipment necessary for the mining operations.

Table 2 is a list of the essential capital equipment estimated to be necessary for this particular mine plan, as well as that equipment associated with the pulverizing plant. Unit and total cost estimates are also provided. All mine yard facilities are considered to be real property for depreciation purposes.

Based on equipment requirements and the mine plan selected, manpower requirements can be determined and labor costs calculated. These estimates are provided in Table 3. Total labor charges include an additional 40% for fringes and benefits.

After manpower requirements are estimated, it is possible to calculate anticipated productivities associated with each aspect of the operation. These are determined as follows:

$$\text{mine:} \quad \frac{476,000 + 476,000 \text{ tpy}}{18 \, m \times 238 \text{ shifts per year}} = 22 \text{ tons per } m \text{ per shift}$$

$$\text{mine and pulverizing plant:} \quad \frac{476,000 + 476,000 \text{ tpy}}{23 \, m \times 238 \text{ shifts per year}} = 174 \text{ tons per } m \text{ per shift}$$

$$\text{project:} \quad \frac{476,000 + 476,000 \text{ tpy}}{30 \, m \times 238 \text{ shifts per year}} = 133 \text{ tons per } m \text{ per shift}$$

where m represents the number of persons assigned to that activity (Table 3).

Operating and Capital Cost Estimates

With labor and capital equipment requirements established, it is possible to estimate the operating and capital costs for the project. Operating costs were calculated as shown in Table 4. These cost estimates were derived from historical internal data available to Limy. The insurance charge was based on 1.5% of capital costs for mine equipment and mine yard facilities, inclusive of allowance for engineering fees and contingencies (Table 5) but exclusive of preproduction development. Also, "total salaried labor" was allocated 75% to the mine and 25% to the pulverizing plant.

Capital cost estimates (Table 5) were derived from the capital equipment list (Table 2). It should be noted that engineering and contingency fees are added to the initial equipment investment in order to estimate the total initial capital investment in the property.

Preproduction development was calculated by taking three months of stripping production multiplied by the stripping cost per ton, derived as follows:

Category	$ per ton, rock
Mining	2.27
Processing (crushing and screening)	1.57
Administration	1.05
Total	$4.89

At a 1:1 stripping ratio, stripping costs are approximately $1.13 per ton.

Working capital was estimated at two months production of limestone rock times the cost for mining ($4.89 per ton of rock).

Spares and supplies were estimated by multiplying the total capital cost by 3% per year and assuming a three-month parts inventory on site.

Project Timing

At this stage a schedule of preproduction activities and associated time requirements can be estimated in order to indicate overall project timing. The project activity schedule for this project is estimated to be that shown in Fig. 1.

Exploration and feasibility activities are estimated to require a two-year period. These activities include mapping, drilling, process testing, and detailed economic studies. Drilling efforts would incorporate core drilling on 500-ft centers to an average depth of 100 ft and fill-in rotary drilling.

To ensure early identification of any environmental road blocks, the permitting process would normally begin once favorable results were obtained from the exploration program. The permitting process should be pursued to completion as

Table 4. Estimated Operating Costs

	Item	$ per year	Salable $ per ton-rock
Mine:	Labor	483,539	1.13
	Maintenance and repairs	355,572	0.83
	Diesel fuel	257,040	0.60
	Explosives and caps	98,532	0.23
	Other (power, supplies, etc.)	235,620	0.55
	Total mine, direct	1,430,303	3.34
	Admin. and supervision (279,451 − 0.75)	209,588	0.49
	Office supplies, etc.	17,000	0.04
	Laboratory supplies (30,000 × 0.5)	15,000	0.04
	Mine insurance (3,631,000 × 0.015)	54,465	0.13
	Shop, warehouse, office insurance (232,000 × 0.15 × 0.75)	2,610	0.01
	Corporate overhead	85,680	0.20
	Subtotal	1,814,646	4.25
	Contingency @ 15%	272,197	0.64
	Total mining cost	2,086,843	4.89*

			$ per ton-plant throughput
Pulverizing plant:	Labor	124,484	1.16
	Maint., repair, power, etc.	363,069	3.39
	Total pulverizing plant, direct†	487,553	4.55
	Admin. and supervision (279,451 × 0.25)	69,863	0.65
	Plant, insurance (3,562,000 × 0.15)	53,430	0.50
	Shop, warehouse, office insurance (232,000 × 0.015 × 0.25)	870	0.01
	Laboratory supplies (30,000 × 0.5)	15,000	0.14
	Corporate overhead	28,000	0.26
	Total pulverizing plant cost†	654,716	6.11
Total operating cost, project:		2,741,559	—

*Rounding error.
†Exclusive of purchase price of feed.

Table 5. Estimated Capital Costs

Mine	$2,906,000	
Engineering, 10% × 2.91 M	291,000	
Contingency, 15% × 2.91 M	434,000	
Total mine		$3,631,000
Pulverizing plant	$2,740,000	
Engineering, 15% × 2.74 M	411,000	
Contingency, 15% × 2.74 M	411,000	
Total pulverizing plant		$3,562,000
Mine yard	$ 185,000	
Engineering, 10% × 0.185 M	19,000	
Contingency, 15% × 0.185 M	28,000	
Total mine yard		$ 232,000
Preproduction development, 476,000 tpy per 12 mon × 3 mon × $1.13 per ton)		$ 134,000
Total capital cost		$7,559,000
Working capital 428,400 tpy × $\frac{2}{12}$ × $4.89 per ton		$ 349,000
Spares and supplies, 5.831 MM × 0.03 per year × 3/12		$ 44,000
Total working capital		$ 393,000

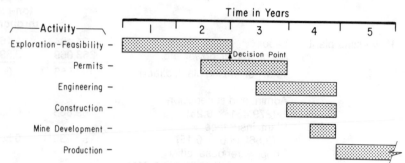

Fig. 1. Simplified project activity schedule.

soon as reasonably possible. Of course, the time required for the permitting process is quite variable and dependent on location and agencies involvement (federal, state, local). The expenditures allocated for environmental studies and permitting are typically classified as *development* expenditures for tax purposes since they are a necessary and vital component of mine development.

Detailed engineering begins when Limy is relatively certain of obtaining the necessary permits. Construction begins at the start of year 4 and is completed within 12 months. The preproduction development campaign occurs during the last three months of year 4 with production beginning in year 5 and continuing for 20 years.

CASE STUDY: CASH FLOW DETERMINATIONS

With the foregoing background information developed on the property, it is possible to calculate net annual cash flows for the project. These cash flows provide insight into the economic viability of the project for Limy management. The results of this economic assessment are presented in the following constant dollar cash flow analysis. A summary of the corresponding escalated dollar analysis is provided in a later section of this chapter.

Table 6 illustrates the magnitude and timing of annual cash flows anticipated during the preproduction period. A total time span of four years is estimated for this activity after the decision is made to acquire the property, assuming the results of the exploration program are as anticipated.

The reader will recall that the owner of the property requires an initial *property payment* of $500,000 if Limy chooses to acquire the property. This expenditure occurs immediately (time 0).

Exploration and feasibility study expenditures are estimates based on the work previously described during the first two years of testing the prospect. The precise allocation of the total estimated expenditure is dependent upon the anticipated timing and type of exploration work scheduled.

Preproduction development expenditure estimates were described previously (Table 5) and occur in year four.

Estimated capital expenditures for the *mine, pulverizing plant, and mine yard facilities* were estimated previously (Table 5). The precise allocation of these expenditures between years three and four is based on engineering estimates of equipment acquisition and needs.

Property tax payments are described in a later section of this chapter in conjunction with production year cash flows.

Table 6. Magnitude and Timing of Annual Cash Flows—Preproduction ($1000, Constant Year 1 Dollars)

Capital expenditure	Time 0	Year 1	Year 2	Year 3	Year 4	Total
Property payment	500	—	—	—	—	500
Exploration and feasibility study	—	180	190	—	—	370
Preproduction development	—	—	—	—	134	134
Mine	—	—	—	1000	2631	3631
Pulverizing plant	—	—	—	1500	2062	3562
Mine yard facilities	—	—	—	100	132	232
Property tax	—	—	—	—	134	134
Working capital	—	—	—	—	44	44
Minimum tax	0	21	22	0	16	59
Total capital expenditures	−500	−201	−212	−2600	−5153	−8666
Cash generated						
Tax savings						
Exploration and feasibility study*	—	72	79	6	5	162
Preproduction development†	—	—	—	—	52	52
Property tax‡	—	—	—	—	62	62
Investment tax credit§	—	—	—	250	469	719
Investment tax credit expl. & devel.¶	—	3	3	0	2	8
Total cash generated	0	75	82	256	590	1003
Net cash flow‖	−500	−126	−130	−2344	−4563	−7663

*See Table 7.
†See Table 7.
‡Amount = property tax × 0.46.
§Investment credit = (mine + pulverizing plant equipment) × 0.10.
¶Investment credit = (capitalized exploration and development) × 0.10.
‖Net cash flow = capital expenditures + cash generated.

Working capital expenditures are for spares and supplies as specified in Table 5.

Minimum tax payments are described in a later section of this chapter in conjunction with production year cash flows.

The bottom portion of Table 6 illustrates the tax savings generated by the corporation as a result of exploration, preproduction development, and property tax expenditures being pretax deductions. These tax savings generated during the preproduction period result from the previous assumption that the corporation is in a position to expense whenever possible and thus shelter other income. It should be noted that investment tax credits also generate tax savings to the corporation. The net impact of these tax savings is to reduce the magnitude of preproduction, pretax capital investment from a total of $8,666,000 to an after-tax cost of $7,663,000. These savings are appreciable and extremely important to project viability because of their impact on early annual cash flows.

The tax savings illustrated on Table 6 were calculated in accordance with the requirements stipulated in the Tax Equity and Fiscal Responsibility Act of 1982 (TEFRA). As discussed in Chapter 8, TEFRA no longer permits expensing of all exploration and development expenditures in the year incurred. Only 85% of the expenditures may be expensed in the year incurred; the remaining 15% is deducted over a five-year period beginning in the year the costs were incurred. The ACRS-type schedule which applies to the remaining 15% expenditure is as follows:

Year 1: 15%
Year 2: 22%
Years 3-5: 21%

The 15% reduction amounts are also eligible for investment tax credits. The appropriate calculations pertaining to exploration and development expenditures and associated investment tax credits are shown on Table 7. The pertinent data determined in Table 7 were transferred to Table 6 for the calculation of tax savings during the preproduction period.

Table 8 represents the estimated net annual cash flows for the project after production beings in project year five.

Revenue estimates are simply the result of multiplying annual production in tons by estimated sales price per ton of product. The revenue calculations are provided in Table 1.

Operating costs are the result of multiplying annual production by the estimated operating cost per ton produced. The estimate of annual operating costs is developed in Table 4 and totals $2,742,000.

The *royalty* obligation is stipulated by the property owner at 10% on all products (f.o.b. mine). The revenue generated from the sale of sugar rock and cement plant feed is provided in Table 1. In addition, however, the remaining 107,100 tons of production is considered to be "sold" to the pulverizing plant at a transfer price of $4.89 per ton (the actual mining cost). Therefore, the royalty calculation is as follows:

[(235,600 tons × $7.25 per ton) + (85,700 tons × $2.50 per ton) + (107,100 tons × $4.89 per ton)] × 10% = $244,600

The calculation of property (ad valorem) tax is a function of specific state tax statutes. The general procedure associated with calculating property taxes is to

Table 7. Calculation of Exploration and Development Tax Savings—Preproduction ($1000)

Year	1	2	3	4	5	6	7	8
Exploration expenditure	180	190						
Allowable expense deduction @ 85%	153	162						
15% reduction amount	27	28						
Year 1:								
15% reduction deduction rate	0.15	0.22	0.21	0.21	0.21			
15% reduction deduction	4	6	6	5	6			
Year 2:								
15% reduction deduction rate		0.15	0.22	0.21	0.21	0.21		
15% reduction deduction		4	6	6	6	6		
Total exploration deduction	157	172	12	11	12	6		
Exploration tax savings @ 46%	72	79	6	5	—	—		
Development expenditures				134				
Allowable expenses deduction @ 85%				114				
15% reduction amount				20				
Year 4:								
15% reduction deduction rate				0.15	0.22	0.21	0.21	0.21
15% reduction deduction				3	4	4	5	4
Total development deductions				117	4	4	5	4
Development tax savings @ 46%				52	—	—	—	—
15% reduction amounts	27	28	0	20	0	0	0	0
Investment tax credit @ 10%	3	3	0	2	0	0	0	0
Exploration and development Deductions carried to production cash flows	—	—	—	—	16	10	5	4

Table 8. Production Cash Flows—Project ($1000)

Year	5	6	7	8	9	10	11	12	13	14
Revenue	4137	4137	4137	4137	4137	4137	4137	4137	4137	4137
Operating costs	2742	2742	2742	2742	2742	2742	2742	2742	2742	2742
Royalty	245	245	245	245	245	245	245	245	245	245
Property tax	145	131	118	104	92	100	85	91	74	78
Net after costs	1005	1019	1032	1046	1058	1050	1065	1059	1076	1072
Explor. & devel. deduction	16	10	5	4	0	0	0	0	0	0
Depreciation—mine	545	782	745	744	740	100	141	221	261	340
Depreciation—pulverizing	508	744	711	710	711	85	124	203	243	321
Net after depreciation	−64	−517	−429	−412	−393	865	800	635	572	411
State income tax @ 5%	0	0	0	0	0	0†	0†	24	29	21
Net after state tax	−64	−517	−429	−412	−393	865	800	611	543	390
Depletion	0	0	0	0	0	0	0	0	0	0
Pretax profit	−64	−517	−429	−412	−393	865	800	611	543	390
Federal income tax @ 46%	29¶	238¶	197¶	190¶	181¶	398	368	281	250	179
Tax credits (ITC)	0	0	0	0	0	120	0	120	0	120
Minimum tax	2	1	1	0	0	0	0	0	0	0
Net profit	−37	−280	−233	−222	−212	587	432	450	293	331
Explor. & devel. deduction	16	10	5	4	0	0	0	0	0	0
Depreciation—mine	545	782	745	744	740	100	141	221	261	340
Depreciation—pulverizing	508	744	711	710	711	85	124	203	243	321
Depletion	0	0	0	0	0	0	0	0	0	0
Operating cash flow	1032	1256	1228	1236	1239	772	697	874	797	992
Capital expenditures	349*	0	0	0	0	1199	0	1199	0	1199
Net cash flow	683	1256	1228	1236	1239	−427	697	−325	797	−207

*Working capital.
†State tax loss carry-forward.
¶Federal corporate tax credit from tax loss.

Table 8. Continued

Year	15	16	17	18	19	20	21	22	23	24
Revenue	4137	4137	4137	4137	4137	4137	4137	4137	4137	4137
Operating costs	2742	2742	2742	2742	2742	2742	2742	2742	2742	2742
Royalty	245	245	245	245	245	245	245	245	245	245
Property tax	72	64	58	73	63	77	68	81	72	62
Net after costs	1078	1086	1092	1077	1087	1073	1082	1069	1078	1088
Explor. & devel. deduction	0	0	0	0	0	0	0	0	0	0
Depreciation—mine	259	254	132	219	139	207	247	328	247	242
Depreciation—pulverizing	243	236	119	203	124	203	243	321	243	236
Net after depreciation	576	596	841	655	824	663	592	420	588	610
State income tax @ 5%	28‡	27	39	30	38	30	27	18	27	28
Net after state tax	548	569	802	625	786	663	565	402	561	582
Depletion	13	54	54	54	54	54	54	54	54	54
Pretax profit	535	515	748	571	732	579	511	348	507	528
Federal income tax @ 46%	246	237	344	263	337	266	235	160	233	243
Tax credits (ITC)	0	0	0	0	0	0	0	0	0	0
Minimum tax	0	0	0	120	0	120	0	120	0	0
Net profit	289	278	404	428	395	433	276	308	274	285
Explor. & devel. deduction	0	0	0	0	0	0	0	0	0	0
Depreciation—mine	259	254	132	219	139	207	247	328	247	242
Depreciation—pulverizing	243	236	119	203	124	203	243	321	243	236
Depletion	13	54	54	54	54	54	54	54	54	54
Operating cash flow	804	822	709	904	712	897	820	1011	818	817
Capital expenditures	0	0	0	1199	0	1199	0	1199	0	393§
Net cash flow	804	822	709	−295	712	−302	820	−188	818	1210

‡(NAD−D) × 0.05, years 15-24
§Recapture of working capital plus spares and supplies.

Table 9. Book Value Calculations—Initial Investment; Assumption: Straight-Line Depreciation

Project year	Depreciation year	Basis, $1000	Rate	Deduction, $1000	Book value, $1000
Mine Yard Facilities (Buildings)—20 Year Life					
4					232
5	1	232	1/20	12	220
6	2	232	1/20	11	209
7	3	232	1/20	12	197
8	4	232	1/20	11	186
9	5	232	1/20	12	174
10	6	232	1/20	11	163
11	7	232	1/20	12	151
12	8	232	1/20	11	140
13	9	232	1/20	12	128
14	10	232	1/20	11	117
15	11	232	1/20	12	105
16	12	232	1/20	11	94
17	13	232	1/20	12	82
18	14	232	1/20	11	71
19	15	232	1/20	12	59
20	16	232	1/20	11	48
21	17	232	1/20	12	36
22	18	232	1/20	11	25
23	19	232	1/20	12	12
24	20	232	1/20	12	0
Mine Equipment—10 Year Life					
4					3631
5	1	3631	1/10	363	3268
6	2	3631	1/10	363	2905
7	3	3631	1/10	363	2542
8	4	3631	1/10	363	2179
9	5	3631	1/10	363	1816
10	6	3631	1/10	363	1452
11	7	3631	1/10	363	1089
12	8	3631	1/10	363	726
13	9	3631	1/10	363	363
14	10	3631	1/10	363	0
Pulverizing Plant—10 Year Life					
4					3562
5	1	3562	1/10	356	3206
6	2	3562	1/10	356	2850
7	3	3562	1/10	356	2493
8	4	3562	1/10	356	2137
9	5	3562	1/10	356	1781
10	6	3562	1/10	356	1425
11	7	3562	1/10	356	1069
12	8	3562	1/10	356	712
13	9	3562	1/10	356	365
14	10	3562	1/10	356	0

Table 10. Book Value Calculations—Replacement; Assumption: Straight-Line Depreciation, 10 Year Life

Depreciation year	Basis, $1000	Rate	Deduction, $1000	Book value, $1000
		Mine Equipment—General		
				605
1	605	1/10	61	544
2	605	1/10	60	484
3	605	1/10	61	423
4	605	1/10	60	363
5	605	1/10	61	302
6	605	1/10	60	242
7	605	1/10	61	181
8	605	1/10	60	121
9	605	1/10	61	60
10	605	1/10	60	0
		Pulverizing Plant Equipment—General		
				594
1	594	1/10	59	536
2	594	1/10	59	476
3	594	1/10	59	417
4	594	1/10	59	359
5	594	1/10	59	299
6	594	1/10	59	240
7	594	1/10	59	181
8	594	1/10	59	122
9	594	1/10	59	63
10	594	1/10	59	0

first determine the fair market value of the property. The property is then assessed at a percentage of its estimated fair market value. The tax liability is determined by applying the appropriate tax rate to the assessed value of the property.

The state of Colorado levies a property tax on real and personal property plus mineral sales. The actual estimate of real and personal property value is based largely on the assessor's judgment; however, it is normally approximated by determining the book value of the assets based on straight-line depreciation. Table 9 shows these book values for the initial investment; Table 10 shows similar values for replacement equipment. Table 11 provides a summary of the calculated book values for the mine, and Table 12 shows summary book values for the pulverizing plant. The mine and pulverizing plant values are segregated for depletion purposes.

The state of Colorado also levies a property tax on ore or mineral sales for each mining operation with gross proceeds of $5000 or more. The appraised value of ore sales is defined as the greater of 25% of gross proceeds or the net proceeds. Gross proceeds are defined as the value of the ore (limestone) after extraction less all costs of treatment, reduction, transportation, and sale of the ore. Net proceeds are defined as gross proceeds less all costs of extraction of the ore. The assessed value is then set at 100% of the appraised value. The calculation procedure for this determination is illustrated in Table 13.

Table 11. Book Value Calculations—Summary: Mine, $1000

Project year	Mine yard facilities	Mine equipment	Replacement A			Replacement B			Total
			Year 10	Year 12	Year 14	Year 18	Year 20	Year 22	
4	232	3631							3863
5	220	3268							3488
6	209	2905							3114
7	197	2542							2739
8	186	2179							2365
9	174	1816							1990
10	163	1452	605						2220
11	151	1089	544						1784
12	140	726	484	605					1955
13	128	363	423	544					1458
14	117		363	484	605				1569
15	105		302	423	544				1374
16	94		242	363	484				1183
17	82		181	302	423				988
18	71		121	242	363	605			1402
19	59		60	181	302	544			1146
20	48			121	242	484	605		1500
21	36			60	181	423	544		1244
22	25				121	363	484	605	1598
23	12				60	302	423	544	1341
24						242	363	484	1089

Table 12. Book Value Calculations—Summary: Pulverizing Plant ($1000)

Project year	Pulverizing plant	Replacement A			Replacement B			Total
		Year 10	Year 12	Year 14	Year 18	Year 20	Year 22	
4	3562							3562
5	3206							3206
6	2850							2850
7	2493							2493
8	2137							2137
9	1781							1781
10	1425	594						2019
11	1069	536						1605
12	712	476	594					1782
13	356	417	536					1309
14		359	476	594				1429
15		299	417	536				1252
16		240	359	417				1075
17		181	299	417				897
18		122	240	359	594			1315
19		63	181	299	536			1079
20			122	240	476	594		1432
21			63	181	359	536		1197
22				122	229	476	536	1551
23				63	240	359	536	1315
24					122	299	476	1075

Table 13. Assessed Value of Rock Sales Calculation

Assumptions: Operating cost breakdown

Mining	$2.27
Processing	1.57
Administration	1.05
Total	$4.89

Gross Value of Rock	2446
Less: Processing costs (428.4 × 1.57)	673
General processing costs (428.4 × (1.57)/(2.27 + 1.57) × 1.05)	184
Gross proceeds	1589
Less: Mining costs (428.4 × 2.27)	972
General mining costs (428.4 × (2.27/(2.27 + 1.57) × 1.05)	266
Net proceeds	351
25% gross proceeds	397
Appraised value	397
Percentage	100%
Assessed value	397

Table 14. Property Tax Calculations ($1000)

Project year	Mine					Pulverizing plant			Total property tax
	Book value	Assessed* value	Assessed value rock sales	Total assessed value	Property† tax	Book value	Assessed value	Property† tax	
4	3863	1159	0	1159	70	3568	1070	64	134
5	3488	1046	397	1443	87	3206	962	58	145
6	3114	934	397	1331	80	2850	855	51	131
7	2739	822	397	1219	73	2493	748	45	118
8	2365	710	397	1107	66	2137	641	38	104
9	1990	597	397	994	60	1781	534	32	92
10	2220	666	397	1063	64	2019	606	36	100
11	1784	535	397	932	56	1605	482	29	85
12	1955	586	397	984	59	1782	535	32	91
13	1458	437	397	834	50	1309	393	24	74
14	1569	471	397	868	52	1429	429	26	78
15	1374	412	397	809	49	1252	376	23	72
16	1183	355	397	752	46	1075	322	19	64
17	988	296	397	693	42	897	269	16	58
18	1402	421	397	818	49	1315	394	24	73
19	1146	344	397	741	44	1079	324	19	63
20	1500	450	397	847	51	1432	430	26	77
21	1244	373	397	770	46	1197	359	22	68
22	1598	479	397	876	53	1551	465	28	81
23	1341	402	397	799	48	1315	394	24	72
24	1089	327	397	724	43	1075	322	19	62

* Assessed value = 30% book value (or appraised value).
† Property tax based on county mil levy of 60 mils.

Table 15. Depreciation Calculations—Initial Investment; Assumptions: ACRS System, Place into Service in Year 5

Project year	Depreciation year	Basis, $1000	ACRS, % per 100	Deduction, $1000
		Mine Equipment		
		$3631 - (0.10 \times 3631 \times 0.50) = 3449$		
5	1	3449	0.15	517
6	2	3449	0.22	759
7	3	3449	0.21	724
8	4	3449	0.21	725
9	5	3449	0.21	724
		Pulverizing Plant		
		$3562 - (0.10 \times 3562 \times 0.50) = 3384$		
5	1	3384	0.15	508
6	2	3384	0.22	744
7	3	3384	0.21	711
8	4	3384	0.21	710
9	5	3384	0.21	711
		Mine Yard Facilities		
5	1	232	0.12	28
6	2	232	0.10	23
7	3	232	0.09	21
8	4	232	0.08	19
9	5	232	0.07	16
10	6	232	0.06	14
11	7	232	0.06	14
12	8	232	0.06	14
13	9	232	0.06	14
14	10	232	0.05	12
15	11	232	0.05	12
16	12	232	0.05	12
17	13	232	0.05	12
18	14	232	0.05	12
19	15	232	0.05	12

For property tax calculations, a mil levy of 60 was assumed. It should be noted, however, that this rate is determined by the county in which the property is located. Table 14 illustrates the entire property tax calculations for the mine and pulverizing plant. The values are summarized under "total property tax" for inclusion into the cash flow analysis.

The state of Colorado does not impose a severance tax on limestone; therefore, a separate line item is not included in the cash flow analysis.

Net after costs refers to the balance remaining from limestone product sales after the above costs and expenditures have been deducted.

Exploration and development deduction expenditures result from the requirement to capitalize 15% of initial exploration and development expenditures. The five-year period over which these expenditures are deducted extends into the production years in this case study. Table 7 shows the calculations for these deduc-

Table 16. Depreciation Calculations

Assumptions:	ACRS system
	50% capital equipment replacement every 8 years
	Expenditure made in 3 equal payments with 2-year interval

Replacement:	Mine	0.50 × 3631 =	1815
	Pulverizing plant	0.50 × 3562 =	1781
	Total		3596

Expenditures:	Mine	1815/3 =	605
		1781/3 =	594
			1199

Expenditure in Years 10, 12, 14, 18, 20, and 22

Depreciation year	Basis, $1000	ACRS, % per 100	Deduction, $1000
Mine Replacement Equipment			
605 − (0.10 × 605 × 0.50) = 575			
1	575	0.15	86
2	575	0.22	127
3	575	0.21	121
4	575	0.21	120
5	575	0.21	121
Pulverizing Plant Replacement Equipment			
594 − (0.10 × 594 × 0.50) = 564			
1	564	0.15	85
2	564	0.22	124
3	564	0.21	118
4	564	0.21	119
5	564	0.21	118

tions. It should be noted that because these deductions are similar to depreciation and depletion allowances, they must also be added back to net profit in the cash flow analysis to arrive at operating cash flow.

Depreciation was calculated using the Accelerated Cost Recovery System (ACRS) as specified in TEFRA (1982) and discussed in Chapter 8. Table 15 shows the depreciation deductions for the initial investment. Note that the depreciable basis for personal property is reduced by 50% of the investment tax credit available as specified by TEFRA. Table 16 provides depreciation calculations for planned replacement equipment. The simple replacement sequence estimated for the project is noted at the top of Table 16. Replacement equipment depreciation is assumed to begin in the year the replacement expenditure is made. Tables 17 and 18 show the total summarized depreciation deductions for the mine and pulverizing plant, respectively. These deductions are segregated because of depletion considerations.

It should be noted that in this particular case no depreciation was declared on any mine equipment or facilities until the first *production* year. Since most, if not all, of the initial equipment was probably placed into service at some time prior to the first production year, a depreciation deduction could have been claimed earlier.

Table 17. Depreciation Deduction Summary—Mine ($1000)

Project year	Mine yard facilities	Mine equipment	Replacement A			Replacement B			Total
			Year 10	Year 12	Year 14	Year 18	Year 20	Year 22	
5	28	517							545
6	23	759							782
7	21	724							745
8	19	725							744
9	16	724							740
10	14		86						100
11	14		127						141
12	14		121	86					261
13	14		120	127					261
14	12		121	121	86				340
15	12			120	127				259
16	12			121	121				254
17	12				120				132
18	12				121	86			219
19	12					127			139
20						121	86		207
21						120	127		247
22						121	121	86	328
23							120	127	247
24							121	121	242

Table 18. Depreciation Deduction Summary—Pulverizing Plant ($1000)

Project year	Pulverizing plant	Replacement A			Replacement B			Total
		Year 10	Year 12	Year 14	Year 18	Year 20	Year 22	
5	508							508
6	744							744
7	711							711
8	710							710
9	711							711
10		85						85
11		124						124
12		118	85					203
13		119	124					243
14		118	118	85				321
15			119	124				243
16			118	118				236
17				119				119
18				118	85			203
19					124			124
20					118	85		203
21					119	124		243
22					118	118	85	321
23						119	124	243
24						118	118	236

Table 19. State Tax—Loss Carry-Forward

Year	10	11	12
Net after depreciation	865	800	635
Loss carry-forward			
Year 5 (64)	64		
Year 6 (517)	517		
Year 7 (429)	284	145	
Year 8 (412)		412	
Year 9 (393)		243	150
State taxable income	0	0	485

However, a more conservative approach was adopted in this case study for illustrative purposes, and no depreciation was declared prior to project start-up.

Net after depreciation is simply the subtraction between "net after costs" and the allowances for exploration and development deductions and depreciation.

State income tax is calculated based on the state of Colorado income tax rate of 5% on net income earned within the state. Since net after depreciation calculations are negative from years five through nine, no state income tax results. Although taxable income does occur in years 10 and 11, no state income taxes are paid because of the carry-forward previous losses in years five through nine (see Table 19). State income taxes are first paid in project year 12 (e.g., 0.05 × 485 = 24). It is interesting to note that in year 15, depletion is first claimed and state income taxes are based on net after depreciation less depletion (e.g., for year 15: (576-13)(0.05))

= 28). This results from the fact that the federal depletion allowance is a deduction for state income tax calculations in the state of Colorado.

Net after state tax is calculated by subtracting the state income tax from net after depreciation. This number is important for calculation of the statutory (percentage) depletion allowance. It is the number used as net income in the 50% of net income constraint on statutory depletion.

The *depletion* allowance is determined as described in Chapter 7. Cost depletion was compared with the allowable statutory depletion and the larger is utilized for the annual deduction. The property acquisition costs are placed in the depletable basis of the cost account. The basis in the cost account is prorated over the property's minable reserves in order to obtain a unit depletion rate. This rate is then multiplied by the annual production in order to obtain the cost depletion allowance. This number is then compared with the statutory depletion calculation which is the lesser of either 50% of net after state tax or 14% of revenue less royalty. This calculation procedure is illustrated in Table 20.

It is interesting to note that the revenue used in the depletion calculations (Table 20) differs from the revenue reported for the project (Table 8). This results from the fact that depletion is based on *gross income from mining*. The revenue derived from only mining activities is determined on the basis of the sale of "first marketable product." In this case study, the revenue from mining must be calculated separately since the products sold from the pulverizing plant do not represent first marketable products. Therefore, it is necessary to estimate the value of the 107,000 tons of limestone delivered to the pulverizing plant. The IRS often uses the "representative field price" method, but in this case would probably require the "proportionate profits" method for deriving the gross income from mining. The simplified transfer price chosen here was the actual mining cost of $4.89 per ton; however, the IRS would allow the addition of a profit margin based upon the proportion of total costs incurred in mining. The mine revenue determination is summarized in Table 21.

Complications can arise, as in the state of Colorado, when depletion is a deduction for state income taxes and is based on 50% of net after state taxes. Under these conditions, a solution using the following calculation procedure is appropriate for the state of Colorado.

$$NAD = \text{Net after costs and depreciation}$$
$$D = \text{Depletion based on 50\% net income}$$
$$ST = \text{State income tax based on 5\% net after depletion}$$

$$D = (NAD - ST) \times 0.5$$
$$ST = (NAD - D) \times 0.05$$

Solve for ST:

$$D = (NAD - ST) \times 0.5 = 0.5\,N - 0.5\,ST$$
$$ST = (NAD - (0.5\,N - 0.5\,ST)) \times 0.05$$
$$ST = 0.0256\,NAD$$

In this case study, net after state tax is always negative so that 50% of net is negative or zero, and this problem does not exist.

It should be noted that no depletion allowance can be claimed until the

Table 20. Mine Depletion Calculations

Year	5	6	7	8	9	10	11	12	13	14
Revenue	2446	2446	2446	2446	2446	2446	2446	2446	2446	2446
Royalty	245	245	245	245	245	245	245	245	245	245
Net after royalty	2201	2201	2201	2201	2201	2201	2201	2201	2201	2201
Operating costs	2087	2087	2087	2087	2087	2087	2087	2087	2087	2087
Property tax	87	80	73	71	60	64	56	59	50	52
Net after costs	27	34	41	43	54	50	58	55	64	62
Explor. & devel. deduction	16	10	5	4	0	0	0	0	0	0
Depreciation	545	782	745	744	740	100	141	221	261	328
Net after depreciation	-534	-758	-709	-705	-686	-50	-83	-166	-197	-266
State Tax	0	0	0	0	0	0	0	0	0	0
Net after state tax	-534	-758	-709	-705	-686	-50	-83	-166	-197	-266
Depletion	0	0	0	0	0	0	0	0	0	0
Depletion basis, $1000	500	500	500	500	500	500	500	500	500	500
Reserves, 1000 per ton	8568.0	8139.6	7711.2	7282.8	6854.4	6426.0	5997.6	5569.2	5140.8	4712.4
Unit depletion, $ per ton	0.058	0.061	0.065	0.069	0.073	0.078	0.083	0.090	0.097	0.106
Production	428.4	428.4	428.4	428.4	428.4	428.4	428.4	428.4	428.4	428.4
Cost depletion	25	26	28	29	31	33	36	38	42	45
50% net	0	0	0	0	0	0	0	0	0	0
14% net after royalty	308	308	308	308	308	308	308	308	308	308
Depletion earned	25	26	28	29	31	33	36	38	42	45
Depletion recaptured	25	26	28	29	31	33	36	38	42	45
Recapture balance	345	319	291	262	231	198	162	124	82	37
Deduction claimed	0	0	0	0	0	0	0	0	0	0

Table 20. Continued

Year	15	16	17	18	19	20	21	22	23	24
Revenue	2446	2446	2446	2446	2446	2446	2446	2446	2446	2446
Royalty	245	245	245	245	245	245	245	245	245	245
Net after royalty	2201	2201	2201	2201	2201	2201	2201	2201	2201	2201
Operating costs	2087	2087	2087	2087	2087	2087	2087	2087	2087	2087
Property tax	70	64	58	73	63	77	67	81	72	62
Net after costs	65	69	72	65	70	63	69	61	66	71
Explor. & devel. deduction	0	0	0	0	0	0	0	0	0	0
Depreciation	259	254	132	219	139	207	247	328	247	242
Net after depreciation	-194	-185	-60	-154	-169	-144	-178	-267	-181	-171
State Tax	0	0	0	0	0	0	0	0	0	0
Net after state tax	-194	-185	-60	-154	-169	-144	-178	-267	-181	-171
Depletion	13	54	54	54	54	54	54	54	54	54
Depletion basis, $1000	500	487	433	379	325	271	217	163	109	55
Reserves, 1000 per ton	4284.0	3855.6	3427.2	2998.8	2570.4	2142.0	1713.6	1285.2	856.8	428.4
Unit depletion, $ per ton	0.117	0.126	0.126	0.126	0.126	0.126	0.126	0.126	0.126	0.126
Production	428.4	428.4	428.4	428.4	428.4	428.4	428.4	428.4	428.4	428.4
Cost depletion	50	54	54	54	54	54	54	54	54	54
50% net	0	0	0	0	0	0	0	0	0	0
14% net after royalty	308	308	308	308	308	308	308	308	308	308
Depletion earned	50	54	54	54	54	54	54	54	54	54
Depletion recaptured	37	0	0	0	0	0	0	0	0	0
Recapture balance	0	0	0	0	0	0	0	0	0	0
Deduction claimed	13	54	54	54	54	54	54	54	54	54

Table 21. Gross Income from Mining

Product	Specifications	Tpy	F.o.b. mine	Revenue, $ per year
Sugar rock	(5½ × 3 in. and 3 × 2 in.: + 95% CaCO₃	235,600	7.25	1,708,000
Cement plant feed	¾ in. −: + 88%CaCO₃)	85,700	2.50	214,200
"Pulverizing" plant feed	(2 × ¾-in.)	107,100	4.89	523,700
Total mine revenue		428,400	5.71	2,446,000

$370,000 expensed exploration expenditure is recaptured. This does not occur until project year 15. Also, the basis in the cost depletion account is altered only by the actual depletion claimed. Therefore, the basis in the cost depletion account does not change until project year 16.

In this case study, cost depletion always applies since the lesser of the statutory depletion calculations is zero (50% of net after state tax). As expected, the cost depletion account is reduced to zero at the end of the project's life.

Pretax profit or taxable income is simply the net amount of income generated by the project which is subject to federal income taxation.

Federal income tax represents taxes imposed on taxable income generated by the project. The 46% rate was used throughout on the assumption that other profit centers in the corporation generated the first $100,000 of taxable income. For project years five through nine, federal corporate tax credits are generated from tax losses due to negative taxable income. It is assumed these tax credits can be used by Limy to offset other corporate taxable income.

Tax credits refer to investment tax credits as described in Chapter 8. These credits simply provide for a reduction in taxes during the year in which an expenditure for a qualifying asset is made. Since buildings do not qualify for investment tax credits, these credits were only generated by expenditures in mine and pulverizing plant equipment. The investment tax credit rate is 10% on qualifying investment, assuming the election is made to reduce the depreciable basis of the asset by 50% of the investment tax credit. If this election is not made, the investment tax credit is limited to 8% on qualifying investment. Since the depreciable basis of the assets was reduced in this case study, the 10% rate was used for all qualifying initial and replacement equipment. In Tables 6 and 8, the investment tax credits are declared the same year in which the capital expenditures are incurred. Table 8 expenditures are associated with replacement equipment in project year 10, 12, 14, 18, 20, and 22.

The investment tax credit is limited by the lesser of the tax liability or $25,000 plus 85% of the tax liability less $25,000. Since this case study assumes the corporate scenario, the assumption is made that the corporate tax liability is adequate to allow declaration of the full tax credit in the year of capital expenditures.

Minimum tax is the tax imposed on preference items in excess of 100% of the corporate tax liability (Chapter 8). The three primary tax preference items relating to mining are:

1) Accelerated depreciation on real property in excess of straight-line deduction.

2) Depletion claimed in excess of adjusted basis at beginning of year.

3) Difference between exploration and development deductions over straight-line deduction for a 10-year period.

Although some uncertainty exists at the present time, it appears that a minimum tax on exploration and development is a function of the corporate structure. It is included here for completeness and illustrative purposes.

Table 22 illustrates the minimum tax calculation procedure for the three preference items for all project years. The minimum tax rate is 15%. These annual minimum tax determinations are transferred to Table 6 and 8 for cash flow calculations. The minimum tax becomes zero after project year nine because the adjusted project tax liability exceeds the tax preference items.

Net profit represents the amount remaining after all taxes are paid from net project earnings. It is simply taxable income less federal income taxes less minimum tax plus investment tax credits. This net profit number represents that component of a cash flow analysis referred to as the return *on* investment.

Because *depreciation, depletion*, and the *capitalized portion of exploration and development expenditures* are noncash items and simply have the effect of reducing taxable income (and therefore taxes paid) for the property, they are added to net profit as a credit in order to determine the true *operating cash flow* from the property for each year. This value actually represents the amount of cash generated by the property, as contrasted to the net profit value. In this regard depreciation, depletion, and capitalized exploration and development represent the return *of* investment component in a cash flow analysis and are of vital interest from a taxation standpoint.

Capital investment represents expenditures for capital assets during a given year. The impact of these capital investments is reflected elsewhere in the pro forma income statement in the line items of depreciation and investment tax credit. In the case study, capital investment represents the periodic replacement of mine and pulverizing plant equipment in production years 10, 12, 14, 18, 20, and 22. Separate depreciation calculations were established for these replacement expenditures and the subsequent yearly depreciation deductions were impacted accordingly (see Tables 16 and 17).

Working capital represents an estimate of the money necessary to cover operating costs until sales proceeds are first received by the project. The estimate should take into consideration the mining methodology, equipment spread, and manpower utilized at the property. In this case study, the estimate was based on the operating cost per ton times the annual production for a two-month period (see Table 5). In theory this account is churned throughout the property life and is available to the owner at the end of the project. Therefore, in project year 24 working capital is *added* to operating cash flow in order to obtain net annual cash flow. The account is larger in project year 24 than that established in project year 5 because it also contains the $44,000 allocated for spares and supplies in preproduction year 4.

The final line item in the spread sheet, *net cash flow*, represents the true cash position of the project for any specific year. It is calculated as net profit plus depreciation plus depletion minus capital investment minus working capital changes. These annual net cash flows are then used in calculating evaluation criteria to test the economic viability of the project.

CASE STUDY: EVALUATION CRITERIA

Following the calculation procedures discussed in Chapter 9 on project evalua-

Table 22. Minimum Tax Calculations ($1000)

Year	1	2	3	4	5	6	7	8	9	10	11	12	13	14
Tax preference items														
Accelerated depreciation: real prop.					28	23	21	19	16	14	14	14	14	12
Straight-line depreciation*					15	16	15	16	15	16	15	16	15	16
Tax preference: depreciation					13	7	6	3	1	0	0	0	0	0
Depletion claimed					0	0	0	0	0	0	0	0	0	0
Adjusted basis					500	500	500	500	500	500	500	500	500	500
Tax preference: depletion					0	0	0	0	0	0	0	0	0	
Exploration and devel. deductions														
Year 1 (180)	157	6	6	5	6	0	0	0	0	0				
Year 1 straight-line	18	18	18	18	18	18	18	18	18	18				
Year 1 tax preference	139	0	0	0	0	0	0	0	0	0				
Year 2 (190)	0	166	6	6	6	6	0	0	0	0	0			
Year 2 straight-line	0	19	19	19	19	19	19	19	19	19	19			
Year 2 tax preference	0	147	0	0	0	0	0	0	0	0	0			
Year 4 (134)	0	0	0	117	4	4	5	4	0	0	0	0	0	
Year 4 straight-line	0	0	0	13	14	13	14	13	14	13	14	13	13	
Year 4 tax preference	0	0	0	104	0	0	0	0	0	0	0	0	0	
Total tax preference items	139	147	0	104	13	7	6	3	1	0	0	0	0	0
Tax liability	0	0	0	0	0	0	0	0	0	398	368	281	250	185
Investment tax credit	—	3	250	471	0	0	0	0	0	120	0	120	0	120
Adjusted tax liability	0	0	0	0	0	0	0	0	0	278	368	161	250	65
Excess	139	147	0	104	13	7	6	3	1	0	0	0	0	0
Minimum tax @ 15%	21	22	0	16	2	1	1	0.45	0.15	0	0	0	0	0

*232 per 15 years = 15.5.

Table 22. Continued

Year	15	16	17	18	19	20	21	22	23	24
Tax preference items										
Accelerated depreciation: real prop.	12	12	12	12	12	0	0	0	0	0
Straight-line depreciation*	15	16	15	16	15	0	0	0	0	0
Tax preference: depreciation	0	0	0	0	0	0	0	0	0	0
Depletion claimed	13	54	54	54	54	54	54	54	54	54
Adjusted basis	500	487	433	379	325	271	217	163	109	55
Tax preference: depletion	0	0	0	0	0	0	0	0	0	0
Exploration and devel. deductions										
Year 1 (180)										
Year 1 straight-line	0	0	0	0	0	0	0	0	0	0
Year 1 tax preference	0	0	0	0	0	0	0	0	0	0
Year 2 (190)										
Year 2 straight-line	0	0	0	0	0	0	0	0	0	0
Year 2 tax preference	0	0	0	0	0	0	0	0	0	0
Year 4 (134)										
Year 4 straight-line	0	0	0	0	0	0	0	0	0	0
Year 4 tax preference	0	0	0	0	0	0	0	0	0	0
Total tax preference items	0	0	0	0	0	0	0	0	0	0
Tax liability	246	237	344	263	337	266	236	160	233	243
Investment tax credit	0	0	0	120	0	120	0	120	0	0
Adjusted tax liability	246	237	344	143	337	146	236	40	233	243
Excess	0	0	0	0	0	0	0	0	0	0
Minimum tax @ 15%	0	0	0	0	0	0	0	0	0	0

*232 per 15 years = 15.5.

tion criteria, some of the more typical criteria were determined for the case study. A required rate of return, or hurdle rate, was used as the discount and/or reinvestment rate wherever appropriate.

Values for the major evaluation criteria at various discount rates follow:

	Discount rate		
Criterion	5%	10%	15%
Net present value ($1000)	− 270	− 1756	− 2304
Profitability index	0.96	0.71	0.55

Payout from project start, years	=	18.8
Payout from first production, years	=	14.8
DCF-ROR, %	=	4.5

From these values it is apparent that Limy Mining Co. would not be overly enthusiastic about acquiring this particular property under the existing economic conditions. While there may be other overriding considerations which support acquisition (e.g., acquisition of potential future reserves, fulfill existing marketing obligations, secure a base of operation in a new market, diversification, etc.), the economics resulting from this preliminary feasibility study suggest that acquisition of this particular property is justified only if the firm's cost of funds is less than 4.5%. It is interesting to note that a similar feasibility study was performed on this same property six years previously. At that time the projected DCR-ROR was almost 11%—a very real example of the relative decline in profitability experienced in the minerals industry in recent years. In this specific case study, the decline in projected return was the result of capital and operating cost increases escalating much faster than the price of salable products.

ESCALATED DOLLAR ANALYSIS

While the preceding constant dollar analysis illustrates most of the principles involved in the financial analysis of a mining property, it ignored, for the sake of simplicity, one important feature—inflation. Chapter 10 described in detail the impact of the declining value of currency on investment analysis and concluded that serious errors could arise if this phenomenon is ignored. Therefore, the analysis was repeated, using the escalation rates shown:

	Annual escalation rates, %			
	Year 1	Year 2	Year 3	Beyond
Revenue	7.0	8.0	8.5	8.5
Operating costs	6.1	8.2	8.5	8.5
Capital costs	6.9	7.6	8.0	8.0
General rate of inflation	7.0	7.7	8.5	8.5

Tables 23 and 24 show the escalated production cash flows for the project. Escalated net annual cash flows were reduced to time zero dollars using the general rate of inflation. Tables 25 through 31 show the salient calculations and backup data for the escalated analysis. For this analysis property taxes were assumed to escalate beyond those determined in the constant dollar scenario at the same rate as revenue.

Table 23. Magnitude and Timing of Annual Cash Flows—Preproduction ($1000, Escalated Dollars)

Capital expenditure	Time 0	Year 1	Year 2	Year 3	Year 4	Total
Property payment	500	—	—	—	—	500
Exploration and feasibility study	—	192	219	—	—	411
Preproduction development	—	—	—	—	180	180
Mine	—	—	—	1242	3531	4773
Pulverizing plant	—	—	—	1863	2769	4632
Mine yard facilities	—	—	—	124	177	301
Property tax	—	—	—	—	175	175
Working capital	—	—	—	—	59	59
Minimum tax	0	22	25	0	21	68
Total capital expenditures	−500	−214	−244	−3229	−6912	−11,099
Cash generated						
Tax savings						
Exploration and feasibility study*	—	77	91	6	6	180
Preproduction development†	—	—	—	—	72	72
Property tax‡	—	—	—	—	81	81
Investment tax credit§	—	—	—	311	630	941
Investment tax credit expl. & devel.¶	—	3	3	0	3	9
Total cash generated	0	80	94	317	792	1283
Net cash flow‖	−500	−134	−150	−2912	−6120	−9816

*See Table 25.
†See Table 25.
‡Amount = property tax × 0.46.
§Investment credit = (mine + pulverizing plant equipment) × 0.10.
¶Investment credit = (capitalized exploration and development) × 0.10.
‖Net cash flow = capital expenditures + cash generated.

Table 24. Production Cash Flows—Project ($1000, Escalated Dollars)

Year	5	6	7	8	9	10	11	12	13	14
Revenue	6106	6623	7186	7798	8460	9180	9958	10,806	11,724	12,721
Operating costs	3981	4319	4686	5084	5517	5986	6493	7047	7645	8295
Royalty	360	391	424	460	499	542	588	638	692	751
Property tax	214	210	205	205	188	222	205	238	210	240
Net after costs	1551	1703	1871	2049	2256	2430	2672	2883	3177	3435
Explor. & devel. deduction	18	13	5	6	0	0	0	0	0	0
Depreciation—mine	716	1027	979	976	974	260	372	638	769	1077
Depreciation—pulverizing	660	968	924	924	924	234	343	601	729	975
Net after depreciation	157	-305	-37	143	358	1936	1957	1644	1679	1383
State income tax @ 5%	8	0	0	0	18†	97	98	82	84	69
Net after state tax	149	-305	-37	143	340	1839	1859	1562	1595	1314
Depletion	0	0	0	0	0	0	0	0	0	0
Pretax profit	149	-305	-37	143	340	1839	1859	1562	1595	1314
Federal income tax @ 46%	69	140‡	17‡	66	156	846	855	719	734	604
Tax credits (ITC)	0	0	0	0	0	0	0	0	0	431
Minimum tax	0	2	1	0	0	0	0	0	0	0
Net profit	80	-167	-21	77	184	1310	1004	1213	861	1141
Explor. & devel. deduction	18	13	5	6	0	0	0	0	0	0
Depreciation—mine	716	1027	979	976	974	260	372	638	769	1077
Depreciation—pulverizing	660	968	924	924	924	234	343	601	729	975
Depletion	0	0	0	0	0	0	0	0	0	0
Operating cash flow	1474	1841	1887	1983	2082	1804	1719	2452	2359	3193
Capital expenditures	507*	0	0	0	0	3171	0	3699	0	4314
Net cash flow — escalated dollars	967	1841	1887	1983	2082	-1367	1719	-1247	2359	-1121
Net cash flow — time zero dollars	657	1153	1089	1055	1021	-618	716	-479	834	-366

*Working capital.
†State tax loss carry-forward.
‡Federal corporate tax credit from tax loss.

Table 24. Continued

Year	15	16	17	18	19	20	21	22	23	24
Revenue	13,801	14,976	16,250	17,628	19,129	20,755	22,518	24,433	26,510	28,760
Operating costs	8999	9764	10,595	11,494	12,473	13,535	14,683	15,931	17,286	18,755
Royalty	814	884	959	1040	1129	1225	1329	1442	1565	1697
Property tax	240	232	228	311	291	386	365	478	461	431
Net after costs	3748	4096	4468	4783	5236	5609	6141	6582	7198	7877
Explor. & devel. deduction	0	0	0	0	0	0	0	0	0	0
Depreciation—mine	891	869	475	922	671	1147	1391	1964	1622	1581
Depreciation—pulverizing	849	829	446	879	636	1113	1348	1905	1573	1534
Net after depreciation	2008	2398	3547	2982	3929	3349	3402	2713	4003	4762
State income tax @ 5%	100	120	177	149	196	167	170	136	200	238
Net after state tax	1908	2278	3370	2833	3733	3182	3232	2577	3803	4524
Depletion	0	28	59	59	59	59	59	59	59	59
Pretax profit	1908	2250	3311	2774	3674	3123	3173	2518	3744	4465
Federal income tax @ 46%	878	1035	1523	1276	1690	1437	1460	1158	1722	2054
Tax credits (ITC)	0	0	0	0	0	685	0	798	0	0
Minimum tax	0	0	0	0	0	0	0	0	0	0
Net profit	1030	1215	1788	2085	1984	2371	1713	2158	2022	2411
Explor. & devel. deduction	0	0	0	0	0	0	0	0	0	0
Depreciation—mine	891	869	475	922	671	1147	1391	1964	1622	1581
Depreciation—pulverizing	849	829	446	879	636	1113	1348	1905	1573	1534
Depletion	0	28	59	59	59	59	59	59	59	59
Operating cash flow	2770	2941	3768	3945	3350	4690	4511	6086	5276	5585
Capital expenditures	0	0	0	5869	0	6846	0	7985	0	2662§
Net cash flow—escalated dollars	2770	2941	2768	−1924	3350	−2156	4511	−1899	5276	8247
Net cash flow—time zero dollars	832	814	706	−453	726	−431	831	−322	825	1189

§Recapture of working capital plus spares and supplies.

Table 25. Calculation of Exploration and Development Tax Savings—Preproduction ($1000, Escalated Dollars)

Year	1	2	3	4	5	6	7	8
Exploration expenditure	192	219						
Allowable expense deduction @ 85%	163	186						
15% reduction amount	29	33						
Year 1:								
15% reduction deduction rate	0.15	0.22	0.21	0.21	0.21			
15% reduction deduction	4	6	6	7	6			
Year 2:								
15% reduction deduction rate		0.15	0.22	0.21	0.21	0.21		
15% reduction deduction		5	7	7	7	7		
Total exploration deduction	167	197	13	14	13	7		
Exploration tax savings @ 46%	77	91	6	6	—	—		
Development expenditures				180				
Allowable expenses deduction @ 85%				153				
15% reduction amount				27				
Year 4:								
15% reduction deduction rate				0.15	0.22	0.21	0.21	0.21
15% reduction deduction				4	6	6	5	6
Total development deductions				157	6	6	5	6
Development tax savings @ 46%				72	—	—	—	—
15% reduction amounts	29	33	0	27	0	0	0	0
Investment tax credit @ 10%	3	3	0	3	0	0	0	0
Exploration and development								
Deductions carried to production cash flows	—	—	—	—	18	13	5	6

Table 26. Depreciation Calculations—Initial Investment
(Escalated Dollars)

Project year	Depreciation year	Basis, $1000	ACRS, % per 100	Deduction, $1000
		Assumptions: ACRS System—Place into Service in Year 5		

Mine Equipment
4773 − (0.10 × 4773 × 0.50) = 4534

Project year	Depreciation year	Basis, $1000	ACRS, % per 100	Deduction, $1000
5	1	4534	0.15	680
6	2	4534	0.22	997
7	3	4534	0.21	952
8	4	4534	0.21	952
9	5	4534	0.21	953

Pulverizing Plant
4632 − (0.10 × 4632 × 0.50) × 4400

5	1	4400	0.15	660
6	2	4400	0.22	968
7	3	4400	0.21	924
8	4	4400	0.21	924
9	5	4400	0.21	924

Mine Yard Facilities

5	1	301	0.12	36
6	2	301	0.10	30
7	3	301	0.09	27
8	4	301	0.08	24
9	5	301	0.07	21
10	6	301	0.06	18
11	7	301	0.06	18
12	8	301	0.06	18
13	9	301	0.06	18
14	10	301	0.05	16
15	11	301	0.05	15
16	12	301	0.05	15
17	13	301	0.05	15
18	14	301	0.05	15
19	15	301	0.05	15

Evaluation criteria for the escalated case are as follows:

Criterion	Discount rate 0%	4%
Net present value	+2184	−2540
Profitability index	1.21	0.74
Payout from project start, years =	19.9	
Payout from first production, years =	15.9	
DCF-ROR, % =	1.9	

The escalation indexes used in this example are arbitrary and track well with each other. The results, therefore, are not unexpected, as the rate of return on the

Table 27. Depreciation Calculations (Escalated Dollars)

Assumptions:	ACRS system
	50% capital equipment replacement every 8 years
	Expenditure made in 3 equal payments with 2-year interval

Replacement:	Mine	$0.50 \times 4773 = 2386$
	Pulverizing plant	$0.50 \times 4632 = 2316$
	Total	4702

Expenditures:	Mine	$2386/3 = 796$
		$2316/3 = 772$
		1568

Expenditure in Years 10, 12, 14, 18, 20, and 22

Depreciation year	Basis, $1000	ACRS, % per 100	Deduction, $1000
Mine Replacement Equipment			
$796 - (0.10 \times 796 \times 0.50) = 756$			
1	756	0.15	113
2	756	0.22	166
3	756	0.21	159
4	756	0.21	159
5	756	0.21	159
Pulverizing Plant Replacement Equipment			
$772 - (0.10 \times 772 \times 0.50) = 733$			
1	733	0.15	110
2	733	0.22	161
3	733	0.21	154
4	733	0.21	154
5	733	0.21	154

investment declined from 4.5% in the constant dollar case to 1.9% in the escalated dollar case. Thus, as mentioned previously, ignoring inflation tends to overvalue a project. Constant dollar cash flows help in developing a technical understanding of a project, but the final investment decision should be based upon an inflation-adjusted analysis.

SUMMARY

It should be apparent from this brief case study that a significant amount of work and backup data are necessary in order to perform even a preliminary cash flow analysis of a mining property. The case study clearly illustrates the importance of tax considerations in a cash flow analysis. Short cuts in this area are rarely warranted.

This case study adequately illustrates the procedure and format associated with compiling a cash flow analysis. Cash flow analyses are difficult and time-consuming and only as accurate as the mine plans and cost estimates used in their formula-

Table 28. Depreciation Deduction Summary—Mine ($1000, Escalated Dollars)

Project year	Mine yard facilities	Mine equipment	Replacement A			Replacement B			Total
			Year 10	Year 12	Year 14	Year 18	Year 20	Year 22	
5	36	680							716
6	30	997							1027
7	27	952							979
8	24	952							976
9	21	953							974
10	18		242						260
11	18		354						372
12	18		338	282					638
13	18		338	413					769
14	16		338	394	329				1077
15	15			394	482				891
16	15			394	460				869
17	15				460				475
18	15				460	447			922
19	15					656			671
20						626	521		1147
21						626	765		1391
22						626	730	608	1964
23							730	892	1622
24							730	851	1581

Table 29. Depreciation Deduction Summary—Pulverizing Plant ($1000, Escalated Dollars)

Project year	Pulverizing plant	Replacement A			Replacement B			Total
		Year 10	Year 12	Year 14	Year 18	Year 20	Year 22	
5	660							660
6	968							968
7	924							924
8	924							924
9	924							924
10		234						234
11		343						343
12		328	273					601
13		328	401					729
14		328	382	319				975
15			382	467				849
16			383	446				829
17				446				446
18				446	433			879
19					636			636
20					607	506		1113
21					607	741		1348
22					607	708	590	1905
23						708	865	1573
24						708	826	1534

tion. Remember, no amount of elegant manipulation can take the place of good data.

SELECTED BIBLIOGRAPHY

Gentry, D.W., and Hrebar, M.J., 1983, "Economic Principles for Industrial Minerals Property Valuations," Short Course, SME-AIME Annual Meeting, Atlanta, GA, March, 172 pp.

Table 30. Mine Depletion Calculations (Escalated Dollars)

Year	5	6	7	8	9	10	11	12	13	14
Revenue	3610	3916	4249	4611	5002	5428	5888	6389	6932	7521
Royalty	360	391	424	460	499	542	588	638	692	751
Net after royalty	3250	3525	3825	4151	4503	4886	5300	5751	6240	6770
Operating costs	3030	3287	3567	3870	4199	4556	4942	5364	5819	6314
Property tax	128	128	127	134	123	142	135	154	142	160
Net after costs	92	110	131	147	181	188	223	233	279	296
Explor. & devel. deduction	18	13	5	6	0	0	0	0	0	0
Depreciation	716	1027	979	976	974	260	372	638	769	1077
Net after depreciation	(642)	(930)	(853)	(835)	(793)	(72)	(149)	(405)	(490)	(781)
State Tax	0	0	0	0	0	0	0	0	0	0
Net after state tax	(642)	(930)	(853)	(835)	(793)	(72)	(149)	(405)	(490)	(781)
Depletion	0	0	0	0	0	0	0	0	0	0
Depletion basis, $1000	500	500	500	500	500	500	500	500	500	500
Reserves, 1000 per ton	8568.0	8139.6	7771.2	7282.8	6854.4	6426.0	5997.6	5569.2	5140.8	4712.4
Unit depletion, $ per ton	0.058	0.061	0.065	0.069	0.073	0.078	0.083	0.090	0.097	0.106
Production	428.4	428.4	428.4	428.4	428.4	428.4	428.4	428.4	428.4	428.4
Cost depletion	25	26	28	29	31	33	36	38	42	45
50% net	0	0	0	0	0	0	0	0	0	0
14% net after royalty	455	494	536	581	630	684	742	805	874	948
Depletion earned	25	26	28	29	31	33	36	38	42	45
Depletion recaptured	25	26	28	29	31	33	36	38	42	45
Recapture balance	386	360	332	303	272	239	203	165	123	78
Deduction claimed	0	0	0	0	0	0	0	0	0	0

Table 30. Continued

Year	15	16	17	18	19	20	21	22	23	24
Revenue	8160	8855	9608	10,423	11,310	12,271	13,314	14,446	15,674	17,004
Royalty	814	884	959	1040	1129	1225	1329	1442	1565	1697
Net after royalty	7346	7971	8649	9383	10,181	11,046	11,985	13,004	14,109	15,307
Operating costs	6849	7432	8064	8748	9493	10,302	11,176	12,125	13,157	14,275
Property tax	234	232	228	311	291	386	365	478	461	431
Net after costs	263	307	357	324	397	358	444	401	491	601
Explor. & devel. deduction	0	0	0	0	0	0	0	0	0	0
Depreciation	891	869	475	922	671	1147	1391	1964	1622	1581
Net after depreciation	(628)	(562)	(118)	(598)	(274)	(789)	(947)	(1563)	(1131)	(980)
State Tax	0	0	0	0	0	0	0	0	0	0
Net after state tax	(628)	(562)	(118)	(598)	(274)	(789)	(947)	(1563)	(1131)	(980)
Depletion	0	26	54	54	54	54	54	54	54	54
Depletion basis, $1000	500	500	472	473	354	295	236	177	118	59
Reserves, 1000 per ton	4284.0	3855.6	3427.2	2998.8	2570.4	2142.0	1713.6	1285.2	856.8	428.4
Unit depletion, $ per ton	0.117	0.130	0.138	0.138	0.138	0.138	0.138	0.138	0.138	0.138
Production	428.4	428.4	428.4	428.4	428.4	428.4	428.4	428.4	428.4	428.4
Cost depletion	50	56	59	59	59	59	59	59	59	59
50% net	0	0	0	0	0	0	0	0	0	0
14% net after royalty	1028	1116	1211	1314	1425	1546	1678	1821	1975	2143
Depletion earned	50	56	59	59	59	59	59	59	59	59
Depletion recaptured	50	28	0	0	0	0	0	0	0	0
Recapture balance	28	0	0	0	0	0	0	0	0	0
Deduction claimed	0	28	59	59	59	59	59	59	59	59

Table 31. Minimum Tax Calculations ($1000, Escalated Dollars)

Year	1	2	3	4	5	6	7	8	9	10	11	12	13	14
Tax preference items														
Accelerated depreciation: real prop.					36	30	27	24	21	18	18	18	18	16
Straight-line depreciation*					20	20	20	20	20	20	20	20	20	20
Tax preference: depreciation					16	10	7	4	1	0	0	0	0	0
Depletion claimed					0	0	0	0	0	0	0	0	0	0
Adjusted basis					500	500	500	500	500	500	500	500	500	500
Tax preference: depletion					0	0	0	0	0	0	0	0	0	0
Exploration and devel. deductions														
Year 1 (192)	167	6	6	7	6	0	0	0	0	0				
Year 1 straight-line	19	19	19	19	19	19	19	20	19	20				
Year 1 tax preference	148	0	0	0	0	0	0	0	0	0				
Year 2 (219)	0	191	7	7	7	7	0	0	0	0	0			
Year 2 straight-line	0	22	22	22	22	22	22	22	22	22	21			
Year 2 tax preference	0	169	0	0	0	0	0	0	0	0	0			
Year 4 (180)	0	0	0	157	6	6	5	6	0	0	0	0	0	
Year 4 straight-line	0	0	0	18	18	18	18	18	18	18	18	18	18	
Year 4 tax preference	0	0	0	139	0	0	0	0	0	0	0	0	0	
Total tax preference items	148	169	0	139	16	10	7	4	1	0	0	0	0	0
Tax liability	0	0	0	0	69	0	0	66	463	846	855	719	734	604
Investment tax credit	3	3	311	633	0	0	0	0	0	317	0	370	0	431
Adjusted tax liability	—	—	—	—	69	0	0	66	463	529	855	349	734	173
Excess	148	169	0	139	0	10	7	0	0	0	0	0	0	0
Minimum tax @ 15%	22	25	0	21	0	2	1	0	0	0	0	0	0	0

*301 per 15 years = 20.

Table 31. Continued

Year	15	16	17	18	19	20	21	22	23	24
Tax preference items										
Accelerated depreciation: real prop.	15	15	15	15	15					
Straight-line depreciation*	20	20	20	20	21					
Tax preference: depreciation	0	0	0	0	0					
Depletion claimed	0	28	59	59	59	59	59	59	59	59
Adjusted basis	500	500	472	413	354	295	236	177	118	59
Tax preference: depletion	0	0	0	0	0	0	0	0	0	0
Exploration and devel. deductions										
Year 1 (192)										
Year 1 straight-line	0	0	0	0	0	0	0	0	0	0
Year 1 tax preference	0	0	0	0	0	0	0	0	0	0
Year 2 (219)										
Year 2 straight-line	0	0	0	0	0	0	0	0	0	0
Year 2 tax preference	0	0	0	0	0	0	0	0	0	0
Year 4 (180)										
Year 4 straight-line	0	0	0	0	0	0	0	0	0	0
Year 4 tax preference	0	0	0	0	0	0	0	0	0	0
Total tax preference items	0	0	0	0	0	0	0	0	0	0
Tax liability	878	1035	1523	1276	1690	1437	1460	1158	1722	2054
Investment tax credit	0	0	0	587	0	685	0	798	0	0
Adjusted tax liability	878	1035	1523	689	1690	752	1460	360	1722	2054
Excess	0	0	0	0	0	0	0	0	0	0
Minimum tax @ 15%	0	0	0	0	0	0	0	0	0	0

*301 per 15 years = 20.

Interest Factors for Discrete Compounding

Table A1. ½% Interest Factors for Annual Compounding Interest

	Single Payment		Equal Payment Series				Uniform gradient-series factor
	Compound-amount factor	Present-worth factor	Compound-amount factor	Sinking-fund factor	Present-worth factor	Capital-recovery factor	
n	To find F Given P F/P i,n	To find P Given F P/F i,n	To find F Given A F/A i,n	To find A Given F A/F i,n	To find P Given A P/A i,n	To find A Given P A/P i,n	To find A Given G A/G i,n
1	1.005	0.9950	1.000	1.0000	0.9950	1.0050	0.0000
2	1.010	0.9901	2.005	0.4988	1.9851	0.5038	0.4988
3	1.015	0.9852	3.015	0.3317	2.9703	0.3367	0.9967
4	1.020	0.9803	4.030	0.2481	3.9505	0.2531	1.4938
5	1.025	0.9754	5.050	0.1980	4.9259	0.2030	1.9900
6	1.030	0.9705	6.076	0.1646	5.8964	0.1696	2.4855
7	1.036	0.9657	7.106	0.1407	6.8621	0.1457	2.9801
8	1.041	0.9609	8.141	0.1228	7.8230	0.1278	3.4738
9	1.046	0.9561	9.182	0.1089	8.7791	0.1139	3.9668
10	1.051	0.9514	10.228	0.0978	9.7304	0.1028	4.4589
11	1.056	0.9466	11.279	0.0887	10.6770	0.0937	4.9501
12	1.062	0.9419	12.336	0.0811	11.6189	0.0861	5.4406
13	1.067	0.9372	13.397	0.0747	12.5562	0.0797	5.9302
14	1.072	0.9326	14.464	0.0691	13.4887	0.0741	6.4190
15	1.078	0.9279	15.537	0.0644	14.4166	0.0694	6.9069
16	1.083	0.9233	16.614	0.0602	15.3399	0.0652	7.3940
17	1.088	0.9187	17.697	0.0565	16.2586	0.0615	7.8803
18	1.094	0.9141	18.786	0.0532	17.1728	0.0582	8.3658
19	1.099	0.9096	19.880	0.0503	18.0824	0.0553	8.8504
20	1.105	0.9051	20.979	0.0477	18.9874	0.0527	9.3342
21	1.110	0.9006	22.084	0.0453	19.8880	0.0503	9.8172
22	1.116	0.8961	23.194	0.0431	20.7841	0.0481	10.2993
23	1.122	0.8916	24.310	0.0411	21.6757	0.0461	10.7806
24	1.127	0.8872	25.432	0.0393	22.5629	0.0443	11.2611
25	1.133	0.8828	26.559	0.0377	23.4456	0.0427	11.7407
26	1.138	0.8784	27.692	0.0361	24.3240	0.0411	12.2195
27	1.144	0.8740	28.830	0.0347	25.1980	0.0397	12.6975
28	1.150	0.8697	29.975	0.0334	26.0677	0.0384	13.1747
29	1.156	0.8653	31.124	0.0321	26.9330	0.0371	13.6510
30	1.161	0.8610	32.280	0.0310	27.7941	0.0360	14.1265
31	1.167	0.8568	33.441	0.0299	28.6508	0.0349	14.6012
32	1.173	0.8525	34.609	0.0289	29.5033	0.0339	15.0750
33	1.179	0.8483	35.782	0.0280	30.3515	0.0330	15.5480
34	1.185	0.8440	36.961	0.0271	31.1956	0.0321	16.0202
35	1.191	0.8398	38.145	0.0262	32.0354	0.0312	16.4915
40	1.221	0.8191	44.159	0.0227	36.1722	0.0277	18.8358
45	1.252	0.7990	50.324	0.0199	40.2072	0.0249	21.1595
50	1.283	0.7793	56.645	0.0177	44.1428	0.0227	23.4624
55	1.316	0.7601	63.126	0.0159	47.9815	0.0209	25.7447
60	1.349	0.7414	69.770	0.0143	51.7256	0.0193	28.0064
65	1.383	0.7231	76.582	0.0131	55.3775	0.0181	30.2475
70	1.418	0.7053	83.566	0.0120	58.9394	0.0170	32.4680
75	1.454	0.6879	90.727	0.0110	62.4137	0.0160	34.6679
80	1.490	0.6710	98.068	0.0102	65.8023	0.0152	36.8474
85	1.528	0.6545	105.594	0.0095	69.1075	0.0145	39.0065
90	1.567	0.6384	113.311	0.0088	72.3313	0.0138	41.1451
95	1.606	0.6226	121.222	0.0083	75.4757	0.0133	43.2633
100	1.647	0.6073	129.334	0.0077	78.5427	0.0127	45.3613

*Thuesen, H.G., Fabrycky, W.J., Thuesen, G.J., *Engineering Economy*, 5th ed., ©1977, pp. 537-557. Reprinted by permission of Prentice-Hall, Inc., Englewood Cliffs, NJ.

Table A2. ¾% Interest Factors for Annual Compounding Interest

n	Single Payment		Equal Payment Series				Uniform gradient-series factor
	Compound-amount factor	Present-worth factor	Compound-amount factor	Sinking-fund factor	Present-worth factor	Capital-recovery factor	
	To find F Given P F/P i,n	To find P Given F P/F i,n	To find F Given A F/A i,n	To find A Given F A/F i,n	To find P Given A P/A i,n	To find A Given P A/P i,n	To find A Given G A/G i,n
1	1.008	0.9926	1.000	1.0000	0.9926	1.0075	0.0000
2	1.015	0.9852	2.008	0.4981	1.9777	0.5056	0.4981
3	1.023	0.9778	3.023	0.3309	2.9556	0.3384	0.9950
4	1.030	0.9706	4.045	0.2472	3.9261	0.2547	1.4907
5	1.038	0.9633	5.076	0.1970	4.8894	0.2045	1.9851
6	1.046	0.9562	6.114	0.1636	5.8456	0.1711	2.4782
7	1.054	0.9491	7.159	0.1397	6.7946	0.1472	2.9701
8	1.062	0.9420	8.213	0.1218	7.7366	0.1293	3.4608
9	1.070	0.9350	9.275	0.1078	8.6716	0.1153	3.9502
10	1.078	0.9280	10.344	0.0967	9.5996	0.1042	4.4384
11	1.086	0.9211	11.422	0.0876	10.5207	0.0951	4.9253
12	1.094	0.9142	12.508	0.0800	11.4349	0.0875	5.4110
13	1.102	0.9074	13.601	0.0735	12.3424	0.0810	5.8954
14	1.110	0.9007	14.703	0.0680	13.2430	0.0755	6.3786
15	1.119	0.8940	15.814	0.0632	14.1370	0.0707	6.8606
16	1.127	0.8873	16.932	0.0591	15.0243	0.0666	7.3413
17	1.135	0.8807	18.059	0.0554	15.9050	0.0629	7.8207
18	1.144	0.8742	19.195	0.0521	16.7792	0.0596	8.2989
19	1.153	0.8677	20.339	0.0492	17.6468	0.0567	8.7759
20	1.161	0.8612	21.491	0.0465	18.5080	0.0540	9.2517
21	1.170	0.8548	22.652	0.0442	19.3628	0.0517	9.7261
22	1.179	0.8484	23.822	0.0420	20.2112	0.0495	10.1994
23	1.188	0.8421	25.001	0.0400	21.0533	0.0475	10.6714
24	1.196	0.8358	26.188	0.0382	21.8892	0.0457	11.1422
25	1.205	0.8296	27.385	0.0365	22.7188	0.0440	11.6117
26	1.214	0.8234	28.590	0.0350	23.5422	0.0425	12.0800
27	1.224	0.8173	29.805	0.0336	24.3595	0.0411	12.5470
28	1.233	0.8112	31.028	0.0322	25.1707	0.0397	13.0128
29	1.242	0.8052	32.261	0.0310	25.9759	0.0385	13.4774
30	1.251	0.7992	33.503	0.0299	26.7751	0.0374	13.9407
31	1.261	0.7932	34.754	0.0288	27.5683	0.0363	14.4028
32	1.270	0.7873	36.015	0.0278	28.3557	0.0353	14.8636
33	1.280	0.7815	37.285	0.0268	29.1371	0.0343	15.3232
34	1.289	0.7757	38.565	0.0259	29.9128	0.0334	15.7816
35	1.299	0.7699	39.854	0.0251	30.6827	0.0326	16.2387
40	1.348	0.7417	46.446	0.0215	34.4469	0.0290	18.5058
45	1.400	0.7145	53.290	0.0188	38.0732	0.0263	20.7421
50	1.453	0.6883	60.394	0.0166	41.5665	0.0241	22.9476
55	1.508	0.6630	67.769	0.0148	44.9316	0.0223	25.1223
60	1.566	0.6387	75.424	0.0133	48.1734	0.0208	27.2665
65	1.625	0.6153	83.371	0.0120	51.2963	0.0195	29.3801
70	1.687	0.5927	91.620	0.0109	54.3046	0.0184	31.4634
75	1.751	0.5710	100.183	0.0100	57.2027	0.0175	33.5163
80	1.818	0.5501	109.073	0.0092	59.9945	0.0167	35.5391
85	1.887	0.5299	118.300	0.0085	62.6838	0.0160	37.5318
90	1.959	0.5105	127.879	0.0078	65.2746	0.0153	39.4946
95	2.034	0.4917	137.823	0.0073	67.7704	0.0148	41.4277
100	2.111	0.4737	148.145	0.0068	70.1746	0.0143	43.3311

Table A3. 1% Interest Factors for Annual Compounding Interest

n	Single Payment		Equal Payment Series				Uniform gradient-series factor
	Compound-amount factor	Present-worth factor	Compound-amount factor	Sinking-fund factor	Present-worth factor	Capital-recovery factor	
	To find F Given P F/P i,n	To find P Given F P/F i,n	To find F Given A F/A i,n	To find A Given F A/F i,n	To find P Given A P/A i,n	To find A Given P A/P i,n	To find A Given G A/G i,n
1	1.010	0.9901	1.000	1.0000	0.9901	1.0100	0.0000
2	1.020	0.9803	2.010	0.4975	1.9704	0.5075	0.4975
3	1.030	0.9706	3.030	0.3300	2.9410	0.3400	0.9934
4	1.041	0.9610	4.060	0.2463	3.9020	0.2563	1.4876
5	1.051	0.9515	5.101	0.1960	4.8534	0.2060	1.9801
6	1.062	0.9421	6.152	0.1626	5.7955	0.1726	2.4710
7	1.072	0.9327	7.214	0.1386	6.7282	0.1486	2.9602
8	1.083	0.9235	8.286	0.1207	7.6517	0.1307	3.4478
9	1.094	0.9143	9.369	0.1068	8.5660	0.1168	3.9337
10	1.105	0.9053	10.462	0.0956	9.4713	0.1056	4.4179
11	1.116	0.8963	11.567	0.0865	10.3676	0.0965	4.9005
12	1.127	0.8875	12.683	0.0789	11.2551	0.0889	5.3815
13	1.138	0.8787	13.809	0.0724	12.1338	0.0824	5.8607
14	1.149	0.8700	14.947	0.0669	13.0037	0.0769	6.3384
15	1.161	0.8614	16.097	0.0621	13.8651	0.0721	6.8143
16	1.173	0.8528	17.258	0.0580	14.7179	0.0680	7.2887
17	1.184	0.8444	18.430	0.0543	15.5623	0.0643	7.7613
18	1.196	0.8360	19.615	0.0510	16.3983	0.0610	8.2323
19	1.208	0.8277	20.811	0.0481	17.2260	0.0581	8.7017
20	1.220	0.8196	22.019	0.0454	18.0456	0.0554	9.1694
21	1.232	0.8114	23.239	0.0430	18.8570	0.0530	9.6354
22	1.245	0.8034	24.472	0.0409	19.6604	0.0509	10.0998
23	1.257	0.7955	25.716	0.0389	20.4558	0.0489	10.5626
24	1.270	0.7876	26.973	0.0371	21.2434	0.0471	11.0237
25	1.282	0.7798	28.243	0.0354	22.0232	0.0454	11.4831
26	1.295	0.7721	29.526	0.0339	22.7952	0.0439	11.9409
27	1.308	0.7644	30.821	0.0325	23.5596	0.0425	12.3971
28	1.321	0.7568	32.129	0.0311	24.3165	0.0411	12.8516
29	1.335	0.7494	33.450	0.0299	25.0658	0.0399	13.3045
30	1.348	0.7419	34.785	0.0288	25.8077	0.0388	13.7557
31	1.361	0.7346	36.133	0.0277	26.5423	0.0377	14.2052
32	1.375	0.7273	37.494	0.0267	27.2696	0.0367	14.6532
33	1.389	0.7201	38.869	0.0257	27.9897	0.0357	15.0995
34	1.403	0.7130	40.258	0.0248	28.7027	0.0348	15.5441
35	1.417	0.7059	41.660	0.0240	29.4086	0.0340	15.9871
40	1.489	0.6717	48.886	0.0205	32.8347	0.0305	18.1776
45	1.565	0.6391	56.481	0.0177	36.0945	0.0277	20.3273
50	1.645	0.6080	64.463	0.0155	39.1961	0.0255	22.4363
55	1.729	0.5785	72.852	0.0137	42.1472	0.0237	24.5049
60	1.817	0.5505	81.670	0.0123	44.9550	0.0223	26.5333
65	1.909	0.5237	90.937	0.0110	47.6266	0.0210	28.5217
70	2.007	0.4983	100.676	0.0099	50.1685	0.0199	30.4703
75	2.109	0.4741	110.913	0.0090	52.5871	0.0190	32.3793
80	2.217	0.4511	121.672	0.0082	54.8882	0.0182	34.2492
85	2.330	0.4292	132.979	0.0075	57.0777	0.0175	36.0801
90	2.449	0.4084	144.863	0.0069	59.1609	0.0169	37.8725
95	2.574	0.3886	157.354	0.0064	61.1430	0.0164	39.6265
100	2.705	0.3697	170.481	0.0059	63.0289	0.0159	41.3426

Table A4. 1¼% Interest Factors for Annual Compounding Interest

	Single Payment		Equal Payment Series				Uniform gradient-series factor
n	Compound-amount factor	Present-worth factor	Compound-amount factor	Sinking-fund factor	Present-worth factor	Capital-recovery factor	
	To find F Given P F/P i,n	To find P Given F P/F i,n	To find F Given A F/A i,n	To find A Given F A/F i,n	To find P Given A P/A i,n	To find A Given P A/P i,n	To find A Given G A/G i,n
1	1.013	0.9877	1.000	1.0001	0.9877	1.0126	0.0000
2	1.025	0.9755	2.013	0.4970	1.9631	0.5095	0.4932
3	1.038	0.9635	3.038	0.3293	2.9265	0.3418	0.9895
4	1.051	0.9516	4.076	0.2454	3.8780	0.2579	1.4830
5	1.064	0.9398	5.127	0.1951	4.8177	0.2076	1.9729
6	1.077	0.9282	6.191	0.1616	5.7459	0.1741	2.4618
7	1.091	0.9168	7.268	0.1376	6.6627	0.1501	2.9491
8	1.105	0.9055	8.359	0.1197	7.5680	0.1322	3.4330
9	1.118	0.8943	9.463	0.1057	8.4623	0.1182	3.9158
10	1.132	0.8832	10.582	0.0946	9.3454	0.1071	4.3960
11	1.147	0.8723	11.714	0.0854	10.2177	0.0979	4.8744
12	1.161	0.8616	12.860	0.0778	11.0792	0.0903	5.3506
13	1.175	0.8509	14.021	0.0714	11.9300	0.0839	5.8248
14	1.190	0.8404	15.196	0.0659	12.7704	0.0784	6.2968
15	1.205	0.8300	16.386	0.0611	13.6004	0.0736	6.7669
16	1.220	0.8198	17.591	0.0569	14.4201	0.0694	7.2350
17	1.235	0.8097	18.811	0.0532	15.2298	0.0657	7.7009
18	1.251	0.7997	20.046	0.0499	16.0293	0.0624	8.1645
19	1.266	0.7898	21.296	0.0470	16.8191	0.0595	8.6264
20	1.282	0.7801	22.563	0.0444	17.5991	0.0569	9.0861
21	1.298	0.7704	23.845	0.0420	18.3695	0.0545	9.5439
22	1.314	0.7609	25.143	0.0398	19.1303	0.0523	9.9993
23	1.331	0.7515	26.457	0.0378	19.8818	0.0503	10.4528
24	1.347	0.7423	27.788	0.0360	20.6240	0.0485	10.9044
25	1.364	0.7331	29.135	0.0344	21.3570	0.0469	11.3539
26	1.381	0.7240	30.499	0.0328	22.0810	0.0453	11.8012
27	1.399	0.7151	31.880	0.0314	22.7960	0.0439	12.2465
28	1.416	0.7063	33.279	0.0301	23.5022	0.0426	12.6898
29	1.434	0.6976	34.695	0.0289	24.1998	0.0414	13.1311
30	1.452	0.6889	36.128	0.0277	24.8886	0.0402	13.5703
31	1.470	0.6804	37.580	0.0267	25.5690	0.0392	14.0074
32	1.488	0.6720	39.050	0.0257	26.2410	0.0382	14.4425
33	1.507	0.6637	40.538	0.0247	26.9047	0.0372	14.8756
34	1.526	0.6555	42.045	0.0238	27.5601	0.0363	15.3066
35	1.545	0.6475	43.570	0.0230	28.2075	0.0355	15.7357
40	1.644	0.6085	51.489	0.0195	31.3266	0.0320	17.8503
45	1.749	0.5718	59.915	0.0167	34.2578	0.0292	19.9144
50	1.861	0.5374	68.880	0.0146	37.0125	0.0271	21.9284
55	1.980	0.5050	78.421	0.0128	39.6013	0.0253	23.8925
60	2.107	0.4746	88.573	0.0113	42.0342	0.0238	25.8072
65	2.242	0.4460	99.375	0.0101	44.3206	0.0226	27.6730
70	2.386	0.4192	110.870	0.0091	46.4693	0.0216	29.4902
75	2.539	0.3939	123.101	0.0082	48.4886	0.0207	31.2594
80	2.702	0.3702	136.116	0.0074	50.3862	0.0199	32.9812
85	2.875	0.3479	149.965	0.0067	52.1696	0.0192	34.6560
90	3.059	0.3270	164.701	0.0061	53.8456	0.0186	36.2844
95	3.255	0.3073	180.382	0.0056	55.4207	0.0181	37.8671
100	3.463	0.2888	197.067	0.0051	56.9009	0.0176	39.4048

Table A5. 1½% Interest Factors for Annual Compounding Interest

n	Single Payment		Equal Payment Series				Uniform gradient-series factor
	Compound-amount factor	Present-worth factor	Compound-amount factor	Sinking-fund factor	Present-worth factor	Capital-recovery factor	
	To find F Given P F/P i,n	To find P Given F P/F i,n	To find F Given A F/A i,n	To find A Given F A/F i,n	To find P Given A P/A i,n	To find A Given P A/P i,n	To find A Given G A/G i,n
1	1.015	0.9852	1.000	1.0000	0.9852	1.0150	0.0000
2	1.030	0.9707	2.015	0.4963	1.9559	0.5113	0.4963
3	1.046	0.9563	3.045	0.3284	2.9122	0.3434	0.9901
4	1.061	0.9422	4.091	0.2445	3.8544	0.2595	1.4814
5	1.077	0.9283	5.152	0.1941	4.7827	0.2091	1.9702
6	1.093	0.9146	6.230	0.1605	5.6972	0.1755	2.4566
7	1.110	0.9010	7.323	0.1366	6.5982	0.1516	2.9405
8	1.127	0.8877	8.433	0.1186	7.4859	0.1336	3.4219
9	1.143	0.8746	9.559	0.1046	8.3605	0.1196	3.9008
10	1.161	0.8617	10.703	0.0934	9.2222	0.1084	4.3772
11	1.178	0.8489	11.863	0.0843	10.0711	0.0993	4.8512
12	1.196	0.8364	13.041	0.0767	10.9075	0.0917	5.3227
13	1.214	0.8240	14.237	0.0703	11.7315	0.0853	5.7917
14	1.232	0.8119	15.450	0.0647	12.5434	0.0797	6.2582
15	1.250	0.7999	16.682	0.0600	13.3432	0.0750	6.7223
16	1.269	0.7880	17.932	0.0558	14.1313	0.0708	7.1839
17	1.288	0.7764	19.201	0.0521	14.9077	0.0671	7.6431
18	1.307	0.7649	20.489	0.0488	15.6726	0.0638	8.0997
19	1.327	0.7536	21.797	0.0459	16.4262	0.0609	8.5539
20	1.347	0.7425	23.124	0.0433	17.1686	0.0583	9.0057
21	1.367	0.7315	24.471	0.0409	17.9001	0.0559	9.4550
22	1.388	0.7207	25.838	0.0387	18.6208	0.0537	9.9018
23	1.408	0.7100	27.225	0.0367	19.3309	0.0517	10.3462
24	1.430	0.6996	28.634	0.0349	20.0304	0.0499	10.7881
25	1.451	0.6892	30.063	0.0333	20.7196	0.0483	11.2276
26	1.473	0.6790	31.514	0.0317	21.3986	0.0467	11.6646
27	1.495	0.6690	32.987	0.0303	22.0676	0.0453	12.0992
28	1.517	0.6591	34.481	0.0290	22.7267	0.0440	12.5313
29	1.540	0.6494	35.999	0.0278	23.3761	0.0428	12.9610
30	1.563	0.6398	37.539	0.0266	24.0158	0.0416	13.3883
31	1.587	0.6303	39.102	0.0256	24.6462	0.0406	13.8131
32	1.610	0.6210	40.688	0.0246	25.2671	0.0396	14.2355
33	1.634	0.6118	42.299	0.0237	25.8790	0.0387	14.6555
34	1.659	0.6028	43.933	0.0228	26.4817	0.0378	15.0731
35	1.684	0.5939	45.592	0.0219	27.0756	0.0369	15.4882
40	1.814	0.5513	54.268	0.0184	29.9159	0.0334	17.5277
45	1.954	0.5117	63.614	0.0157	32.5523	0.0307	19.5074
50	2.105	0.4750	73.683	0.0136	34.9997	0.0286	21.4277
55	2.268	0.4409	84.530	0.0118	37.2715	0.0268	23.2894
60	2.443	0.4093	96.215	0.0104	39.3803	0.0254	25.0930
65	2.632	0.3799	108.803	0.0092	41.3378	0.0242	26.8392
70	2.835	0.3527	122.364	0.0082	43.1549	0.0232	28.5290
75	3.055	0.3274	136.973	0.0073	44.8416	0.0223	30.1631
80	3.291	0.3039	152.711	0.0066	46.4073	0.0216	31.7423
85	3.545	0.2821	169.665	0.0059	47.8607	0.0209	33.2676
90	3.819	0.2619	187.930	0.0053	49.2099	0.0203	34.7399
95	4.114	0.2431	207.606	0.0048	50.4622	0.0198	36.1602
100	4.432	0.2256	228.803	0.0044	51.6247	0.0194	37.5295

Table A6. 2% Interest Factors for Annual Compounding Interest

	Single Payment		Equal Payment Series				Uniform gradient-series factor
	Compound-amount factor	Present-worth factor	Compound-amount factor	Sinking-fund factor	Present-worth factor	Capital-recovery factor	
n	To find F Given P $F/P\ i,n$	To find P Given F $P/F\ i,n$	To find F Given A $F/A\ i,n$	To find A Given F $A/F\ i,n$	To find P Given A $P/A\ i,n$	To find A Given P $A/P\ i,n$	To find A Given G $A/G\ i,n$
1	1.020	0.9804	1.000	1.0000	0.9804	1.0200	0.0000
2	1.040	0.9612	2.020	0.4951	1.9416	0.5151	0.4951
3	1.061	0.9423	3.060	0.3268	2.8839	0.3468	0.9868
4	1.082	0.9239	4.122	0.2426	3.8077	0.2626	1.4753
5	1.104	0.9057	5.204	0.1922	4.7135	0.2122	1.9604
6	1.126	0.8880	6.308	0.1585	5.6014	0.1785	2.4423
7	1.149	0.8706	7.434	0.1345	6.4720	0.1545	2.9208
8	1.172	0.8535	8.583	0.1165	7.3255	0.1365	3.3961
9	1.195	0.8368	9.755	0.1025	8.1622	0.1225	3.8681
10	1.219	0.8204	10.950	0.0913	8.9826	0.1113	4.3367
11	1.243	0.8043	12.169	0.0822	9.7869	0.1022	4.8021
12	1.268	0.7885	13.412	0.0746	10.5754	0.0946	5.2643
13	1.294	0.7730	14.680	0.0681	11.3484	0.0881	5.7231
14	1.319	0.7579	15.974	0.0626	12.1063	0.0826	6.1786
15	1.346	0.7430	17.293	0.0578	12.8493	0.0778	6.6309
16	1.373	0.7285	18.639	0.0537	13.5777	0.0737	7.0799
17	1.400	0.7142	20.012	0.0500	14.2919	0.0700	7.5256
18	1.428	0.7002	21.412	0.0467	14.9920	0.0667	7.9681
19	1.457	0.6864	22.841	0.0438	15.6785	0.0638	8.4073
20	1.486	0.6730	24.297	0.0412	16.3514	0.0612	8.8433
21	1.516	0.6598	25.783	0.0388	17.0112	0.0588	9.2760
22	1.546	0.6468	27.299	0.0366	17.6581	0.0566	9.7055
23	1.577	0.6342	28.845	0.0347	18.2922	0.0547	10.1317
24	1.608	0.6217	30.422	0.0329	18.9139	0.0529	10.5547
25	1.641	0.6095	32.030	0.0312	19.5235	0.0512	10.9745
26	1.673	0.5976	33.671	0.0297	20.1210	0.0497	11.3910
27	1.707	0.5859	35.344	0.0283	20.7069	0.0483	11.8043
28	1.741	0.5744	37.051	0.0270	21.2813	0.0470	12.2145
29	1.776	0.5631	38.792	0.0258	21.8444	0.0458	12.6214
30	1.811	0.5521	40.568	0.0247	22.3965	0.0447	13.0251
31	1.848	0.5413	42.379	0.0236	22.9377	0.0436	13.4257
32	1.885	0.5306	44.227	0.0226	23.4683	0.0426	13.8230
33	1.922	0.5202	46.112	0.0217	23.9886	0.0417	14.2172
34	1.961	0.5100	48.034	0.0208	24.4986	0.0408	14.6083
35	2.000	0.5000	49.994	0.0200	24.9986	0.0400	14.9961
40	2.208	0.4529	60.402	0.0166	27.3555	0.0366	16.8885
45	2.438	0.4102	71.893	0.0139	29.4902	0.0339	18.7034
50	2.692	0.3715	84.579	0.0118	31.4236	0.0318	20.4420
55	2.972	0.3365	98.587	0.0102	33.1748	0.0302	22.1057
60	3.281	0.3048	114.052	0.0088	34.7609	0.0288	23.6961
65	3.623	0.2761	131.126	0.0076	36.1975	0.0276	25.2147
70	4.000	0.2500	149.978	0.0067	37.4986	0.0267	26.6632
75	4.416	0.2265	170.792	0.0059	38.6771	0.0259	28.0434
80	4.875	0.2051	193.772	0.0052	39.7445	0.0252	29.3572
85	5.383	0.1858	219.144	0.0046	40.7113	0.0246	30.6064
90	5.943	0.1683	247.157	0.0041	41.5869	0.0241	31.7929
95	6.562	0.1524	278.085	0.0036	42.3800	0.0236	32.9189
100	7.245	0.1380	312.232	0.0032	43.0984	0.0232	33.9863

Table A7. 3% Interest Factors for Annual Compounding Interest

n	Single Payment		Equal Payment Series				Uniform gradient-series factor
	Compound-amount factor	Present-worth factor	Compound-amount factor	Sinking-fund factor	Present-worth factor	Capital-recovery factor	
	To find F Given P $F/P\ i,n$	To find P Given F $P/F\ i,n$	To find F Given A $F/A\ i,n$	To find A Given F $A/F\ i,n$	To find P Given A $P/A\ i,n$	To find A Given P $A/P\ i,n$	To find A Given G $A/G\ i,n$
1	1.030	0.9709	1.000	1.0000	0.9709	1.0300	0.0000
2	1.061	0.9426	2.030	0.4926	1.9135	0.5226	0.4926
3	1.093	0.9152	3.091	0.3235	2.8286	0.3535	0.9803
4	1.126	0.8885	4.184	0.2390	3.7171	0.2690	1.4631
5	1.159	0.8626	5.309	0.1884	4.5797	0.2184	1.9409
6	1.194	0.8375	6.468	0.1546	5.4172	0.1846	2.4138
7	1.230	0.8131	7.662	0.1305	6.2303	0.1605	2.8819
8	1.267	0.7894	8.892	0.1125	7.0197	0.1425	3.3450
9	1.305	0.7664	10.159	0.0984	7.7861	0.1284	3.8032
10	1.344	0.7441	11.464	0.0872	8.5302	0.1172	4.2565
11	1.384	0.7224	12.808	0.0781	9.2526	0.1081	4.7049
12	1.426	0.7014	14.192	0.0705	9.9540	0.1005	5.1485
13	1.469	0.6810	15.618	0.0640	10.6350	0.0940	5.5872
14	1.513	0.6611	17.086	0.0585	11.2961	0.0885	6.0211
15	1.558	0.6419	18.599	0.0538	11.9379	0.0838	6.4501
16	1.605	0.6232	20.157	0.0496	12.5611	0.0796	6.8742
17	1.653	0.6050	21.762	0.0460	13.1661	0.0760	7.2936
18	1.702	0.5874	23.414	0.0427	13.7535	0.0727	7.7081
19	1.754	0.5703	25.117	0.0398	14.3238	0.0698	8.1179
20	1.806	0.5537	26.870	0.0372	14.8775	0.0672	8.5229
21	1.860	0.5376	28.676	0.0349	15.4150	0.0649	8.9231
22	1.916	0.5219	30.537	0.0328	15.9369	0.0628	9.3186
23	1.974	0.5067	32.453	0.0308	16.4436	0.0608	9.7094
24	2.033	0.4919	34.426	0.0291	16.9356	0.0591	10.0954
25	2.094	0.4776	36.459	0.0274	17.4132	0.0574	10.4768
26	2.157	0.4637	38.553	0.0259	17.8769	0.0559	10.8535
27	2.221	0.4502	40.710	0.0246	18.3270	0.0546	11.2256
28	2.288	0.4371	42.931	0.0233	18.7641	0.0533	11.5930
29	2.357	0.4244	45.219	0.0221	19.1885	0.0521	11.9558
30	2.427	0.4120	47.575	0.0210	19.6005	0.0510	12.3141
31	2.500	0.4000	50.003	0.0200	20.0004	0.0500	12.6678
32	2.575	0.3883	52.503	0.0191	20.3888	0.0491	13.0169
33	2.652	0.3770	55.078	0.0182	20.7658	0.0482	13.3616
34	2.732	0.3661	57.730	0.0173	21.1318	0.0473	13.7018
35	2.814	0.3554	60.462	0.0165	21.4872	0.0465	14.0375
40	3.262	0.3066	75.401	0.0133	23.1148	0.0433	15.6502
45	3.782	0.2644	92.720	0.0108	24.5187	0.0408	17.1556
50	4.384	0.2281	112.797	0.0089	25.7298	0.0389	18.5575
55	5.082	0.1968	136.072	0.0074	26.7744	0.0374	19.8600
60	5.892	0.1697	163.053	0.0061	27.6756	0.0361	21.0674
65	6.830	0.1464	194.333	0.0052	28.4529	0.0352	22.1841
70	7.918	0.1263	230.594	0.0043	29.1234	0.0343	23.2145
75	9.179	0.1090	272.631	0.0037	29.7018	0.0337	24.1634
80	10.641	0.0940	321.363	0.0031	30.2008	0.0331	25.0354
85	12.336	0.0811	377.857	0.0027	30.6312	0.0327	25.8349
90	14.300	0.0699	443.349	0.0023	31.0024	0.0323	26.5667
95	16.578	0.0603	519.272	0.0019	31.3227	0.0319	27.2351
100	19.219	0.0520	607.288	0.0017	31.5989	0.0317	27.8445

Table A8. 4% Interest Factors for Annual Compounding Interest

n	Single Payment		Equal Payment Series				Uniform gradient-series factor
	Compound-amount factor	Present-worth factor	Compound-amount factor	Sinking-fund factor	Present-worth factor	Capital-recovery factor	
	To find F Given P $F/P\ i,n$	To find P Given F $P/F\ i,n$	To find F Given A $F/A\ i,n$	To find A Given F $A/F\ i,n$	To find P Given A $P/A\ i,n$	To find A Given P $A/P\ i,n$	To find A Given G $A/G\ i,n$
1	1.040	0.9615	1.000	1.0000	0.9615	1.0400	0.0000
2	1.082	0.9246	2.040	0.4902	1.8861	0.5302	0.4902
3	1.125	0.8890	3.122	0.3204	2.7751	0.3604	0.9739
4	1.170	0.8548	4.246	0.2355	3.6299	0.2755	1.4510
5	1.217	0.8219	5.416	0.1846	4.4518	0.2246	1.9216
6	1.265	0.7903	6.633	0.1508	5.2421	0.1908	2.3857
7	1.316	0.7599	7.898	0.1266	6.0021	0.1666	2.8433
8	1.369	0.7307	9.214	0.1085	6.7328	0.1485	3.2944
9	1.423	0.7026	10.583	0.0945	7.4353	0.1345	3.7391
10	1.480	0.6756	12.006	0.0833	8.1109	0.1233	4.1773
11	1.539	0.6496	13.486	0.0742	8.7605	0.1142	4.6090
12	1.601	0.6246	15.026	0.0666	9.3851	0.1066	5.0344
13	1.665	0.6006	16.627	0.0602	9.9857	0.1002	5.4533
14	1.732	0.5775	18.292	0.0547	10.5631	0.0947	5.8659
15	1.801	0.5553	20.024	0.0500	11.1184	0.0900	6.2721
16	1.873	0.5339	21.825	0.0458	11.6523	0.0858	6.6720
17	1.948	0.5134	23.698	0.0422	12.1657	0.0822	7.0656
18	2.026	0.4936	25.645	0.0390	12.6593	0.0790	7.4530
19	2.107	0.4747	27.671	0.0361	13.1339	0.0761	7.8342
20	2.191	0.4564	29.778	0.0336	13.5903	0.0736	8.2091
21	2.279	0.4388	31.969	0.0313	14.0292	0.0713	8.5780
22	2.370	0.4220	34.248	0.0292	14.4511	0.0692	8.9407
23	2.465	0.4057	36.618	0.0273	14.8569	0.0673	9.2973
24	2.563	0.3901	39.083	0.0256	15.2470	0.0656	9.6479
25	2.666	0.3751	41.646	0.0240	15.6221	0.0640	9.9925
26	2.772	0.3607	44.312	0.0226	15.9828	0.0626	10.3312
27	2.883	0.3468	47.084	0.0212	16.3296	0.0612	10.6640
28	2.999	0.3335	49.968	0.0200	16.6631	0.0600	10.9909
29	3.119	0.3207	52.966	0.0189	16.9837	0.0589	11.3121
30	3.243	0.3083	56.085	0.0178	17.2920	0.0578	11.6274
31	3.373	0.2965	59.328	0.0169	17.5885	0.0569	11.9371
32	3.508	0.2851	62.701	0.0160	17.8736	0.0560	12.2411
33	3.648	0.2741	66.210	0.0151	18.1477	0.0551	12.5396
34	3.794	0.2636	69.858	0.0143	18.4112	0.0543	12.8325
35	3.946	0.2534	73.652	0.0136	18.6646	0.0536	13.1199
40	4.801	0.2083	95.026	0.0105	19.7928	0.0505	14.4765
45	5.841	0.1712	121.029	0.0083	20.7200	0.0483	15.7047
50	7.107	0.1407	152.667	0.0066	21.4822	0.0466	16.8123
55	8.646	0.1157	191.159	0.0052	22.1086	0.0452	17.8070
60	10.520	0.0951	237.991	0.0042	22.6235	0.0442	18.6972
65	12.799	0.0781	294.968	0.0034	23.0467	0.0434	19.4909
70	15.572	0.0642	364.290	0.0028	23.3945	0.0428	20.1961
75	18.945	0.0528	448.631	0.0022	23.6804	0.0422	20.8206
80	23.050	0.0434	551.245	0.0018	23.9154	0.0418	21.3719
85	28.044	0.0357	676.090	0.0015	24.1085	0.0415	21.8569
90	34.119	0.0293	817.983	0.0012	24.2673	0.0412	22.2826
95	41.511	0.0241	1012.785	0.0010	24.3978	0.0410	22.6550
100	50.505	0.0198	1237.624	0.0008	24.5050	0.0408	22.9800

Table A9. 5% Interest Factors for Annual Compounding Interest

n	Single Payment		Equal Payment Series				Uniform gradient-series factor
	Compound-amount factor	Present-worth factor	Compound-amount factor	Sinking-fund factor	Present-worth factor	Capital-recovery factor	
	To find F Given P F/P i,n	To find P Given F P/F i,n	To find F Given A F/A i,n	To find A Given F A/F i,n	To find P Given A P/A i,n	To find A Given P A/P i,n	To find A Given G A/G i,n
1	1.050	0.9524	1.000	1.0000	0.9524	1.0500	0.0000
2	1.103	0.9070	2.050	0.4878	1.8594	0.5378	0.4878
3	1.158	0.8638	3.153	0.3172	2.7233	0.3672	0.9675
4	1.216	0.8227	4.310	0.2320	3.5460	0.2820	1.4391
5	1.276	0.7835	5.526	0.1810	4.3295	0.2310	1.9025
6	1.340	0.7462	6.802	0.1470	5.0757	0.1970	2.3579
7	1.407	0.7107	8.142	0.1228	5.7864	0.1728	2.8052
8	1.477	0.6768	9.549	0.1047	6.4632	0.1547	3.2445
9	1.551	0.6446	11.027	0.0907	7.1078	0.1407	3.6758
10	1.629	0.6139	12.587	0.0795	7.7217	0.1295	4.0991
11	1.710	0.5847	14.207	0.0704	8.3064	0.1204	4.5145
12	1.796	0.5568	15.917	0.0628	8.8633	0.1128	4.9219
13	1.886	0.5303	17.713	0.0565	9.3936	0.1065	5.3215
14	1.980	0.5051	19.599	0.0510	9.8987	0.1010	5.7133
15	2.079	0.4810	21.579	0.0464	10.3797	0.0964	6.0973
16	2.183	0.4581	23.658	0.0423	10.8378	0.0923	6.4736
17	2.292	0.4363	25.840	0.0387	11.2741	0.0887	6.8423
18	2.407	0.4155	28.132	0.0356	11.6896	0.0856	7.2034
19	2.527	0.3957	30.539	0.0328	12.0853	0.0828	7.5569
20	2.653	0.3769	33.066	0.0303	12.4622	0.0803	7.9030
21	2.786	0.3590	35.719	0.0280	12.8212	0.0780	8.2416
22	2.925	0.3419	38.505	0.0260	13.1630	0.0760	8.5730
23	3.072	0.3256	41.430	0.0241	13.4886	0.0741	8.8971
24	3.225	0.3101	44.502	0.0225	13.7987	0.0725	9.2140
25	3.386	0.2953	47.727	0.0210	14.0940	0.0710	9.5238
26	3.556	0.2813	51.113	0.0196	14.3752	0.0696	9.8266
27	3.733	0.2679	54.669	0.0183	14.6430	0.0683	10.1224
28	3.920	0.2551	58.403	0.0171	14.8981	0.0671	10.4114
29	4.116	0.2430	62.323	0.0161	15.1411	0.0661	10.6936
30	4.322	0.2314	66.439	0.0151	15.3725	0.0651	10.9691
31	4.538	0.2204	70.761	0.0141	15.5928	0.0641	11.2381
32	4.765	0.2099	75.299	0.0133	15.8027	0.0633	11.5005
33	5.003	0.1999	80.064	0.0125	16.0026	0.0625	11.7566
34	5.253	0.1904	85.067	0.0118	16.1929	0.0618	12.0063
35	5.516	0.1813	90.320	0.0111	16.3742	0.0611	12.2498
40	7.040	0.1421	120.800	0.0083	17.1591	0.0583	13.3775
45	8.985	0.1113	159.700	0.0063	17.7741	0.0563	14.3644
50	11.467	0.0872	209.348	0.0048	18.2559	0.0548	15.2233
55	14.636	0.0683	272.713	0.0037	18.6335	0.0537	15.9665
60	18.679	0.0535	353.584	0.0028	18.9293	0.0528	16.6062
65	23.840	0.0420	456.798	0.0022	19.1611	0.0522	17.1541
70	30.426	0.0329	588.529	0.0017	19.3427	0.0517	17.6212
75	38.833	0.0258	756.654	0.0013	19.4850	0.0513	18.0176
80	49.561	0.0202	971.229	0.0010	19.5965	0.0510	18.3526
85	63.254	0.0158	1245.087	0.0008	19.6838	0.0508	18.6346
90	80.730	0.0124	1594.607	0.0006	19.7523	0.0506	18.8712
95	103.035	0.0097	2040.694	0.0005	19.8059	0.0505	19.0689
100	131.501	0.0076	2610.025	0.0004	19.8479	0.0504	19.2337

Table A10. 6% Interest Factors for Annual Compounding Interest

	Single Payment		Equal Payment Series				Uniform gradient-series factor
n	Compound-amount factor	Present-worth factor	Compound-amount factor	Sinking-fund factor	Present-worth factor	Capital-recovery factor	
	To find F Given P F/P i,n	To find P Given F P/F i,n	To find F Given A F/A i,n	To find A Given F A/F i,n	To find P Given A P/A i,n	To find A Given P A/P i,n	To find A Given G A/G i,n
1	1.060	0.9434	1.000	1.0000	0.9434	1.0600	0.0000
2	1.124	0.8900	2.060	0.4854	1.8334	0.5454	0.4854
3	1.191	0.8396	3.184	0.3141	2.6730	0.3741	0.9612
4	1.262	0.7921	4.375	0.2286	3.4651	0.2886	1.4272
5	1.338	0.7473	5.637	0.1774	4.2124	0.2374	1.8836
6	1.419	0.7050	6.975	0.1434	4.9173	0.2034	2.3304
7	1.504	0.6651	8.394	0.1191	5.5824	0.1791	2.7676
8	1.594	0.6274	9.897	0.1010	6.2098	0.1610	3.1952
9	1.689	0.5919	11.491	0.0870	6.8017	0.1470	3.6133
10	1.791	0.5584	13.181	0.0759	7.3601	0.1359	4.0220
11	1.898	0.5268	14.972	0.0668	7.8869	0.1268	4.4213
12	2.012	0.4970	16.870	0.0593	8.3839	0.1193	4.8113
13	2.133	0.4688	18.882	0.0530	8.8527	0.1130	5.1920
14	2.261	0.4423	21.015	0.0476	9.2950	0.1076	5.5635
15	2.397	0.4173	23.276	0.0430	9.7123	0.1030	5.9260
16	2.540	0.3937	25.673	0.0390	10.1059	0.0990	6.2794
17	2.693	0.3714	28.213	0.0355	10.4773	0.0955	6.6240
18	2.854	0.3504	30.906	0.0324	10.8276	0.0924	6.9597
19	3.026	0.3305	33.760	0.0296	11.1581	0.0896	7.2867
20	3.207	0.3118	36.786	0.0272	11.4699	0.0872	7.6052
21	3.400	0.2942	39.993	0.0250	11.7641	0.8520	7.9151
22	3.604	0.2775	43.392	0.0231	12.0416	0.0831	8.2166
23	3.820	0.2618	46.996	0.0213	12.3034	0.0813	8.5099
24	4.049	0.2470	50.816	0.0197	12.5504	0.0797	8.7951
25	4.292	0.2330	54.865	0.0182	12.7834	0.0782	9.0722
26	4.549	0.2198	59.156	0.0169	13.0032	0.0769	9.3415
27	4.822	0.2074	63.706	0.0157	13.2105	0.0757	9.6030
28	5.112	0.1956	68.528	0.0146	13.4062	0.0746	9.8568
29	5.418	0.1846	73.640	0.0136	13.5907	0.0736	10.1032
30	5.744	0.1741	79.058	0.0127	13.7648	0.0727	10.3422
31	6.088	0.1643	84.802	0.0118	13.9291	0.0718	10.5740
32	6.453	0.1550	90.890	0.0110	14.0841	0.0710	10.7988
33	6.841	0.1462	97.343	0.0103	14.2302	0.0703	11.0166
34	7.251	0.1379	104.184	0.0096	14.3682	0.0696	11.2276
35	7.686	0.1301	111.435	0.0090	14.4983	0.0690	11.4319
40	10.286	0.0972	154.762	0.0065	15.0463	0.0665	12.3590
45	13.765	0.0727	212.744	0.0047	15.4558	0.0647	13.1413
50	18.420	0.0543	290.336	0.0035	15.7619	0.0635	13.7964
55	24.650	0.0406	394.172	0.0025	15.9906	0.0625	14.3411
60	32.988	0.0303	533.128	0.0019	16.1614	0.0619	14.7910
65	44.145	0.0277	719.083	0.0014	16.2891	0.0614	15.1601
70	59.076	0.0169	967.932	0.0010	16.3846	0.0610	15.4614
75	79.057	0.0127	1300.949	0.0008	16.4559	0.0608	15.7058
80	105.796	0.0095	1746.600	0.0006	16.5091	0.0606	15.9033
85	141.579	0.0071	2342.982	0.0004	16.5490	0.0604	16.0620
90	189.465	0.0053	3141.075	0.0003	16.5787	0.0603	16.1891
95	253.546	0.0040	4209.104	0.0002	16.6009	0.0602	16.2905
100	339.302	0.0030	5638.368	0.0002	16.6176	0.0602	16.3711

Table A11. 7% Interest Factors for Annual Compounding Interest

	Single Payment		Equal Payment Series				Uniform gradient-series factor
	Compound-amount factor	Present-worth factor	Compound-amount factor	Sinking-fund factor	Present-worth factor	Capital-recovery factor	
n	To find F Given P F/P i,n	To find P Given F P/F i,n	To find F Given A F/A i,n	To find A Given F A/F i,n	To find P Given A P/A i,n	To find A Given P A/P i,n	To find A Given G A/G i,n
1	1.070	0.9346	1.000	1.0000	0.9346	1.0700	0.0000
2	1.145	0.8734	2.070	0.4831	1.8080	0.5531	0.4831
3	1.225	0.8163	3.215	0.3111	2.6243	0.3811	0.9549
4	1.311	0.7629	4.440	0.2252	3.3872	0.2952	1.4155
5	1.403	0.7130	5.751	0.1739	4.1002	0.2439	1.8550
6	1.501	0.6664	7.153	0.1398	4.7665	0.2098	2.3032
7	1.606	0.6228	8.654	0.1156	5.3893	0.1856	2.7304
8	1.718	0.5820	10.260	0.0975	5.9713	0.1675	3.1466
9	1.838	0.5439	11.978	0.0835	6.5152	0.1535	3.5517
10	1.967	0.5084	13.816	0.0724	7.0236	0.1424	3.9461
11	2.105	0.4751	15.784	0.0634	7.4987	0.1334	4.3296
12	2.252	0.4440	17.888	0.0559	7.9427	0.1259	4.7025
13	2.410	0.4150	20.141	0.0497	8.3577	0.1197	5.0649
14	2.579	0.3878	22.550	0.0444	8.7455	0.1144	5.4167
15	2.759	0.3625	25.129	0.0398	9.1079	0.1098	5.7583
16	2.952	0.3387	27.888	0.0359	9.4467	0.1059	6.0897
17	3.159	0.3166	30.840	0.0324	9.7632	0.1024	6.4110
18	3.380	0.2959	33.999	0.0294	10.0591	0.0994	6.7225
19	3.617	0.2765	37.379	0.0268	10.3356	0.0968	7.0242
20	3.870	0.2584	40.996	0.0244	10.5940	0.0944	7.3163
21	4.141	0.2415	44.865	0.0223	10.8355	0.0923	7.5990
22	4.430	0.2257	49.006	0.0204	11.0613	0.0904	7.8725
23	4.741	0.2110	53.436	0.0187	11.2722	0.0887	8.1369
24	5.072	0.1972	58.177	0.0172	11.4693	0.0872	8.3923
25	5.427	0.1843	63.249	0.0158	11.6536	0.0858	8.6391
26	5.807	0.1722	68.676	0.0146	11.8258	0.0846	8.8773
27	6.214	0.1609	74.484	0.0134	11.9867	0.0834	9.1072
28	6.649	0.1504	80.698	0.0124	12.1371	0.0824	9.3290
29	7.114	0.1406	87.347	0.0115	12.2777	0.0815	9.4527
30	7.612	0.1314	94.461	0.0106	12.4091	0.0806	9.7487
31	8.145	0.1228	102.073	0.0098	12.5318	0.0798	9.9471
32	8.715	0.1148	110.218	0.0091	12.6466	0.0791	10.1381
33	9.325	0.1072	118.933	0.0084	12.7538	0.0784	10.3219
34	9.978	0.1002	128.259	0.0078	12.8540	0.0778	10.4987
35	10.677	0.0937	138.237	0.0072	12.9477	0.0772	10.6687
40	14.974	0.0668	199.635	0.0050	13.3317	0.0750	11.4234
45	21.002	0.0476	285.749	0.0035	13.6055	0.0735	12.0360
50	29.457	0.0340	406.529	0.0025	13.8008	0.0725	12.5287
55	41.315	0.0242	575.929	0.0017	13.9399	0.0717	12.9215
60	57.946	0.0173	813.520	0.0012	14.0392	0.0712	13.2321
65	81.273	0.0123	1146.755	0.0009	14.1099	0.0709	13.4760
70	113.989	0.0088	1614.134	0.0006	14.1604	0.0706	13.6662
75	159.876	0.0063	2269.657	0.0005	14.1964	0.0705	13.8137
80	224.234	0.0045	3189.063	0.0003	14.2220	0.0703	13.9274
85	314.500	0.0032	4478.576	0.0002	14.2403	0.0702	14.0146
90	441.103	0.0023	6287.185	0.0002	14.2533	0.0702	14.0812
95	618.670	0.0016	8823.854	0.0001	14.2626	0.0701	14.1319
100	867.716	0.0012	12381.662	0.0001	14.2693	0.0701	14.1703

MINE INVESTMENT ANALYSIS

Table A12. 8% Interest Factors for Annual Compounding Interest

	Single Payment		Equal Payment Series				Uniform gradient-series factor
n	Compound-amount factor	Present-worth factor	Compound-amount factor	Sinking-fund factor	Present-worth factor	Capital-recovery factor	
	To find F Given P F/P i,n	To find P Given F P/F i,n	To find F Given A F/A i,n	To find A Given F A/F i,n	To find P Given A P/A i,n	To find A Given P A/P i,n	To find A Given G A/G i,n
1	1.080	0.9259	1.000	1.0000	0.9259	1.0800	0.0000
2	1.166	0.8573	2.080	0.4808	1.7833	0.5608	0.4808
3	1.260	0.7938	3.246	0.3080	2.5771	0.3880	0.9488
4	1.360	0.7350	4.506	0.2219	3.3121	0.3019	1.4040
5	1.469	0.6806	5.867	0.1705	3.9927	0.2505	1.8465
6	1.587	0.6302	7.336	0.1363	4.6229	0.2163	2.2764
7	1.714	0.5835	8.923	0.1121	5.2064	0.1921	2.6937
8	1.851	0.5403	10.637	0.0940	5.7466	0.1740	2.0985
9	1.999	0.5003	12.488	0.0801	6.2469	0.1601	3.4910
10	2.159	0.4632	14.487	0.0690	6.7101	0.1490	3.8713
11	2.332	0.4289	16.645	0.0601	7.1390	0.1401	4.2395
12	2.518	0.3971	18.977	0.0527	7.5361	0.1327	4.5958
13	2.720	0.3677	21.495	0.0465	7.9038	0.1265	4.9402
14	2.937	0.3405	24.215	0.0413	8.2442	0.1213	5.2731
15	3.172	0.3153	27.152	0.0368	8.5595	0.1168	5.5945
16	3.426	0.2919	30.324	0.0330	8.8514	0.1130	5.9046
17	3.700	0.2703	33.750	0.0296	9.1216	0.1096	6.2038
18	3.996	0.2503	37.450	0.0267	9.3719	0.1067	6.4920
19	4.316	0.2317	41.446	0.0241	9.6036	0.1041	6.7697
20	4.661	0.2146	45.762	0.0219	9.8182	0.1019	7.0370
21	5.034	0.1987	50.423	0.0198	10.0168	0.0998	7.2940
22	5.437	0.1840	55.457	0.0180	10.2008	0.0980	7.5412
23	5.871	0.1703	60.893	0.0164	10.3711	0.0964	7.7786
24	6.341	0.1577	66.765	0.0150	10.5288	0.0950	8.0066
25	6.848	0.1460	73.106	0.0137	10.6748	0.0937	8.2254
26	7.396	0.1352	79.954	0.0125	10.8100	0.0925	8.4352
27	7.988	0.1252	87.351	0.0115	10.9352	0.0915	8.6363
28	8.627	0.1159	95.339	0.0105	11.0511	0.0905	8.8289
29	9.317	0.1073	103.966	0.0096	11.1584	0.0896	9.0133
30	10.063	0.0994	113.283	0.0088	11.2578	0.0888	9.1897
31	10.868	0.0920	123.346	0.0081	11.3498	0.0881	9.3584
32	11.737	0.0852	134.214	0.0075	11.4350	0.0875	9.5197
33	12.676	0.0789	145.951	0.0069	11.5139	0.0869	9.6737
34	13.690	0.0731	158.627	0.0063	11.5869	0.0863	9.8208
35	14.785	0.0676	172.317	0.0058	11.6546	0.0858	9.9611
40	21.725	0.0460	259.057	0.0039	11.9246	0.0839	10.5699
45	31.920	0.0313	386.506	0.0026	12.1084	0.0826	11.0447
50	46.902	0.0213	573.770	0.0018	12.2335	0.0818	11.4107
55	68.914	0.0145	848.923	0.0012	12.3186	0.0812	11.6902
60	101.257	0.0099	1253.213	0.0008	12.3766	0.0808	11.9015
65	148.780	0.0067	1847.248	0.0006	12.4160	0.0806	12.0602
70	218.606	0.0046	2720.080	0.0004	12.4428	0.0804	12.1783
75	321.205	0.0031	4002.557	0.0003	12.4611	0.0803	12.2658
80	471.955	0.0021	5886.935	0.0002	12.4735	0.0802	12.3301
85	693.456	0.0015	8655.706	0.0001	12.4820	0.0801	12.3773
90	1018.915	0.0010	12723.939	0.0001	12.4877	0.0801	12.4116
95	1497.121	0.0007	18701.507	0.0001	12.4917	0.0801	12.4365
100	2199.761	0.0005	27484.516	0.0001	12.4943	0.0800	12.4545

Table A13. 9% Interest Factors for Annual Compounding Interest

n	Single Payment		Equal Payment Series				Uniform gradient-series factor
	Compound-amount factor	Present-worth factor	Compound-amount factor	Sinking-fund factor	Present-worth factor	Capital-recovery factor	
	To find F Given P F/P i,n	To find P Given F P/F i,n	To find F Given A F/A i,n	To find A Given F A/F i,n	To find P Given A P/A i,n	To find A Given P A/P i,n	To find A Given G A/G i,n
1	1.090	0.9174	1.000	1.0000	0.9174	1.0900	0.0000
2	1.188	0.8417	2.090	0.4785	1.7591	0.5685	0.4785
3	1.295	0.7722	3.278	0.3051	2.5313	0.3951	0.9426
4	1.412	0.7084	4.573	0.2187	3.2397	0.3087	1.3925
5	1.539	0.6499	5.985	0.1671	3.8897	0.2571	1.8282
6	1.677	0.5963	7.523	0.1329	4.4859	0.2229	2.2498
7	1.828	0.5470	9.200	0.1087	5.0330	0.1987	2.6574
8	1.993	0.5019	11.028	0.0907	5.5348	0.1807	3.0512
9	2.172	0.4604	13.021	0.0768	5.9953	0.1668	3.4312
10	2.367	0.4224	15.193	0.0658	6.4177	0.1558	3.7978
11	2.580	0.3875	17.560	0.0570	6.8052	0.1470	4.1510
12	2.813	0.3555	20.141	0.0497	7.1607	0.1397	4.4910
13	3.066	0.3262	22.953	0.0436	7.4869	0.1336	4.8182
14	3.342	0.2993	26.019	0.0384	7.7862	0.1284	5.1326
15	3.642	0.2745	29.361	0.0341	8.0607	0.1241	5.4346
16	3.970	0.2519	33.003	0.0303	8.3126	0.1203	5.7245
17	4.328	0.2311	36.974	0.0271	8.5436	0.1171	6.0024
18	4.717	0.2120	41.301	0.0242	8.7556	0.1142	6.2687
19	5.142	0.1945	46.018	0.0217	8.9501	0.1117	6.5236
20	5.604	0.1784	51.160	0.0196	9.1286	0.1096	6.7675
21	6.109	0.1637	56.765	0.0176	9.2923	0.1076	7.0006
22	6.659	0.1502	62.873	0.0159	9.4424	0.1059	7.2232
23	7.258	0.1378	69.532	0.0144	9.5802	0.1044	7.4358
24	7.911	0.1264	76.790	0.0130	9.7066	0.1030	7.6384
25	8.623	0.1160	84.701	0.0118	9.8226	0.1018	7.8316
26	9.399	0.1064	93.324	0.0107	9.9290	0.1007	8.0156
27	10.245	0.0976	102.723	0.0097	10.0266	0.0997	8.1906
28	11.167	0.0896	112.968	0.0089	10.1161	0.0989	8.3572
29	12.172	0.0822	124.135	0.0081	10.1983	0.0981	8.5154
30	13.268	0.0754	136.308	0.0073	10.2737	0.0973	8.6657
31	14.462	0.0692	149.575	0.0067	10.3428	0.0967	8.8083
32	15.763	0.0634	164.037	0.0061	10.4063	0.0961	8.9436
33	17.182	0.0582	179.800	0.0056	10.4645	0.0956	9.0718
34	18.728	0.0534	196.982	0.0051	10.5178	0.0951	9.1933
35	20.414	0.0490	215.711	0.0046	10.5668	0.0946	9.3083
40	31.409	0.0318	337.882	0.0030	10.7574	0.0930	9.7957
45	48.327	0.0207	525.859	0.0019	10.8812	0.0919	10.1603
50	74.358	0.0135	815.084	0.0012	10.9617	0.0912	10.4295
55	114.408	0.0088	1260.092	0.0008	11.0140	0.0908	10.6261
60	176.031	0.0057	1944.792	0.0005	11.0480	0.0905	10.7683
65	270.846	0.0037	2998.288	0.0003	11.0701	0.0903	10.8702
70	416.730	0.0024	4619.223	0.0002	11.0845	0.0902	10.9427
75	641.191	0.0016	7113.232	0.0002	11.0938	0.0902	10.9940
80	986.552	0.0010	10950.574	0.0001	11.0999	0.0901	11.0299
85	1517.932	0.0007	16854.800	0.0001	11.1038	0.0901	11.0551
90	2335.527	0.0004	25939.184	0.0001	11.1064	0.0900	11.0726
95	3593.497	0.0003	39916.635	0.0000	11.1080	0.0900	11.0847
100	5529.041	0.0002	61422.675	0.0000	11.1091	0.0900	11.0930

Table A14. 10% Interest Factors for Annual Compounding Interest

n	Single Payment		Equal Payment Series				Uniform gradient-series factor
	Compound-amount factor	Present-worth factor	Compound-amount factor	Sinking-fund factor	Present-worth factor	Capital-recovery factor	
	To find F Given P F/P i,n	To find P Given F P/F i,n	To find F Given A F/A i,n	To find A Given F A/F i,n	To find P Given A P/A i,n	To find A Given P A/P i,n	To find A Given G A/G i,n
1	1.100	0.9091	1.000	1.0000	0.9091	1.1000	0.0000
2	1.210	0.8265	2.100	0.4762	1.7355	0.5762	0.4762
3	1.331	0.7513	3.310	0.3021	2.4869	0.4021	0.9366
4	1.464	0.6830	4.641	0.2155	3.1699	0.3155	1.3812
5	1.611	0.6209	6.105	0.1638	3.7908	0.2638	1.8101
6	1.772	0.5645	7.716	0.1296	4.3553	0.2296	2.2236
7	1.949	0.5132	9.487	0.1054	4.8684	0.2054	2.6216
8	2.144	0.4665	11.436	0.0875	5.3349	0.1875	3.0045
9	2.358	0.4241	13.579	0.0737	5.7590	0.1737	3.3724
10	2.594	0.3856	15.937	0.0628	6.1446	0.1628	3.7255
11	2.853	0.3505	18.531	0.0540	6.4951	0.1540	4.0641
12	3.138	0.3186	21.384	0.0468	6.8137	0.1468	4.3884
13	3.452	0.2897	24.523	0.0408	7.1034	0.1408	4.6988
14	3.798	0.2633	27.975	0.0358	7.3667	0.1358	4.9955
15	4.177	0.2394	31.772	0.0315	7.6061	0.1315	5.2789
16	4.595	0.2176	35.950	0.0278	7.8237	0.1278	5.5493
17	5.054	0.1979	40.545	0.0247	8.0216	0.1247	5.8071
18	5.560	0.1799	45.599	0.0219	8.2014	0.1219	6.0526
19	6.116	0.1635	51.159	0.0196	8.3649	0.1196	6.2861
20	6.728	0.1487	57.275	0.0175	8.5136	0.1175	6.5081
21	7.400	0.1351	64.003	0.0156	8.6487	0.1156	6.7189
22	8.140	0.1229	71.403	0.0140	8.7716	0.1140	6.9189
23	8.954	0.1117	79.453	0.0126	8.8832	0.1126	7.1085
24	9.850	0.1015	88.497	0.0113	8.9848	0.1113	7.2881
25	10.835	0.0923	98.347	0.0102	9.0771	0.1102	7.4580
26	11.918	0.0839	109.182	0.0092	9.1610	0.1092	7.6187
27	13.110	0.0763	121.100	0.0083	9.2372	0.1083	7.7704
28	14.421	0.0694	134.210	0.0075	9.3066	0.1075	7.9137
29	15.863	0.0630	148.631	0.0067	9.3696	0.1067	8.0489
30	17.449	0.0573	164.494	0.0061	9.4269	0.1061	8.1762
31	19.194	0.0521	181.943	0.0055	9.4790	0.1055	8.2962
32	21.114	0.0474	201.138	0.0050	9.5264	0.1050	8.4091
33	23.225	0.0431	222.252	0.0045	9.5694	0.1045	8.5152
34	25.548	0.0392	245.477	0.0041	9.6086	0.1041	8.6149
35	28.102	0.0356	271.024	0.0037	9.6442	0.1037	8.7086
40	45.259	0.0221	442.593	0.0023	9.7791	0.1023	9.0962
45	72.890	0.0137	718.905	0.0014	9.8628	0.1014	9.3741
50	117.391	0.0085	1163.909	0.0009	9.9148	0.1009	9.5704
55	189.059	0.0053	1880.591	0.0005	9.9471	0.1005	9.7075
60	304.482	0.0033	3034.816	0.0003	9.9672	0.1003	9.8023
65	490.371	0.0020	4893.707	0.0002	9.9796	0.1002	9.8672
70	789.747	0.0013	7887.470	0.0001	9.9873	0.1001	9.9113
75	1271.895	0.0008	12708.954	0.0001	9.9921	0.1001	9.9410
80	2048.400	0.0005	20474.002	0.0001	9.9951	0.1001	9.9609
85	3298.969	0.0003	32979.690	0.0000	9.9970	0.1000	9.9742
90	5313.023	0.0002	53120.226	0.0000	9.9981	0.1000	9.9831
95	8556.676	0.0001	85556.760	0.0000	9.9988	0.1000	9.9889
100	13780.612	0.0001	137796.123	0.0000	9.9993	0.1000	9.9928

Table A15. 12% Interest Factors for Annual Compounding Interest

n	Single Payment		Equal Payment Series				Uniform gradient-series factor
	Compound-amount factor	Present-worth factor	Compound-amount factor	Sinking-fund factor	Present-worth factor	Capital-recovery factor	
	To find F Given P $F/P\ i,n$	To find P Given F $P/F\ i,n$	To find F Given A $F/A\ i,n$	To find A Given F $A/F\ i,n$	To find P Given A $P/A\ i,n$	To find A Given P $A/P\ i,n$	To find A Given G $A/G\ i,n$
1	1.120	0.8929	1.000	1.0000	0.8929	1.1200	0.0000
2	1.254	0.7972	2.120	0.4717	1.6901	0.5917	0.4717
3	1.405	0.7118	3.374	0.2964	2.4018	0.4164	0.9246
4	1.574	0.6355	4.779	0.2092	3.0374	0.3292	1.3589
5	1.762	0.5674	6.353	0.1574	3.6048	0.2774	1.7746
6	1.974	0.5066	8.115	0.1232	4.1114	0.2432	2.1721
7	2.211	0.4524	10.089	0.0991	4.5638	0.2191	2.5515
8	2.476	0.4039	12.300	0.0813	4.9676	0.2013	2.9132
9	2.773	0.3606	14.776	0.0677	5.3283	0.1877	3.2574
10	3.106	0.3220	17.549	0.0570	5.6502	0.1770	3.5847
11	3.479	0.2875	20.655	0.0484	5.9377	0.1684	3.8953
12	3.896	0.2567	24.133	0.0414	6.1944	0.1614	4.1897
13	4.364	0.2292	28.029	0.0357	6.4236	0.1557	4.4683
14	4.887	0.2046	32.393	0.0309	6.6282	0.1509	4.7317
15	5.474	0.1827	37.280	0.0268	6.8109	0.1468	4.9803
16	6.130	0.1631	42.753	0.0234	6.9740	0.1434	5.2147
17	6.866	0.1457	48.884	0.0205	7.1196	0.1405	5.4353
18	7.690	0.1300	55.750	0.0179	7.2497	0.1379	5.6427
19	8.613	0.1161	63.440	0.0158	7.3658	0.1358	5.8375
20	9.646	0.1037	72.052	0.0139	7.4695	0.1339	6.0202
21	10.804	0.0926	81.699	0.0123	7.5620	0.1323	6.1913
22	12.100	0.0827	92.503	0.0108	7.6447	0.1308	6.3514
23	13.552	0.0738	104.603	0.0096	7.7184	0.1296	6.5010
24	15.179	0.0659	118.155	0.0085	7.7843	0.1285	6.6407
25	17.000	0.0588	133.334	0.0075	7.8431	0.1275	6.7708
26	19.040	0.0525	150.334	0.0067	7.8957	0.1267	6.8921
27	21.325	0.0469	169.374	0.0059	7.9426	0.1259	7.0049
28	23.884	0.0419	190.699	0.0053	7.9844	0.1253	7.1098
29	26.750	0.0374	214.583	0.0047	8.0218	0.1247	7.2071
30	29.960	0.0334	241.333	0.0042	8.0552	0.1242	7.2974
31	33.555	0.0298	217.293	0.0037	8.0850	0.1237	7.3811
32	37.582	0.0266	304.848	0.0033	8.1116	0.1233	7.4586
33	42.092	0.0238	342.429	0.0029	8.1354	0.1229	7.5303
34	47.143	0.0212	384.521	0.0026	8.1566	0.1226	7.5965
35	52.800	0.0189	431.664	0.0023	8.1755	0.1223	7.6577
40	93.051	0.0108	767.091	0.0013	8.2438	0.1213	7.8988
45	163.988	0.0061	1358.230	0.0007	8.2825	0.1207	8.0572
50	289.002	0.0035	2400.018	0.0004	8.3045	0.1204	8.1597

Table A16. 15% Interest Factors for Annual Compounding Interest

	Single Payment		Equal Payment Series				Uniform gradient-series factor
n	Compound-amount factor	Present-worth factor	Compound-amount factor	Sinking-fund factor	Present-worth factor	Capital-recovery factor	
	To find F Given P F/P i,n	To find P Given F P/F i,n	To find F Given A F/A i,n	To find A Given F A/F i,n	To find P Given A P/A i,n	To find A Given P A/P i,n	To find A Given G A/G i,n
1	1.150	0.8696	1.000	1.0000	0.8696	1.1500	0.0000
2	1.323	0.7562	2.150	0.4651	1.6257	0.6151	0.4651
3	1.521	0.6575	3.473	0.2880	2.2832	0.4380	0.9071
4	1.749	0.5718	4.993	0.2003	2.8550	0.3503	1.3263
5	2.011	0.4972	6.742	0.1483	3.3522	0.2983	1.7228
6	2.313	0.4323	8.754	0.1142	3.7845	0.2642	2.0972
7	2.660	0.3759	11.067	0.0904	4.1604	0.2404	2.4499
8	3.059	0.3269	13.727	0.0729	4.4873	0.2229	2.7813
9	3.518	0.2843	16.786	0.0596	4.7716	0.2096	3.0922
10	4.046	0.2472	20.304	0.0493	5.0188	0.1993	3.3832
11	4.652	0.2150	24.349	0.0411	5.2337	0.1911	3.6550
12	5.350	0.1869	29.002	0.0345	5.4206	0.1845	3.9082
13	6.153	0.1625	34.352	0.0291	5.5832	0.1791	4.1438
14	7.076	0.1413	40.505	0.0247	5.7245	0.1747	4.3624
15	8.137	0.1229	47.580	0.0210	5.8474	0.1710	4.5650
16	9.358	0.1069	55.717	0.0180	5.9542	0.1680	4.7523
17	10.761	0.0929	65.075	0.0154	6.0472	0.1654	4.9251
18	12.375	0.0808	75.836	0.0132	6.1280	0.1632	5.0843
19	14.232	0.0703	88.212	0.0113	6.1982	0.1613	5.2307
20	16.367	0.0611	102.444	0.0098	6.2593	0.1598	5.3651
21	18.822	0.0531	118.810	0.0084	6.3125	0.1584	5.4883
22	21.645	0.0462	137.632	0.0073	6.3587	0.1573	5.6010
23	24.891	0.0402	159.276	0.0063	6.3988	0.1563	5.7040
24	28.625	0.0349	184.168	0.0054	6.4338	0.1554	5.7979
25	32.919	0.0304	212.793	0.0047	6.4642	0.1547	5.8834
26	37.857	0.0264	245.712	0.0041	6.4906	0.1541	5.9612
27	43.535	0.0230	283.569	0.0035	6.5135	0.1535	6.0319
28	50.066	0.0200	327.104	0.0031	6.5335	0.1531	6.0960
29	57.575	0.0174	377.170	0.0027	6.5509	0.1527	6.1541
30	66.212	0.0151	434.745	0.0023	6.5660	0.1523	6.2066
31	76.144	0.0131	500.957	0.0020	6.5791	0.1520	6.2541
32	87.565	0.0114	577.100	0.0017	6.5905	0.1517	6.2970
33	100.700	0.0099	664.666	0.0015	6.6005	0.1515	6.3357
34	115.805	0.0086	765.365	0.0013	6.6091	0.1513	6.3705
35	133.176	0.0075	881.170	0.0011	6.6166	0.1511	6.4019
40	267.864	0.0037	1779.090	0.0006	6.6418	0.1506	6.5168
45	538.769	0.0019	3585.128	0.0003	6.6543	0.1503	6.5830
50	1083.657	0.0009	7217.716	0.0002	6.6605	0.1501	6.6205

Table A17. 20% Interest Factors for Annual Compounding Interest

n	Single Payment		Equal Payment Series				Uniform gradient-series factor
	Compound-amount factor	Present-worth factor	Compound-amount factor	Sinking-fund factor	Present-worth factor	Capital-recovery factor	
	To find F Given P F/P i,n	To find P Given F P/F i,n	To find F Given A F/A i,n	To find A Given F A/F i,n	To find P Given A P/A i,n	To find A Given P A/P i,n	To find A Given G A/G i,n
1	1.200	0.8333	1.000	1.0000	0.8333	1.2000	0.0000
2	1.440	0.6945	2.200	0.4546	1.5278	0.6546	0.4546
3	1.728	0.5787	3.640	0.2747	2.1065	0.4747	0.8791
4	2.074	0.4823	5.368	0.1863	2.5887	0.3863	1.2742
5	2.488	0.4019	7.442	0.1344	2.9906	0.3344	1.6405
6	2.986	0.3349	9.930	0.1007	3.3255	0.3007	1.9788
7	3.583	0.2791	12.916	0.0774	3.6046	0.2774	2.2902
8	4.300	0.2326	16.499	0.0606	3.8372	0.2606	2.5756
9	5.160	0.1938	20.799	0.0481	4.0310	0.2481	2.8364
10	6.192	0.1615	25.959	0.0385	4.1925	0.2385	3.0739
11	7.430	0.1346	32.150	0.0311	4.3271	0.2311	3.2893
12	8.916	0.1122	39.581	0.0253	4.4392	0.2253	3.4841
13	10.699	0.0935	48.497	0.0206	4.5327	0.2206	3.6597
14	12.839	0.0779	59.196	0.0169	4.6106	0.2169	3.8175
15	15.407	0.0649	72.035	0.0139	4.6755	0.2139	3.9589
16	18.488	0.0541	87.442	0.0114	4.7296	0.2114	4.0851
17	22.186	0.0451	105.931	0.0095	4.7746	0.2095	4.1976
18	26.623	0.0376	128.117	0.0078	4.8122	0.2078	4.2975
19	31.948	0.0313	154.740	0.0065	4.8435	0.2065	4.3861
20	38.338	0.0261	186.688	0.0054	4.8696	0.2054	4.4644
21	46.005	0.0217	225.026	0.0045	4.8913	0.2045	4.5334
22	55.206	0.0181	271.031	0.0037	4.9094	0.2037	4.5942
23	66.247	0.0151	326.237	0.0031	4.9245	0.2031	4.6475
24	79.497	0.0126	392.484	0.0026	4.9371	0.2026	4.6943
25	95.396	0.0105	471.981	0.0021	4.9476	0.2021	4.7352
26	114.475	0.0087	567.377	0.0018	4.9563	0.2018	4.7709
27	137.371	0.0073	681.853	0.0015	4.9636	0.2015	4.8020
28	164.845	0.0061	819.223	0.0012	4.9697	0.2012	4.8291
29	197.814	0.0051	984.068	0.0010	4.9747	0.2010	4.8527
30	237.376	0.0042	1181.882	0.0009	4.9789	0.2009	4.8731
31	284.852	0.0035	1419.258	0.0007	4.9825	0.2007	4.8908
32	341.822	0.0029	1704.109	0.0006	4.9854	0.2006	4.9061
33	410.186	0.0024	2045.931	0.0005	4.9878	0.2005	4.9194
34	492.224	0.0020	2456.118	0.0004	4.9899	0.2004	4.9308
35	590.668	0.0017	2948.341	0.0003	4.9915	0.2003	4.9407
40	1469.772	0.0007	7343.858	0.0002	4.9966	0.2001	4.9728
45	3657.262	0.0003	18281.310	0.0001	4.9986	0.2001	4.9877
50	9100.438	0.0001	45497.191	0.0000	4.9995	0.2000	4.9945

Table A18. 25% Interest Factors for Annual Compounding Interest

n	Single Payment		Equal Payment Series				Uniform gradient-series factor
	Compound-amount factor	Present-worth factor	Compound-amount factor	Sinking-fund factor	Present-worth factor	Capital-recovery factor	
	To find F Given P F/P i,n	To find P Given F P/F i,n	To find F Given A F/A i,n	To find A Given F A/F i,n	To find P Given A P/A i,n	To find A Given P A/P i,n	To find A Given G A/G i,n
1	1.250	0.8000	1.000	1.0000	0.8000	1.2500	0.0000
2	1.563	0.6400	2.250	0.4445	1.4400	0.6945	0.4445
3	1.953	0.5120	3.813	0.2623	1.9520	0.5123	0.8525
4	2.441	0.4096	5.766	0.1735	2.3616	0.4235	1.2249
5	3.052	0.3277	8.207	0.1219	2.6893	0.3719	1.5631
6	3.815	0.2622	11.259	0.0888	2.9514	0.3388	1.8683
7	4.768	0.2097	15.073	0.0664	3.1611	0.3164	2.1424
8	5.960	0.1678	19.842	0.0504	3.3289	0.3004	2.3873
9	7.451	0.1342	25.802	0.0388	3.4631	0.2888	2.6048
10	9.313	0.1074	33.253	0.0301	3.5705	0.2801	2.7971
11	11.642	0.0859	42.566	0.0235	3.6564	0.2735	2.9663
12	14.552	0.0687	54.208	0.0185	3.7251	0.2685	3.1145
13	18.190	0.0550	68.760	0.0146	3.7801	0.2646	3.2438
14	22.737	0.0440	86.949	0.0115	3.8241	0.2615	3.3560
15	28.422	0.0352	109.687	0.0091	3.8593	0.2591	3.4530
16	35.527	0.0282	138.109	0.0073	3.8874	0.2573	3.5366
17	44.409	0.0225	173.636	0.0058	3.9099	0.2558	3.6084
18	55.511	0.0180	218.045	0.0046	3.9280	0.2546	3.6698
19	69.389	0.0144	273.556	0.0037	3.9424	0.2537	3.7222
20	86.736	0.0115	342.945	0.0029	3.9539	0.2529	3.7667
21	108.420	0.0092	429.681	0.0023	3.9631	0.2523	3.8045
22	135.525	0.0074	538.101	0.0019	3.9705	0.2519	3.8365
23	169.407	0.0059	673.626	0.0015	3.9764	0.2515	3.8634
24	211.758	0.0047	843.033	0.0012	3.9811	0.2512	3.8861
25	264.698	0.0038	1054.791	0.0010	3.9849	0.2510	3.9052
26	330.872	0.0030	1319.489	0.0008	3.9879	0.2508	3.9212
27	413.590	0.0024	1650.361	0.0006	3.9903	0.2506	3.9346
28	516.988	0.0019	2063.952	0.0005	3.9923	0.2505	3.9457
29	646.235	0.0016	2580.939	0.0004	3.9938	0.2504	3.9551
30	807.794	0.0012	3227.174	0.0003	3.9951	0.2503	3.9628
31	1009.742	0.0010	4034.968	0.0003	3.9960	0.2503	3.9693
32	1262.177	0.0008	5044.710	0.0002	3.9968	0.2502	3.9746
33	1577.722	0.0006	6306.887	0.0002	3.9975	0.2502	3.9791
34	1972.152	0.0005	7884.609	0.0001	3.9980	0.2501	3.9828
35	2465.190	0.0004	9856.761	0.0001	3.9984	0.2501	3.9858

Table A19. 30% Interest Factors for Annual Compounding Interest

	Single Payment		Equal Payment Series				Uniform gradient-series factor
	Compound-amount factor	Present-worth factor	Compound-amount factor	Sinking-fund factor	Present-worth factor	Capital-recovery factor	
n	To find F Given P F/P i,n	To find P Given F P/F i,n	To find F Given A F/A i,n	To find A Given F A/F i,n	To find P Given A P/A i,n	To find A Given P A/P i,n	To find A Given G A/G i,n
1	1.300	0.7692	1.000	1.0000	0.7692	1.3000	0.0000
2	1.690	0.5917	2.300	0.4348	1.3610	0.7348	0.4348
3	2.197	0.4552	3.990	0.2506	1.8161	0.5506	0.8271
4	2.856	0.3501	6.187	0.1616	2.1663	0.4616	1.1783
5	3.713	0.2693	9.043	0.1106	2.4356	0.4106	1.4903
6	4.827	0.2072	12.756	0.0784	2.6428	0.3784	1.7655
7	6.275	0.1594	17.583	0.0569	2.8021	0.3569	2.0063
8	8.157	0.1226	23.858	0.0419	2.9247	0.3419	2.2156
9	10.605	0.0943	32.015	0.0312	3.0190	0.3312	2.3963
10	13.786	0.0725	42.620	0.0235	3.0915	0.3235	2.5512
11	17.922	0.0558	56.405	0.0177	3.1473	0.3177	2.6833
12	23.298	0.0429	74.327	0.0135	3.1903	0.3135	2.7952
13	30.288	0.0330	97.625	0.0103	3.2233	0.3103	2.8895
14	39.374	0.0254	127.913	0.0078	3.2487	0.3078	2.9685
15	51.186	0.0195	167.286	0.0060	3.2682	0.3060	3.0345
16	66.542	0.0150	218.472	0.0046	3.2832	0.3046	3.0892
17	86.504	0.0116	285.014	0.0035	3.2948	0.3035	3.1345
18	112.455	0.0089	371.518	0.0027	3.3037	0.3027	3.1718
19	146.192	0.0069	483.973	0.0021	3.3105	0.3021	3.2025
20	190.050	0.0053	630.165	0.0016	3.3158	0.3016	3.2276
21	247.065	0.0041	820.215	0.0012	3.3199	0.3012	3.2480
22	321.184	0.0031	1067.280	0.0009	3.3230	0.3009	3.2646
23	417.539	0.0024	1388.464	0.0007	3.3254	0.3007	3.2781
24	542.801	0.0019	1806.003	0.0006	3.3272	0.3006	3.2890
25	705.641	0.0014	2348.803	0.0004	3.3286	0.3004	3.2979
26	917.333	0.0011	3054.444	0.0003	3.3297	0.3003	3.3050
27	1192.533	0.0008	3971.778	0.0003	3.3305	0.3003	3.3107
28	1550.293	0.0007	5164.311	0.0002	3.3312	0.3002	3.3153
29	2015.381	0.0005	6714.604	0.0002	3.3317	0.3002	3.3189
30	2619.996	0.0004	8729.985	0.0001	3.3321	0.3001	3.3219
31	3405.994	0.0003	11349.981	0.0001	3.3324	0.3001	3.3242
32	4427.793	0.0002	14755.975	0.0001	3.3326	0.3001	3.3261
33	5756.130	0.0002	19183.768	0.0001	3.3328	0.3001	3.3276
34	7482.970	0.0001	24939.899	0.0001	3.3329	0.3001	3.3288
35	9727.860	0.0001	32422.868	0.0000	3.3330	0.3000	3.3297

Table A20. 40% Interest Factors for Annual Compounding Interest

n	Single Payment		Equal Payment Series				Uniform gradient-series factor
	Compound-amount factor	Present-worth factor	Compound-amount factor	Sinking-fund factor	Present-worth factor	Capital-recovery factor	
	To find F Given P F/P i,n	To find P Given F P/F i,n	To find F Given A F/A i,n	To find A Given F A/F i,n	To find P Given A P/A i,n	To find A Given P A/P i,n	To find A Given G A/G i,n
1	1.400	0.7143	1.000	1.0001	0.7143	1.4001	0.0000
2	1.960	0.5103	2.400	0.4167	1.2245	0.8167	0.4167
3	2.744	0.3645	4.360	0.2294	1.5890	0.6294	0.7799
4	3.842	0.2604	7.104	0.1408	1.8493	0.5408	1.0924
5	5.378	0.1860	10.946	0.0914	2.0352	0.4914	1.3580
6	7.530	0.1329	16.324	0.0613	2.1680	0.4613	1.5811
7	10.541	0.0949	23.853	0.0420	2.2629	0.4420	1.7664
8	14.758	0.0678	34.395	0.0291	2.3306	0.4291	1.9186
9	20.661	0.0485	49.153	0.0204	2.3790	0.4204	2.0423
10	28.925	0.0346	69.814	0.0144	2.4136	0.4144	2.1420
11	40.496	0.0247	98.739	0.0102	2.4383	0.4102	2.2215
12	56.694	0.0177	139.234	0.0072	2.4560	0.4072	2.2846
13	79.371	0.0126	195.928	0.0052	2.4686	0.4052	2.3342
14	111.120	0.0090	275.299	0.0037	2.4775	0.4037	2.3729
15	155.568	0.0065	386.419	0.0026	2.4840	0.4026	2.4030
16	217.794	0.0046	541.986	0.0019	2.4886	0.4019	2.4262
17	304.912	0.0033	759.780	0.0014	2.4918	0.4014	2.4441
18	426.877	0.0024	1064.691	0.0010	2.4942	0.4010	2.4578
19	597.627	0.0017	1491.567	0.0007	2.4959	0.4007	2.4682
20	836.678	0.0012	2089.195	0.0005	2.4971	0.4005	2.4761
21	1171.348	0.0009	2925.871	0.0004	2.4979	0.4004	2.4821
22	1639.887	0.0007	4097.218	0.0003	2.4985	0.4003	2.4866
23	2295.842	0.0005	5737.105	0.0002	2.4990	0.4002	2.4900
24	3214.178	0.0004	8032.945	0.0002	2.4993	0.4002	2.4926
25	4499.847	0.0003	11247.110	0.0001	2.4995	0.4001	2.4945
26	6299.785	0.0002	15746.960	0.0001	2.4997	0.4001	2.4959
27	8819.695	0.0002	22046.730	0.0001	2.4998	0.4001	2.4970
28	12347.570	0.0001	30866.430	0.0001	2.4998	0.4001	2.4978
29	17286.590	0.0001	43213.990	0.0001	2.4999	0.4001	2.4984
30	24201.230	0.0001	60500.580	0.0001	2.4999	0.4001	2.4988

Table A21. 50% Interest Factors for Annual Compounding Interest

	Single Payment		Equal Payment Series				Uniform gradient-series factor
	Compound-amount factor	Present-worth factor	Compound-amount factor	Sinking-fund factor	Present-worth factor	Capital-recovery factor	
n	To find F Given P F/P i,n	To find P Given F P/F i,n	To find F Given A F/A i,n	To find A Given F A/F i,n	To find P Given A P/A i,n	To find A Given P A/P i,n	To find A Given G A/G i,n
1	1.500	0.6667	1.000	1.0000	0.6667	1.5000	0.0001
2	2.250	0.4445	2.500	0.4000	1.1112	0.9001	0.4001
3	3.375	0.2963	4.750	0.2106	1.4075	0.7106	0.7369
4	5.063	0.1976	8.125	0.1231	1.6050	0.6231	1.0154
5	7.594	0.1317	13.188	0.0759	1.7367	0.5759	1.2418
6	11.391	0.0878	20.781	0.0482	1.8245	0.5482	1.4226
7	17.086	0.0586	32.172	0.0311	1.8830	0.5311	1.5649
8	25.629	0.0391	49.258	0.0204	1.9220	0.5204	1.6752
9	38.443	0.0261	74.887	0.0134	1.9480	0.5134	1.7597
10	57.665	0.0174	113.330	0.0089	1.9654	0.5089	1.8236
11	86.498	0.0116	170.995	0.0059	1.9769	0.5059	1.8714
12	129.746	0.0078	257.493	0.0039	1.9846	0.5039	1.9068
13	194.620	0.0052	387.239	0.0026	1.9898	0.5026	1.9329
14	291.929	0.0035	581.858	0.0018	1.9932	0.5018	1.9519
15	437.894	0.0023	873.788	0.0012	1.9955	0.5012	1.9657
16	656.841	0.0016	1311.681	0.0008	1.9970	0.5008	1.9757
17	985.261	0.0011	1968.522	0.0006	1.9980	0.5006	1.9828
18	1477.891	0.0007	2953.783	0.0004	1.9987	0.5004	1.9879
19	2216.837	0.0005	4431.671	0.0003	1.9991	0.5003	1.9915
20	3325.256	0.0004	6648.511	0.0002	1.9994	0.5002	1.9940
21	4987.882	0.0003	9973.765	0.0002	1.9996	0.5002	1.9958
22	7481.824	0.0002	14961.640	0.0001	1.9998	0.5001	1.9971
23	11222.730	0.0001	22443.470	0.0001	1.9999	0.5001	1.9980
24	16834.100	0.0001	33666.210	0.0001	1.9999	0.5001	1.9986
25	25251.160	0.0001	50500.330	0.0001	2.0000	0.5001	1.9991

APPENDIX **B***

Effective Interest Rates Corresponding to Nominal Rate

Table B1. Effective Interest Rates Corresponding to Nominal Rate r

| r | Compounding Frequency | | | | | |
	Semi-annually $\left(1 + \frac{r}{2}\right)^2 - 1$	Quarterly $\left(1 + \frac{r}{4}\right)^4 - 1$	Monthly $\left(1 + \frac{r}{12}\right)^{12} - 1$	Weekly $\left(1 + \frac{r}{52}\right)^{52} - 1$	Daily $\left(1 + \frac{r}{365}\right)^{365} - 1$	Continuously $\left(1 + \frac{r}{\infty}\right)^{\infty} - 1$
0.01	0.010025	0.010038	0.010046	0.010049	0.010050	0.010050
0.02	0.020100	0.020151	0.020184	0.020197	0.020200	0.020201
0.03	0.030225	0.030339	0.030416	0.030444	0.030451	0.030455
0.04	0.040400	0.040604	0.040741	0.040793	0.040805	0.040811
0.05	0.050625	0.050945	0.051161	0.051244	0.051261	0.051271
0.06	0.060900	0.061364	0.061678	0.061797	0.061799	0.061837
0.07	0.071225	0.071859	0.072290	0.072455	0.072469	0.072508
0.08	0.081600	0.082432	0.082999	0.083217	0.083246	0.083287
0.09	0.092025	0.093083	0.093807	0.094085	0.094132	0.094174
0.10	0.102500	0.103813	0.104713	0.105060	0.105126	0.105171
0.11	0.113025	0.114621	0.115718	0.116144	0.116231	0.116278
0.12	0.123600	0.125509	0.126825	0.127336	0.127447	0.127497
0.13	0.134225	0.136476	0.138032	0.138644	0.138775	0.138828
0.14	0.144900	0.147523	0.149341	0.150057	0.150217	0.150274
0.15	0.155625	0.158650	0.160755	0.161582	0.161773	0.161834
0.16	0.166400	0.169859	0.172270	0.173221	0.173446	0.173511
0.17	0.177225	0.181148	0.183891	0.184974	0.185235	0.185305
0.18	0.188100	0.192517	0.195618	0.196843	0.197142	0.197217
0.19	0.199025	0.203917	0.207451	0.208828	0.209169	0.209250
0.20	0.210000	0.215506	0.219390	0.220931	0.221316	0.221403
0.21	0.221025	0.227124	0.231439	0.233153	0.233584	0.233678
0.22	0.232100	0.238825	0.243596	0.245494	0.245976	0.246077
0.23	0.243225	0.250609	0.255863	0.257957	0.258492	0.258600
0.24	0.254400	0.262477	0.268242	0.270542	0.271133	0.271249
0.25	0.265625	0.274429	0.280731	0.283250	0.283901	0.284025
0.26	0.276900	0.286466	0.293333	0.296090	0.296796	0.296930
0.27	0.288225	0.298588	0.306050	0.309049	0.309821	0.309964
0.28	0.299600	0.310796	0.318880	0.322135	0.322976	0.323130
0.29	0.311025	0.323089	0.331826	0.335350	0.336264	0.336428
0.30	0.322500	0.335469	0.344889	0.348693	0.349684	0.349859
0.31	0.334025	0.347936	0.358068	0.362168	0.363238	0.363425
0.32	0.345600	0.360489	0.371366	0.375775	0.376928	0.377128
0.33	0.357225	0.373130	0.384784	0.389515	0.390756	0.390968
0.34	0.368900	0.385859	0.398321	0.403389	0.404722	0.404948
0.35	0.380625	0.398676	0.411979	0.417399	0.418827	0.419068

*Thuesen, H.G., Fabrycky, W.J., Thuesen, G.J., *Engineering Economy*, 5th ed., ©1977, pp. 559. Reprinted by permission of Prentice-Hall, Inc., Englewood Cliffs, NJ.

APPENDIX C^*

Interest Factors for Continuous Compounding Interest

Table C1.1% Interest Factors for Continuous Compounding Interest

n	Single Payment		Equal Payment Series				Uniform gradient-series factor
	Compound-amount factor	Present-worth factor	Compound-amount factor	Sinking-fund factor	Present-worth factor	Capital-recovery factor	
	To find F Given P F/P r,n	To find P Given F P/F r,n	To find F Given A F/A r,n	To find A Given F A/F r,n	To find P Given A P/A r,n	To find A Given P A/P r,n	To find A Given G A/G r,n
1	1.010	0.9901	1.000	1.0000	0.9901	1.0101	0.0000
2	1.020	0.9802	2.010	0.4975	1.9703	0.5076	0.4975
3	1.030	0.9705	3.030	0.3300	2.9407	0.3401	0.9933
4	1.041	0.9608	4.061	0.2463	3.9015	0.2563	1.4875
5	1.051	0.9512	5.102	0.1960	4.8527	0.2061	1.9800
6	1.062	0.9418	6.153	0.1625	5.7945	0.1726	2.4708
7	1.073	0.9324	7.215	0.1386	6.7269	0.1487	2.9600
8	1.083	0.9231	8.287	0.1207	7.6500	0.1307	3.4475
9	1.094	0.9139	9.370	0.1067	8.5639	0.1168	3.9334
10	1.105	0.9048	10.465	0.0956	9.4688	0.1056	4.4175
11	1.116	0.8958	11.570	0.0864	10.3646	0.0965	4.9000
12	1.128	0.8869	12.686	0.0788	11.2515	0.0889	5.3809
13	1.139	0.8781	13.814	0.0724	12.1296	0.0825	5.8600
14	1.150	0.8694	14.952	0.0669	12.9990	0.0769	6.3376
15	1.162	0.8607	16.103	0.0621	13.8597	0.0722	6.8134
16	1.174	0.8522	17.264	0.0579	14.7118	0.0680	7.2876
17	1.185	0.8437	18.438	0.0542	15.5555	0.0643	7.7601
18	1.197	0.8353	19.623	0.0510	16.3908	0.0610	8.2310
19	1.209	0.8270	20.821	0.0480	17.2177	0.0581	8.7002
20	1.221	0.8187	22.030	0.0454	18.0365	0.0555	9.1677
21	1.234	0.8106	23.251	0.0430	18.8470	0.0531	9.6336
22	1.246	0.8025	24.485	0.0409	19.6496	0.0509	10.0978
23	1.259	0.7945	25.731	0.0389	20.4441	0.0489	10.5604
24	1.271	0.7866	26.990	0.0371	21.2307	0.0471	11.0213
25	1.284	0.7788	28.261	0.0354	22.0095	0.0454	11.4806
26	1.297	0.7711	29.545	0.0339	22.7806	0.0439	11.9381
27	1.310	0.7634	30.842	0.0324	23.5439	0.0425	12.3941
28	1.323	0.7558	32.152	0.0311	24.2997	0.0412	12.8484
29	1.336	0.7483	33.475	0.0299	25.0480	0.0399	13.3010
30	1.350	0.7408	34.811	0.0287	25.7888	0.0388	13.7520
31	1.363	0.7335	36.161	0.0277	26.5223	0.0377	14.2013
32	1.377	0.7262	37.525	0.0267	27.2484	0.0367	14.6490
33	1.391	0.7189	38.902	0.0257	27.9673	0.0358	15.0950
34	1.405	0.7118	40.293	0.0248	28.6791	0.0349	15.5394
35	1.419	0.7047	41.698	0.0240	29.3838	0.0340	15.9821
40	1.492	0.6703	48.937	0.0204	32.8034	0.0305	18.1711
45	1.568	0.6376	56.548	0.0177	36.0563	0.0277	20.3190
50	1.649	0.6065	64.548	0.0155	39.1505	0.0256	22.4261
55	1.733	0.5770	72.959	0.0137	42.0939	0.0238	24.4926
60	1.822	0.5488	81.802	0.0122	44.8936	0.0223	26.5187
65	1.916	0.5221	91.097	0.0110	47.5569	0.0210	28.5045
70	2.014	0.4966	100.869	0.0099	50.0902	0.0200	30.4505
75	2.117	0.4724	111.142	0.0090	52.5000	0.0191	32.3567
80	2.226	0.4493	121.942	0.0082	54.7922	0.0183	34.2235
85	2.340	0.4274	133.296	0.0075	56.9727	0.0176	36.0513
90	2.460	0.4066	145.232	0.0069	59.0463	0.0169	37.8402
95	2.586	0.3868	157.779	0.0063	61.0198	0.0164	39.5907
100	2.718	0.3679	170.970	0.0059	62.8965	0.0159	41.3032

*Thuesen, H.G., Fabrycky, W.J., Thuesen, G.J., *Engineering Economy*, 5th ed., ©1977, pp. 561-575. Reprinted by permission of Prentice-Hall, Inc., Englewood Cliffs, NJ.

Table C2. 2% Interest Factors for Continuous Compounding Interest

n	Single Payment		Equal Payment Series				Uniform gradient-series factor
	Compound-amount factor	Present-worth factor	Compound-amount factor	Sinking-fund factor	Present-worth factor	Capital-recovery factor	
	To find F Given P $F/P\ r,n$	To find P Given F $P/F\ r,n$	To find F Given A $F/A\ r,n$	To find A Given F $A/F\ r,n$	To find P Given A $P/A\ r,n$	To find A Given P $A/P\ r,n$	To find A Given G $A/G\ r,n$
1	1.020	0.9802	1.000	1.0000	0.9802	1.0202	0.0000
2	1.041	0.9608	2.020	0.4950	1.9410	0.5152	0.4950
3	1.062	0.9418	3.061	0.3267	2.8828	0.3469	0.9867
4	1.083	0.9231	4.123	0.2426	3.8059	0.2628	1.4750
5	1.105	0.9048	5.206	0.1921	4.7107	0.2123	1.9600
6	1.128	0.8869	6.311	0.1585	5.5976	0.1787	2.4417
7	1.150	0.8694	7.439	0.1344	6.4670	0.1546	2.9200
8	1.174	0.8522	8.589	0.1164	7.3191	0.1366	3.3951
9	1.197	0.8353	9.763	0.1024	8.1544	0.1226	3.8667
10	1.221	0.8187	10.960	0.0913	8.9731	0.1115	4.3351
11	1.246	0.8025	12.181	0.0821	9.7757	0.1023	4.8002
12	1.271	0.7866	13.427	0.0745	10.5623	0.0947	5.2619
13	1.297	0.7711	14.699	0.0680	11.3333	0.0882	5.7203
14	1.323	0.7558	15.995	0.0625	12.0891	0.0827	6.1754
15	1.350	0.7408	17.319	0.0578	12.8299	0.0780	6.6272
16	1.377	0.7262	18.668	0.0536	13.5561	0.0738	7.0757
17	1.405	0.7118	20.046	0.0499	14.2679	0.0701	7.5209
18	1.433	0.6977	21.451	0.0466	14.9655	0.0668	7.9628
19	1.462	0.6839	22.884	0.0437	15.6494	0.0639	8.4015
20	1.492	0.6703	24.346	0.0411	16.3197	0.0613	8.8368
21	1.522	0.6571	25.838	0.0387	16.9768	0.0589	9.2688
22	1.553	0.6440	27.360	0.0366	17.6208	0.0568	9.6976
23	1.584	0.6313	28.913	0.0346	18.2521	0.0548	10.1231
24	1.616	0.6188	30.497	0.0328	18.8709	0.0530	10.5453
25	1.649	0.6065	32.113	0.0312	19.4774	0.0514	10.9643
26	1.682	0.5945	33.762	0.0296	20.0719	0.0498	11.3801
27	1.716	0.5828	35.444	0.0282	20.6547	0.0484	11.7925
28	1.751	0.5712	37.160	0.0269	21.2259	0.0471	12.2018
29	1.786	0.5599	38.910	0.0257	21.7858	0.0459	12.6078
30	1.822	0.5488	40.696	0.0246	22.3346	0.0448	13.0106
31	1.859	0.5380	42.518	0.0235	22.8725	0.0437	13.4102
32	1.896	0.5273	44.377	0.0225	23.3998	0.0427	13.8065
33	1.935	0.5169	46.274	0.0216	23.9167	0.0418	14.1997
34	1.974	0.5066	48.209	0.0208	24.4233	0.0410	14.5897
35	2.014	0.4966	50.182	0.0199	24.9199	0.0401	14.9765
40	2.226	0.4493	60.666	0.0165	27.2591	0.0367	16.8630
45	2.460	0.4066	72.253	0.0139	29.3758	0.0341	18.6714
50	2.718	0.3679	85.058	0.0118	31.2910	0.0320	20.4028
55	3.004	0.3329	99.210	0.0101	33.0240	0.0303	22.0588
60	3.320	0.3012	114.850	0.0087	34.5921	0.0289	23.6409
65	3.669	0.2725	132.135	0.0076	36.0109	0.0278	25.1507
70	4.055	0.2466	151.238	0.0066	37.2947	0.0268	26.5899
75	4.482	0.2231	172.349	0.0058	38.4564	0.0260	27.9604
80	4.953	0.2019	195.682	0.0051	39.5075	0.0253	29.2640
85	5.474	0.1827	221.468	0.0045	40.4585	0.0247	30.5028
90	6.050	0.1653	249.966	0.0040	41.3191	0.0242	31.6786
95	6.686	0.1496	281.461	0.0036	42.0978	0.0238	32.7937
100	7.389	0.1353	316.269	0.0032	42.8024	0.0234	33.8499

Table C3. 3% Interest Factors for Continuous Compounding Interest

n	Single Payment		Equal Payment Series				Uniform gradient-series factor
	Compound-amount factor	Present-worth factor	Compound-amount factor	Sinking-fund factor	Present-worth factor	Capital-recovery factor	
	To find F Given P F/P r,n	To find P Given F P/F r,n	To find F Given A F/A r,n	To find A Given F A/F r,n	To find P Given A P/A r,n	To find A Given P A/P r,n	To find A Given G A/G r,n
1	1.030	0.9705	1.000	1.0000	0.9705	1.0305	0.0000
2	1.062	0.9418	2.030	0.4925	1.9122	0.5230	0.4925
3	1.094	0.9139	3.092	0.3234	2.8262	0.3538	0.9800
4	1.128	0.8869	4.186	0.2389	3.7131	0.2693	1.4625
5	1.162	0.8607	5.314	0.1882	4.5738	0.2186	1.9400
6	1.197	0.8353	6.476	0.1544	5.4090	0.1849	2.4126
7	1.234	0.8106	7.673	0.1303	6.2196	0.1608	2.8801
8	1.271	0.7866	8.907	0.1123	7.0063	0.1427	3.3427
9	1.310	0.7634	10.178	0.0983	7.7696	0.1287	3.8003
10	1.350	0.7408	11.488	0.0871	8.5105	0.1175	4.2529
11	1.391	0.7189	12.838	0.0779	9.2294	0.1084	4.7006
12	1.433	0.6977	14.229	0.0703	9.9271	0.1007	5.1433
13	1.477	0.6771	15.662	0.0639	10.6041	0.0943	5.5811
14	1.522	0.6571	17.139	0.0584	11.2612	0.0888	6.0139
15	1.568	0.6376	18.661	0.0536	11.8988	0.0841	6.4419
16	1.616	0.6188	20.229	0.0494	12.5176	0.0799	6.8650
17	1.665	0.6005	21.845	0.0458	13.1181	0.0762	7.2831
18	1.716	0.5828	23.511	0.0425	13.7008	0.0730	7.6964
19	1.768	0.5655	25.227	0.0397	14.2663	0.0701	8.1049
20	1.822	0.5488	26.995	0.0371	14.8152	0.0675	8.5085
21	1.878	0.5326	28.817	0.0347	15.3477	0.0652	8.9072
22	1.935	0.5169	30.695	0.0326	15.8646	0.0630	9.3012
23	1.994	0.5016	32.629	0.0307	16.3662	0.0611	9.6904
24	2.054	0.4868	34.623	0.0289	16.8529	0.0593	10.0748
25	2.117	0.4724	36.678	0.0273	17.3253	0.0577	10.4545
26	2.181	0.4584	38.795	0.0258	17.7837	0.0562	10.8294
27	2.248	0.4449	40.976	0.0244	18.2286	0.0549	11.1996
28	2.316	0.4317	43.224	0.0231	18.6603	0.0536	11.5652
29	2.387	0.4190	45.540	0.0220	19.0792	0.0524	11.9261
30	2.460	0.4066	47.927	0.0209	19.4858	0.0513	12.2823
31	2.535	0.3946	50.387	0.0199	19.8803	0.0503	12.6339
32	2.612	0.3829	52.921	0.0189	20.2632	0.0494	12.9810
33	2.691	0.3716	55.533	0.0180	20.6348	0.0485	13.3235
34	2.773	0.3606	58.224	0.0172	20.9954	0.0476	13.6614
35	2.858	0.3499	60.998	0.0164	21.3453	0.0469	13.9948
40	3.320	0.3012	76.183	0.0131	22.9459	0.0436	15.5953
45	3.857	0.2593	93.826	0.0107	24.3235	0.0411	17.0874
50	4.482	0.2231	114.324	0.0088	25.5092	0.0392	18.4750
55	5.207	0.1921	138.140	0.0072	26.5297	0.0377	19.7623
60	6.050	0.1653	165.809	0.0060	27.4081	0.0365	20.9538
65	7.029	0.1423	197.957	0.0051	28.1642	0.0355	22.0540
70	8.166	0.1225	235.307	0.0043	28.8149	0.0347	23.0677
75	9.488	0.1054	278.702	0.0036	29.3750	0.0341	23.9996
80	11.023	0.0907	329.119	0.0030	29.8570	0.0335	24.8543
85	12.807	0.0781	387.696	0.0026	30.2720	0.0330	25.6368
90	14.880	0.0672	455.753	0.0022	30.6291	0.0327	26.3516
95	17.288	0.0579	534.823	0.0019	30.9365	0.0323	27.0033
100	20.086	0.0498	626.690	0.0016	31.2010	0.0321	27.5963

Table C4. 4% Interest Factors for Continuous Compounding Interest

n	Single Payment		Equal Payment Series				Uniform gradient-series factor
	Compound-amount factor	Present-worth factor	Compound-amount factor	Sinking-fund factor	Present-worth factor	Capital-recovery factor	
	To find F Given P F/P r,n	To find P Given F P/F r,n	To find F Given A F/A r,n	To find A Given F A/F r,n	To find P Given A P/A r,n	To find A Given P A/P r,n	To find A Given G A/G r,n
1	1.041	0.9608	1.000	1.0000	0.9608	1.0408	0.0000
2	1.083	0.9231	2.041	0.4900	1.8839	0.5308	0.4900
3	1.128	0.8869	3.124	0.3201	2.7708	0.3609	0.9734
4	1.174	0.8522	4.252	0.2352	3.6230	0.2760	1.4500
5	1.221	0.8187	5.425	0.1843	4.4417	0.2251	1.9201
6	1.271	0.7866	6.647	0.1505	5.2283	0.1913	2.3835
7	1.323	0.7558	7.918	0.1263	5.9841	0.1671	2.8402
8	1.377	0.7262	9.241	0.1082	6.7103	0.1490	3.2904
9	1.433	0.6977	10.618	0.0942	7.4079	0.1350	3.7339
10	1.492	0.6703	12.051	0.0830	8.0783	0.1238	4.1709
11	1.553	0.6440	13.543	0.0738	8.7223	0.1147	4.6013
12	1.616	0.6188	15.096	0.0663	9.3411	0.1071	5.0252
13	1.682	0.5945	16.712	0.0598	9.9356	0.1007	5.4425
14	1.751	0.5712	18.394	0.0544	10.5068	0.0952	5.8534
15	1.822	0.5488	20.145	0.0497	11.0556	0.0905	6.2578
16	1.896	0.5273	21.967	0.0455	11.5829	0.0863	6.6558
17	1.974	0.5066	23.863	0.0419	12.0895	0.0827	7.0474
18	2.054	0.4868	25.837	0.0387	12.5763	0.0795	7.4326
19	2.138	0.4677	27.892	0.0359	13.0440	0.0767	7.8114
20	2.226	0.4493	30.030	0.0333	13.4933	0.0741	8.1840
21	2.316	0.4317	32.255	0.0310	13.9250	0.0718	8.5503
22	2.411	0.4148	34.572	0.0289	14.3398	0.0697	8.9105
23	2.509	0.3985	36.983	0.0270	14.7383	0.0679	9.2644
24	2.612	0.3829	39.492	0.0253	15.1212	0.0661	9.6122
25	2.718	0.3679	42.104	0.0238	15.4891	0.0646	9.9539
26	2.829	0.3535	44.822	0.0223	15.8425	0.0631	10.2896
27	2.945	0.3396	47.651	0.0210	16.1821	0.0618	10.6193
28	3.065	0.3263	50.596	0.0198	16.5084	0.0606	10.9431
29	3.190	0.3135	53.661	0.0186	16.8219	0.0595	11.2609
30	3.320	0.3012	56.851	0.0176	17.1231	0.0584	11.5730
31	3.456	0.2894	60.171	0.0166	17.4125	0.0574	11.8792
32	3.597	0.2780	63.626	0.0157	17.6905	0.0565	12.1797
33	3.743	0.2671	67.223	0.0149	17.9576	0.0557	12.4746
34	3.896	0.2567	70.966	0.0141	18.2143	0.0549	12.7638
35	4.055	0.2466	74.863	0.0134	18.4609	0.0542	13.0475
40	4.953	0.2019	96.862	0.0103	19.5562	0.0511	14.3845
45	6.050	0.1653	123.733	0.0081	20.4530	0.0489	15.5918
50	7.389	0.1353	156.553	0.0064	21.1872	0.0472	16.6775
55	9.025	0.1108	196.640	0.0051	21.7883	0.0459	17.6498
60	11.023	0.0907	245.601	0.0041	22.2805	0.0449	18.5172
65	13.464	0.0743	305.403	0.0033	22.6834	0.0441	19.2882
70	16.445	0.0608	378.445	0.0027	23.0133	0.0435	19.9710
75	20.086	0.0498	467.659	0.0021	23.2834	0.0430	20.5737
80	24.533	0.0408	576.625	0.0017	23.5045	0.0426	21.1038
85	29.964	0.0334	709.717	0.0014	23.6856	0.0422	21.5687
90	36.598	0.0273	872.275	0.0012	23.8338	0.0420	21.9751
95	44.701	0.0224	1070.825	0.0009	23.9552	0.0418	22.3295
100	54.598	0.0183	1313.333	0.0008	24.0545	0.0416	22.6376

Table C5. 5% Interest Factors for Continuous Compounding Interest

n	Single Payment Compound-amount factor	Single Payment Present-worth factor	Equal Payment Series Compound-amount factor	Equal Payment Series Sinking-fund factor	Equal Payment Series Present-worth factor	Equal Payment Series Capital-recovery factor	Uniform gradient-series factor
	To find F Given P F/P r,n	To find P Given F P/F r,n	To find F Given A F/A r,n	To find A Given F A/F r,n	To find P Given A P/A r,n	To find A Given P A/P r,n	To find A Given G A/G r,n
1	1.051	0.9512	1.000	1.0000	0.9512	1.0513	0.0000
2	1.105	0.9048	2.051	0.4875	1.8561	0.5388	0.4875
3	1.162	0.8607	3.156	0.3168	2.7168	0.3681	0.9667
4	1.221	0.8187	4.318	0.2316	3.5355	0.2829	1.4376
5	1.284	0.7788	5.540	0.1805	4.3143	0.2318	1.9001
6	1.350	0.7408	6.824	0.1466	5.0551	0.1978	2.3544
7	1.419	0.7047	8.174	0.1224	5.7598	0.1736	2.8004
8	1.492	0.6703	9.593	0.1043	6.4301	0.1555	3.2382
9	1.568	0.6376	11.084	0.0902	7.0678	0.1415	3.6678
10	1.649	0.6065	12.653	0.0790	7.6743	0.1303	4.0892
11	1.733	0.5770	14.301	0.0699	8.2513	0.1212	4.5025
12	1.822	0.5488	16.035	0.0624	8.8001	0.1136	4.9077
13	1.916	0.5221	17.857	0.0560	9.3221	0.1073	5.3049
14	2.014	0.4966	19.772	0.0506	9.8187	0.1019	5.6941
15	2.117	0.4724	21.786	0.0459	10.2911	0.0972	6.0753
16	2.226	0.4493	23.903	0.0418	10.7404	0.0931	6.4487
17	2.340	0.4274	26.129	0.0383	11.1678	0.0896	6.8143
18	2.460	0.4066	28.468	0.0351	11.5744	0.0864	7.1721
19	2.586	0.3868	30.928	0.0323	11.9611	0.0836	7.5222
20	2.718	0.3679	33.514	0.0298	12.3290	0.0811	7.8646
21	2.858	0.3499	36.232	0.0276	12.6789	0.0789	8.1996
22	3.004	0.3329	39.090	0.0256	13.0118	0.0769	8.5270
23	3.158	0.3166	42.094	0.0238	13.3284	0.0750	8.8471
24	3.320	0.3012	45.252	0.0221	13.6296	0.0734	9.1599
25	3.490	0.2865	48.572	0.0206	13.9161	0.0719	9.4654
26	3.669	0.2725	52.062	0.0192	14.1887	0.0705	9.7638
27	3.857	0.2593	55.732	0.0180	14.4479	0.0692	10.0551
28	4.055	0.2466	59.589	0.0168	14.6945	0.0681	10.3395
29	4.263	0.2346	63.644	0.0157	14.9291	0.0670	10.6170
30	4.482	0.2231	67.907	0.0147	15.1522	0.0660	10.8877
31	4.711	0.2123	72.389	0.0138	15.3645	0.0651	11.1517
32	4.953	0.2019	77.101	0.0130	15.5664	0.0643	11.4091
33	5.207	0.1921	82.054	0.0122	15.7584	0.0635	11.6601
34	5.474	0.1827	87.261	0.0115	15.9411	0.0627	11.9046
35	5.755	0.1738	92.735	0.0108	16.1149	0.0621	12.1429
40	7.389	0.1353	124.613	0.0080	16.8646	0.0593	13.2435
45	9.488	0.1054	165.546	0.0061	17.4485	0.0573	14.2024
50	12.183	0.0821	218.105	0.0046	17.9032	0.0559	15.0329
55	15.643	0.0639	285.592	0.0035	18.2573	0.0548	15.7480
60	20.086	0.0498	372.247	0.0027	18.5331	0.0540	16.3604
65	25.790	0.0388	483.515	0.0021	18.7479	0.0533	16.8822
70	33.115	0.0302	626.385	0.0016	18.9152	0.0529	17.3245
75	42.521	0.0235	809.834	0.0012	19.0455	0.0525	17.6979
80	54.598	0.0183	1045.387	0.0010	19.1469	0.0522	18.0116
85	70.105	0.0143	1347.843	0.0008	19.2260	0.0520	18.2742
90	90.017	0.0111	1736.205	0.0006	19.2875	0.0519	18.4931
95	115.584	0.0087	2234.871	0.0005	19.3354	0.0517	18.6751
100	148.413	0.0067	2875.171	0.0004	19.3728	0.0516	18.8258

Table C6. 6% Interest Factors for Continuous Compounding Interest

n	Single Payment		Equal Payment Series				Uniform gradient-series factor
	Compound-amount factor	Present-worth factor	Compound-amount factor	Sinking-fund factor	Present-worth factor	Capital-recovery factor	
	To find F Given P $F/P\ r,n$	To find P Given F $P/F\ r,n$	To find F Given A $F/A\ r,n$	To find A Given F $A/F\ r,n$	To find P Given A $P/A\ r,n$	To find A Given P $A/P\ r,n$	To find A Given G $A/G\ r,n$
1	1.062	0.9418	1.000	1.0000	0.9418	1.0618	0.0000
2	1.128	0.8869	2.062	0.4850	1.8287	0.5469	0.4850
3	1.197	0.8353	3.189	0.3136	2.6640	0.3754	0.9600
4	1.271	0.7866	4.387	0.2280	3.4506	0.2898	1.4251
5	1.350	0.7408	5.658	0.1768	4.1914	0.2386	1.8802
6	1.433	0.6977	7.008	0.1427	4.8891	0.2045	2.3254
7	1.522	0.6571	8.441	0.1185	5.5461	0.1803	2.7607
8	1.616	0.6188	9.963	0.1004	6.1649	0.1622	3.1862
9	1.716	0.5828	11.579	0.0864	6.7477	0.1482	3.6020
10	1.822	0.5488	13.295	0.0752	7.2965	0.1371	4.0080
11	1.935	0.5169	15.117	0.0662	7.8133	0.1280	4.4044
12	2.054	0.4868	17.052	0.0587	8.3001	0.1205	4.7912
13	2.181	0.4584	19.106	0.0523	8.7585	0.1142	5.1685
14	2.316	0.4317	21.288	0.0470	9.1902	0.1088	5.5363
15	2.460	0.4066	23.604	0.0424	9.5968	0.1042	5.8949
16	2.612	0.3829	26.064	0.0384	9.9797	0.1002	6.2442
17	2.773	0.3606	28.676	0.0349	10.3403	0.0967	6.5845
18	2.945	0.3396	31.449	0.0318	10.6799	0.0936	6.9157
19	3.127	0.3198	34.393	0.0291	10.9997	0.0909	7.2379
20	3.320	0.3012	37.520	0.0267	11.3009	0.0885	7.5514
21	3.525	0.2837	40.840	0.0245	11.5845	0.0863	7.8562
22	3.743	0.2671	44.366	0.0225	11.8517	0.0844	8.1525
23	3.975	0.2516	48.109	0.0208	12.1032	0.0826	8.4403
24	4.221	0.2369	52.084	0.0192	12.3402	0.0810	8.7199
25	4.482	0.2231	56.305	0.0178	12.5633	0.0796	8.9913
26	4.759	0.2101	60.786	0.0165	12.7734	0.0783	9.2546
27	5.053	0.1979	65.545	0.0153	12.9713	0.0771	9.5101
28	5.366	0.1864	70.598	0.0142	13.1577	0.0760	9.7578
29	5.697	0.1755	75.964	0.0132	13.3332	0.0750	9.9980
30	6.050	0.1653	81.661	0.0123	13.4985	0.0741	10.2307
31	6.424	0.1557	87.711	0.0114	13.6542	0.0732	10.4561
32	6.821	0.1466	94.135	0.0106	13.8008	0.0725	10.6743
33	7.243	0.1381	100.956	0.0099	13.9389	0.0718	10.8855
34	7.691	0.1300	108.198	0.0093	14.0689	0.0711	11.0899
35	8.166	0.1225	115.889	0.0086	14.1914	0.0705	11.2876
40	11.023	0.0907	162.091	0.0062	14.7046	0.0680	12.1809
45	14.880	0.0672	224.458	0.0045	15.0849	0.0663	12.9295
50	20.086	0.0498	308.645	0.0032	15.3665	0.0651	13.5519
55	27.113	0.0369	422.285	0.0024	15.5752	0.0642	14.0654
60	36.598	0.0273	575.683	0.0017	15.7298	0.0636	14.4862
65	49.402	0.0203	782.748	0.0013	15.8443	0.0631	14.8288
70	66.686	0.0150	1062.257	0.0010	15.9292	0.0628	15.1060
75	90.017	0.0111	1439.555	0.0007	15.9920	0.0625	15.3291
80	121.510	0.0082	1948.854	0.0005	16.0386	0.0624	15.5078
85	164.022	0.0061	2636.336	0.0004	16.0731	0.0622	15.6503
90	221.406	0.0045	3564.339	0.0003	16.0986	0.0621	15.7633
95	298.867	0.0034	4817.012	0.0002	16.1176	0.0621	15.8527
100	403.429	0.0025	6507.944	0.0002	16.1316	0.0620	15.9232

Table C7. 7% Interest Factors for Continuous Compounding Interest

n	Single Payment		Equal Payment Series				Uniform gradient-series factor
	Compound-amount factor	Present-worth factor	Compound-amount factor	Sinking-fund factor	Present-worth factor	Capital-recovery factor	
	To find F Given P F/P r,n	To find P Given F P/F r,n	To find F Given A F/A r,n	To find A Given F A/F r,n	To find P Given A P/A r,n	To find A Given P A/P r,n	To find A Given G A/G r,n
1	1.073	0.9324	1.000	1.0000	0.9324	1.0725	0.0000
2	1.150	0.8694	2.073	0.4825	1.8018	0.5550	0.4825
3	1.234	0.8106	3.223	0.3103	2.6123	0.3828	0.9534
4	1.323	0.7558	4.456	0.2244	3.3681	0.2969	1.4126
5	1.419	0.7047	5.780	0.1730	4.0728	0.2455	1.8603
6	1.522	0.6571	7.199	0.1389	4.7299	0.2114	2.2965
7	1.632	0.6126	8.721	0.1147	5.3425	0.1872	2.7211
8	1.751	0.5712	10.353	0.0966	5.9137	0.1691	3.1344
9	1.878	0.5326	12.104	0.0826	6.4463	0.1551	3.5364
10	2.014	0.4966	13.981	0.0715	6.9429	0.1440	3.9272
11	2.160	0.4630	15.995	0.0625	7.4059	0.1350	4.3069
12	2.316	0.4317	18.155	0.0551	7.8376	0.1276	4.6756
13	2.484	0.4025	20.471	0.0489	8.2401	0.1214	5.0334
14	2.664	0.3753	22.955	0.0436	8.6154	0.1161	5.3804
15	2.858	0.3499	25.620	0.0390	8.9654	0.1161	5.7168
16	3.065	0.3263	28.478	0.0351	9.2917	0.1076	6.0428
17	3.287	0.3042	31.542	0.0317	9.5959	0.1042	6.3585
18	3.525	0.2837	34.829	0.0287	9.8795	0.1012	6.6640
19	3.781	0.2645	38.355	0.0261	10.1440	0.0986	6.9596
20	4.055	0.2466	42.136	0.0237	10.3906	0.0963	7.2453
21	4.349	0.2299	46.191	0.0217	10.6205	0.0942	7.5215
22	4.665	0.2144	50.540	0.0198	10.8349	0.0923	7.7882
23	5.003	0.1999	55.205	0.0181	11.0348	0.0906	8.0456
24	5.366	0.1864	60.208	0.0166	11.2212	0.0891	8.2940
25	5.755	0.1738	65.573	0.0153	11.3949	0.0878	8.5335
26	6.172	0.1620	71.328	0.0140	11.5570	0.0865	8.7643
27	6.619	0.1511	77.500	0.0129	11.7080	0.0854	8.9867
28	7.099	0.1409	84.119	0.0119	11.8489	0.0844	9.2009
29	7.614	0.1313	91.218	0.0110	11.9802	0.0835	9.4070
30	8.166	0.1225	98.833	0.0101	12.1027	0.0826	9.6052
31	8.758	0.1142	106.999	0.0094	12.2169	0.0819	9.7958
32	9.393	0.1065	115.757	0.0086	12.3233	0.0812	9.9790
33	10.047	0.0993	125.150	0.0080	12.4226	0.0805	10.1550
34	10.805	0.0926	135.225	0.0074	12.5151	0.0799	10.3239
35	11.588	0.0863	146.030	0.0069	12.6014	0.0794	10.4860
40	16.445	0.0608	213.006	0.0047	12.9529	0.0772	11.2017
45	23.336	0.0429	308.049	0.0033	13.2006	0.0758	11.7769
50	33.115	0.0302	442.922	0.0023	13.3751	0.0748	12.2347
55	46.993	0.0213	634.316	0.0016	13.4981	0.0741	12.5957
60	66.686	0.0150	905.916	0.0011	13.5847	0.0736	12.8781
65	94.632	0.0106	1291.336	0.0008	13.6458	0.0733	13.0974
70	134.290	0.0075	1838.272	0.0006	13.6889	0.0731	13.2664
75	190.566	0.0053	2614.412	0.0004	13.7192	0.0729	13.3959
80	270.426	0.0037	3715.807	0.0003	13.7406	0.0728	13.4946
85	383.753	0.0026	5278.761	0.0002	13.7556	0.0727	13.5695
90	544.572	0.0019	7496.698	0.0001	13.7662	0.0727	13.6260
95	772.784	0.0013	10644.100	0.0001	13.7737	0.0726	13.6685
100	1096.633	0.0009	15110.476	0.0001	13.7790	0.0726	13.7003

Table C8. 8% Interest Factors for Continuous Compounding Interest

n	Single Payment		Equal Payment Series				Uniform gradient-series factor
	Compound-amount factor	Present-worth factor	Compound-amount factor	Sinking-fund factor	Present-worth factor	Capital-recovery factor	
	To find F Given P F/P r,n	To find P Given F P/F r,n	To find F Given A F/A r,n	To find A Given F A/F r,n	To find P Given A P/A r,n	To find A Given P A/P r,n	To find A Given G A/G r,n
1	1.083	0.9231	1.000	1.0000	0.9231	1.0833	0.0000
2	1.174	0.8522	2.083	0.4800	1.7753	0.5633	0.4800
3	1.271	0.7866	3.257	0.3071	2.5619	0.3903	0.9467
4	1.377	0.7262	4.528	0.2209	3.2880	0.3041	1.4002
5	1.492	0.6703	5.905	0.1694	3.9584	0.2526	1.8405
6	1.616	0.6188	7.397	0.1352	4.5772	0.2185	2.2676
7	1.751	0.5712	9.013	0.1110	5.1484	0.1942	2.6817
8	1.896	0.5273	10.764	0.0929	5.6757	0.1762	3.0829
9	2.054	0.4868	12.660	0.0790	6.1624	0.1623	3.4713
10	2.226	0.4493	14.715	0.0680	6.6117	0.1513	3.8470
11	2.411	0.4148	16.940	0.0590	7.0265	0.1423	4.2102
12	2.612	0.3829	19.351	0.0517	7.4094	0.1350	4.5611
13	2.829	0.3535	21.963	0.0455	7.7629	0.1288	4.8998
14	3.065	0.3263	24.792	0.0403	8.0891	0.1236	5.2265
15	3.320	0.3012	27.857	0.0359	8.3903	0.1192	5.5415
16	3.597	0.2780	31.177	0.0321	8.6684	0.1154	5.8449
17	3.896	0.2567	34.774	0.0288	8.9250	0.1121	6.1369
18	4.221	0.2369	38.670	0.0259	9.1620	0.1092	6.4178
19	4.572	0.2187	42.891	0.0233	9.3807	0.1066	6.6879
20	4.953	0.2019	47.463	0.0211	9.5826	0.1044	6.9473
21	5.366	0.1864	52.416	0.0191	9.7689	0.1024	7.1963
22	5.812	0.1721	57.781	0.0173	9.9410	0.1006	7.4352
23	6.297	0.1588	63.594	0.0157	10.0998	0.0990	7.6642
24	6.821	0.1466	69.890	0.0143	10.2464	0.0976	7.8836
25	7.389	0.1353	76.711	0.0130	10.3818	0.0963	8.0937
26	8.004	0.1249	84.100	0.0119	10.5067	0.0952	8.2948
27	8.671	0.1153	92.105	0.0109	10.6220	0.0942	8.4870
28	9.393	0.1065	100.776	0.0099	10.7285	0.0932	8.6707
29	10.176	0.0983	110.169	0.0091	10.8267	0.0924	8.8461
30	11.023	0.0907	120.345	0.0083	10.9175	0.0916	9.0136
31	11.941	0.0838	131.368	0.0076	11.0012	0.0909	9.1734
32	12.936	0.0773	143.309	0.0070	11.0785	0.0903	9.3257
33	14.013	0.0714	156.245	0.0064	11.1499	0.0897	9.4708
34	15.180	0.0659	170.258	0.0059	11.2157	0.0892	9.6090
35	16.445	0.0608	185.439	0.0054	11.2765	0.0887	9.7405
40	24.533	0.0408	282.547	0.0035	11.5173	0.0868	10.3069
45	36.598	0.0273	427.416	0.0023	11.6786	0.0856	10.7426
50	54.598	0.0183	643.535	0.0016	11.7868	0.0849	11.0738
55	81.451	0.0123	965.947	0.0010	11.8593	0.0843	11.3230
60	121.510	0.0082	1446.928	0.0007	11.9079	0.0840	11.5088
65	181.272	0.0055	2164.469	0.0005	11.9404	0.0838	11.6461
70	270.426	0.0037	3234.913	0.0003	11.9623	0.0836	11.7469
75	403.429	0.0025	4831.828	0.0002	11.9769	0.0835	11.8203
80	601.845	0.0017	7214.146	0.0002	11.9867	0.0834	11.8735
85	897.847	0.0011	10768.146	0.0001	11.9933	0.0834	11.9119
90	1339.431	0.0008	16070.091	0.0001	11.9977	0.0834	11.9394
95	1998.196	0.0005	23979.664	0.0001	12.0007	0.0833	11.9591
100	2980.958	0.0004	35779.360	0.0000	12.0026	0.0833	11.9731

Table C9. 9% Interest Factors for Continuous Compounding Interest

	Single Payment		Equal Payment Series				Uniform gradient-series factor
	Compound-amount factor	Present-worth factor	Compound-amount factor	Sinking-fund factor	Present-worth factor	Capital-recovery factor	
n	To find F Given P F/P r,n	To find P Given F P/F r,n	To find F Given A F/A r,n	To find A Given F A/F r,n	To find P Given A P/A r,n	To find A Given P A/P r,n	To find A Given G A/G r,n
1	1.094	0.9139	1.000	1.0000	0.9139	1.0942	0.0000
2	1.197	0.8353	2.094	0.4775	1.7492	0.5717	0.4775
3	1.310	0.7634	3.291	0.3038	2.5126	0.3980	0.9401
4	1.433	0.6977	4.601	0.2173	3.2103	0.3115	1.3878
5	1.568	0.6376	6.305	0.1657	3.8479	0.2599	1.8206
6	1.716	0.5828	7.603	0.1315	4.4306	0.2257	2.2388
7	1.878	0.5326	9.319	0.1073	4.9632	0.2015	2.6424
8	2.054	0.4868	11.197	0.0893	5.4500	0.1835	3.0316
9	2.248	0.4449	13.251	0.0755	5.8948	0.1697	3.4065
10	2.460	0.4066	15.499	0.0645	6.3014	0.1587	3.7674
11	2.691	0.3716	17.959	0.0557	6.6730	0.1499	4.1145
12	2.945	0.3396	20.650	0.0484	7.0126	0.1426	4.4479
13	3.222	0.3104	23.594	0.0424	7.3230	0.1366	4.7680
14	3.525	0.2837	26.816	0.0373	7.6066	0.1315	5.0750
15	3.857	0.2593	30.342	0.0330	7.8658	0.1271	5.3691
16	4.221	0.2369	34.199	0.0293	8.1028	0.1234	5.6507
17	4.618	0.2165	38.420	0.0260	8.3193	0.1202	5.9201
18	5.053	0.1979	43.038	0.0232	8.5172	0.1174	6.1776
19	5.529	0.1809	48.091	0.0208	8.6981	0.1150	6.4234
20	6.050	0.1653	53.620	0.0187	8.8634	0.1128	6.6579
21	6.619	0.1511	59.670	0.0168	9.0144	0.1109	6.8815
22	7.243	0.1381	66.289	0.0151	9.1525	0.1093	7.0945
23	7.925	0.1262	73.532	0.0136	9.2787	0.1078	7.2972
24	8.671	0.1153	81.457	0.0123	9.3940	0.1065	7.4900
25	9.488	0.1054	90.128	0.0111	9.4994	0.1053	7.6732
26	10.381	0.0963	99.616	0.0100	9.5958	0.1042	7.8471
27	11.359	0.0880	109.997	0.0091	9.6838	0.1033	8.0122
28	12.429	0.0805	121.356	0.0083	9.7643	0.1024	8.1686
29	13.599	0.0735	133.784	0.0075	9.8378	0.1017	8.3169
30	14.880	0.0672	147.383	0.0068	9.9050	0.1010	8.4572
31	16.281	0.0614	162.263	0.0062	9.9664	0.1003	8.5900
32	17.814	0.0561	178.544	0.0056	10.0225	0.0998	8.7155
33	19.492	0.0513	196.358	0.0051	10.0739	0.0993	8.8341
34	21.328	0.0469	215.850	0.0046	10.1207	0.0988	8.9460
35	23.336	0.0429	237.178	0.0042	10.1636	0.0984	9.0516
40	36.598	0.0273	378.004	0.0027	10.3285	0.0968	9.4950
45	57.397	0.0174	598.863	0.0017	10.4336	0.0959	9.8207
50	90.017	0.0111	945.238	0.0011	10.5007	0.0952	10.0569
55	141.175	0.0071	1488.463	0.0007	10.5434	0.0949	10.2263
60	221.406	0.0045	2340.410	0.0004	10.5707	0.0946	10.3464
65	347.234	0.0029	3676.528	0.0003	10.5880	0.0945	10.4309
70	544.572	0.0019	5771.978	0.0002	10.5991	0.0944	10.4898
75	854.059	0.0012	9058.298	0.0001	10.6062	0.0943	10.5307
80	1339.431	0.0008	14212.274	0.0001	10.6107	0.0943	10.5588
85	2100.646	0.0005	22295.318	0.0001	10.6136	0.0942	10.5781
90	3294.468	0.0003	34972.053	0.0000	10.6154	0.0942	10.5913
95	5166.754	0.0002	54853.132	0.0000	10.6166	0.0942	10.6002
100	8103.084	0.0001	86032.870	0.0000	10.6173	0.0942	10.6063

Table C10. 10% Interest Factors for Continuous Compounding Interest

n	Single Payment		Equal Payment Series				Uniform gradient-series factor
	Compound-amount factor	Present-worth factor	Compound-amount factor	Sinking-fund factor	Present-worth factor	Capital-recovery factor	
	To find F Given P $F/P\ r,n$	To find P Given F $P/F\ r,n$	To find F Given A $F/A\ r,n$	To find A Given F $A/F\ r,n$	To find P Given A $P/A\ r,n$	To find A Given P $A/P\ r,n$	To find A Given G $A/G\ r,n$
1	1.105	0.9048	1.000	1.0000	0.9048	1.1052	0.0000
2	1.221	0.8187	2.105	0.4750	1.7236	0.5802	0.4750
3	1.350	0.7408	3.327	0.3006	2.4644	0.4058	0.9335
4	1.492	0.6703	4.676	0.2138	3.1347	0.3190	1.3754
5	1.649	0.6065	6.168	0.1621	3.7412	0.2673	1.8009
6	1.822	0.5488	7.817	0.1279	4.2901	0.2331	2.2101
7	2.014	0.4966	9.639	0.1038	4.7866	0.2089	2.6033
8	2.226	0.4493	11.653	0.0858	5.2360	0.1910	2.9806
9	2.460	0.4066	13.878	0.0721	5.6425	0.1772	3.3423
10	2.718	0.3679	16.338	0.0612	6.0104	0.1664	3.6886
11	3.004	0.3329	19.056	0.0525	6.3433	0.1577	4.0198
12	3.320	0.3012	22.060	0.0453	6.6445	0.1505	4.3362
13	3.669	0.2725	25.381	0.0394	6.9170	0.1446	4.6381
14	4.055	0.2466	29.050	0.0344	7.1636	0.1396	4.9260
15	4.482	0.2231	33.105	0.0302	7.3867	0.1354	5.2001
16	4.953	0.2019	37.587	0.0266	7.5886	0.1318	5.4608
17	5.474	0.1827	42.540	0.0235	7.7713	0.1287	5.7086
18	6.050	0.1653	48.014	0.0208	7.9366	0.1260	5.9437
19	6.686	0.1496	54.063	0.0185	8.0862	0.1237	6.1667
20	7.389	0.1353	60.749	0.0165	8.2215	0.1216	6.3780
21	8.166	0.1225	68.138	0.0147	8.3440	0.1199	6.5779
22	9.025	0.1108	76.305	0.0131	8.4548	0.1183	6.7669
23	9.974	0.1003	85.330	0.0117	8.5550	0.1169	6.9454
24	11.023	0.0907	95.304	0.0105	8.6458	0.1157	7.1139
25	12.183	0.0821	106.327	0.0094	8.7279	0.1146	7.2727
26	13.464	0.0743	118.509	0.0084	8.8021	0.1136	7.4223
27	14.880	0.0672	131.973	0.0076	8.8693	0.1128	7.5631
28	16.445	0.0608	146.853	0.0068	8.9301	0.1120	7.6954
29	18.174	0.0550	163.297	0.0061	8.9852	0.1113	7.8198
30	20.086	0.0498	181.472	0.0055	9.0349	0.1107	7.9365
31	22.198	0.0451	201.557	0.0050	9.0800	0.1101	8.0459
32	24.533	0.0408	223.755	0.0045	9.1208	0.1097	8.1485
33	27.113	0.0369	248.288	0.0040	9.1576	0.1092	8.2446
34	29.964	0.0334	275.400	0.0036	9.1910	0.1088	8.3345
35	33.115	0.0302	305.364	0.0033	9.2212	0.1085	8.4185
40	54.598	0.0183	509.629	0.0020	9.3342	0.1071	8.7620
45	90.017	0.0111	846.404	0.0012	9.4027	0.1064	9.0028
50	148.413	0.0067	1401.653	0.0007	9.4443	0.1059	9.1692
55	244.692	0.0041	2317.104	0.0004	9.4695	0.1056	9.2826
60	403.429	0.0025	3826.427	0.0003	9.4848	0.1054	9.3592
65	665.142	0.0015	6314.879	0.0002	9.4940	0.1053	9.4105
70	1096.633	0.0009	10417.644	0.0001	9.4997	0.1053	9.4445
75	1808.042	0.0006	17181.959	0.0001	9.5031	0.1052	9.4668
80	2980.958	0.0004	28334.430	0.0001	9.5052	0.1052	9.4815
85	4914.769	0.0002	46721.745	0.0000	9.5064	0.1052	9.4910
90	8103.084	0.0001	77037.303	0.0000	9.5072	0.1052	9.4972
95	13359.727	0.0001	127019.209	0.0000	9.5076	0.1052	9.5012
100	22026.466	0.0001	209425.440	0.0000	9.5079	0.1052	9.5038

Table C11. 12% Interest Factors for Continuous Compounding Interest

	Single Payment		Equal Payment Series				Uniform gradient-series factor
	Compound-amount factor	Present-worth factor	Compound-amount factor	Sinking-fund factor	Present-worth factor	Capital-recovery factor	
n	To find F Given P F/P r,n	To find P Given F P/F r,n	To find F Given A F/A r,n	To find A Given F A/F r,n	To find P Given A P/A r,n	To find A Given P A/P r,n	To find A Given G A/G r,n
1	1.128	0.8869	1.000	1.0000	0.8869	1.1275	0.0000
2	1.271	0.7866	2.128	0.4700	1.6736	0.5975	0.4700
3	1.433	0.6977	3.399	0.2942	2.3712	0.4217	0.9202
4	1.616	0.6188	4.832	0.2070	2.9900	0.3345	1.3506
5	1.822	0.5488	6.448	0.1551	3.5388	0.2826	1.7615
6	2.054	0.4868	8.270	0.1209	4.0256	0.2484	2.1531
7	2.316	0.4317	10.325	0.0969	4.4573	0.2244	2.5257
8	2.612	0.3829	12.641	0.0791	4.8402	0.2066	2.8796
9	2.945	0.3396	15.253	0.0656	5.1798	0.1931	3.2153
10	3.320	0.3012	18.197	0.0550	5.4810	0.1825	3.5332
11	3.743	0.2671	21.518	0.0465	5.7481	0.1740	3.8337
12	4.221	0.2369	25.261	0.0396	5.9850	0.1671	4.1174
13	4.759	0.2101	29.482	0.0339	6.1952	0.1614	4.3848
14	5.366	0.1864	34.241	0.0292	6.3815	0.1567	4.6364
15	6.050	0.1653	39.606	0.0253	6.5468	0.1528	4.8728
16	6.821	0.1466	45.656	0.0219	6.6935	0.1494	5.0947
17	7.691	0.1300	52.477	0.0191	6.8235	0.1466	5.3025
18	8.671	0.1153	60.167	0.0166	6.9388	0.1441	5.4969
19	9.777	0.1023	68.838	0.0145	7.0411	0.1420	5.6785
20	11.023	0.0907	78.615	0.0127	7.1318	0.1402	5.8480
21	12.429	0.0805	89.638	0.0112	7.2123	0.1387	6.0058
22	14.013	0.0714	102.067	0.0098	7.2836	0.1373	6.1528
23	15.800	0.0633	116.080	0.0086	7.3469	0.1361	6.2893
24	17.814	0.0561	131.880	0.0076	7.4031	0.1351	6.4160
25	20.086	0.0498	149.694	0.0067	7.4528	0.1342	6.5334
26	22.646	0.0442	169.780	0.0059	7.4970	0.1334	6.6422
27	25.534	0.0392	192.426	0.0052	7.5362	0.1327	6.7428
28	28.789	0.0347	217.960	0.0046	7.5709	0.1321	6.8358
29	32.460	0.0308	246.749	0.0041	7.6017	0.1316	6.9215
30	36.598	0.0273	279.209	0.0036	7.6290	0.1311	7.0006
31	41.264	0.0242	315.807	0.0032	7.6533	0.1307	7.0734
32	46.525	0.0215	357.071	0.0028	7.6748	0.1303	7.1404
33	52.457	0.0191	403.597	0.0025	7.6938	0.1300	7.2020
34	59.145	0.0169	456.054	0.0022	7.7107	0.1297	7.2586
35	66.686	0.0150	515.200	0.0020	7.7257	0.1294	7.3105
40	121.510	0.0082	945.203	0.0011	7.7788	0.1286	7.5114
45	221.406	0.0045	1728.720	0.0006	7.8079	0.1281	7.6392
50	403.429	0.0025	3156.382	0.0003	7.8239	0.1278	7.7191

Table C12. 15% Interest Factors for Continuous Compounding Interest

	Single Payment		Equal Payment Series				Uniform gradient-series factor
	Compound-amount factor	Present-worth factor	Compound-amount factor	Sinking-fund factor	Present-worth factor	Capital-recovery factor	
n	To find F Given P $F/P\ r,n$	To find P Given F $P/F\ r,n$	To find F Given A $F/A\ r,n$	To find A Given F $A/F\ r,n$	To find P Given A $P/A\ r,n$	To find A Given P $A/P\ r,n$	To find A Given G $A/G\ r,n$
1	1.162	0.8607	1.000	1.0000	0.8607	1.1618	0.0000
2	1.350	0.7408	2.162	0.4626	1.6015	0.6244	0.4626
3	1.568	0.6376	3.512	0.2848	2.2392	0.4466	0.9004
4	1.822	0.5488	5.080	0.1969	2.7880	0.3587	1.3137
5	2.117	0.4724	6.902	0.1449	3.2603	0.3067	1.7029
6	2.460	0.4066	9.019	0.1109	3.6669	0.2727	2.0685
7	2.858	0.3499	11.479	0.0871	4.0168	0.2490	2.4110
8	3.320	0.3012	14.336	0.0698	4.3180	0.2316	2.7311
9	3.857	0.2593	17.657	0.0566	4.5773	0.2185	3.0295
10	4.482	0.2231	21.514	0.0465	4.8004	0.2083	3.3070
11	5.207	0.1921	25.996	0.0385	4.9925	0.2003	3.5645
12	6.050	0.1653	31.203	0.0321	5.1578	0.1939	3.8028
13	7.029	0.1423	37.252	0.0269	5.3000	0.1887	4.0228
14	8.166	0.1225	44.281	0.0226	5.4225	0.1844	4.2255
15	9.488	0.1054	52.447	0.0191	5.5279	0.1089	4.4119
16	11.023	0.0907	61.935	0.0162	5.6186	0.1780	4.5829
17	12.807	0.0781	72.958	0.0137	5.6967	0.1756	4.7394
18	14.880	0.0672	85.765	0.0117	5.7639	0.1735	4.8823
19	17.288	0.0579	100.645	0.0099	5.8217	0.1718	5.0127
20	20.086	0.0498	117.933	0.0085	5.8715	0.1703	5.1313
21	23.336	0.0429	138.018	0.0073	5.9144	0.1691	5.2390
22	27.113	0.0369	161.354	0.0062	5.9513	0.1680	5.3367
23	31.500	0.0318	188.467	0.0053	5.9830	0.1672	5.4251
24	36.598	0.0273	219.967	0.0046	6.0103	0.1664	5.5050
25	42.521	0.0235	256.566	0.0339	6.0339	0.1657	5.5771
26	49.402	0.0203	299.087	0.0034	6.0541	0.1652	5.6420
27	57.397	0.0174	348.489	0.0029	6.0715	0.1647	5.7004
28	66.686	0.0150	405.886	0.0025	6.0865	0.1643	5.7529
29	77.478	0.0129	472.573	0.0021	6.0994	0.1640	5.8000
30	90.017	0.0111	550.051	0.0018	6.1105	0.1637	5.8422
31	104.585	0.0096	640.068	0.0016	6.1201	0.1634	5.8799
32	121.510	0.0082	744.653	0.0014	6.1283	0.1632	5.9136
33	141.175	0.0071	866.164	0.0012	6.1354	0.1630	5.9438
34	164.022	0.0061	1007.339	0.0010	6.1415	0.1628	5.9706
35	190.566	0.0053	1171.361	0.0009	6.1467	0.1627	5.9945
40	403.429	0.0025	2486.673	0.0004	6.1639	0.1622	6.0798
45	854.059	0.0012	5271.188	0.0002	6.1719	0.1620	6.1264
50	1808.042	0.0006	11166.008	0.0001	6.1758	0.1619	6.1515

Table C13. 20% Interest Factors for Continuous Compounding Interest

n	Single Payment		Equal Payment Series				Uniform gradient-series factor
	Compound-amount factor	Present-worth factor	Compound-amount factor	Sinking-fund factor	Present-worth factor	Capital-recovery factor	
	To find F Given P F/P r,n	To find P Given F P/F r,n	To find F Given A F/A r,n	To find A Given F A/F r,n	To find P Given A P/A r,n	To find A Given P A/P r,n	To find A Given G A/G r,n
1	1.221	0.8187	1.000	1.0000	0.8187	1.2214	0.0000
2	1.492	0.6703	2.221	0.4502	1.4891	0.6716	0.4502
3	1.822	0.5488	3.713	0.2693	2.0379	0.4907	0.8676
4	2.226	0.4493	5.535	0.1807	2.4872	0.4021	1.2528
5	2.718	0.3679	7.761	0.1289	2.8551	0.3503	1.6068
6	3.320	0.3012	10.479	0.0954	3.1563	0.3168	1.9306
7	4.055	0.2466	13.799	0.0725	3.4029	0.2939	2.2255
8	4.953	0.2019	17.854	0.0560	3.6048	0.2774	2.4929
9	6.050	0.1653	22.808	0.0439	3.7701	0.2653	2.7344
10	7.389	0.1353	28.857	0.0347	3.9054	0.2561	2.9515
11	9.025	0.1108	36.246	0.0276	4.0162	0.2490	3.1460
12	11.023	0.0907	45.271	0.0221	4.1069	0.2435	3.3194
13	13.464	0.0743	56.294	0.0178	4.1812	0.2392	3.4736
14	16.445	0.0608	69.758	0.0143	4.2420	0.2357	3.6102
15	20.086	0.0498	86.203	0.0116	4.2918	0.2330	3.7307
16	24.533	0.0408	106.288	0.0094	4.3326	0.2308	3.8368
17	29.964	0.0334	130.821	0.0077	4.3659	0.2291	3.9297
18	36.598	0.0273	160.785	0.0062	4.3933	0.2276	4.0110
19	44.701	0.0224	197.383	0.0051	4.4156	0.2265	4.0819
20	54.598	0.0183	242.084	0.0041	4.4339	0.2255	4.1435
21	66.686	0.0150	296.683	0.0034	4.4489	0.2248	4.1970
22	81.451	0.0123	363.369	0.0028	4.4612	0.2242	4.2432
23	99.484	0.0101	444.820	0.0023	4.4713	0.2237	4.2831
24	121.510	0.0082	544.304	0.0018	4.4795	0.2232	4.3175
25	148.413	0.0067	665.814	0.0015	4.4862	0.2229	4.3471
26	181.272	0.0055	814.228	0.0012	4.4917	0.2226	4.3724
27	221.406	0.0045	995.500	0.0010	4.4963	0.2224	4.3942
28	270.426	0.0037	1216.906	0.0008	4.5000	0.2222	4.4127
29	330.300	0.0030	1487.333	0.0007	4.5030	0.2221	4.4286
30	403.429	0.0025	1817.632	0.0006	4.5055	0.2220	4.4421
31	492.749	0.0020	2221.061	0.0005	4.5075	0.2219	4.4536
32	601.845	0.0017	2713.810	0.0004	4.5092	0.2218	4.4634
33	735.095	0.0014	3315.655	0.0003	4.5105	0.2217	4.4717
34	897.847	0.0011	4050.750	0.0003	4.5116	0.2217	4.4788
35	1096.633	0.0009	4948.598	0.0002	4.5125	0.2216	4.4847
40	2980.958	0.0004	13459.444	0.0001	4.5152	0.2215	4.5032
45	8103.084	0.0001	36594.322	0.0000	4.5161	0.2214	4.5111
50	22026.466	0.0001	99481.443	0.0000	4.5165	0.2214	4.5144

Table C14. 25% Interest Factors for Continuous Compounding Interest

n	Single Payment		Equal Payment Series				Uniform gradient-series factor
	Compound-amount factor	Present-worth factor	Compound-amount factor	Sinking-fund factor	Present-worth factor	Capital-recovery factor	
	To find F Given P F/P r,n	To find P Given F P/F r,n	To find F Given A F/A r,n	To find A Given F A/F r,n	To find P Given A P/A r,n	To find A Given P A/P r,n	To find A Given G A/G r,n
1	1.284	0.7788	1.000	1.0000	0.7788	1.2840	0.0000
2	1.649	0.6065	2.284	0.4378	1.3853	0.7219	0.4378
3	2.117	0.4724	3.933	0.2543	1.8577	0.5383	0.8351
4	2.718	0.3679	6.050	0.1653	2.2256	0.4493	1.1929
5	3.490	0.2865	8.768	0.1141	2.5121	0.3981	1.5131
6	4.482	0.2231	12.258	0.0816	2.7352	0.3656	1.7975
7	5.755	0.1738	16.740	0.0597	2.9090	0.3438	2.0486
8	7.389	0.1353	22.495	0.0445	3.0443	0.3285	2.2687
9	9.488	0.1054	29.884	0.0335	3.1497	0.3175	2.4605
10	12.183	0.0821	39.371	0.0254	3.2318	0.3094	2.6266
11	15.643	0.0639	51.554	0.0194	3.2957	0.3034	2.7696
12	20.086	0.0498	67.197	0.0149	3.3455	0.2989	2.8921
13	25.790	0.0388	87.282	0.0115	3.3843	0.2955	2.9964
14	33.115	0.0302	113.072	0.0089	3.4145	0.2929	3.0849
15	42.521	0.0235	146.188	0.0069	3.4380	0.2909	3.1596
16	54.598	0.0183	188.709	0.0053	3.4563	0.2893	3.2223
17	70.105	0.0143	243.307	0.0041	3.4706	0.2881	3.2748
18	90.017	0.0111	313.413	0.0032	3.4817	0.2872	3.3186
19	115.584	0.0087	403.430	0.0025	3.4904	0.2865	3.3550
20	148.413	0.0067	519.014	0.0019	3.4971	0.2860	3.3851
21	190.566	0.0053	667.427	0.0015	3.5023	0.2855	3.4100
22	244.692	0.0041	857.993	0.0012	3.5064	0.2852	3.4305
23	314.191	0.0032	1102.685	0.0009	3.5096	0.2849	3.4474
24	403.429	0.0025	1416.876	0.0007	3.5121	0.2847	3.4612
25	518.013	0.0019	1820.305	0.0006	3.5140	0.2846	3.4725
26	665.142	0.0015	2338.318	0.0004	3.5155	0.2845	3.4817
27	854.059	0.0012	3008.459	0.0003	3.5167	0.2844	3.4892
28	1096.633	0.0009	3857.518	0.0003	3.5176	0.2843	3.4953
29	1408.105	0.0007	4954.151	0.0002	3.5183	0.2842	3.5002
30	1808.042	0.0006	6362.256	0.0002	3.5189	0.2842	3.5042
31	2321.572	0.0004	8170.298	0.0001	3.5193	0.2842	3.5075
32	2980.958	0.0004	10491.871	0.0001	3.5196	0.2841	3.5101
33	3827.626	0.0003	13472.829	0.0001	3.5199	0.2841	3.5122
34	4914.769	0.0002	17300.455	0.0001	3.5201	0.2841	3.5139
35	6310.688	0.0002	22215.223	0.0001	3.5203	0.2841	3.5153

Table C15. 30% Interest Factors for Continuous Compounding Interest

n	Single Payment		Equal Payment Series				Uniform gradient-series factor
	Compound-amount factor	Present-worth factor	Compound-amount factor	Sinking-fund factor	Present-worth factor	Capital-recovery factor	
	To find F Given P F/P r,n	To find P Given F P/F r,n	To find F Given A F/A r,n	To find A Given F A/F r,n	To find P Given A P/A r,n	To find A Given P A/P r,n	To find A Given G A/G r,n
1	1.350	0.7408	1.000	1.0000	0.7408	1.3499	0.0000
2	1.822	0.5488	2.350	0.4256	1.2896	0.7754	0.4256
3	2.460	0.4066	4.172	0.2397	1.6962	0.5896	0.8030
4	3.320	0.3012	6.632	0.1508	1.9974	0.5007	1.1343
5	4.482	0.2231	9.952	0.1005	2.2205	0.4504	1.4222
6	6.050	0.1653	14.433	0.0693	2.3858	0.4192	1.6701
7	8.166	0.1225	20.483	0.0488	2.5083	0.3987	1.8815
8	11.023	0.0907	28.649	0.0349	2.5990	0.3848	2.0602
9	14.880	0.0672	39.672	0.0252	2.6662	0.3751	2.2099
10	20.086	0.0498	54.552	0.0183	2.7160	0.3682	2.3343
11	27.113	0.0369	74.638	0.0134	2.7529	0.3633	2.4371
12	36.598	0.0273	101.750	0.0098	2.7802	0.3597	2.5212
13	49.402	0.0203	138.349	0.0072	2.8004	0.3571	2.5897
14	66.686	0.0150	187.751	0.0053	2.8154	0.3552	2.6452
15	90.017	0.0111	254.437	0.0039	2.8266	0.3538	2.6898
16	121.510	0.0082	344.454	0.0029	2.8348	0.3528	2.7255
17	164.022	0.0061	465.965	0.0022	2.8409	0.3520	2.7540
18	221.406	0.0045	629.987	0.0016	2.8454	0.3515	2.7766
19	298.867	0.0034	851.393	0.0012	2.8487	0.3510	2.7945
20	403.429	0.0025	1150.261	0.0009	2.8512	0.3507	2.8086
21	544.572	0.0018	1553.689	0.0007	2.8531	0.3505	2.8197
22	735.095	0.0014	2098.261	0.0005	2.8544	0.3503	2.8283
23	992.275	0.0010	2833.356	0.0004	2.8554	0.3502	2.8351
24	1339.431	0.0008	3825.631	0.0003	2.8562	0.3501	2.8404
25	1808.042	0.0006	5165.062	0.0002	2.8567	0.3501	2.8445
26	2440.602	0.0004	6973.104	0.0002	2.8571	0.3500	2.8476
27	3294.468	0.0003	9413.706	0.0001	2.8574	0.3500	2.8501
28	4447.067	0.0002	12708.174	0.0001	2.8577	0.3499	2.8520
29	6002.912	0.0002	17155.241	0.0001	2.8578	0.3499	2.8535
30	8103.084	0.0001	23158.153	0.0001	2.8580	0.3499	2.8546
31	10938.019	0.0001	31261.237	0.0000	2.8580	0.3499	2.8555
32	14764.782	0.0001	42199.257	0.0000	2.8581	0.3499	2.8561
33	19930.370	0.0001	56964.038	0.0000	2.8582	0.3499	2.8566
34	26903.186	0.0001	76894.409	0.0000	2.8582	0.3499	2.8570
35	36315.503	0.0000	103797.595	0.0000	2.8582	0.3499	2.8573

APPENDIX **D***

Funds Flow Conversion Factors

Table D1. Funds Flow Conversion Factors

r	$\dfrac{e^r - 1}{r}$ $A/\bar{A}\,r$ ( )
1	1.005020
2	1.010065
3	1.015150
4	1.020270
5	1.025422
6	1.030608
7	1.035831
8	1.041088
9	1.046381
10	1.051709
11	1.057073
12	1.062474
13	1.067910
14	1.073384
15	1.078894
16	1.084443
17	1.090028
18	1.095652
19	1.101313
20	1.107014
21	1.112752
22	1.118530
23	1.124347
24	1.130204
25	1.136101
26	1.142038
27	1.148016
28	1.154035
29	1.160094
30	1.166196
31	1.172339
32	1.178524
33	1.184751
34	1.191022
35	1.197335
36	1.203692
37	1.210093
38	1.216538
39	1.223027
40	1.229561

*Thuesen, H.G., Fabrycky, W.J., Thuesen, G.J., *Engineering Economy*, 5th ed., ©1977, pp. 577. Reprinted by permission of Prentice-Hall, Inc., Englewood Cliffs, NJ.

Geometric Gradient Interest Factors

Table E1. Discrete Compounding: i = 5%

n	Geometric series present worth factor, (P/A,i,j,n)				
	j = 4%	j = 6%	j = 8%	j = 10%	j = 15%
1	0.9524	0.9524	0.9524	0.9524	0.9524
2	1.8957	1.9138	1.9320	1.9501	1.9955
3	2.8300	2.8844	2.9396	2.9954	3.1379
4	3.7554	3.8643	3.9759	4.0904	4.3891
5	4.6721	4.8535	5.0419	5.2375	5.7595
6	5.5799	5.8521	6.1383	6.4393	7.2604
7	6.4792	6.8602	7.2661	7.6983	8.9043
8	7.3699	7.8779	8.4261	9.0173	10.7047
9	8.2521	8.9053	9.6192	10.3991	12.6765
10	9.1258	9.9425	10.8464	11.8467	14.8362
11	9.9913	10.9896	12.1087	13.3632	17.2016
12	10.8485	12.0466	13.4070	14.9519	19.7922
13	11.6976	13.1137	14.7425	16.6163	22.6295
14	12.5386	14.1910	16.1161	18.3599	25.7371
15	13.3715	15.2785	17.5289	20.1866	29.1407
16	14.1966	16.3764	18.9821	22.1002	32.8683
17	15.0137	17.4848	20.4769	24.1050	36.9510
18	15.8231	18.5037	22.0143	26.2052	41.4226
19	16.6248	19.7332	23.5956	28.4055	46.3200
20	17.4189	20.8736	25.2222	30.7105	51.6838
21	18.2054	22.0247	26.8952	33.1253	57.5584
22	18.9844	23.1869	28.6160	35.5550	63.9925
23	19.7559	24.3601	30.3860	38.3053	71.0394
24	20.5202	25.5445	32.2006	41.0817	78.7575
25	21.2771	26.7401	34.0791	43.9904	87.2106
26	22.0269	27.9472	36.0052	47.0375	96.4637
27	22.7695	29.1657	37.9863	50.2298	106.6086
28	23.5050	30.3959	40.0240	53.5741	117.7142
29	24.2335	31.6377	42.1199	57.0776	129.8774
30	24.9551	32.8914	44.2757	60.7480	143.1991
31	24.6698	34.1571	46.4931	64.5931	157.7895
32	26.3777	35.4348	48.7739	68.6213	173.7695
33	27.0789	36.7248	51.1198	72.8414	191.2713
34	27.7734	38.0267	53.5328	77.2624	210.4400
35	28.4612	39.3413	56.0146	81.8940	231.4343
36	29.1426	40.6583	58.5674	86.7461	254.4280
37	29.8174	42.0080	61.1932	91.8292	279.6116
38	30.4858	43.3605	63.8939	97.1544	307.1937
39	31.1478	44.7258	66.6719	102.7332	337.4026
40	31.8036	46.1042	69.5291	108.5776	370.4886
41	32.4531	47.4957	72.4681	114.7004	406.7256
42	33.0964	48.9004	75.4910	121.1147	446.4138
43	33.7335	50.3185	78.6002	127.8344	489.8817
44	34.3647	51.7501	81.7983	134.8742	537.4895
45	34.9898	53.1953	85.0878	142.2491	589.6314
46	35.6089	54.6543	88.4713	149.9753	646.7391
47	36.2221	56.1272	91.9514	158.0693	709.2857
48	36.8296	57.6141	95.5310	166.5488	777.7891
49	37.4312	59.1152	99.2128	175.4321	852.8167
50	38.0271	60.6306	102.9998	184.7384	934.9897

Principles of Engineering Economic Analysis, J.A. White, M.H. Agee, and K.E. Case, © 1977, John Wiley & Sons, Inc., New York. Reprinted by permission of John Wiley & Sons, Inc.

Table E1. Discrete Compounding: i = 5%

	Geometric series future worth factor, (F/A,i,j,n)				
n	j = 4%	j = 6%	j = 8%	j = 10%	j = 15%
1	1.0000	1.0000	1.0000	1.0000	1.0000
2	2.0900	2.1100	2.1300	2.1500	2.2000
3	3.2761	3.3391	3.4029	3.4675	3.6325
4	4.5648	4.6971	4.8328	4.9719	5.3350
5	5.9629	6.1944	6.4349	6.6346	7.3508
6	7.4777	7.8423	8.2260	8.6293	9.7297
7	9.1169	9.6530	10.2241	10.8323	12.5292
8	10.8886	11.6393	12.4492	13.3227	15.8157
9	12.8016	13.8151	14.9225	16.1324	19.6655
10	14.8850	16.1953	17.6677	19.2970	24.1666
11	17.0885	18.7959	20.7100	22.8555	29.4205
12	19.4824	21.6340	24.0771	26.8514	35.5439
13	22.0576	24.7279	27.7992	31.3324	42.6714
14	24.8255	28.0972	31.9087	36.3513	50.9577
15	27.7985	31.7630	36.4414	41.9664	60.5813
16	30.9893	35.7477	41.4356	48.2420	71.7475
17	34.4118	40.0754	46.9333	55.2490	84.6925
18	38.0803	44.7720	52.9800	63.0660	99.6883
19	42.0101	49.8649	59.6250	71.7792	117.0482
20	46.2175	55.3838	66.9220	81.4840	137.1324
21	50.7195	61.3601	74.9290	92.2857	160.3556
22	55.5342	67.8277	83.7093	104.3003	187.1948
23	60.6808	74.8226	93.3313	117.6556	218.1993
24	65.1796	82.3835	103.8694	132.4927	254.0008
25	72.0619	90.5516	115.4040	148.9670	295.3260
26	78.3203	99.3710	128.0227	167.2501	343.0112
27	85.0088	108.8890	141.8202	187.5008	398.0186
28	92.1426	119.1558	156.8992	210.0173	461.4548
29	99.7484	130.2252	173.3713	234.9391	534.5932
30	107.8545	142.1549	191.3572	252.5492	618.8983
31	115.4906	155.0061	210.9877	293.1261	716.0550
32	125.6883	168.8445	232.4047	326.9767	828.0013
33	135.4807	183.7401	255.7620	364.4393	956.9664
34	145.9032	199.7677	281.2262	405.8864	1105.5146
35	156.9926	217.0071	308.9776	451.7284	1276.5951
36	168.7884	235.5436	339.2119	502.4173	1473.6004
37	181.3317	255.4680	372.1406	558.4508	1700.4322
38	194.6664	276.8775	407.9933	620.3773	1961.5785
39	208.8385	299.8756	447.0182	688.6005	2262.2007
40	223.8968	324.5729	489.4844	764.3853	2608.2356
41	239.8927	351.0873	535.6832	847.8839	3006.5109
42	258.8804	379.5445	585.9298	940.0422	3464.8795
43	274.9172	410.0783	640.5658	1041.8080	3992.3730
44	294.0635	442.8332	699.9607	1154.1385	4599.3787
45	314.3832	477.9603	764.5147	1278.1095	5297.8426
46	335.9435	515.6229	834.6609	1414.9055	6101.5040
47	358.8153	555.9946	910.8680	1565.8303	7026.1639
48	383.0741	509.2602	993.6434	1732.3193	8089.9944
49	406.7984	645.6171	1083.5362	1915.9525	9313.8949
50	436.0716	695.2754	1181.1404	2118.4691	10721.9004

Table E2. Discrete Compounding: i = 8%

	Geometric series present worth factor, (P/A,i,j,n)				
n	j = 4%	j = 6%	j = 8%	j = 10%	j = 15%
1	0.9259	0.9259	0.9259	0.9259	0.9259
2	1.8176	1.8347	1.8519	1.8690	1.9119
3	2.6762	2.7267	2.7778	2.8295	2.9617
4	3.5030	3.6021	3.7037	3.8079	4.0796
5	4.2992	4.4613	4.6296	4.8043	5.2699
6	5.0659	5.3046	5.5556	5.8192	6.5374
7	5.8042	6.1323	6.4815	6.8529	7.8871
8	6.5151	6.9447	7.4074	7.9057	9.3242
9	7.1997	7.7420	8.3333	8.9780	10.8545
10	7.8590	8.5246	9.2593	10.0702	12.4839
11	8.4939	9.2926	10.1852	11.1826	14.2190
12	9.1052	10.0465	11.1111	12.3157	16.0665
13	9.6939	10.7863	12.0370	13.4696	18.0338
14	10.2608	11.5125	12.9630	14.6450	20.1286
15	10.8067	12.2252	13.8889	15.8421	22.3592
16	11.3324	12.9248	14.8148	17.0014	24.7343
17	11.8386	13.6114	15.7407	18.3033	27.2634
18	12.3260	14.2852	16.6667	19.5682	29.9564
19	12.7954	14.9466	17.5926	20.8565	32.8239
20	13.2475	15.5957	18.5185	22.1687	35.8773
21	13.6827	16.2329	19.4444	23.5051	39.1286
22	14.1019	16.8582	20.3704	24.8663	42.5906
23	14.5055	17.4719	21.2963	26.2527	46.2771
24	14.8942	18.0743	22.2222	27.6648	50.2024
25	15.2685	18.6655	23.1481	29.1031	54.3822
26	15.6289	19.2458	24.0741	30.5679	58.8329
27	15.9760	19.8153	25.0000	32.0599	63.5721
28	16.3102	20.3743	25.9259	33.5796	68.6184
29	16.6321	20.9229	26.8519	35.1273	73.9919
30	16.9420	21.4614	27.7778	36.7038	79.7136
31	17.2404	21.9899	28.7037	38.3094	85.8061
32	17.5278	22.5086	29.6296	39.9447	92.2935
33	17.8046	23.0177	30.5556	41.6104	99.2015
34	18.0711	23.5173	31.4815	43.3069	106.5571
35	18.3277	24.0078	32.4074	45.0348	114.3895
36	18.5748	24.4891	33.3333	46.7947	122.7296
37	18.8128	24.9615	34.2593	48.5872	131.6102
38	19.0419	25.4252	35.1852	50.4129	141.0664
39	19.2626	25.8803	36.1111	52.2724	151.1355
40	19.4751	26.3269	37.0370	54.1663	161.8573
41	19.6797	26.7653	37.9630	56.0953	173.2739
42	19.8768	27.1956	38.8889	58.0600	185.4306
43	20.0665	27.6179	39.8148	60.0611	198.3752
44	20.2493	28.0324	40.7407	62.0993	212.1587
45	20.4252	28.4392	41.6667	64.1752	226.8357
46	20.5946	28.8385	42.5926	66.2896	242.4639
47	20.7578	29.2304	43.5185	68.4431	259.1051
48	20.9149	29.6150	44.4444	70.6365	276.8249
49	21.0662	29.9925	45.3704	72.8705	295.6932
50	21.2119	30.3630	46.2963	75.1459	315.7844

Table E2. Discrete Compounding: i = 8%

	Geometric series future worth factor, (F/A,i,j,n)				
n	j = 4%	j = 6%	j = 8%	j = 10%	j = 15%
1	1.0000	1.0000	1.0000	1.0000	1.0000
2	2.1200	2.1400	2.1600	2.1800	2.2300
3	3.3712	3.4348	3.4992	3.5644	3.7309
4	4.7658	4.9006	5.0388	5.1806	5.5502
5	6.3169	6.5551	6.8024	7.0591	7.7433
6	8.0389	8.4178	8.8160	9.2343	10.3741
7	9.9473	10.5097	11.1081	11.7446	13.5171
8	12.0590	12.8541	13.7106	14.6329	17.2585
9	14.3923	15.4763	16.6584	17.9472	21.6982
10	16.9670	18.4039	19.9900	21.7409	25.9519
11	19.8046	21.6670	23.7482	26.0739	33.1536
12	22.9284	25.2967	27.9797	31.0129	40.4583
13	26.3638	29.3348	32.7362	36.6324	49.0452
14	30.1379	33.8145	38.0747	43.0152	59.1216
15	34.2806	38.7805	44.0579	50.2540	70.9270
16	38.8240	44.2795	50.7547	58.4515	84.7383
17	43.8029	50.3623	58.2410	67.7226	100.8749
18	49.2551	57.0840	66.6003	78.1949	119.7062
19	55.2213	64.5051	75.9244	90.0104	141.6582
20	61.7459	72.6911	86.3140	103.3271	167.2226
21	68.8766	81.7135	97.8801	118.3208	196.9669
22	76.6655	91.6501	110.7443	135.1867	231.5458
23	85.1687	102.5857	125.0404	154.1419	271.7142
24	94.4469	114.6123	140.9151	175.4276	318.3428
25	104.5660	127.8302	158.5295	199.3115	372.4354
26	115.5972	142.3485	178.0604	226.0912	435.1492
27	127.6173	158.2858	199.7015	256.0966	507.8179
28	140.7101	175.7710	223.6657	289.6944	591.9787
29	154.9656	194.9443	250.1861	327.2909	689.4026
30	170.4815	215.9583	279.5182	369.3373	802.1302
31	187.3634	238.9784	311.9424	416.3337	932.5124
32	205.7256	264.1848	347.7654	468.8347	1083.2569
33	225.6917	291.7730	387.3237	527.4552	1257.4826
34	247.3954	321.9554	430.9857	592.8768	1458.7810
35	270.9814	354.9629	479.1547	665.8546	1691.2883
36	296.6060	391.0460	532.2724	747.2254	1959.7669
37	324.4384	430.4769	590.8224	837.9162	2269.7001
38	354.6616	473.5512	655.3338	938.9534	2627.4007
39	387.4733	520.5895	726.3857	1051.4740	3040.1361
40	423.0875	571.9402	804.6119	1176.7367	3516.2718
41	461.7355	627.9811	890.7054	1316.1349	4065.4371
42	503.6674	689.1225	985.4243	1471.2109	4698.7151
43	549.1536	755.8093	1089.5977	1643.6714	5428.8619
44	598.4864	828.5245	1204.1322	1835.4052	6270.5578
45	651.9818	907.7919	1330.0187	2048.5017	7240.6974
46	709.9816	994.1799	1468.3407	2285.2723	8358.7225
47	772.8549	1088.3048	1620.2820	2548.2737	9647.0049
48	841.0011	1190.8351	1787.1366	2840.3330	11131.2877
49	914.8517	1302.4957	1970.3181	3164.5769	12841.1914
50	994.8732	1424.0729	2171.3709	3524.4620	14810.7976

Table E3. Discrete Compounding: i = 10%

	Geometric series present worth factor, $(P/A, i, j, n)$				
n	*j* = 4%	*j* = 6%	*j* = 8%	*j* = 10%	*j* = 15%
1	0.9091	0.9091	0.9091	0.9091	0.9091
2	1.7685	1.7851	1.8017	1.8182	1.8595
3	2.5812	2.6293	2.6780	2.7273	2.8531
4	3.3495	3.4428	3.5384	3.6364	3.8919
5	4.0759	4.2267	4.3831	4.5455	4.9779
6	4.7627	4.9821	5.2125	5.4545	6.1133
7	5.4120	5.7100	6.0269	6.3636	7.3002
8	6.0259	6.4115	6.8254	7.2727	8.5411
9	6.6063	7.0874	7.6113	8.1818	9.8385
10	7.1550	7.7388	8.3820	9.0909	11.1948
11	7.6738	8.3664	9.1387	10.0000	12.6127
12	8.1644	8.9713	9.8817	10.9091	14.0951
13	8.6281	9.5542	10.6111	11.8182	15.6449
14	9.0566	10.1158	11.3273	12.7273	17.2651
15	9.4811	10.6571	12.0304	13.6364	18.9590
16	9.8731	11.1786	12.7208	14.5455	20.7298
17	10.2436	11.6812	13.3986	15.4545	22.5812
18	10.5940	12.1656	14.0640	16.3636	24.5167
19	10.9252	12.6323	14.7174	17.2727	26.5402
20	11.2384	13.0820	15.3589	18.1818	28.6556
21	11.5345	13.5154	15.9888	19.0909	30.8672
22	11.8144	13.9330	16.6071	20.0000	33.1794
23	12.0791	14.3354	17.2143	20.9091	35.5966
24	12.3293	14.7232	17.8104	21.8182	38.1238
25	12.5659	15.0969	18.3957	22.7273	40.7658
26	12.7896	15.4570	18.9703	23.6364	43.5278
27	13.0011	15.8041	19.5345	24.5455	46.4155
28	13.2010	16.1385	20.0884	25.4545	49.4343
29	13.3900	16.4607	20.6322	26.3636	52.5905
30	13.5688	16.7712	21.1662	27.2727	55.8900
31	13.7377	17.0704	21.6904	28.1818	59.3396
32	13.8975	17.3588	22.2052	29.0909	62.9459
33	14.0485	17.6367	22.7105	30.0000	66.7162
34	14.1913	17.9044	23.2067	30.9091	70.6578
35	14.3264	18.1624	23.6938	31.8182	74.7786
36	14.4540	18.4111	24.1721	32.7273	79.0867
37	14.5747	18.6507	24.6417	33.6364	83.5907
38	14.6888	18.8816	25.1028	34.5455	88.2994
39	14.7967	19.1040	25.5555	35.4545	93.2221
40	14.8987	19.3184	25.9999	36.3636	98.3685
41	14.9951	19.5250	26.4363	37.2727	103.7489
42	15.0853	19.7241	26.8647	38.1818	109.3739
43	15.1725	19.9160	27.2854	39.0909	115.2545
44	15.2540	20.1009	27.6983	40.0000	121.4024
45	15.3311	20.2790	28.1038	40.9091	127.8298
46	15.4039	20.4507	28.5019	41.8182	134.5493
47	15.4728	20.6161	28.8928	42.7273	141.5743
48	15.5379	20.7755	29.2766	43.6364	148.9186
49	15.5995	20.9291	29.6534	44.5455	156.5967
50	15.6577	21.0772	30.0233	45.4545	164.6238

Table E3. Discrete Compounding: i = 10%

	Geometric series future worth factor, (F/A,i,j,n)				
n	j = 4%	j = 6%	j = 8%	j = 10%	j = 15%
1	1.0000	1.0000	1.0000	1.0000	1.0000
2	2.1400	2.1600	2.1800	2.2000	2.2500
3	3.4356	3.4996	3.5644	3.6300	3.7975
4	4.9040	5.0406	5.1806	5.3240	5.6981
5	6.5643	6.8071	7.0591	7.3205	8.0169
6	8.4374	8.8260	9.2343	9.6631	10.8300
7	10.5464	11.1272	11.7446	12.4009	14.2261
8	12.9170	13.7435	14.6329	15.5897	18.3087
9	15.5773	16.7117	17.9472	19.2923	23.1986
10	18.5583	20.0724	21.7409	23.5795	29.0363
11	21.8944	23.8705	26.0739	28.5312	35.9855
12	25.6233	28.1558	31.0129	34.2374	44.2364
13	29.7866	32.9836	36.6324	40.7996	54.0103
14	34.4304	38.4149	43.0152	48.3318	65.5641
15	39.6051	44.5172	50.2540	56.9625	79.1963
16	45.3665	51.3655	58.4515	66.8360	95.2530
17	51.7762	59.0424	67.7226	78.1145	114.1359
18	58.9017	67.5395	78.1949	90.9805	136.3107
19	66.8177	77.2577	90.0104	105.6384	162.3173
20	75.8063	88.0091	103.3271	122.3182	192.7807
21	85.3580	100.0172	118.3208	141.2775	228.4254
22	96.1726	113.4184	135.1867	162.8055	270.0894
23	108.1598	128.3638	154.1419	187.2263	318.7431
24	121.4405	145.0200	175.4276	214.9033	375.5089
25	136.1478	163.5709	199.3115	245.2433	441.6849
26	152.4284	184.2198	226.0912	281.7024	518.7724
27	170.4438	207.1912	256.0966	321.7908	608.5064
28	190.3715	232.7327	289.5944	367.0798	712.8924
29	212.4074	261.1176	327.2909	418.2088	834.2472
30	236.7657	292.6478	369.3373	475.8928	975.2474
31	263.6868	327.6560	416.3337	540.9315	1138.9839
32	293.4286	366.5098	468.8347	614.2190	1329.0258
33	326.2796	409.6141	527.4552	696.7546	1549.4935
34	362.5559	457.4161	592.8768	789.6553	1805.1427
35	402.6058	510.4088	665.8546	894.1684	2101.4617
36	446.8125	569.1357	747.2254	1011.6877	2444.7834
37	495.5976	634.1965	837.9162	1143.7692	2842.4136
38	549.4255	706.2523	938.9534	1292.1500	3302.7796
39	668.8069	786.0318	1051.4740	1458.7694	3835.6009
40	674.3039	874.3384	1176.7367	1645.7911	4452.0858
41	746.5353	972.0580	1316.1349	1855.6295	5165.1579
42	826.1819	1080.1667	1471.2109	2090.9776	5989.7163
43	913.9929	1199.7404	1643.6714	2354.8391	6942.9380
44	1010.7927	1331.9649	1835.4052	2650.5630	8044.6188
45	1117.4385	1478.1468	2048.5017	2981.8834	9317.5757
46	1235.0785	1639.7261	2255.2723	3352.9622	10788.1025
47	1364.6612	1818.2892	2548.2737	3768.4380	12486.4975
48	1507.4451	2015.5841	2840.3330	4233.4793	14447.6696
49	1664.7601	2233.5363	3164.5769	4753.8445	16711.8372
50	1838.0695	2474.2675	3524.4620	5335.9479	19325.3318

Table E4. Discrete Compounding: i = 15%

		Geometric series present worth factor, (P/A,i,j,n)			
n	j = 4%	j = 6%	j = 8%	j = 10%	j = 15%
1	0.8696	0.8696	0.8696	0.8696	0.8696
2	1.6560	1.6711	1.6862	1.7013	1.7391
3	2.3671	2.4099	2.4531	2.4969	2.6087
4	3.0103	3.0908	3.1734	3.2579	3.4783
5	3.5919	3.7185	3.8498	3.9858	4.3478
6	4.1179	4.2971	4.4850	4.6821	5.2174
7	4.5936	4.8303	5.0816	5.3481	6.0870
8	5.0237	5.3219	5.6418	5.9851	6.9565
9	5.4128	5.7749	6.1680	6.5945	7.8261
10	5.7646	6.1926	6.6621	7.1773	8.6957
11	6.0828	6.5775	7.1261	7.7348	9.5652
12	6.3705	6.9323	7.5619	8.2681	10.4348
13	6.6307	7.2593	7.9712	8.7782	11.3043
14	6.8660	7.5608	8.3556	9.2661	12.1739
15	7.0789	7.8386	8.7165	9.7328	13.0435
16	7.2713	8.0947	9.0555	10.1792	13.9130
17	7.4454	8.3308	9.3739	10.6062	14.7826
18	7.6028	8.5484	9.6729	11.0146	15.6522
19	7.7451	8.7489	9.9537	11.4053	16.5217
20	7.8738	8.9338	10.2173	11.7790	17.3913
21	7.9903	9.1042	10.4650	12.1364	18.2609
22	8.0955	9.2613	10.6976	12.4783	19.1304
23	8.1907	9.4060	10.9160	12.8053	20.0000
24	8.2768	9.5395	11.1211	13.1181	20.8696
25	8.3547	9.6625	11.3137	13.4173	21.7391
26	8.4251	9.7759	11.4946	13.7035	22.6087
27	8.4888	9.8803	11.6645	13.9773	23.4783
28	8.5464	9.9767	11.8241	14.2392	24.3478
29	8.5985	10.0655	11.9739	14.4896	25.2174
30	8.6456	10.1473	12.1146	14.7292	26.0870
31	8.6882	10.2227	12.2468	14.9584	26.9565
32	8.7267	10.2922	12.3709	15.1776	27.8261
33	8.7615	10.3563	12.4874	15.3873	28.6957
34	8.7930	10.4154	12.5969	15.5878	29.5652
35	8.8215	10.4698	12.6997	15.7796	30.4348
36	8.8473	10.5200	12.7962	15.9631	31.3043
37	8.8706	10.5663	12.8869	16.1386	32.1739
38	8.8917	10.6089	12.9720	16.3065	33.0435
39	8.9107	10.6482	13.0520	16.4671	33.9130
40	8.9280	10.6845	13.1271	16.6207	34.7826
41	8.9436	10.7178	13.1976	16.7676	35.6522
42	8.9576	10.7486	13.2639	16.9082	36.5217
43	8.9704	10.7770	13.3261	17.0426	37.3913
44	8.9819	10.8031	13.3845	17.1712	38.2609
45	8.9923	10.8272	13.4393	17.2942	39.1304
46	9.0018	10.8495	13.4908	17.4118	40.0000
47	9.0103	10.8699	13.5392	17.5244	40.8696
48	9.0180	10.8888	13.5847	17.6320	41.7391
49	9.0250	10.9062	13.6273	17.7350	42.6087
50	9.0313	10.9222	13.6674	17.8334	43.4783

Table E4. Discrete Compounding: i = 15%

	Geometric series future worth factor, (P/A,i,j,n)				
n	j = 4%	j = 6%	j = 8%	j = 10%	j = 15%
1	1.0000	1.0000	1.0000	1.0000	1.0000
2	2.1900	2.2100	2.2300	2.2500	2.3000
3	3.6001	3.6651	3.7309	3.7975	3.9675
4	5.2650	5.4059	5.5502	5.6981	6.0835
5	7.2246	7.4792	7.7433	8.0169	8.7450
6	9.5249	9.9394	10.3741	10.8300	12.0681
7	12.2190	12.8488	13.5171	14.2261	16.1914
8	15.3678	16.2797	17.2585	18.3087	21.2802
9	19.0415	20.3155	21.6982	23.1986	27.5312
10	23.3210	25.0523	26.9519	29.0363	35.1788
11	28.2994	30.6010	33.1536	35.9855	44.5011
12	34.0838	37.0895	40.4583	44.2364	55.8287
13	40.7974	44.6651	49.0452	54.0103	69.5533
14	48.5821	53.4978	59.1216	65.5641	86.1390
15	57.6011	63.7834	70.9270	79.1963	106.1356
16	68.0422	75.7474	84.7383	95.2530	130.1930
17	80.1215	89.6499	100.8749	114.1359	159.0796
18	94.0876	105.7902	119.7062	136.3107	193.7028
19	110.2266	124.5130	141.6582	162.3173	235.1336
20	128.8674	146.2156	167.2226	192.7807	284.6354
21	150.3886	171.3550	196.9669	228.4254	343.6973
22	175.2257	200.4579	231.5458	270.0894	414.0734
23	203.8795	234.1301	271.7142	318.7431	497.8292
24	236.9261	273.0694	318.3428	375.5089	597.3950
25	275.0283	318.0787	372.4354	441.6849	715.6294
26	318.9484	370.0824	435.1492	518.7724	855.8928
27	369.5631	430.1441	507.8179	608.5064	1022.1335
28	427.8810	499.4881	591.9787	712.8924	1218.9888
29	495.0618	579.5230	689.4026	834.2472	1451.9027
30	572.4398	671.8698	802.1302	975.2474	1727.2636
31	681.5491	778.3937	932.5124	1138.9839	2052.5649
32	784.1546	901.2409	1083.2569	1329.0258	2436.5932
33	882.2859	1042.8804	1257.4826	1549.4935	2889.6473
34	1018.2772	1206.1531	1458.7810	1805.1427	3423.7942
35	1174.8130	1394.3271	1691.2883	2101.4617	4053.1681
36	1354.9811	1611.1622	1959.7669	2444.7834	4794.3188
37	1562.3322	1860.9838	2269.7001	2842.4136	5666.6185
38	1800.9501	2148.7675	2627.4007	3302.7796	6692.7359
39	2075.5314	2480.2368	3040.1361	3835.6009	7899.1896
40	2391.4775	2861.9759	3516.2718	4452.0858	9316.9929
41	2755.0002	3301.5580	4065.4371	5165.1579	10982.4054
42	3173.2432	3807.6945	4698.7151	5989.7168	12937.8093
43	3654.4225	4390.4057	5428.8619	6942.9380	15232.7302
44	4207.9864	5061.2171	6270.5578	8044.6188	17925.0267
45	4844.8008	5833.3851	7240.6974	9317.5757	21082.2757
46	5577.3622	6722.1575	8358.7225	10788.1025	24783.3864
47	6420.0413	7745.0716	9647.0049	12486.4975	29120.4790
48	7389.3653	8922.2982	11131.2877	14447.6696	34201.0732
49	8504.3406	10277.0368	12841.1914	16711.8372	40150.6349
50	9786.8251	11835.9699	14810.7976	19325.3318	47115.5409

Table E5. Continuous Compounding: r = 5%

	Geometric series present worth factor, $(P/A,r,c,n)_{\infty}$				
n	$c = 4\%$	$c = 6\%$	$c = 8\%$	$c = 10\%$	$c = 15\%$
1	0.9512	0.9512	0.9512	0.9512	0.9512
2	1.8930	1.9120	1.9314	1.9512	2.0025
3	2.8254	2.8825	2.9415	3.0025	3.1643
4	3.7485	3.8627	3.9823	4.1077	4.4484
5	4.6824	4.8527	5.0548	5.2695	5.8674
6	5.5673	5.8527	6.1600	6.4909	7.4357
7	6.4631	6.8628	7.2988	7.7749	9.1690
8	7.3500	7.8830	8.4723	9.1248	11.0845
9	8.2281	8.9134	9.6816	10.5439	13.2015
10	9.0975	9.9542	10.9276	12.0357	15.5412
11	9.9582	11.0055	12.2117	13.6040	18.1269
12	10.8103	12.0673	13.5348	15.2527	20.9845
13	11.6540	13.1398	14.8982	16.9860	24.1427
14	12.4893	14.2231	16.3032	18.8081	27.6331
15	13.3162	15.3173	17.7509	20.7236	31.4905
16	14.1350	16.4225	19.2427	22.7374	35.7538
17	14.9455	17.5388	20.7800	24.8544	40.4651
18	15.7481	18.6663	22.3641	27.0799	45.6721
19	16.5426	19.8051	23.9964	29.4196	51.4267
20	17.3292	20.9554	25.6784	31.8792	57.7865
21	18.1080	22.1172	27.4116	34.4649	64.8152
22	18.8791	23.2907	29.1977	37.1832	72.5831
23	19.6425	24.4760	31.0381	40.0408	81.1679
24	20.3982	25.6732	32.9346	43.0450	90.6557
25	21.1465	26.8825	34.8888	46.2032	101.1412
26	21.8873	28.1039	36.9026	49.5233	112.7296
27	22.6208	29.3376	38.9777	53.0136	125.5367
28	23.3469	30.5836	41.1159	56.6829	139.6907
29	24.0658	31.8422	43.3193	60.5404	155.3334
30	24.7776	33.1135	45.5898	64.5956	172.6211
31	25.4823	34.3975	47.9295	68.8587	191.7271
32	26.1800	35.6944	50.3404	73.3404	212.8424
33	26.8707	37.0044	52.8247	78.0518	236.1785
34	27.5546	38.3275	55.3847	83.0049	261.9688
35	28.2316	39.6640	58.0226	88.2118	290.4716
36	28.9019	41.0138	60.7409	93.6858	321.9720
37	29.5656	42.3772	63.5420	99.4404	356.7853
38	30.2226	43.7544	66.4284	105.4900	395.2600
39	30.8732	45.1453	69.4026	111.8499	437.7810
40	31.5172	46.5503	72.4675	118.5358	484.7741
41	32.1548	47.9694	75.6257	125.5644	536.7095
42	32.7861	49.4027	78.8301	132.9535	594.1069
43	33.4111	50.8504	82.2335	140.7214	657.5409
44	34.0299	52.3127	85.6892	148.8876	727.6463
45	34.6425	53.7897	89.2500	157.4724	805.1248
46	35.2490	55.2815	92.9193	166.4974	890.7517
47	35.8495	56.7883	96.7003	175.9852	985.3842
48	36.4441	58.3103	100.5965	185.9594	1089.9691
49	37.0327	59.8475	104.6114	196.4449	1205.5534
50	37.6154	61.4002	108.7485	207.4681	1333.2938

Table E5. Continuous Compounding: r = 5%

	Geometric series future worth factor, $(F/A,r,c,n)_{\infty}$				
n	c = 4%	c = 6%	c = 8%	c = 10%	c = 15%
1	1.0000	1.0000	1.0000	1.0000	1.0000
2	2.0921	2.1131	2.1346	2.1564	2.2131
3	3.2826	3.3489	3.4175	3.4884	3.6764
4	4.5784	4.7179	4.8640	5.0171	5.4332
5	5.9567	6.2310	6.4905	6.7662	7.5339
6	7.5150	7.9003	8.3151	8.7618	10.0372
7	9.1716	9.7387	10.3575	11.0332	13.0114
8	10.9650	11.7600	12.6392	13.6126	16.5362
9	12.9043	13.9790	15.1837	16.5361	20.7041
10	14.9992	16.4118	18.0166	19.8435	25.5231
11	17.2601	19.0753	21.1659	23.5792	31.4185
12	19.6977	21.9881	24.6620	27.7923	38.2363
13	22.3237	25.1699	28.5381	32.5373	46.2464
14	25.1503	28.6419	32.8305	37.8748	55.6462
15	28.1905	32.4267	37.5786	43.8719	66.6654
16	31.4579	36.5489	42.8255	50.6030	79.5711
17	34.9673	41.0345	48.6178	58.1505	94.6740
18	38.7340	45.9116	55.0067	66.6059	112.3352
19	42.7743	51.2102	62.0476	76.0705	132.9744
20	47.1057	56.9626	69.8011	86.6566	157.0800
21	51.7464	63.2032	78.3329	99.4886	185.2192
22	56.7159	69.9691	87.7147	111.7044	218.0516
23	62.0347	77.2999	98.0244	126.4566	256.3440
24	67.7245	85.2381	109.3467	142.9144	300.9874
25	73.8085	93.8290	121.7740	161.2649	353.0176
26	80.3111	103.1215	135.4066	181.7157	413.6383
27	87.2579	113.1674	150.3535	204.4962	484.2484
28	94.6764	124.0227	166.7334	229.8606	566.4738
29	102.5954	135.7471	184.6753	258.0905	662.2039
30	111.0455	148.4043	204.3195	289.4972	773.6343
31	120.0591	162.0628	225.8184	324.4256	903.3165
32	129.6703	176.7957	249.3376	363.2572	1054.2155
33	139.9152	192.6812	275.0572	406.4143	1229.7767
34	150.8323	209.8029	303.1729	454.3643	1434.0037
35	162.4618	228.2503	333.8972	507.6241	1671.5485
36	174.8466	248.1191	367.4612	566.7660	1947.8169
37	183.0319	269.5116	404.1156	632.4230	2269.0901
38	202.0654	292.5371	444.1330	705.2953	2642.6664
39	216.9977	317.3125	487.8094	786.1577	3077.0262
40	232.8823	343.9627	535.4663	875.8673	3582.0230
41	249.7754	372.6212	587.4528	975.3722	4169.1061
42	267.7369	403.4307	644.1479	1085.7209	4851.5781
43	286.8295	436.5436	705.9633	1208.0733	5644.8957
44	307.1202	472.1228	773.3457	1343.7123	6567.0180
45	328.6790	510.3423	846.7805	1494.0568	7638.8114
46	351.5804	551.3878	926.7940	1660.6759	8884.5204
47	375.9028	595.4579	1013.9582	1845.3049	10332.3143
48	401.7293	642.7645	1108.8934	2049.8628	12014.9221
49	429.1474	693.5341	1212.2730	2276.4720	13970.3711
50	458.2495	748.0082	1324.8280	2527.4790	16242.8438

Table E6. Continuous Compounding: r = 8%

	Geometric series present worth factor, $(P/A,r,c,n)_\infty$				
n	c = 4%	c = 6%	c = 8%	c = 10%	c = 15%
1	0.9231	0.9231	0.9231	0.9231	0.9231
2	1.8100	1.8280	1.8462	1.8649	1.9132
3	2.6622	2.7149	2.7693	2.8257	2.9750
4	3.4809	3.5842	3.6925	3.8059	4.1138
5	4.2675	4.4364	4.6156	4.8059	5.3352
6	5.0233	5.2716	5.5387	5.8261	6.6452
7	5.7495	6.0904	6.4618	6.8669	8.0501
8	6.4471	6.8929	7.3849	7.9287	9.5570
9	7.1175	7.6795	8.3080	9.0120	11.1730
10	7.7615	8.4506	9.2312	10.1172	12.9063
11	8.3803	9.2064	10.1543	11.2447	14.7652
12	8.9748	9.9472	11.0774	12.3949	16.7589
13	9.5460	10.6733	12.0005	13.5685	18.8972
14	10.0948	11.3851	12.9236	14.7657	21.1905
15	10.6221	12.0828	13.8467	15.9871	23.6501
16	11.1287	12.7666	14.7699	17.2332	26.2881
17	11.6155	13.4370	15.6930	18.5044	29.1173
18	12.0832	14.0940	16.6161	19.8013	32.1517
19	12.5325	14.7380	17.5392	21.1245	35.4060
20	12.9642	15.3693	18.4623	22.4743	38.8964
21	13.3790	15.9881	19.3854	23.8514	42.6398
22	13.7775	16.5946	20.3086	25.2564	46.6546
23	14.1604	17.1892	21.2317	26.6897	50.9606
24	14.5283	17.7719	22.1548	28.1520	55.5788
25	14.8817	18.3431	23.0779	29.6438	60.5318
26	15.2213	18.9030	24.0010	31.1658	65.8440
27	15.5476	19.4518	24.9241	32.7185	71.5413
28	15.8611	19.9898	25.8473	34.3026	77.6518
29	16.1623	20.5171	26.7704	35.9187	84.2053
30	16.4517	21.0339	27.6935	37.5674	91.2340
31	16.7297	21.5405	28.6166	39.2494	98.7723
32	16.9968	22.0371	29.5397	40.9654	106.8572
33	17.2535	22.5239	30.4628	42.7161	115.5283
34	17.5001	23.0010	31.3860	44.5021	124.8282
35	17.7370	23.4686	32.3091	46.3242	134.8024
36	17.9647	23.9271	33.2322	48.1832	145.4998
37	18.1834	24.3764	34.1553	50.0796	156.9728
38	18.3935	24.8168	35.0784	52.0144	169.2778
39	18.5954	25.2485	36.0015	53.9883	182.4749
40	18.7894	25.6717	36.9247	56.0021	196.6289
41	18.9758	26.0865	37.8478	58.0565	211.8092
42	19.1548	26.4930	38.7709	60.1524	228.0903
43	19.3269	26.8916	39.6940	62.2907	245.5518
44	19.4922	27.2822	40.6171	64.4722	264.2794
45	19.6510	27.6651	41.5402	66.6977	284.3650
46	19.8036	28.0404	42.4634	68.9682	305.9069
47	19.9502	28.4083	43.3865	71.2846	329.0107
48	20.0910	28.7689	44.3096	73.6478	353.7898
49	20.2264	29.1223	45.2327	76.0587	380.3656
50	20.3564	29.4688	46.1558	78.5183	408.8683

Table E6. Continuous Compounding: r = 8%

		Geometric series future worth factor, $(F/A,r,c,n)_{\infty}$			
n	c = 4%	c = 6%	c = 8%	c = 10%	c = 15%
1	1.0000	1.0000	1.0000	1.0000	1.0000
2	2.1241	2.1451	2.1666	2.1885	2.2451
3	3.3843	3.4513	3.5205	3.5921	3.7820
4	4.7937	4.9359	5.0850	5.2412	5.6653
5	6.3664	6.6183	6.8856	7.1695	7.9592
6	8.1181	8.5194	8.9509	9.4154	10.7391
7	10.0654	10.6623	11.3125	12.0217	14.0932
8	12.2269	13.0722	14.0054	15.0367	18.1246
9	14.6224	15.7771	17.0683	18.5146	22.9543
10	17.2735	18.8071	20.5443	22.5162	28.7235
11	20.2040	22.1956	24.4810	27.1098	35.5975
12	23.4395	25.9790	28.9308	32.3718	43.7693
13	27.0078	30.1972	33.9521	38.3881	53.4643
14	30.9392	34.8937	39.6090	45.2546	64.9459
15	35.2667	40.1162	45.9728	53.0790	78.5212
16	40.0261	45.9170	53.1219	61.9814	94.5487
17	45.2562	52.3530	61.1429	72.0967	113.4466
18	50.9993	59.4865	70.1315	83.5754	135.7023
19	57.3014	67.3856	80.1932	96.5858	161.8843
20	64.2121	76.1247	91.4445	111.3160	192.6550
21	71.7857	85.7851	104.0137	127.9763	228.7862
22	80.0809	96.4553	118.0422	146.8012	271.1772
23	89.1614	108.2322	133.6861	168.0529	320.8754
24	99.0967	121.2214	151.1169	192.0237	379.1005
25	109.9619	135.5383	170.5240	219.0400	447.2729
26	121.8386	151.3086	192.1155	249.4657	527.0461
27	134.8154	168.6694	216.1207	283.7067	620.3446
28	148.9884	187.7705	242.7919	322.2155	729.4088
29	164.4621	208.7749	272.4066	365.4965	856.8454
30	181.3496	231.8605	305.2702	414.1118	1005.6880
31	199.7738	257.2211	341.7185	468.6875	1179.4660
32	219.8680	285.0681	382.1205	529.9210	1382.2852
33	241.7768	315.6315	426.8820	598.5891	1618.9221
34	265.6571	349.1623	476.4489	675.5565	1894.9324
35	291.6791	385.9336	531.3113	761.7857	2216.7776
36	320.0274	426.2430	592.0073	858.3481	2591.9728
37	350.9022	470.4147	659.1281	966.4356	3029.2571
38	384.5208	518.8015	733.3229	1087.3745	3538.7925
39	421.1186	571.7877	815.3045	1222.6399	4132.3956
40	460.9512	629.7914	905.8552	1373.8725	4823.8051
41	504.2955	693.2681	1005.8337	1542.8964	5628.9945
42	551.4520	762.7131	1116.1825	1731.7400	6566.5343
43	602.7463	838.6659	1237.9352	1942.6579	7658.0136
44	658.5318	921.7130	1372.2262	2178.1560	8928.5294
45	719.1914	1012.4930	1520.2993	2441.0191	10407.2556
46	785.1404	1111.7003	1683.5188	2734.3416	12128.1042
47	856.8290	1220.0904	1863.3805	3061.5612	14130.4931
48	934.7453	1338.4850	2061.5244	3426.4968	16460.2392
49	1019.4185	1467.7778	2279.7482	3833.3901	19170.5950
50	1111.4222	1608.9405	2520.0222	4286.9517	22323.4542

Table E7. Continuous Compounding: r = 10%

			Geometric series present worth factor, $(P/A,i,j,n)_\infty$		
n	c = 4%	c = 6%	c = 8%	c = 10%	c = 15%
1	0.9048	0.9048	0.9048	0.9048	0.9048
2	1.7570	1.7742	1.7918	1.8097	1.8561
3	2.5595	2.6095	2.6611	2.7145	2.8561
4	3.3153	3.4120	3.5133	3.6193	3.9073
5	4.0271	4.1830	4.3485	4.5242	5.0125
6	4.6974	4.9239	5.1673	5.4290	6.1743
7	5.3287	5.6356	5.9698	6.3339	7.3957
8	5.9232	6.3195	6.7564	7.2387	8.6798
9	6.4831	6.9785	7.5275	8.1435	10.0296
10	7.0104	7.6078	8.2832	9.0484	11.4487
11	7.5070	8.2143	9.0241	9.9532	12.9405
12	7.9746	8.7971	9.7502	10.8580	14.5088
13	8.4151	9.3570	10.4620	11.7629	16.1578
14	8.8298	9.8949	11.1597	12.6677	17.8908
15	9.2205	10.4118	11.8435	13.5726	19.7129
16	9.5883	10.9084	12.5138	14.4774	21.6285
17	9.9348	11.3855	13.1709	15.3822	23.6422
18	10.2611	11.8439	13.8149	16.2871	25.7592
19	10.5684	12.2843	14.4462	17.1919	27.9843
20	10.8577	12.7075	15.0650	18.0967	30.3244
21	11.1303	13.1141	15.6715	19.0016	32.7840
22	11.3869	13.5047	16.2660	19.9064	35.3697
23	11.6286	13.8800	16.8488	20.8113	38.0880
24	11.8563	14.2406	17.4200	21.7161	40.9457
25	12.0707	14.5870	17.9799	22.6209	43.9498
26	12.2728	14.9199	18.5287	23.5258	47.1080
27	12.4627	15.2397	19.0667	24.4306	50.4281
28	12.6418	15.5470	19.5939	25.3354	53.9185
29	12.8104	15.8422	20.1108	26.2403	57.5878
30	12.9692	16.1259	20.6174	27.1451	61.4452
31	13.1188	16.3984	21.1140	28.0500	65.5004
32	13.2597	16.6603	21.6007	28.9548	69.7635
33	13.3923	16.9119	22.0779	29.8596	74.2452
34	13.5172	17.1536	22.5455	30.7645	78.9567
35	13.6349	17.3858	23.0039	31.6693	83.9097
36	13.7457	17.6089	23.4533	32.5741	89.1167
37	13.8500	17.8233	23.8937	33.4790	94.5906
38	13.9483	18.0293	24.3254	34.3838	100.3452
39	14.0409	18.2272	24.7486	35.2887	106.3949
40	14.1280	18.4173	25.1634	36.1935	112.7547
41	14.2101	18.6000	25.5699	37.0983	119.4406
42	14.2874	18.7755	25.9684	38.0032	126.4693
43	14.3602	18.9442	26.3591	39.9080	133.8583
44	14.4288	19.1052	26.7420	39.8128	141.6262
45	14.4934	19.2619	27.1173	40.7177	149.7924
46	14.5542	19.4114	27.4852	41.6225	158.3773
47	14.6114	19.5551	27.8457	42.5274	167.4023
48	14.6654	19.6932	28.1992	43.4322	176.8900
49	14.7162	19.8259	28.5457	44.3370	186.8642
50	14.7640	19.9533	28.8853	45.2419	197.3498

Table E7. Continuous Compounding: r = 10%

	Geometric series future worth factor, $(F/A,r,c,n)_\infty$				
n	c = 4%	c = 6%	c = 8%	c = 10%	c = 15%
1	1.0000	1.0000	1.0000	1.0000	1.0000
2	2.1460	2.1670	2.1835	2.2103	2.2670
3	3.4550	3.5224	3.5921	3.6642	3.8553
4	4.9458	5.0901	5.2412	5.3994	5.8291
5	6.6395	6.8967	7.1695	7.4591	8.2642
6	8.5592	8.9718	9.4154	9.8923	11.2504
7	10.7306	11.3488	12.0217	12.7548	14.8932
8	13.1823	14.0643	15.0367	16.1100	19.3172
9	15.9458	17.1595	18.5146	20.0299	24.6689
10	19.0562	20.6802	22.5162	24.5960	31.1208
11	22.5521	24.6773	27.1098	29.9011	38.8755
12	26.4767	29.2074	32.3718	36.0500	48.1710
13	30.8773	34.3336	38.3881	43.1615	59.2869
14	35.8067	40.1260	45.2546	51.3702	72.5508
15	41.3232	46.6624	53.0790	60.8280	88.3472
16	47.4914	54.0295	61.9814	71.7070	107.1265
17	54.3826	62.3236	72.0967	84.2016	129.4163
18	62.0759	71.6514	83.5754	98.5311	155.8342
19	70.6589	82.1317	96.5858	114.9433	187.1032
20	80.2285	93.8963	111.3160	133.7179	224.0688
21	90.8917	107.0916	127.9763	155.1702	267.7198
22	102.7672	121.8800	146.8012	179.6557	319.2122
23	115.9863	138.4416	168.0529	207.5753	379.8967
24	130.6939	156.9766	192.0237	239.3804	451.3512
25	147.0505	177.7066	219.0400	275.5794	535.4184
26	165.2346	200.8779	249.4657	316.7448	634.2500
27	185.4417	226.7632	283.7067	363.5209	750.3571
28	207.8894	255.6652	322.2155	416.6325	886.6703
29	232.8182	287.9193	365.4965	476.8948	1046.6085
30	260.4938	323.8974	414.1118	545.2244	1234.1597
31	291.2103	364.0116	468.6875	622.6516	1453.9746
32	325.2928	408.7188	529.9210	710.3344	1711.4754
33	363.1008	458.5251	598.5891	809.5735	2012.9833
34	405.0318	513.9913	675.5565	921.8297	2365.8655
35	451.5256	575.7389	761.7857	1048.7435	2778.7077
36	503.0682	644.4560	858.3481	1192.1563	3261.5132
37	560.1970	720.9052	966.4356	1354.1347	3825.9360
38	623.5064	805.9308	1087.3745	1536.9976	4485.5507
39	693.6533	900.4679	1222.6399	1743.3462	5256.1676
40	771.3643	1005.5522	1373.8725	1976.0980	6156.1980
41	857.4424	1122.3303	1542.8964	2238.5242	7207.0797
42	952.7756	1252.0716	1731.7400	2534.2921	8433.7723
43	1058.3454	1396.1817	1942.6579	2867.5122	9365.3318
44	1175.2371	1556.2165	2178.1560	3242.7909	11535.5801
45	1304.6503	1733.8984	2441.0191	3665.2891	13483.8828
46	1447.9113	1931.1339	2734.3416	4140.7880	15756.0539
47	1606.4860	2150.0328	3061.5612	4675.7628	18405.4073
48	1781.9951	2392.9306	3426.4968	5277.4643	21493.9796
49	1976.2301	2662.4116	3833.3901	5954.0105	25093.9520
50	2191.1713	2961.3357	4286.9517	6714.4890	29289.3025

Table E8. Continuous Compounding: r = 15%

		Geometric series present worth factor, $(P/A,r,c,n)_\infty$			
n	c = 4%	c = 6%	c = 8%	c = 10%	c = 15%
1	0.8607	0.8607	0.8607	0.8607	0.8607
2	1.6318	1.6473	1.6632	1.6794	1.7214
3	2.3225	2.3663	2.4115	2.4582	2.5821
4	2.9413	3.0233	3.1092	3.1991	3.4428
5	3.4956	3.6238	3.7597	3.9037	4.3035
6	3.9922	4.1726	4.3662	4.5741	5.1642
7	4.4370	4.6742	4.9317	5.2117	6.0250
8	4.8356	5.1326	5.4590	5.8182	6.8857
9	5.1926	5.5515	5.9507	6.3952	7.7464
10	5.5124	5.9344	6.4091	6.9440	8.6071
11	5.7989	6.2844	6.8365	7.4660	9.4678
12	6.0556	6.6042	7.2350	7.9626	10.3285
13	6.2855	6.8965	7.6066	8.4350	11.1892
14	6.4915	7.1636	7.9530	8.8843	12.0499
15	6.6760	7.4078	8.2761	9.3117	12.9106
16	6.8413	7.6309	8.5773	9.7183	13.7713
17	6.9894	7.8348	8.8581	10.1050	14.6320
18	7.1220	8.0212	9.1199	10.4729	15.4927
19	7.2409	8.1915	9.3641	10.8229	16.3535
20	7.3473	8.3472	9.5917	11.1557	17.2142
21	7.4427	8.4895	9.8040	11.4724	18.0749
22	7.5281	8.6195	10.0019	11.7736	18.9356
23	7.6046	8.7383	10.1864	12.0601	19.7963
24	7.6732	8.8470	10.3584	12.3326	20.6570
25	7.7346	8.9462	10.5189	12.5918	21.5177
26	7.7897	9.0369	10.6684	12.8384	22.3784
27	7.8389	9.1198	10.8079	13.0730	23.2391
28	7.8831	9.1956	10.9379	13.2961	24.0998
29	7.9227	9.2649	11.0591	13.5084	24.9605
30	7.9581	9.3282	11.1722	13.7103	25.8212
31	7.9898	9.3860	11.2776	13.9023	26.6819
32	8.0183	9.4389	11.3759	14.0850	27.5427
33	8.0438	9.4872	11.4675	14.2588	28.4034
34	8.0666	9.5313	11.5529	14.4241	29.2641
35	8.0870	9.5717	11.6326	14.5813	30.1248
36	8.1053	9.6086	11.7069	14.7309	30.9855
37	8.1218	9.6423	11.7761	14.8732	31.8462
38	8.1365	9.6731	11.8407	15.0085	32.7069
39	8.1496	9.7013	11.9009	15.1372	33.5676
40	8.1614	9.7270	11.9570	15.2597	34.4283
41	8.1720	9.7505	12.0094	15.3762	35.2890
42	8.1814	9.7720	12.0582	15.4870	36.1497
43	8.1899	9.7916	12.1037	15.5924	37.0104
44	8.1975	9.8096	12.1461	15.6926	37.8712
45	8.2043	9.8260	12.1856	15.7880	38.7319
46	8.2104	9.8410	12.2225	15.8787	39.5926
47	8.2159	9.8547	12.2569	15.9650	40.4533
48	8.2208	9.8672	12.2890	16.0471	41.3140
49	8.2252	9.8787	12.3189	16.1252	42.1747
50	8.2291	9.8891	12.3468	16.1995	43.0354

Table E8. Continuous Compounding: r = 15%

	Geometric series future worth factor, $(F/A,r,c,n)_\infty$				
n	c = 4%	c = 6%	c = 8%	c = 10%	c = 15%
1	1.0000	1.0000	1.0000	1.0000	1.0000
2	2.2026	2.2237	2.2451	2.2670	2.3237
3	3.6424	3.7110	3.7820	3.8553	4.0496
4	5.3594	5.5088	5.6653	5.8291	6.2732
5	7.4002	7.6716	7.9592	8.2642	9.1106
6	9.8192	10.2630	10.7391	11.2504	12.7020
7	12.6795	13.3572	14.0932	14.8932	17.2172
8	16.0546	17.0408	18.1246	19.3172	22.8612
9	20.0300	21.4147	22.9543	24.6689	29.8811
10	24.7048	26.5963	28.7235	31.1208	38.5743
11	30.1947	32.7226	35.5975	38.8755	49.2986
12	36.6340	39.9531	43.7693	48.1710	62.4838
13	44.1787	48.4733	53.4643	59.2869	78.6454
14	53.0104	58.4994	64.9459	72.5508	98.4016
15	63.3399	70.2829	78.5212	88.3472	122.4925
16	75.4126	84.1167	94.5487	107.1265	151.8038
17	89.5134	100.3414	113.4466	129.4163	187.3940
18	105.9736	119.3533	135.7023	155.8342	230.5279
19	125.1782	141.6134	161.8843	187.1032	282.7149
20	147.5746	167.6581	192.6550	224.0688	345.7556
21	173.6828	198.1110	228.7862	267.7198	421.7963
22	204.1070	233.6976	271.1772	319.2122	513.3934
23	239.5494	275.2612	320.8754	379.8967	623.5907
24	280.8260	323.7828	379.1005	451.3512	756.0094
25	328.8850	380.4027	447.2729	535.4184	914.9559
26	384.8281	446.4466	527.0461	634.2500	1105.5481
27	449.9357	523.4557	620.3446	750.3571	1333.8661
28	525.6954	613.2219	729.4088	886.6703	1607.1288
29	613.8357	717.8277	856.8454	1046.6085	1933.9036
30	716.3653	839.6942	1005.6880	1234.1597	2324.3539
31	835.6179	981.6351	1179.4660	1453.9746	2790.5311
32	974.3051	1146.9210	1382.2852	1711.4754	3346.7195
33	1135.5776	1339.3531	1618.9221	2012.9833	4009.8438
34	1323.0964	1563.3490	1894.9324	2365.8655	4799.9488
35	1541.1149	1824.0430	2216.7776	2778.7077	5740.7668
36	1794.5752	2127.4018	2591.9728	3261.5132	6860.3857
37	2089.2197	2480.3594	3029.2571	3825.9360	8192.0374
38	2431.7199	2890.9738	3538.7925	4485.5507	9775.0271
39	2829.8277	3368.6090	4132.3956	5256.1676	11655.8286
40	3292.5495	3924.1466	4823.8051	6156.1980	13889.3752
41	3830.3498	4570.2310	5628.9945	7207.0797	16540.5805
42	4455.3867	5321.5557	6566.5343	8433.7723	19686.1302
43	5181.7864	6195.1943	7658.0136	9865.3318	23416.5921
44	6025.9614	7210.9860	8928.5294	11535.5801	27838.9009
45	7006.9808	8391.9836	10407.2556	13483.8828	33079.2835
46	8146.9998	9764.9737	12128.1042	15756.0539	39286.7031
47	9471.7599	11361.0807	14130.4931	18405.4073	46636.9116
48	11011.1685	13216.4694	16460.2392	21493.9796	55337.2197
49	12799.9736	15373.1610	19170.5950	25093.9520	65632.1075
50	14878.5469	17879.9807	22323.4542	29289.3025	77809.8264

Index

A

AACE. *See* American Association of Cost Engineers.
Accelerated cost recovery system (ACRS), 231, 412
Accelerated depreciation, 223
Accounting
  and investment analysis, 25
  cost, 26
  department of, 111
  financial
    unit depletion used, 157
    vs. tax, 246
  for risk in mining investments, 375, 377
  rate of return, 255
  statements, 25, 26
Accounts
  classified, 227
  component, 226
  composite, 227
  money market, 338
  multiple asset, 226
  payable, receivable, 113
  single item, 226
  types of, 226, 227
  vintage, 228
Acquisition
  lands, costs of, 206
  of mineral properties, 8
ACRS. *See* Accelerated Cost Recovery System.
Ad valorem property tax. *See* Taxes, ad valorem.
After tax
  cost, preferred stock, 335
  returns, mining, 351
Aggregation of properties, 160
Agreements, operating deficiency, 366
Alabama tax bases, 192, 195, 198
Alaska mineral royalty system, 203
Allocation vs. valuation, 30
Alternative evaluating criteria, 284
AMAX, Mt. Tolman project, 68, 69
American Association of Cost Engineers (AACE), 104

Amortization, 32
  of discount, 327
  tax deduction, 180
Anisotropies, geologic, 71
Annual value (cost), 262
Annuities, 39, 46
Area of influence, 74
Arizona
  capitalization method, 182
  Lakeshore mine, 353
  price copper, 84
  tax bases, 192, 194, 195, 203
Arkansas tax bases, 195, 201
Arm's-length transactions, 13
Assays, 72
  contract clause, 86
  limits, 61
Assessment
  rate, taxes, 184
  work, 356
Assets, 26
  accounts, multiple, 226
  capital gains taxes on, 308
  expenditures for, 180, 181
  portfolio, 2
  recouping value, 162
Australia
  tax system, 161
  uranium discoveries, 349

B

Balance
  of payments, 359
  sheet, 334, 349
Banks, commercial, 350, 361
Base opportunity cost, 323
Bases
  capitalized account, 180
  project comparison, 254
  property depreciation, 221
Benefit/cost ratio, 264
Benefits
  from investments, 254

maximizing, 33
Bidding, competitive, 8
Binomial distribution, 110
Black Lung Revenue Act of 1978, 168
Block method, underground, 73
Bonanza image, mining, 169
Bonds, 53
  investments, 328
  valuation problem, 53
Book value weights, 333
Budgeting. *See* Capital, budgeting.
Budgets, 53
  cash, for lenders, 369
  ceilings, 19
  methodology, capital, 317
Bureau of Labor Statistics, 111, 137, 303
Business-risk profile, 294
Buy-out arrangements, 348
Byproduct credits, 95

C

California mineral royalty system, 204
Canada
  copper production cost, 95
  discovery Saskatchewan potash, 349
  tax laws benefit mining, 161
Capacity, capital costs and, 115
Capital
  asset pricing model, 331
  attrition of, 6
  budgeting, 2, 3
    defined, 285
    methodology, 317
    theory, 9
  corporate cost of, 325
  costs. *See* Costs.
  gains taxation, 308
  intensity, mining, 17, 171, 352
    inflation impact, 301
  investment, 18, 419
    decision, 1
  maintenance, working, 367
  rationing, 286
  resources, 2
    available, 350
  structures, optimum, 343
  testing exploration target, 348
  working, 113, 419
Capitalization
  future net proceeds, 182
  mining ventures, 349
Capitalized earnings approach, 182
Captive theory, taxation, 166
Carried interests, 349, 356
Carved out production payment, 215

Cash
  flows, 1, 30, 37, 46, 50, 393
    affect mineral depreciation, 151
    analyses, 14, 30, 391
    approach to evaluation, 247
    calculation, components of annual, 31
    component, variable, 50
    constant, current dollars, 308, 309
    determinations, 400
    net, 31
    new mining project, 258
    on hand, working capital, 113
    surplus, outlay, 32
Cathode price, copper, 84
Caving, block, 67
CDF. *See* Cumulative distribution function.
CE. *See Chemical Engineering* Plant Construction.
Charles River Associates, 91
Chemical Engineering (CE) Plant Construction, 137
  measuring inflation, 305
Chile, 93
CLADR. *See* Class life asset depreciation range.
Claims, ownership of, 348
Class life asset depreciation range system (CLADR), 228
Classification system, international, 62
Co-ownerships. *See* Joint ventures.
Coal
  depletion allowance, 152
  prices, 85
  state severance taxes, 195-197
Coal Week, 85, 90
Colorado
  aggregating uranium properties, 160
  limestone mining property in, case study, 392
  tax bases, 192, 195, 196, 198, 201, 204
  valuation approach, 183
Comex. *See* New York Commodity Exchange.
Commodities
  exchanges, 84
  fundible, 83
  place-value, 393
Common stock, 1
  cost, financing, 330
  market price of, 253
  price indicates firm wealth, 321
Competition affects price, 98
Completion guarantee, 366
Component cost ratio method, estimating, 126, 130
Compound growth rates, 317
Compounding
  continuous, 329
  see also Interest, compounding.
Concentrating, cost of, 158
Conference method, estimating, 113

Confidence intervals, 110
Conservation
  and inflation, 302
  mineral, taxation, 187
  mining, 175
Constant dollars, 308-310
  analysis, role of, 317
  need for analyses, 324
Consumer patterns, measuring, 303, 304
Consumer Price Index (CPI), 85, 141, 302
Consumption, per capita minerals, 7
Contingencies
  estimating fee for, 128
  part estimation, 105, 136
Contingent proposals. 286
Continuous cast copper rod, price, 84
Contracts, 85, 86
  fixed-price, 311
  purchase, 85
  sales, 366
    inflation impact, 310
  title in, 86
Conventions, depreciation, 230
Convertible securities, cost of, 332
Copper
  byproduct price, 95
  depletion allowance, 155
  econometric modeling, 91
  joint ventures for, 353
  porphyry deposit, 178
Corporate Franchise Act of 1909, 151
Corporations
  allocating profit, 159
  federal income tax, 239
  forms of, 350
Cost of living indicator, 302, 303
Costs, 11, 15
  accelerated recovery system, 231
  approach, taxes, 181
  benefit ratio, 264
  bonus, 217-219
  capacity
    curves, 115
    relationship method, 129
  capital, 18, 19
    estimates, 397
    estimating, 103, 112
    fixed, 112, 113
    intensive, mining, 171
    requirements, 25
    selecting discount rate, 321
  cash vs. noncash, 19
  composite, 333
  depletion, 155
  depreciated replacement, 12
  direct, 15
    vs. indirect, 17
  discovery, 152

estimates
  special features, mining, 347
  types of, 104
estimation techniques, 112
  first, 18
  future vs. historic, 333
  increases, inflationary, 50
  indices of, 117, 124
    trends, price, 305
    types, 136
  information, kinds, 108
  measured, 108, 109
  minimizing, 33
  monitoring changes, 303
  normalized, 333
  of capital calculation, 337
  operating, 15, 397, 402
    direct, indirect, 128
    estimating, 103, 129
    requirements, 25
  operation, taxes as, 174
  opportunity, 19, 35, 322
  policy, 108, 109
  ratio method, estimating, 120
  schedules, 94
  see alsoOverhead costs.
  see also History, costs.
  specific, 333
  spot, 333
  taxes part, 165
  transactions, 35, 322
  underwriting, 326
  variable, fixed, 17
Coupon rate, 53, 328
CPI Detailed Reporter, 304
CPI-U. See Urban consumer index.
CPI-W. See Urban wage earners' index.
CPI. See Consumer Price Index.
Criteria
  evaluation
    decision, 3
    project, 253
  of taxes, 174
Cumulative distribution function (CDF), 385
Currency
  depreciating, 358
  exchange, 358
Current dollars, 308-310
Curves
  cost-capacity, 115
  estimating, 143
Cut-and-fill mining, cost breakdown, 135
Cutoff
  grade, 65
    taxes affect, 175
  point, determining gross income, 158

D

DB. *See* Declining-balance method.
DCF. *See* Discounted cash flow.
DCFROI. *See* Discounted cash flow-return on investment.
DCR. *See* Debt coverage ratio.
Debt
  cost of, 327
  financing, 315, 326, 334
  repayment schedules, 349
  senior, 349
  subordinated, 349
  to equity ratio, 364
Debt coverage ratio (DCR), 364
Decisions
  financing, investment, 324
  making, cash-flow-based, 15
  mine investment, 58
    evaluating projects for, 254
  separating investment, financing, 338
  theory, 375
  variable, 33
Decisions under uncertainty, risk, 375
Declining balance (DB) method of depreciation, 223
Deductions
  deferred
    development, 213
    expense, 180
  depletion as, 157
  development expenditure, 211-215
  pretax, 32, 180
  recapture, 160, 180, 207, 208
Default, risk of, 8
Definitions
  basic tax, 180
  inflation, 301
  taxes, 167
  uniform in estimation, 103
Degree of necessity, investments, 255
Demand
  schedules, 94
  shifts affect price, 93
Density, 78
Depletion, 31
  allowance, minerals, 151
  basis of, 155
  earned, 161
  federal statutes, 194, 206
  foreign rates, 359
  percentage, 152
Deposits
  configurations of, 177-179
  massive low-grade, 61
  specifications for model, 93
  stratabound, vein-type, 61
  uncertainty in determining value, 171, 172
  withdrawls, intraperiod, 47

Depreciation, 31, 412
  additional first-year, 222
  as tax deduction, 180
  declining tax shelter, 307
  foreign, 359
  methods of, 223
  systems of, 226
  tax treatment of, 220
Descartes' Rule of Signs, 271
Detailed cost breakdown method, estimating, 132
Development
  economic, taxes for, 168
  expenditure deductions, 211-215
  international institutions, 350
  requirements for, 107
Dilution, 66, 67
Dip Slip Mines, 379
Disbursements, 41
Discount rate
  adjustment, 308
  risk-adjusted, 378
  selecting, 321
    components of, 322
Discounted cash flow (DCF) analyses, 267, 325
Discounted cash flow-return on investment (DCFROI), 163
  inflationary adjustments, 313
Discounted payback period, 258
Discounting, intuitive approach, 321
Dividend
  decision, 1
  corporate, 253
  valuation model, 329
DOE. *See* US Dept. of Energy.
Dollar
  analysis
    escalated, 422
    role of, constant, current, 308
  see also Current dollars.
Drilling
  methods, 70
  program, exploration, 9

E

E/P ratio, 330
Earning power, 36
Earnings
  capitalized approach, 182
  future, mining, 172
  per share, maximizing, 253
  retained, 322
Economic
  compliance, administration of taxes, 174
  evaluation, taxes part, 165

feasibility, degrees of, 61, 65
rent, taxes as, 170
size of plant, 117
variables, 25
viability, potential, 9
Economic Recovery Tax Act of 1981, 231
Economic studies, 368
Economy of scale, 117, 353
El Paso Natural Gas Co., 353
Employment, full, tax objective, 168
Engineering News-Record (ENR) Cost Indices, 137
as factor cost index, 305
Engineering
department, 111
estimating expenses, 128
England. *See* United Kingdom.
Environmental
awareness, 6
affects taxes, 166
legislation, 7
matters, contracts, 88
regulations on mining, 349
Environmental Protection Agency (EPA), 111
Equations
for estimating curves, 143
interest, 37-40
Equipment
and materials-labor index, 137, 306
cost ratio method, 123
replacement problem, 33
selection, 396
variable curves, 120
Equitable, taxes, 174
Equity
capital, cost of, 328
financing, 326
funds, 36
Equivalence, 52
Equivalent uniform annual value (EUAV), 262
Error
estimating, 14
standard, sample percentage, 110
Escalation
definition, 302
indexes, 427
measuring, 305
selecting rates of, 316
Essex Wire and Cable Co., 353
Estimates
accuracy, degree of, 104
definitive, 104, 107
detailed, 104, 107
factored, 104, 126
in situ geological, 67
order of magnitude cost, 104, 105
quality of, 108
types of, 104

Estimating
capital, operating costs, 103
cost, handbook, 143
curves, equations for, 143
error, 14
quotations for, 127
rules-of-thumb in, 129
EUAV. *See* Equivalent uniform annual value.
Evaluation criteria, project, 253
methods related to, 280
special problems, mining, 347
Exchange rates, fluctuating currency, 358
Excise taxes. *See* Taxes, excise.
Exhaustibility, minerals, 171
Expenditures, 180
exploration
adjusted, 209
development deduction, 411
feasibility study, 400
tax treatment, 206
for tax purposes, 180, 181
see also Costs, Expenses.
tax base of, 169
Expenses
general
administrative, marketing, 15
estimating, 128, 129
noncash, 32
tax deduction, 180
Expertise, acquisition of, 353
Exploration
allocating costs to, 255
as current deduction, 207, 208
capital for, 348
compensation for, 170
expenditures
adjusted, 209
development deduction, 411
feasibility study, 408
tax treatment, 206
limited partnership for, 352
part estimation process, 104
program, reconnaissance, 10
recapture of deductions, 160
Exponential capacity adjustment (scaling) method, 114, 115
Export
credit agencies, 350
see also Taxes, export.
Expropriation, 5
risk of, 353

F

Factor cost indices, 136, 305
Factors, sinking fund, 40
Farm-out, 357

FASB. *See* Financial Accounting Standards Board.
FCLAA. *See* Federal Coal Leasing Amendments Act.
Feasibility studies, 4, 392
  estimating costs, 103
Federal Coal Leasing Amendments Act (FCLAA) of 1976, 8
Federal Reserve Bank, 111
Federal taxes, 166
  depletion allowance, 151, 194
  income, 238
Financial
  agreements, multilayered, 361
  analysis, 11
    of mining projects, 7
  statements, 26
  tax books, 26, 215
Financial Accounting Standards Board (FASB), 26, 68
Financing
  complexity of, 349
  decision, 1
  international, repayment schedules, 365
  mix, optimal, 2
  multilayer, 349
  new mining ventures, 349
  production payment, 361
  project, 350, 361, 363
Floating cone concept, 76
Florida tax bases, 198, 202
Force majeure, 87
Foreign mining ventures, 357-361
France, tax system, 161
Franchise taxes. *See* Taxes, franchise.
Front-end loader, ownership costs, 132-135
Funding, external sources, 349
Funds, continuous flow of, 48
Future
  net proceeds, 182
  values, 262
Futures trading, 84

G

Gao. *See* US General Accounting Office.
General Electric Co., 353
Generators, open pit limits, 76
Geologic
  risk, 353
  uncertainties, deposits, 172
  variability, natural, 70
Geological assurance, degrees of, 61, 64
Geology department, 111
Geomechanics features, mines, 172
Geometric series, present, future value factors, 51
Geostatistics, 71

Geothermal steam depletion allowance, 155
GNPIPD. *See* Gross Natl. Product Implicit Price.
Gold, depletion allowance, 155
Government
  federal
    depletion allowance, revenue loss, 151
    taxes, 166-169, 193, 206-219
  states
    statutes, taxes, 191-205
    taxes, 179-191
Grades
  affect revenue, 82
  control, 70
  cutoff, 65
    taxes affect, 175
  deposit, 59, 77, 78
  distribution, 60
  estimation, 67, 70
  mill-head, 66
Gradient, 50
Graphs, cost capacity, 129, 130
Gross Natl. Product Implicit Price Deflator (GNPIPD), 305
Gross proceeds method, valuation, 183
Group
  accounts, 227
  organization, 351
Growth rate of return (GRR), 275

H

Hecla Mining Co., 353
History
  costs
    trends measured, 142
    updating, 103
  depletion allowance, 151, 152
  long, CPI, 303
  mining, bonanza image, 169
  vs, future costs, 333
Horizontal arrangement, ventures, 354
Hoskold valuation premise, 182
  method explained, 279
House Ways and Means Committee, 209
Hurdle rate, 259, 325
Hydrometallurgical extraction, 158

I

Idaho tax bases, 195, 198, 202, 204
Implied reinvestment question, 270
In situ geological estimate, 67
Income
  capitalized future, 14
  determining percentage depletion, 156

partnership distribution, 351
statement, pro forma, 14, 30, 368, 391, 419
taxable, 31, 180
taxes, 169, 173
    defined, 184
    impact of, 188, 190
    state laws, 195
Incremental (marginal) analysis, 292
Indifference, level of, 100
Industrial minerals
    depletion allowance reduced, 155
    prices, 85
    severance taxes on, 201-203
Inflation, 35
    adjusting CPI for, 141
    defining, measuring, 301
    differential, 315
    in mine investment decision, 301
    increment for, 324
    pressures on mining, 349
    see also Uniform inflation.
    working capital estimates, 113
Information
    available for estimates, 105
    filing returns, 355
    management systems, 104
    sources of
        estimate, 110
        price, 90
Infrastructure, 5
    taxes to develop, 359
Interest
    and inflation, 315
    compounding, 37, 43, 44, 46
    continuous, 44, 45, 49
    equations, 37-40
    factor relationships, 51, 52
    large payment obligations, 328
    lender's viewpoint, 35
    market rate of, 54
    nominal, effective, rate of, 37
    payment, compounding intervals, 43
    problems, 41, 49
    rates, 35, 36, 43, 47
        prime, 334
    simple, 36
    tables, 38
Internal rate of return (IRR), 265, 267, 312
Internal Revenue Service (IRS)
    depletion allowance calculations, 155
    tax litigation, 167
    tests, partnerships, joint ventures, 352
    treats exploration expenditures, 207
International
    agencies, 111
    markets, 349
    ventures, 357
International Tin Council, 89

Inventories, working capital for, 113
Investment
    alternatives, evaluating, 284
    analysis, 3
        accounting, 25
        effect inflation on, 307, 310
    decision, inflation in, 301
    environment, 5
    goals, 9, 10
    projects with unequal, 290
    proposals
        evaluating, 254
        mine-related, 347
    rate of return on, 14, 162, 321
    recovering mining, 152
    risk mining, accounting for, 375
    tax credit reduced, 235
    what constitutes, 296
Investment tax credit (ITC), 236
Iron depletion allowance, 155
IRR. *See Internal rate of return.*
Irreducibles, 10
IRS. *See Internal Revenue Service.*
ITC. *See Investment tax credit.*
Iterative process, 4

J

Japan, smelter contract, 86, 87
Joint ventures, 349, 352, 369
    explanation of, 352-356
    horizontal, vertical agreements, 354

K

Kentucky severance tax, 196
Kriging block, 77

L

Labor
    and materials-equipment index, 137, 306
    estimating escalation, 118
    increasing skill levels, 359
Lakeshore mine, 353
Land as asset, 308
Lang factor, 120
LDCs. *See* Lesser developed countries.
Leach-electrowinning extraction, 158
Lead times, mining, 171, 310
Lead, depletion allowance reduced, 155
Lease agreements, 348
Leasing federal coal, 8
Legal department, 111
Legislation, tax, 151, 152, 155, 175

Lending institutions, 8
Lesser developed countries (LDCs), 350
Level of indifference, 100
Leverage
  financial, 326
  negative, 315
Liabilities, 26
  partnership distribution of, 351
Lignite depletion allowance, 155
Limestone mining property, Colorado, case
  study of, 392
Limy Mining Co., 392
Litigation
  over minerals depletion allowance, 158
  over taxes, 167, 182
LME. *See* London Metal Exchange.
Loading
  cost of, 158
  unloading, contract clause, 86
Loans, term, 349
London Metal Exchange (LME), 84
Long-run supply curves, 97
Louisiana tax bases, 196, 198, 202

M

M&S. *See* Marshall and Swift Cost Indices.
Maintenance allowance, 230
Man-hour reports, 110
Management
  committee of, 352
  estimates for budgeting, 107
  responsibilities, 1
  wealth maximization goal, 253
Manpower requirements, 396
Marginal analysis. *See* Incremental.
Marginal weighted average cost, capital, 332
Market
  value
    approaches to estimating, 12, 13
    as tax basis, 181
    common stock, 253, 329
    fair, mine property, 152
    weights, 333
Market-determined cost of capital, 324, 325
Marketing
  department, 111
  guarantee, 366
  requirements, 393
Markets, 11, 13
  and pricing, 83
  business-risk profile in, 294
  international, 5, 24, 349
  measuring price changes, 303
  response to signals, 93
  risk of, mining, 172
  spot, 85

MARR. *See* Minimum acceptable rate of re-
  turn.
Marshall & Stevens. *See* Marshall and Swift.
Marshall and Swift Cost Indices (M&S), 137,
  305
MAS. *See* Minerals Availability System.
Materials-equipment-labor index, 137, 306
Maturity date, 53
McKelvey diagram, 61
Metals
  indestructibility of, 172
  state severance taxes, 198-201
Metals Week, 90
Mill recovery, 386
Milling, custom, smelting, 86
Mine
  evaluation
    activity, 3
    ore reserves for, 64
  impact depletion allowance, 162
  investment decision, 58
  inflation in, 301
  operation, tax impact, 173
  planning variables, 58
  plans, 67
  valuation studies, 11, 57
Minerals
  conservation, taxation, 175, 187
  depletion allowance, 6, 151
  development, 6
  future prices, estimating, 81
  inventory, 66
  markets, 83
  occurrence, 58
  ownership rights, 348
  properties
    acquisition of, 8
    valuation of, 24
  resource classification, 61
  substitutability of, 63
Minerals Availability System (MAS), USBM,
  98, 142
Mines
  investment opportunities, evaluating, 4
  metals, depletion allowance, 152
  taxation of, 165
    state, 179
  underground vein, 74
  valuation, 3
    effect taxes on, 189
Minimum acceptable rate of return (MARR),
  324
Mining
  bonanza image of, 169
  cave, waste dilution in, 67
  characteristics, 5, 6
  cut-and-fill, cost breakdown, 135
  method selection, 394
  problems, evaluating special, 369

projects, financial analysis of, 7
special
  evaluation problems, 347
  features of, 194
  transactions, 13
Mining and Mineral Policy Act of 1970, 191
Minnesota severance tax, 198
Mississippi severance tax, 202
Missouri tax bases, 192, 195, 204
Modeling
  econometric, 91
  no change, price, 90
  price forecasting, 90
Models
  analytical financial, 10
  appraiser's, 13
  mine property, 392
  price forecasting, 91
  same change, pricing, 90
  Stochastic risk analysis, 379
Modified half-year convention, 230
Module method, estimating, 126
Moisture, contract clause, 86
Molybdenum
  depletion allowance reduced, 155
  prices affect demand, 98
Monetary exchange rates, 356
Money, time value of, 26, 35
Montana tax bases, 193, 195, 196, 199, 202, 204
Mortgages, 357, 367
Motion-time data, 110
Multilayer package financing, 350
Multiple
  asset accounts, 226
  roots question, 271
Mutually exclusive proposals, 286

N

Natural heritage theory, taxation, 6, 166
Nelson Refinery Construction Cost Index
  (NR), 137, 305
Net present value (NPV), 260
Net proceeds evaluation methods, 183
Net smelter return (NSR), 88, 159
Nevada tax bases, 193, 199, 202
New Mexico
  aggregating uranium properties, 160
  tax bases, 193, 195, 197, 199, 202, 205
New York Commodity Exchange, 84
No change modeling, price, 90
Nonmetals, severance taxes on, 201-203
Nonmining processes listed, 158
North Dakota severance tax, 197
NPV. *See Net present value.*
NR. *See* Nelson Refinery Construction Cost
  Index.
NSR. *See* Net smelter return.

O

O'Hara approach, estimating system, 146
Ocean mining, 353
Ohio tax bases, 197, 203
Oklahoma tax bases, 200, 203
Oligopolies and prices, 89
OPEC. *See* Organ. of Petroleum Exporting
  Countries.
Open pit mines, estimates for, 147
Operating
  deficiency agreement, 366
  environment, 25
  interests, 356
Order of magnitude cost estimates, 104, 105
Ore reserves
  assessment, 393
  base, 65
  calculating, 59, 65, 75
  definitions, 64, 65
  estimation, 57, 71, 73, 74
    geostatistical, 77
    magnitude, quality, 24
  mine evaluation, 64
  quantity of, 58
  reporting practice, 67
  system
    computer, 75
    proven, probable, possible, 64
  warehouse analogy of, 60
Ore-grade material, 65
Ore-waste boundary, 58, 61
Oregon mineral royalty system, 204
Ores, 24
  bodies, 58, 59, 64
  grades, 385
Organization of Petroleum Exporting Countries (OPEC), 5
Organizations, forms of, 350
Overhaul allowance, 230
Overhead costs
  estimating, 128, 129
  of large company, 354
Ownership
  complexity of, 348
  complicate evaluation, 348
  costs, front-end loader, 132-135
  indigenous, 359
  surface, 348

P

Parameters, risk-adjusted input, 378
Partnerships, 349, 351
  limited, mining, 352
Payable units produced, 82
Payback (payout) period, 256
  using constant dollars to determine, 310

Payee's, payer's tax treatment, 217
Payment intervals, varying, 43
Payout. *See* Payback period.
Percentage
    depletion, 152, 156
    recovery, 82
    repair allowance, 230
Permitting
    process, 398
    requirements, 5
Personal property, 220
PI. *See* Profitability index.
Planning commissions, 6
Pleasure, postponement of, 35
Political process, taxes part of, 170
Polygons, triangles, methods of, 74
Pool organization, 351
Population mean, 77
Potash, Saskatchewan discovery, 349
PPI. *See* Producer Price Indices.
Preferred stock, 326
    cost of, 328
Preliminary cost estimate, 105, 107
Preproduction, 5, 31, 33
    affects payback period, 258
    development, 397
    engineering for estimates, 105
Present value (worth), 260
    profile, 262, 263
Pretax net income, 156
Prices
    controls, 349
    estimating future mineral, 81
    forecasting model, limitations of, 91
    forecasting, 103
        future, 142
    historic series, extrapolating, 302
    purchase, 14
    stabilization by taxes, 168
    transfer, 354
    unit, 83
Pricing, rational, 92
Prime rate, interest, 334
Principal, 36
Principle of substitution, 12
Probability
    density function, 384
    distributions
        continuous, 384
        discrete, 380
Proceeds
    future net, capitalizing, 182
    gross, net methods, valuation, 183
Producer
    expenditure patterns, 136, 304
    price, 89
        measuring, 304, 305
Producer Prices and Price Indices, 137, 305
Producer Price Index (PPI), 85, 136, 304

Product requirements, 393
Production, 31, 33
    capital cost per ton, 114
    cost estimates, 30
    department of, 111
    domestic minerals, 7
    effect taxes on, 191
    increasing US costs, 349
    level, break-even, 18
    payments, 361
    requirements, 393
    return on capital cost of, 170
    revenue component, 82
    technology, 25
Productivity
    effect inflation, 302
    levels decreasing, US, 349
Profit, 32, 58
    allocating, 158, 159
    effect taxes on mineral, 191
    Hoskold method, 279
    maximization, 1, 253
    net, 419
    repatriation rules, 356
    sharing arrangements, 348
Profitability index (PI), 264
Proforma income statement. *See* Income, statement.
Project
    cost
        estimates, 103
        indices, 136, 137, 305
    dependent vs. independent, 286
    evaluation, 3
        criteria, 253, 419
        special problems, 347
    perceived risk of, 322
    timing, 398
    unequal life of, 287
    viability, 19
Projections, price, 90
Property
    aggregation of, 160
    depreciable, gain on disposition of, 236
    owners, 348
    percentage repair allowance, 230
    real, 220
    tangible, intangible, real, personal, 220
    taxes, 173
        affect prices, 85
        defined, 181
        impact of, 187, 188
        see also Taxes, ad valorem property.
        state bases, 192, 193
    transactions, mineral, 13
    value, 8
Proportional profits method, 159
Purchase contracts, 85

Purchasing power
  inflation affects, 113
  of dollar, 302

R

Ranking investment alternatives, 285
Rate
  compound growth, 317
  discount, selecting, 321
  inflation, 301
  of return
    accounting, 255
    external, 272
    investments, 162
    minimum acceptable, 309
  see also Interest, rates.
  tax, 184
Real property, 220
Real world cost schedule, 96
Recapture deductions, 160, 180, 207, 208
Receding face theory, 221
Records, collecting costs, 104
Regional indices, price, 303
Regulations, environmental, 5
Regulatory
  medium, taxes as, 168
  requirements, 8
Reinvestment question, implied, 270
Rent
  agreements, 348
  economic, 170
  payments, 172
Repairs allowance, 230
Repayment schedules, international financing, 365
Required rate of return (RRR), 310
Research & development, 255
Reserves, 66, 68
Resources, 61, 62
  calculations, 65
  sterilization of, 179
Retained
  earnings, 329
  production payment, 215
Revenue Act of 1913, 152
Revenue, 81, 82
  estimates, 402
  fluctuations, 25
  from tax programs, 167, 180
  inflation impacts, 308
  loss, depletion allowance, 151
  sales, 24
  special features, mining, 347
Risk, 35
  analysis model, Stochastic, 379
  business, 324
    financial, 253

facing lenders, 365
  financial, business, 376
  in mining investments, accounting for, 375, 377
  increment of, 322
  joint ventures, 353
  methods to compensate for
    continuous probability distributions, 384
    discrete probability distributions, 380
    risk-adjusted discount rate, 378
    risk-adjusted input parameters, 378
    risk-adjusted payout period, 377
  mining
    characteristics of, 376
    high, 169, 310
    political, 350
    severe, in mining industry, 377
Royalties, 348, 402
  from mining, 172
  income as depletion, 159
  obligation to leasors, 348
  overriding, 356
  relation to price, 85
  standard, 357
  systems for state lands, 203-205
RRR. *See* Required rate of return.
Rule of nearest points, 74
Rules-of-thumb in estimating, 129

S

Salaries, partnerships, 351
Sales
  contracts, inflation impact, 311
  taxes, 169
Salvage value, 222
Same change model, pricing, 90
Samples, 71, 72, 74, 77
Sampling, 69-72, 76, 77
  programs, 57
  components of, 69
Scaling. *See* Exponential capacity adjustment method.
Schedules
  payments, production, 348
  production cost, 94
  supply, demand, 94, 96, 98
Securities, 367
  cost of convertible, 332
  investments in, 40
Semi-fabrication, nonmining cost, 158
Senate Finance Committee, 216
Senior debt, 349
Severance taxes
  definition of, 173, 174, 185
  impact of, 187-189
  on industrial minerals, 201, 203
  on nonmetals, 201-203

tabulation by states, 195-203
to curtail mineral development, 168
Shareholders, 2
  maximizing wealth, 253, 321
  reports to, 111
  unit depletion reported, 157
Short-run supply schedules, 96
Shortage, minerals, 166
Similar project method, estimating, 129
Sinking fund, 40
  debentures, 334
  establishment of, 280
Size reduction, 72
Smelters
  custom contracts, 86, 302
  net return, 88, 159
Social effects, achieving by taxes, 175
Socioeconomic impact, mining, 186
Solvent extraction, 158
South Africa tax system, 161
South Dakota tax bases, 197, 200, 203
Splitting limits, 86
Spread sheet, 391, 419
Standard error, sample percentage, 110
State taxes, 166, 179
  types of, 181, 192, 193, 195-205
Statistics
  applied, 69, 77
  techniques, work sampling as, 109
Statutory depletion, 155, 156
Stochastic risk analysis models, 379
Stock. *See* Preferred stock and Common stock.
Stockholders. *See* Shareholders.
Straight-line (SL) method of depreciation, 223
Stripping
  preproduction, 397
  ratio, 60, 394
Structural limits, 60
Subore-grade material, 65
Subsidiaries, 350
Sum-of-the-years-digits (SYD) method, depreciation, 226
Supply schedules, 94, 96-98
Surface Mining and Control Reclamation Act of 1977, 168
Survey of Current Business, , 305
Syndicate, 351
Systems
  approach, estimating, 142
  of depreciation, 226

T

Tax Equity & Fiscal Responsibility Act (TE-FRA), 402
  investment tax credit reduced, 235
  reduces depletion allowances, 151, 206, 208
Tax Reduction Act of 1975, 151, 359

Tax Reform Act of 1969, 151, 207, 360
Taxable income, 180
Taxation, 8
  and mineral conservation, 187
  mine, 165
  mining ventures, 349
  state mine, 179
Taxes
  achieving social effects by, 175
  ad valorem property, 8, 14, 173, 402
    defined, 181
  adequacy of, 175
  basic definitions, 180
  books, 26
  credits, 418
    investment, 402
  depletion allowance deduction, 151
  depreciation, codes define, 307
  excise, 169
    definition, 174
  export, 174, 360
  federal income, 238
    minimum, 239
  foreign, impact of, 358
  franchise, 169
    definition of, 174
  holiday, Canadian, 161
  import, 174
  interrelationships between US and foreign, 359
  litigation over, 167, 182
  local, 165
  loophole, 151
  marginal rate, 327
  minimum, 239, 418
  net after state, 415
  neutrality of, 175
  on partnership, 351
  payer's, payee's treatment, 217
  policies, 6
    affect mining, 347
  preference items, three primary, 418
  privilege, 169
  profit, pretax, 418
  program objectives, 168
  provisions, 8
  rates, effective, 242
  sales, 169
  see also Income, taxes.
  social achievement, economic effects of, 175
  state income, calculating, 414
  to be convenient, 174
  types of, 173, 174
Technological advancements, 25
Technology, sophisticated, 353
TEFRA. *See* Tax Equity & Fiscal Responsibility Act.
Tennessee tax bases, 195, 197
Testing, analytical, 72

Texas tax bases, 203, 205
Theory of regionalized variables, 77
Time
  cards, 110
  expected returns, 253, 310
  long lead, mining, 171, 322
  requirements, estimates, 105
  studies, unit operations, 109
  value of money, 261
Title in contracts, 86
Tolling, smelter, 86
Tombstone, Arizona, 169
Tonnage, 66, 78
  revenue component, 82
  value curve, 176
Trade associations, 111
Trading companies, 353
Transactions taxes, 169, 174
  Privilege, Arizona, 194
Trust, investment, 351
Turnover ratio method, estimating, 114

U

UN. *See* United Nations.
Underground mining methods, labor costs of,
  131
Unemployment
  reducing, 358
  taxes, 174
Unequal investment, projects, 290
Uniform inflation
  no income taxes, 312
  with income taxes, 314
Unit
  cost method, estimating, 113
  depletion, 155
  operations, cost of, 132
  sales price, 83
  valuation methods, 14
    for taxation, 183
Unit of production (UOP) method, deprecia-
  tion, 223
Unitary taxation statutes, 184
United Kingdom tax system, 161
United Nations (UN), 62, 111
United States
  aluminum producers, prices, 89
  foreign discoveries impact, 349
UOP. *See* Unit of production method.
Uranium
  Canadian, Australian discoveries, 349
  depletion allowance reduced, 155
  prices, 85
Urban consumer index (CPI-U), 303
Urban wage earners' index (CPI-W), 303
US Bureau of Mines (USBM), 111
  Minerals Availability System (MAS), 98, 142

US Dept. of Commerce, 111
US Dept. of Energy (DOE), 111
US Dept. of Labor, Bureau of Labor Statistics,
  137, 302
US General Accounting Office (GAO), 191
US Geological Survey, 64
USBM. *See* US Bureau of Mines.
Useful life, property, 221
USGS. *See* US Geological Survey.
Utah
  aggregating uranium properties, 160
  tax bases, 193, 195, 200, 205
  valuation approach, 183

V

Valuation
  calculations, 25
  criteria, project, 253
  mine, effect taxes, 189
  problem, bond, 53
  studies, 7, 9
  vs. allocation, 30
Values, 11, 12, 15
  discovery, 152
  face, 53
  intrinsic, 8
  market
    as tax basis, 181
    fair, 8, 152
  present, future, annual, 260
  salvage, 14
  single, 8
  threshold, 61
  uncertainty in determining, 171
  variations in, 13
Variables
  decision, 30, 33
  input, 31
  mine planning, 58
  project, 4
  theory of regionalized, 77
Variogram development, 77
Vertical agreement, 354
Virginia City, Nevada, 169
Volumes, converting to weights, 78

W

Wall rock, 66
Wall Street Journal, 334
Waste products, 7
Wealth growth rate (WGR), 273
Wealth, 1
  maximizing owners, 253
  ownership of, 169

redistribution by taxation, 168
Weighing, contract clause, 86
Weighting, inverse distance, 75
Weights, converting from volumes, 78
WGR. *See* Wealth growth rate.
Wholesale Price Index. *See* Producer Price Indices.
Wisconsin
  severance tax, 200
  taxation by, 165
Work, measured costs of, 109
Working
  capital estimates, 112, 113
    drain, inflation, 308
  interests, 356
World Bank, 111
Wyoming
  tax bases, 193, 197, 201, 203, 205
  valuation approach, 183

Y

Yellow cake, 24

Z

Zambian copper, price, 84
Zinc
  depletion allowance reduced, 155